The Ecology
of Natural Disturbance
and Patch Dynamics

CONTRIBUTORS

Nicholas V. L. Brokaw

Charles D. Canham

Norman L. Christensen

B. S. Collins

Joseph H. Connell

Julie Sloan Denslow

K. P. Dunne

Kathryn E. Freemark

Subodh Jain

James R. Karr

Michael J. Keough

Orie L. Loucks

P. L. Marks

S. T. A. Pickett

Mary L. Plumb-Mentjes

Kevin Rice

Deborah Rogers

James R. Runkle

Timothy D. Schowalter

S. W. Seagle

H. H. Shugart

Wayne P. Sousa

Douglas G. Sprugel

John N. Thompson

Thomas T. Veblen

Peter M. Vitousek

Robert C. Vrijenhoek

P. S. White

John A. Wiens

The Ecology of Natural Disturbance and Patch Dynamics

Edited by

S. T. A. PICKETT

Department of Biological Sciences
and Bureau of Biological Research
Rutgers University
New Brunswick, New Jersey

P. S. WHITE

Uplands Field Research Laboratory
Great Smoky Mountains National Park
Twin Creeks Area
Gatlinburg, Tennessee
and Graduate Program in Ecology
University of Tennessee
Knoxville, Tennessee

ACADEMIC PRESS, INC.
Harcourt Brace Jovanovich, Publishers
San Diego   New York   Berkeley   Boston
London   Sydney   Tokyo   Toronto

ACADEMIC PRESS, INC.
San Diego, California 92101

United Kingdom Edition published by
ACADEMIC PRESS LIMITED
24-28 Oval Road, London NW1 7DX

Library of Congress Cataloging in Publication Data
Main entry under title:

The Ecology of natural disturbance and patch dynamics.

   Bibliography: p.
   Includes index.
   1. Natural disasters–Environmental aspects.
2. Biotic communities. I. Pickett, S. T. A.
II. White, P. S. III. Title: Patch dynamics.
QH545.N3E28 1985   574.5'22   84-18599
ISBN 0–12–554520–7 (hardcover) (alk. paper)
ISBN 0–12–554521–5 (paperback) (alk. paper)

PRINTED IN THE UNITED STATES OF AMERICA
89 90 91 92 93   9 8 7 6 5 4

Contents

20 Modeling Forest Landscapes and the Role of Disturbance in Ecosystems and Communities
H. H. SHUGART AND S. W. SEAGLE

Part V   SYNTHESIS

21 Patch Dynamics: A Synthesis
S. T. A. PICKETT AND P. S. WHITE

Bibliography   385

Index   457

Contributors

Numbers in parentheses indicate the pages on which the authors' contributions begin.

Nicholas V. L. Brokaw[1] (53), Department of Biology, Kenyon College, Gambier, Ohio 43022, and Smithsonian Tropical Research Institute, Balboa, Panama

Charles D. Canham (197), Section of Ecology and Systematics, Cornell University, Ithaca, New York 14853

Norman L. Christensen (85), Department of Botany, Duke University, Durham, North Carolina 27706

B. S. Collins (217), Department of Biological Sciences and Bureau of Biological Research, Rutgers University, New Brunswick, New Jersey 08901

Joseph H. Connell (125), Department of Biological Sciences, University of California, Santa Barbara, California 93106

Julie Sloan Denslow (307), Departments of Zoology and Botany, University of Wisconsin, Madison, Wisconsin 53706

K. P. Dunne (217), Department of Biological Sciences and Bureau of Biological Research, Rutgers University, New Brunswick, New Jersey 08901

Kathryn E. Freemark (153), Department of Biology, Carleton University, Ottawa, Ontario, Canada K1S 5B6

Subodh Jain (287), Department of Agronomy and Range Science, University of California, Davis, California 95617

James R. Karr[2] (153), Department of Ecology, Ethology, and Evolution, University of Illinois, Champaign, Illinois 61820

Michael J. Keough[3] (125), Department of Biological Sciences, University of California, Santa Barbara, California 93106

Orie L. Loucks (71), Butler University, Indianapolis, Indiana 46208

[1]Present address: Manomet Bird Observatory, Manomet, Massachusetts 02345.

[2]Present address: Smithsonian Tropical Research Institute, P.O. Box 2072, Balboa, Panama.

[3]Present address: Department of Biological Science, Florida State University, Tallahassee, Florida 32306.

P. L. Marks (197), Section of Ecology and Systematics, Cornell University, Ithaca, New York 14853

S. T. A. Pickett[4] (3, 217, 371), Department of Biological Sciences and Bureau of Biological Research, Rutgers University, New Brunswick, New Jersey 08901

Mary L. Plumb-Mentjes (71), U.S. Army Corps of Engineers, New Orleans, Louisiana 70160-0267

Kevin Rice[5] (287), Department of Agronomy and Range Science, University of California, Davis, California 95617

Deborah Rogers (71), Technical Information Project, Pierre, South Dakota 57501

James R. Runkle (17), Department of Biological Sciences, Wright State University, Dayton, Ohio 45435

Timothy D. Schowalter (235), Department of Entomology, Oregon State University, Corvallis, Oregon 97331

S. W. Seagle (353), Graduate Program in Ecology, University of Tennessee, Knoxville, Tennessee 37996

H. H. Shugart[6] (353), Environmental Sciences Division, Oak Ridge National Laboratory, Oak Ridge, Tennessee 37831

Wayne P. Sousa (101), Department of Zoology, University of California, Berkeley, California 94720

Douglas G. Sprugel[7] (335), Department of Forestry, Michigan State University, East Lansing, Michigan 48824

John N. Thompson (253), Departments of Botany and Zoology, Washington State University, Pullman, Washington 99164

Thomas T. Veblen (35), Department of Geography, University of Colorado, Boulder, Colorado 80309

Peter M. Vitousek[8] (325), Department of Biology, University of North Carolina, Chapel Hill, North Carolina 27514

Robert C. Vrijenhoek (265), Department of Biological Sciences and Bureau of Biological Research, Rutgers University, New Brunswick, New Jersey 08901

P. S. White[9] (3, 371), Uplands Field Research Laboratory, Great Smoky Mountains National Park, Twin Creeks Area, Gatlinburg, Tennessee 37738, and Graduate Program in Ecology, University of Tennessee, Knoxville, Tennessee 37916

John A. Wiens (169), Department of Biology, University of New Mexico, Albuquerque, New Mexico 87131

[4]Present address: Institute of Ecosystem Studies, New York Botanical Garden, Mary Flagler Cary Arboretum, Box AB, Millbrook, New York 12545.

[5]Present address: Department of Botany, Washington State University, Pullman, Washington 99164-4230.

[6]Present address: Department of Environmental Sciences, University of Virginia, Charlottesville, Virginia 22903.

[7]Present address: College of Forest Resources, AR-10, University of Washington, Seattle, Washington 98195.

[8]Present address: Department of Biological Sciences, Stanford University, Stanford, California 94305.

[9]Present address: Department of Biology, University of North Carolina, Chapel Hill, North Carolina 27514.

Preface

Ecologists have always been aware of the importance of natural dynamics in ecosystems, but historically, the focus has been on successional development of equilibrium communities. While this approach has generated appreciable understanding of the composition and functioning of ecosystems, recently many workers have turned their attention to processes of disturbance themselves and to the evolutionary significance of such events. This shifted emphasis has inspired studies in diverse systems. We use the phrase "patch dynamics" (Thompson, 1978) to describe their common focus.

Focus on patch dynamics leads workers to explicit studies of disturbance-related phenomena—the conditions created by disturbance; the frequency, severity, intensity, and predictability of such events; and the responses of organisms to disturbance regimes. The phrase "patch dynamics" embraces disturbances external to the community as well as internal processes of change. Patch dynamics includes not only such coarse-scale, infrequent events as hurricanes, but also such fine-scale events as the shifting mosaic of badger mounds in a prairie. The scope of this book includes population, community, ecosystem, and landscape levels. The most basic theme is an evolutionary one: How does the dynamic setting of populations influence their evolution? What are the implications for communities and ecosystems?

This book seeks to bring together the findings and ideas of workers studying such varied systems as marine invertebrate communities; grasslands; and boreal, temperate, and tropical forests. Our primary goal is to present a synthesis of diverse individual contributions. The book is divided into three main sections: (1) examples of patch dynamics in diverse systems; (2) adaptations of organisms and evolution of populations in patch dynamic environments; and (3) implications of patch dynamics for the organization of communities and the functioning of ecosystems. We feel this approach demonstrates the commonality of disturbance-generated phenomena over a wide range of scales and levels of organization and thus validates the broad applicability of the patch dynamic viewpoint. We seek to present clearly a framework that can stimulate the generation of explicit hypotheses and theory and thus form an alternative to equilibrium concepts of the evolution of populations, composition of communities, and functioning of ecosystems. We hope, in addition, that this volume will help identify areas of future research.

This book draws principally on terrestrial and marine systems, in which most work on patch dynamics has been done. Freshwater environments have received less emphasis, because studies of patch dynamics have been rare in these systems. By contrast, there is a rich, and largely recent, literature on a wide variety of terrestrial systems, examining both biotic and abiotic components and a variety of trophic levels. This book provides a common focus for this growing body of work, and aquatic patch dynamics have been included in this synthesis where possible.

Although this book is replete with detailed examples of community dynamics and organism adaptation, no book of this length could exhaustively treat all systems; even some terrestrial systems are missing here. For example, descriptions of boreal forest patch dynamics were omitted in reliance on recent, widely available reviews. We have sought, rather, to emphasize building a theoretical framework in which to view disturbance in natural systems. The stress here is on the processes of dynamics and emerging generalizations from patch dynamic systems, and not a mere catalog of examples. Treating certain systems in depth, while drawing connections and parallels with others, will best serve an understanding of processes and successful generalization.

A state-of-the-art conference on succession held in 1980 and summarized by West *et al.* (1981) convinced us that although there is much scattered information on the subject of disturbance and patch dynamics and many workers are currently interested in the topic, there is no compilation and synthesis of this information; nor is there any work which develops a broad framework or theory incorporating disturbance and its effect. We feel that this survey, drawing on workers familiar with many different systems and having interests at various levels of organization, is an ideal vehicle for meeting these needs.

This book is aimed at ecologists and advanced students working in this rapidly expanding field. Because we incorporate theoretical, empirical, and applied approaches to the effects of disturbance on plants, animals, and entire ecosystems, the book should be useful to workers of diverse backgrounds and interest.

S. T. A. Pickett
P. S. White

The Ecology
of Natural Disturbance
and Patch Dynamics

Part I

INTRODUCTION

Chapter 1

Natural Disturbance and Patch Dynamics: An Introduction

P. S. WHITE

Uplands Field Research Laboratory
Great Smoky Mountains National Park
Twin Creeks Area
Gatlinburg, Tennessee
and
Graduate Program in Ecology
University of Tennessee
Knoxville, Tennessee

S. T. A. PICKETT

Department of Biological Sciences and Bureau of Biological Research
Rutgers University
New Brunswick, New Jersey

I. THE DYNAMICS OF BIOLOGICAL SYSTEMS

The processes of growth, death, and replacement ensure that biological systems are dynamic, if only on a local scale. The extreme case in which these dynamics are entirely a function of individual mortality is illustrated by the monocarpic tropical

THE ECOLOGY
OF NATURAL DISTURBANCE
AND PATCH DYNAMICS

tree *Tachigalia versicolor* (Foster, 1977) and by the cyclic successions of some shrub-dominated communities (Watt, 1947; Christensen, Chapter 6, this volume). In most biological systems, however, other factors besides individual growth and death contribute to dynamics. Foremost among these factors are natural disturbances (White, 1979). The catalog of ecosystems in which natural disturbances play a fundamental role is a long one, as several reviews attest (Knapp, 1974; Grubb, 1977; Miles, 1979; White, 1979; Pickett, 1980; Oliver, 1981).

Natural disturbances and patch dynamics occur on a wide variety of spatial and temporal scales (Delcourt *et al.*, 1983). In this volume, dynamics are described that span a temporal scale of 10^0 to 10^3 years and a spatial scale of 10^{-4} to 10^6 m^2. The range of these scales is wide because the biological systems treated in this volume are varied and because disturbance effects occur on a variety of scales within each system. Our goal in this volume is the exploration of general themes in this diversity; in this chapter we foster that goal by defining three central concepts: patch dynamics, perturbation, and disturbance. These definitions lead to a discussion of endogenous and exogenous factors in community pattern.

II. DEFINITIONS: PATCH DYNAMICS, PERTURBATION, AND DISTURBANCE

The sources of variation in disturbances include differences in ecosystem scale, differences in kinds of disturbances, and differences in disturbance regimes. Even for a single ecosystem and disturbance event, effects vary at different trophic levels and occur over a wide range of biological levels from suborganismal (e.g., physiological effects; Sousa, Chapter 7, this volume) and organismal (e.g., behavioral changes; Wiens, Chapter 10, this volume) to ecosystem-wide (e.g., nutrient availability; Vitousek, Chapter 18, this volume). Most disturbances produce heterogeneous and patchy effects; these effects may themselves depend on the state of the community prior to the disturbance. The consequences of a given disturbance are strongly dependent on a variety of biotic and physical factors (e.g., regional climatic gradients, topographic gradients, and substrate types). Despite this plurality of "disturbances," a definition must be sought in a search for generality.

A. Patch Dynamics

We have adopted the term "patch dynamics" (Thompson, 1978; Pickett and Thompson, 1978) to label the central organizing theme of this book for the following reasons:

1. "Patch" implies a relatively discrete spatial pattern, but does not establish any constraint on patch size, internal homogeneity, or discreteness.
2. "Patch" implies a relationship of one patch to another in space and to the surrounding, unaffected or less affected matrix.
3. "Patch dynamics" emphasizes patch change.

We do not suggest that patches are spatially discrete in an absolute sense (note the descriptions of gap edge in Chapters 2 by Runkle, 4 by Brokaw, and 8 by Connell and Keough, this volume), nor that patches are internally homogeneous (note the range of fire effects described by Christensen, Chapter 6, this volume), nor that patches are necessarily temporally discrete in origin (note that treefalls on gap edges increase the original gap area; Runkle, Chapter 2, this volume). Particular definitions of "patch" will always be relative to the system at hand. But within a particular system, we do suggest that community structure and behavior vary locally and in a relatively patch-wise manner. Organisms, by their very nature, take up space and use resources; biological systems, on some level, are patchy.

A concept similar to "patch dynamics" is "shifting mosaic" (Bormann and Likens, 1979; Heinselman, 1981a). This concept connotes, however, a uniformity of patch distribution in time and space such that an overall landscape equilibrium of patches applies. Such equilibria are to be expected where feedback occurs between community characteristics and disturbance events (e.g., when susceptibility to disturbance increases with the time since the last disturbance), where the patch size is small relative to the homogeneous landscape unit, and where disturbance regimes are stable. We prefer "patch dynamics" for more general situations in which such an equilibrium has not been demonstrated and for situations in which patch scale is large relative to the scale of the relevant landscape (Romme and Knight, 1981). Further, environmental fluctuation, which may cause a shift in the disturbance regime, occurs on time scales similar to those of disturbances operating in the same system (see, e.g., Neilson and Wullstein, 1983). Equilibrium landscapes would therefore seem to be the exception, rather than the rule (for example, most North American landscapes have probably been influenced by changing disturbance regimes in the last several thousand to tens of thousands of years [Delcourt et al., 1983]).

Despite the difficulty in demonstrating the existence of patch equilibria, the concept of the "shifting mosaic" is important on a theoretical level. For example, all successional states would be predictable and permanent (if locally fleeting) features of steady state landscapes under a stable disturbance regime. Size class distributions of disturbance patches would be invariant. Disturbance regimes could be calculated from either the temporal or spatial distribution of disturbance patches because these two aspects of disturbance would be directly linked. The simultaneous occurrence of local dynamics and broad-scale equilibrium also underscores the central importance of scale hierarchies in the interpretation of natural systems (see, e.g., Allen and Starr, 1982; see also Zedler and Goff, 1973). Dynamics on one scale determine stasis on another.

B. Perturbation

"Disturbance" and "perturbation" have been used synonymously by some authors and yet possess particular meanings in the work of others. In general, "perturbation" has been used with a whole system orientation in the sense of any change in a parameter that defines a system (Loucks, 1975). The problem with applying this term to natural systems is that it is difficult to separate "perturbation" from the background variance in system parameters. The problem of defining the "normal" for the environmental settings of natural systems has led to semantic debate [e.g., is recurrent fire part of the "normal" environment of a boreal forest? and are there fire "climaxes" or not? (Whittaker, 1953; see the discussion in White, 1979)].

We feel, therefore, that the term "perturbation" is most useful in three relatively narrow contexts: (a) when the parameters or behaviors that define a system have been explicitly defined, (b) when a given disturbance is known to be new to the system at hand (e.g., some kinds of human disturbances), and (c) when the disturbance is under direct experimenter control. In these cases, a "perturbation" is a departure (explicitly defined) from a normal state, behavior, or trajectory (also explicitly defined). It is unlikely that natural ecosystems can be characterized in enough detail to warrant frequent use of "perturbation." However, this remains a possibility in the context of experimental treatments, model simulations, microcosms under researcher control, and components of natural systems. In other cases, we feel that there is no compelling reason to substitute "perturbation" for "disturbance."

C. Disturbance

Based on the discussion above, we define "disturbance" not in a sense that is relative to the "normal" environment but rather in a more tractable and physical sense. Thus, our definition of "disturbance" includes environmental fluctuations and destructive events, whether or not these are perceived as "normal" for a particular system. This definition must be applicable in a wide variety of systems; it must allow for the fact that "disturbance" is relative to the spatial (e.g., organism size) and temporal (e.g., organism lifespan) dimensions of the system at hand. Disturbance to bryophyte communities on streamside boulders can occur on a spatial (e.g., 10^{-4} m^2) and temporal scale (e.g., annual) that is irrelevant to the disturbance regime of the forest community growing on the same site. A definition of "disturbance" must be explicitly derived from relevant community dimensions (Allen and Starr, 1982): What is the ratio of the disturbance patch size to the mean size of the dominant organisms? What is the ratio of the disturbance return interval to the mean lifespan of the organism? Conversion of disturbance characteristics to relative scales in this way may be more meaningful than simple comparisons of absolute measurements (e.g., disturbance patch size in square meters).

Our final caveat is that although the formulation of definitions is a necessary human activity, it is ultimately arbitrary. Where the spatial and temporal properties of the system at hand can be measured, this measurement is obviously to be preferred over any classification scheme. The appropriate way to measure disturbance regimes is only now emerging (Table 1; see Paine and Levin (1981) for an approach based on patch demography). In this context, we propose the following statement:

> A disturbance is any relatively discrete event in time that disrupts ecosystem, community, or population structure and changes resources, substrate availability, or the physical environment.

This is a purposely generalized definition, and matters of scale and process will have to be specified in each case. Two general kinds of disturbance can be distinguished: destructive events and environmental fluctuation (see, e.g., Neilson and Wullstein, 1983).

Disturbances merge with secular environmental change over various spatial and time frames (MacMahon, 1981; Delcourt *et al.*, 1983; Delcourt and Delcourt, 1983). Most often, a disturbance results in ''open space''—for example, gaps in forests (Runkle, Chapter 2; Veblen, Chapter 3; and Brokaw, Chapter 4, this volume) and marine invertebrate communities (Sousa, Chapter 7; and Connell and Keough, Chapter 8, this volume). Resources are often made more available by disturbance (Canham and Marks, Chapter 11, this volume), but this is not always the case (Vitousek, Chapter 18, this volume). Disturbances often create patchiness;

TABLE 1

Definitions of Disturbance Regime Descriptors[a]

Descriptor	Definition
Distribution	Spatial distribution, including relationship to geographic, topographic, environmental, and community gradients
Frequency	Mean number of events per time period. Frequency is often used for probability of disturbance when expressed as a decimal fraction of events per year
Return interval, cycle, or turnover time	The inverse of frequency; mean time between disturbances
Rotation period	Mean time needed to disturb an area equivalent to the study area (the study area is arbitrarily defined; some sites may be disturbed several times in this period and others not at all—thus, ''study area'' must be explicitly defined)
Predictability	A scaled inverse function of variance in the return interval
Area or size	Area disturbed. This can be expressed as area per event, area per time period, area per event per time period, or total area per disturbance type per time period. Frequently given as a percentage of total available area
Magnitude	
intensity	Physical force of the event per area per time (e.g., heat released per area per time period for fire and windspeed for hurricanes)
severity	Impact on the organism, community, or ecosystem (e.g., basal area removed)
Synergism	Effects on the occurrence of other disturbances (e.g., drought increases fire intensity and insect damage increases susceptibility to windstorm)

[a] For a given disturbance descriptor, measures of central tendency and dispersion, as well as frequency distributions, are of interest.

however, some disturbances result in such diffuse effects that patches are difficult to define.

Two related terms were used by Harper (1977): "disaster" and "catastrophe." Harper contrasts these as follows: a "disaster" occurs so frequently that it is likely to occur within the life cycles of successive generations, while a "catastrophe" occurs rarely, so that it is unlikely to be experienced as a repeated, selective force. Harper expressed these differences in evolutionary terms: a disaster would be likely to increase fitness through selection, while a catastrophe would generally decrease fitness. Nearly all of the disturbances discussed in this book are disasters in Harper's sense, and his formulation makes a key distinction. However, as with the narrow use of "perturbation" that we propose, his terms require additional knowledge about the system at hand—knowledge that is likely to develop through the patch dynamics perspective, but that is not universally present when study is initiated.

D. Endogenous and Exogenous Causes of Disturbance

Early in the development of the field of ecology, two sorts of community change were recognized: autogenesis and allogenesis (see the discussion in White, 1979). In the former, change is driven by the biological properties of the system at hand, while in the latter an outside driving environmental "forcing" function is present. Factors responsible for change were divided into endogenous (within the community) and exogenous (outside the community). Classically, natural disturbances were treated as exogenous. In other situations, during disturbance-free periods, succession was driven by endogenous factors and autogenesis occurred.

Several chapters in this volume note that these distinctions are difficult to make in natural systems. For example, the creation of open space in marine invertebrate communities can result from the operation of an outside, periodic, and acute disturbance but can also result from the operation of an internal, periodic, and chronic biological disturbance, like predation (Chapters 7, by Sousa; and 8, by Connell and Keough, this volume). Senesence and death of overstory trees is an intrinsic community rhythm, but the proximal cause is often windstorm (see Chapters 2, by Runkle; 3, by Veblen; and 4, by Brokaw, this volume). Where the successional state of the community influences the likelihood of disturbance and where the community possesses disturbance-promoting traits (see, e.g., Loucks *et al.*, Chapter 5, and Christensen, Chapter 6, this volume), the classification of disturbances as endogenous or exogenous becomes problematic.

It seems logical to regard endogenous and exogenous factors as endpoints of a continuum. There are, in fact, several other continua in the definition of "disturbance" stated above: the relative discreteness of the disturbance event in time, the relative discreteness of disturbance patches, and the relative effect on ecosystem resources. Based on these continua, we can establish classical "disturbance" as a special case (we label this "class I disturbance") of the general phenomenon of disturbance, that of the demonstrably exogenous disturbance that acts at a single point in time, creates abrupt patch boundaries, and increases resource availability through decreased biological use of resources or increased decomposition or both (Sprugel, Chapter 19, this volume). This narrow definition permits the traditional interpretation, namely, that disturbance acts to "reset" the successional clock of the ecosystem without influencing the ultimate, predictable trajectory of that change or the potential biomass supportable on that site. In such a situation, the dominant organisms are removed, but the physical and biotic parameters of the site return to their previous state through autogenesis. We should make it clear that such a situation is restricted in nature. At the very least, disturbance itself introduces stochastic influences on community composition that reduce predictability of response.

Recognizing a second range on the exogenous–endogenous continuum, we establish endogenous community dynamics (e.g., the classic pattern and process of Watt, 1947) as a second special case ("class II disturbance"). Our definition of "disturbance" does not require that either of our special cases holds. As a result, we see no difficulty in including both physical and biological kinds of disturbance (e.g., Sousa, Chapter 7, this volume).

The inherent problems in defining normalcy, as discussed above, recur in terms of the concept of autogenesis. Autogenesis can be defined as competitive replacement only within a defined environmental context. The definition of "environmental context" is assumed but rarely specified in detail. Similarly, predictions about the outcome of competition experiments, such as those of Gause (Gause and Witt, 1935), can be made only with reference to controlled experimental conditions. Based on autogenesis and competitive exclusion, it has been predicted by some plant ecologists that uninterrupted successions lead to dominance by a single autotroph (see, e.g., Langford and Buell, 1969).

We can have no basic argument with this view; it is internally consistent. We do argue, however, with the realism of the view and with whether the competitive development of equilibrium communities characterizes the evolutionary setting of organisms (Pickett, 1980). Fluctuating environments lead to multiple resetting of the local successional trajectory (Botkin, 1981). Further, disturbance is a recurrent feature of many systems (leading us back to semantic arguments on the nature of the normal environmental setting for competition). The importance of the disturbance regime itself in species interactions was shown by Armstrong (1976), who performed experiments with two species of fungi, one a superior competitor and the other a faster colonizer. The two species achieved stable coexistence when the availability of new substrates (the disturbance regime) was regulated. Zedler and Goff (1973) established the role of scale in population stability and coexistence in shade-tolerant and shade-intolerant temperate trees.

III. NATURAL DISTURBANCE: THE PATCH DYNAMICS PERSPECTIVE

The chapters that follow describe the dynamics of natural systems and the consequences of these dynamics for the understanding of populations, communities, and

TABLE 2

An Index to Recent Literature on Major Natural Disturbances and to the Chapters of this Volume[a]

Disturbance	Ecosystem or geographic area	Chapters in this volume	Other representative literature
Fire	Boreal forest	19	Heinselman (1973, 1981a)
	Temperate forest	2, 3	Ahlgren (1974)
	Coastal plain		Komarek (1974); Forman and Boerner (1981)
	Western montane forest		Romme (1982); Romme and Knight (1981)
	Grasslands	5	Daubenmire (1968); Vogl (1974)
	Chaparral	6	Zedler (1981); Mooney et al. (1981)
Hurricane	Marine	7, 8	Spurr (1956); Reiners and Reiners (1965); Chabrek and Palmisano (1973); Webb (1958); Whitmore (1974)
	Terrestrial		
Other windstorms	Temperate forest	2, 3, 19	Jones (1945); Henry and Swan (1974); Sprugel (1976); Bormann and Likens (1979)
Gap dynamics	Mesic forests	2–4, 12, 17, 20	Watt (1925); Bray (1956); Forcier (1975); Fox (1977); Grubb (1977); Foster (1977)
	Marine	7, 8	Paine and Levin (1981)
Ice storm	Temperate forest	2	Lemon (1961)
Ice push on shores	Temperate and boreal		Raup (1975)
Cryogenesis	Arctic tundra		Sigafoos (1952); Churchill and Hanson (1958)
	Alpine tundra		Johnson and Billings (1962)
Freeze damage	Various		Silberbauer-Gottsberger et al. (1977)
Fluctuating water levels in basins	Various		Raup (1975); Buell and Buell (1975)

Disturbance		Vegetation/habitat	References
Rare rainstorms		Desert	Zedler (1981)
Droughts	5	Grasslands	Visher (1949); Weaver (1968)
		Temperate forest	Hough and Forbes (1943)
Alluvial processes		Various	Johnson et al. (1976); Keeley (1979); Nanson and Beach (1977); Bell and del Moral (1977)
Flash floods		Desert	S. G. Fisher et al. (1982)
Coastal processes		Various	Olson (1958); Martin (1959); Schroeder et al. (1976)
		Mangroves	Thom (1967)
Salinity changes		Various	Chabrek and Palmisano (1973); Barbour and DeJong (1977)
Miscellaneous marine processes	7, 8	Temperate and tropical	Leviten and Kohn (1980); van Blaricom (1982)
Landslides and other earth movements	3	Steep topography	Flaccus (1959); Garwood et al. (1979)
Lava flows	3	Various	Eggler (1971)
Karst processes			
Biotic			
insect outbreaks	13	Various	Blais (1954)
disease		Deciduous forest	Anderson and Anderson (1963)
predation	7, 8	Marine	Platt (1975)
burrowing animals	5	Grassland	Ives (1942)
beaver		Temperate and boreal	Siccama et al. (1976); Strong (1977)
vascular plants		Temperate and tropical	
Human-caused	17	Various	Harmon et al. (1984)

[a] This list was primarily developed with reference to North American terrestrial vegetation (see White, 1979), but other systems are also included.

ecosystems. The first eight chapters develop these themes in detail for particular kinds of systems: north temperate forests (Runkle, Chapter 2), south temperate montane forests (Veblen, Chapter 3), tropical forests (Brokaw, Chapter 4), grasslands (Loucks *et al.,* Chapter 5), shrublands (Christensen, Chapter 6), and marine intertidal (Sousa, Chapter 7) and subtidal (Connell and Keough, Chapter 8) communities. The subsequent five chapters discuss particular groups of organisms in relation to disturbance: vertebrates (Karr and Freemark, Chapter 9, and Wiens, Chapter 10), woody plants (Canham and Marks, Chapter 11), forest herbs (Collins *et al.,* Chapter 12), and insects (Schowalter, Chapter 13). The final eight chapters are concerned with research implications of disturbance and patch dynamics for a variety of levels of biological organization: interacting plant and herbivore populations (Thompson, Chapter 14), animal population genetics (Vrijenhoek, Chapter 15), plant population genetics (Rice and Jain, Chapter 16), species coexistence (Denslow, Chapter 17), nutrient dynamics (Vitousek, Chapter 18), ecosystem energetics (Sprugel, Chapter 19), modeling (Shugart and Seagle, Chapter 20), and future research directions (Pickett and White, Chapter 21).

The disturbances and dynamics described in these chapters have fundamental implications for the discipline of ecology. There are frequent references to the heterogeneity and lack of equilibrium in the natural world; both factors have made the derivation of appropriate theory and models difficult and contribute to frequent dissatisfaction with existing approaches (e.g., see Chapters 3, 4, 9, and 10, this volume). Both nonequilibrium and patchiness characterize the setting of evolution itself (e.g., Chapter 15, this volume). The equivalence or nonequivalence of spatial and temporal gradients in communities is of central importance in this regard (Chapter 10, this volume), as is the contrast between spatial and temporal sources of landscape patches (Chapter 9, this volume).

Patch dynamics has implications for applied ecology. Several chapters discuss resource management and conservation implications. Preservation of natural systems necessarily involves a paradox: we seek to preserve systems that change (White and Bratton, 1980). Success in a conservation effort thus requires an understanding of landscape patch structure and dynamics (Pickett and Thompson, 1978). A final applied problem is that the existence of natural change, whether consisting of cyclic replacements or successional trajectories, complicates the testing of hypotheses about human impacts in these systems.

No single volume can treat all ecosystems for which disturbance is significant. Fire has been well summarized in previous work (Ahlgren and Ahlgren, 1960; Kozlowski and Ahlgren, 1974; Mooney *et al.,* 1981); in particular, the importance of fire in the boreal forest has been fully described in the published literature (see, e.g., Heinselman, 1981a, and references therein). We present a catalog of disturbances and an index to representative literature in Table 2. Some of the disturbances listed in that table are associated with particular climates (e.g., cryogenesis), particular geographic situations (e.g., coastal erosion), particular kinds of topography (e.g., debris avalanches), or particular kinds of terrestrial substrates (e.g., dune movement). The literature cited in Table 2 and in the succeeding chapters of this volume, plus the several review papers we have cited, provides access to a wealth of disturbance studies. Our purpose in organizing this volume was not to provide coverage of all disturbances but rather to choose examples that would aid in the development of generality. From the standpoint of understanding the role of disturbances, we were drawn to systems in which small-scale and frequent disturbances were the rule. Nonetheless, in almost every case, the systems that experience such disturbances experience disturbances on other scales as well.

Part II

PATCH DYNAMICS IN NATURE

Chapter **2**

Disturbance Regimes in Temperate Forests

JAMES R. RUNKLE

Department of Biological Sciences
Wright State University
Dayton, Ohio

I. INTRODUCTION

Different forest types can be characterized by the mortality patterns of their canopy trees. This chapter will begin by defining the parameters necessary to characterize the pattern of death of dominant individuals (canopy trees) in a community, also referred to as that community's ''disturbance regime.'' ''Disturbance'' is defined here as a force that kills at least one canopy tree. The disturbance regimes of two particular forest types will then be described. Finally, descriptions of natural disturbance regimes will be compared with the results of manipulative studies or artificial disturbance regimes. Special attention will be given to the relative importance of large-scale versus small-scale disturbance.

II. COMPONENTS OF A DISTURBANCE REGIME
FOR FORESTS

A. Average Disturbance Rates

The average rate at which trees die can have important consequences for the species composition and structure of a forest. High disturbance rates should select for fast-growing species that reproduce early and are short-lived (Grime, 1974, 1977). If disturbance rates are too high, the tree lifeform is no longer viable and community dominance switches to shrub or herb lifeforms. Natural disturbance rates for forests also have theoretical minimum values set by the maximum age and size limits of tree species. As a tree ages and increases in size, its efficiency in transporting water, nutrients, and photosynthate usually decreases (Spurr and Barnes, 1973; Oldeman, 1978). Its roots must support a proportionally greater aboveground biomass (Borchert, 1976), and its photosynthetic tissues must support a proportionally greater mass of nonphotosynthetic tissue (Harper, 1977). These factors, plus the tendency of the tree to develop a more massive crown, render it increasingly susceptible to smaller and more common disturbances. This relationship between the external environment (frequency of disturbances, e.g., wind speeds of a certain magnitude) and the plant itself (the rate at which aging increases its susceptibility to disturbances of smaller magnitude) diminishes the usefulness of terms such as "allogenic" or "autogenic" in connection with natural disturbances.

As a result of the above factors, forest disturbance rates seem to be constrained to a fairly small range of potential values. As one consequence, forest dominants in most parts of the world have a range of life spans of 100–1000 years (Budowski, 1965; Fowells, 1965; Ashton, 1969). For temperate deciduous forests the normal range is even smaller: 300 years is the age often reached by dominants with few individuals living more than 500 years (Jones, 1945).

The average rate of forest disturbance also shows fairly little variation, despite wide differences in vegetation and types of disturbance. Northern conifer forests affected primarily by fire (Heinselman, 1973; Zackrisson, 1977) and temperate and tropical forests affected primarily by the death of scattered individuals (Leigh, 1975; Abrell and Jackson, 1977; Hartshorn, 1978; Naka, 1982; Runkle, 1982) all show average rates of disturbance of $\approx 1\%$/year (ranging from ≈ 0.5 to ≈ 2.0/year in large samples). Although these forests are different from each other in many ways, they are similar in that most canopy individuals die due to the one mechanism studied—fire or wind throw. Disturbance rates for some specific agents of tree death (e.g., 0.02–0.16%/year for tropical landslides; Garwood et al., 1979) may be lower than $\approx 1\%$ because many trees die due to factors other than the one studied.

Disturbance rates of 0.5–2.0%/year give natural return intervals (average time between disturbances for a given site) of 50–200 years. These values can be reconciled with 300- to 500-year average tree longevities for the following reasons. First, certain trees live longer than average due to their presence in more protected locations or to chance deviations from normal weather conditions. Second, many important forest dominants often persist for many years under a closed canopy, growing very slowly. For instance, using age–size (diameter at breast height, dbh) regressions, trees in mesic sites in the Great Smoky Mountains National Park averaged about 91 years to reach 25 cm dbh, the approximate minimum size at which they reach the canopy (Runkle, 1982). The average time spent by individuals in the canopy, again using age–size regressions, was 127 years, in good agreement with the natural rotation periods noted above.

The somewhat surprising conclusion is that different mesic forests probably do not show very great differences in their average rates of disturbance. Therefore, important differences among the disturbance regimes of different forests are more likely to occur in the distribution of tree deaths in time and space and in the severity of the disturbance.

B. Distribution of Disturbance in Space

Over a broad geographic area, a given level of disturbance can affect either many adjacent individuals, creating a few large disturbed patches, or many scattered individuals, creating many small disturbed patches. Because patch size affects the nature of the vegetation's response to the disturbance, these two alternatives should yield different results (see Section II,B,2).

Before proceeding, one note on terminology is useful. The term ''gap'' was used by Watt (1947) to refer to a site at which a canopy individual had died and at which active recruitment of new individuals into the canopy was occurring. The emphasis was on relatively small within-community disturbance patches. This emphasis has generally been maintained in later usages of the term (see, e.g., Bray, 1956; Williamson, 1975; Whitmore, 1978; Ehrenfeld, 1980; Barden, 1981; Runkle, 1981, 1982; Shugart and West, 1981; Nakashizuka and Numata, 1982a,b). This chapter will retain this usage, although clearly a gradient exists between disturbances affecting a single tree and those affecting many square kilometers of forest.

1. Relation of the Environment to the Size of the Disturbed Area

The physical environment within a small open area surrounded by forest differs from that under the canopy or in a large open area. In a small opening, temperatures fluctuate more and light and soil moisture are both more abundant than under a closed canopy. As the opening size decreases, humidity increases, wind speed decreases, and temperatures remain more constant (Geiger, 1965). Opening size is frequently quantified as the D/H, ratio, where D is the diameter of the open area and H is the mean height of the surrounding stand (Geiger, 1965). Several studies (Jackson, 1959; Minckler, 1961; Berry, 1964; Minckler and Woerheide, 1965; Minckler et al., 1973) have shown light to increase with increased opening size, reaching a maximum when $D/H \approx 2$. March and Skeen (1976) found that differences in light between a small opening and a closed forest persist throughout the growing season. Minckler et al. (1973) found the opening size to determine the number of years the increase in soil moisture persists, although not the size of the initial difference. Tomanek (1960) found that the shape and orientation of openings, as well as their size, can be important in determining their microclimate.

2. Relation of Species Composition to the Size of the Disturbed Area

Many forestry studies and general reviews state that the selective cutting of individual trees will favor shade-tolerant species such as American beech (*Fagus grandifolia*), eastern hemlock (*Tsuga canadensis*), and sugar maple (*Acer saccharum*) (see, e.g., U.S. Forest Service, 1973; McCauley and Trimble, 1975; Leak and Filip, 1977; Tubbs, 1977). However, openings as small as 400 m² have been found to be sufficient for tuliptree (*Liriodendron tulipifera*) and yellow birch (*Betula alleghaniensis*) to maintain themselves in a forest (Merz and Boyce, 1958; Tubbs, 1969; Trimble, 1970; Schlesinger, 1976; Beck and Della-Bianca, 1981). Tryon and Trimble (1969) found a 1000-m² opening sufficient to regenerate several intolerant species, with relatively few adverse affects of border trees on the growth of saplings near the edge of the opening. Runkle (1982, 1984) found significant differences in the response of potential canopy species to differences in gap size for naturally formed gaps ≤1000 m² and generally ≤400 m². Williamson (1975) found evidence that gaps 50–250 m² were sufficient to regenerate tuliptree and white ash (*Fraxinus americana*).

C. Distribution of Disturbance in Time

A given average annual disturbance rate can be achieved by a low level of disturbance occurring in most years or by occasional years of very high disturbance followed by many years of few or no disturbances. Forests at either extreme are known. On a local level, differences in the periodicity of disturbance often parallel differences in the spatial distribution of disturbance. If most tree mortality is concentrated in a few years, then probably much tree mortality is concentrated in large openings. Therefore, species composition at sites where disturbance is concentrated in time should resemble species composition at sites where disturbance is concentrated in space. The temporal distribution of disturbance is more important on a landscape level than on a local level because it determines the synchrony of the regeneration processes occurring over a broad area. The level of synchrony of regrowth is important because of the close relationship between tree population dynamics and ecosystem changes in biomass and production (Peet and Christensen, 1980; Peet, 1981).

D. Severity of Disturbance

In addition to varying in temporal and spatial distributions, disturbances can vary in their severity. "Severity" measures the degree to which the predisturbance vegetation has been damaged and ecosystem properties have been disrupted. It is equivalent to the term "magnitude" used by White (1979). The vegetation of a site will develop more slowly after a severe disturbance than after a mild disturbance. The size and severity of a disturbance are two different properties. It is possible to have a small, severe disturbance or a large, mild one.

Several compilations of species regeneration strategies have been made (e.g., Bormann and Likens, 1979; Oliver, 1981; Canham and Marks, Chapter 11, this volume). In general, individuals growing after disturbance are present at the time of disturbance as suppressed seedlings and saplings, as seeds buried in the soil, or as seeds newly dispersed into the area. The severity of disturbance determines which of these strategies is most likely to succeed. A mild disturbance, e.g., windthrow of just the canopy trees, probably favors the suppressed sapling strategy. A more severe disturbance may eliminate suppressed saplings but leave the soil intact and favor species such as pin cherry (*Prunus pensylvanica*), which are well represented in the seed pool (Marks, 1974). A disturbance that is both very severe, e.g., long-term agriculture (eliminating saplings and the seed pool) and is conducted over a large area (greatly diminishing the potential seed rain) can result in a very protracted recovery time.

Some types of disturbances can enhance the success of certain regeneration strategies through the creation of special microhabitats. Uprooting of trees creates pits and mounds that differ in several properties from soils that have not been overturned (Lyford and MacLean, 1966; Armson and Fessenden, 1973; Stone, 1975). In particular, pits have more litter and standing water and mounds have less than do other soils. Some species differ in the part of the pit and mound surface on which they grow (Hutnik, 1952). Decomposing logs also provide a specialized habitat on which some species, such as yellow birch and eastern hemlock, reproduce (Fowells, 1965). Other examples in which the type of disturbance determines the pattern of species replacement are given by Grubb (1977).

The severity of disturbance can also be measured as the effect on ecosystem functioning. The primary influence is on soil properties and long-term nutrient dynamics. A severe disturbance results in substantial erosion and nutrient losses, which may take decades for the ecosystem to replace (Bormann and Likens, 1979). For example, low-intensity fires may have no long-term effect on ecosystem properties, but intense fires can volatilize much nitrogen, cause severe erosion, and greatly diminish future productivity at the site (Wells *et al.*, 1979).

E. Rates of Recovery from Disturbance

The rate at which a community recovers from disturbance depends upon the characteristics of the disturbance discussed above. For small and mild disturbances, recovery is determined primarily by the rates at which bordering canopy trees expand into the opened area and seedlings and saplings grow into the canopy. For larger and more severe disturbances, a more varied and elaborate process of vegetation development occurs.

In general, the latter process has been studied as ecological succession, while the former has been considered as characterizing gap dynamics. The division between the two processes is arbitrary, with "succession" being used primarily when whole communities change and "gap dynamics" when the disturbances occur within a single community. "Community" here refers to a site of sufficient size to be studied by itself. Dynamics within a gap caused by a single treefall are usually studied in relation to the surrounding forest, whereas recovery of a large area in which trees were blown down by a large storm or burned in a fire is often studied without mention of the surrounding areas.

This section will concentrate on the recovery processes most important in gap dynamics, as defined above, that is, lateral growth of canopy trees bordering the gap and height growth of seedlings and saplings within the gap.

1. Lateral Extension Growth

Several studies differing in location, species, and technique have measured reasonably similar rates of branch growth by trees bordering openings (Trimble and Tryon, 1966; Phares and Williams, 1971; Erdmann et al., 1975; Hibbs, 1982; Runkle, 1982). Average rates generally range from 4 to 14 cm/year. Some trees expand at rates of up to 20 to 26 cm/year. The impact of these branch growth rates on gap regeneration depends upon the rate of sapling height growth and the size of the gap. Smaller gaps have a large ratio of edge to interior. Therefore, lateral extension growth should be proportionally more important in small gaps than in big ones.

2. Sapling Height Growth

The rate at which a gap closes due to the height growth of saplings depends on both the rate of height growth of saplings and the heights of the saplings at the time the gap was formed.

Many species from different areas in the eastern deciduous forest show average growth rates of 0.5–1.0 m/year following cutting or in naturally created (usually large) openings (e.g., Kramer, 1943; Downs, 1946; Kozlowski and Ward, 1957; Tryon and Trimble, 1969; Marks, 1975). Minckler et al. (1973) found species height growth to range from 9 to 73 cm/year near the centers of gaps of different sizes (less than or equal to two times the height of surrounding trees). Hibbs (1982) measured sapling height growth (the average of the three largest stems) in a hemlock–hardwood stand in Massachusetts. In small gaps (≤5-m radius), saplings of different species grew 10–50 cm/year; in open field conditions, species grew 25–50 cm/year. Hibbs (1982) related the rates of sapling growth to the rates of canopy branch growth calculated for the same woods and concluded that few or no tree seedlings will reach the canopy in openings with a radius of <5 m. Some seedlings may reach the canopy in larger gaps because of the increased time until canopy closure occurs via branch growth.

Small gaps may still close primarily due to sapling height growth if sufficiently large, suppressed saplings are present in the gap when it is formed. Good descriptions of the height distribution of saplings in gaps immediately after formation are not available. However, many forests contain large numbers of suppressed saplings. If the disturbance is mild, then one of these saplings may grow to become the next canopy tree in less time than it would take for a new seedling to reach the canopy, especially if taller individuals grow faster than small ones, as at least occasionally occurs (Laufersweiler, 1955; Burton et al., 1969; Tubbs, 1977).

To include the effects of both initial sapling size distribution and growth rates, Runkle (1982) compared the rate at which the total gap area disappeared for small gaps (average, 100 m^2) with the rate expected if branch growth by canopy trees were the only mechanism of gap closure. After the fifth year, height growth was the primary mechanism of gap closure. Even small gaps can result in successful tree regeneration if they include a large, formerly suppressed individual.

F. Importance of Multiple-Gap Episodes

One last component of a forest disturbance regime is the extent to which a tree may be affected by two or more different gaps in the course of its growth into the canopy. Such multiple gap effects should be more common after mild disturbances than after severe ones, which might kill the regenerating individual. Such multiple episodes are also more important when disturbances are small and scattered rather than clustered. Small, scattered disturbances have the greatest ratio of edge (areas affected but not injured) to internal area.

Such multiple gap episodes may be fairly common. Individuals of several species, notably hemlock, often show multiple release and suppression of ring widths, implying several episodes of gap formation and closure (Henry and Swan, 1974; Oliver and Stephens, 1977). Also, if, as mentioned earlier, average tree mortality rates are approximately 1%/year, repeated disturbances should be fairly common. For example, 36 gaps examined in Hueston Woods State Park, Ohio (Runkle, 1981, 1982) had 257 border trees, or 7.1 border trees per gap. Given those values (1%/year, mortality; 7.1 border trees per gap), about half of the gaps should be affected by a new disturbance (death of at least one border tree) within 10 years of initial gap formation. The Hueston Woods gaps were revisited 4 years after their original census (Runkle, 1984); 11 border trees had died or become moribund during that time, a value close to the 10 predicted from average rates of disturbance. Therefore, for this forest, return rates of disturbance may be common and generated primarily by deaths not influenced by the proximity of a previous tree death. In other forests, e.g., high-elevation forests of balsam fir (Abies balsamea) (Sprugel and Bormann, 1981), repeated disturbances are even more common because the environment next to a disturbed area is more severe than elsewhere, and so new tree deaths occur primarily among border trees.

That such multiple-gap episodes are common for at least some forest types may be very important for forest regeneration and species evolution. Species may be able to reach the canopy fairly often by using a series of small gaps rather than a single large one.

Species specializing in this mode of reproduction should be able to take advantage of temporary openings in the canopy and then should suffer only slightly after the canopy closes, thereby increasing the chance that they will still be able to respond while awaiting a new gap in the vicinity. High rates of multiple gap occurrence can also imply that individuals of tolerant species will usually be exposed to one or more gaps at some stage before reaching the canopy.

The question of whether understory-tolerant species can occasionally reach the canopy without benefiting from gaps is not resolved. To my knowledge, no species under a closed canopy has been shown to have a steady increase in height to reach canopy status. Several lines of evidence suggest that this phenomenon will rarely if ever occur. Seedlings and saplings in complete shade grow very slowly. For example, Morey (1936) found that on average it took beech 12 years and hemlock 29 years to reach a height of 1.2 m. Sugar maple and beech seedlings in Ontario grew only 2–4 cm/year both under shaded conditions and in a 200-m^2 gap (Cypher and Boucher, 1982). In small gaps (10–50 m^2), Hibbs (1982) found hemlock to grow only 10–20 cm/year. Presumably growth under a closed canopy would be even less. As a result of these very slow growth rates, these species would take $\geq$200–300 years to reach the canopy without occasional spurts of faster growth. This time interval is at the outer limit of the lifespan for most of these species (Fowells, 1965). Similar conclusions can be reached for diameter growth. Many tolerant trees show little or no diameter growth under shaded conditions. For example, one study in Pennsylvania found a 9-cm dbh beech missing rings for 46 of the previous 70 years and a hemlock missing rings for 39 of the previous 70 years (Turberville and Hough, 1939). Given the rates of disturbance that occur in these forests, however, the probability that a gap will affect one spot at least once within a 100- to 300-year period is extremely high.

III. NATURAL DISTURBANCE REGIMES FOR SPECIFIC TEMPERATE FORESTS

Workers in the eastern deciduous forest of North America have had several advantages in the determination of natural disturbance regimes. Although the majority of the original forest has been logged or severely disturbed, several remnants do remain on which the formerly widespread processes of forest regeneration can be studied. Historical records of other primeval forests and natural disturbances exist. Some of these historical records are remarkably quantitative, such as those of the General Land Office Survey (Bourdo, 1956). Also, North American plant ecologists have long been interested in the processes of forest disturbance and succession, and so much information is available in the literature.

This section will describe the disturbance regime associated with two different forest types and locations. The cove forests of the southern Appalachians are affected almost entirely by small-scale, mild disturbances. The forests of the Allegheny Plateau, in Pennsylvania, are affected by both small-scale and large-scale, usually mild, disturbances. The description of the Allegheny forests will also include the distribution of disturbances over the landscape, including the effect of topographic position on both disturbance regime and vegetation.

A third type of disturbance regime exists in the white pine (*Pinus strobus*)–northern hardwoods section of northern Minnesota and adjacent Canada. For these forests, fire is the primary source of large-scale forest disturbance (Frissell, 1973; Heinselman, 1973) and has been important for 10,000 years (Swain, 1973). Heinselman (1973) found average rates of burning of ≈1%/year before widespread fire suppression was adopted. Disturbance was clumped in time and space, with most of the area burned in one of only a few major fire years. All present stands within the 415,782-ha Boundary Waters Canoe Area owe their origin to fire. Therefore, small-scale disturbances seem to be relatively unimportant. Because this disturbance regime has been adequately summarized elsewhere (Heinselman, 1973, 1981a), it will not be discussed further here.

A. Cove Forests of the Southern Appalachians

Cove forests occur in sheltered areas near creeks at middle elevations throughout much of the southern Appalachian mountains (Braun, 1950; Whittaker, 1956; Golden, 1981). They are dominated by differing combinations of mesophytic tree species, particularly sugar maple, yellow buckeye (*Aesculus octandra*), yellow birch, American beech, silverbell (*Halesia carolina*), white basswood (*Tilia heterophylla*), and eastern hemlock.

The disturbance regime for the cove forests is determined by their regional and local topographic positions. Fire occurrence rates on a county basis are very low to moderate for most of the mountainous counties of eastern Tennessee and western North Carolina, in contrast to higher rates nearer the coast (Nelson and Zillgitt, 1969). Within the mountains, fires are uncommon, occurring primarily on south-facing slopes near ridge tops, especially on lower ridges (Barden and Woods, 1976; Harmon, 1982). North-facing lower slopes and sheltered ravines have the lowest incidence of fire (Harmon, 1982).

Wind-related disturbance tends to be dominated by small-scale events. Glaze storms are more common than large-scale, damaging tropical storms (Nelson and Zillgitt, 1969). Tornadoes are not as common or severe as they are in most of the rest of the eastern deciduous forest (Fujita, 1976). Occasional tornadoes do occur, however.

Human disturbance of most sites once dominated by mixed mesophytic species has been extensive. Therefore, most work on the long-term dynamics of mixed mesophytic forests has been done in one of the remaining old-growth remnants, either the Great Smoky Mountains National Park (GSMNP) of Tennessee and North Carolina or unlogged coves in one of the nearby national forests. These areas were protected from extensive logging by their regional inaccessibility until about 1900 and by the formation of the GSMNP in 1940. Between 1900 and 1940, however, virgin timber was removed by commercial loggers from most of the present-day park. Also, substantial areas of the GSMNP had been cleared and selectively cut by local people (Frome, 1966). Despite these human influences, enough undisturbed forest remains at middle and high elevations to allow a meaningful characterization of the natural disturbance regime.

The exact locations of the sites to be discussed below were further constrained by two additional factors. Most sampling was done far enough away from streams or near small enough streams so that *Rhododendron maximum* was nearly or entirely absent. The presence of a dense shrub layer of rhododendron influences regeneration by greatly diminishing the success of an advance sapling regeneration strategy. As a consequence, cove forests with rhododendron have more red maple (*Acer rubrum*) and more hemlock and *Betula* spp., which regenerate on fallen logs, and less sugar maple, yellow buckeye, beech, silverbell, and basswood, all of which depend on advance sapling growth than do cove forests without abundant rhododendron (Oosting and Bourdeau, 1955; Barden, 1979, 1980; Lorimer, 1980). The second local restriction in sampling was to avoid slope communities in which American chestnut (*Castanea dentata*) had been important before its demise (Woods and Shanks, 1959).

The disturbance regimes of the cove forests are thus influenced by their regional and local positions. Deaths of canopy trees occur primarily as scattered small-scale disturbances affecting only one or a few trees at a time in any one location. Likely causes of tree mortality are glaze storms, lightning strikes, or occasional very high winds. Disturbances are not very severe. Surrounding vegetation diminishes the loss of nutrients from the site, so that long-term ecosystem functioning should not be harmed. Many saplings and other advance regeneration are present, so vegetation recovery should proceed rapidly.

The following data on the disturbance regime parameters for cove forests are summarized primarily from Runkle (1982), unless otherwise stated.

Overall, in the cove forests, 0.5–2.0% of the land surface area in individual sites was converted from forest to new treefall gaps per year. The average for all sites studied was 1.2–1.3%/year, in agreement with figures from other forest types, as discussed above. Romme and Martin (1982) found lower disturbance rates (0.25–1.0%/year, depending on the method of calculation) for an old-growth mixed mesophytic forest in Kentucky. They did not include very small gaps created by parts of still living canopy trees, so the two results are not strictly comparable.

Gap areas followed a lognormal distribution, with many small and a few large gaps. The average size of a canopy opening was ~31 m^2 if the very small gaps caused by the fall of large branches or small canopy trees were included. Canopy opening sizes ranged up to 1490 m^2, with ≈1% of the total land area in gaps of >400 m^2. Most gaps were created by the death of single trees, but multiple treefalls accounted for most of the larger gaps. Similar values for gap size in these forests are given in Barden (1981). Similar values for gap size were also found in a climax stand of Japanese beech (*Fagus crenata*) and other mixed mesophytic species in Japan (Nakashizuka and Numata, 1982a,b). Because the gaps were fairly small, with diameter/canopy height ratios <1, the difference in environmental conditions between the gap and the forest understory is smaller than for a forest dominated by large-scale disturbances. However, small gaps have greater edge/area ratios than do larger openings. Therefore, cove forests should contain very large fractions of land area partly affected by disturbance.

Yearly fluctuations in the rate at which gaps are formed occur but are minor. For example, for 10 different sites in the southern Appalachians and for 15 years per site, the maximum fraction of land area per year in gaps was only 7.4%. Every year, several storms in the general area down enough trees to cause notable economic damage (Environmental Data Service, 1975). The rugged topography results in different areas having different peak years of disturbance, with no sign of regional synchrony in gap formation.

A disturbance regime characterized by many small gaps with a large ratio of edge to area might be expected to show high rates of repeat disturbance. New gaps should often form close enough to old gaps to maintain the changed environmental conditions associated with gaps and to slow the processes of gap closure. In one study designed to test this hypothesis (Runkle, unpublished data), high rates of repeat disturbance were found. For 273 gaps revisited 6–7 years after originally being sampled, one or more canopy trees surrounding the gap had died or been severely injured in 114 gaps, a former large stump from the tree creating the gap had fallen in 62 gaps, and new gaps were created near but not immediately adjacent to the original gap 35 times. In only 112 gaps did none of these new disturbances occur. Canopy trees surrounding gaps died at about the same rate as canopy trees in general. Multiplying the number of original surrounding canopy trees by a 1%/year disturbance rate by 6 or 7 years gives a predicted number of deaths of 151 trees versus 164 deaths or severe injuries actually recorded. For these forests, therefore, repeat disturbances are common and are a property of the size and age distributions of gaps. The evolutionarily important consequence of this result is that tree species should be favored whose saplings are able to alternate between periods of moderate to rapid growth while in gaps and periods of slow growth during the times between gaps.

Gaps close both by the branch extension growth of trees surrounding the gap and by the height growth of saplings in the gap. For these small gaps, both processes are important in gap closure. Small gaps close primarily by lateral extension growth, except where large, previously suppressed saplings are present. Large gaps close primarily by sapling height growth.

Species responses to gap size form a gradient. At one extreme are tolerant species, whose life cycle usually includes a lengthy suppressed sapling stage. These species, e.g., sugar maple, yellow buckeye, beech, and hemlock, are adapted to alternating periods of growth and suppression, and therefore seem especially able to benefit from small but repeated disturbances. They can also grow well in some larger gaps. At the other extreme are intolerant species, e.g., tuliptree, which can grow very rapidly in large gaps but cannot grow in small gaps and cannot withstand suppression. These species are therefore restricted to gaps large enough to preclude closure by lateral extension growth or by previously suppressed saplings.

Given these species differences, are processes presently occurring in the range of gap sizes studied sufficient to account for the canopy composition of the stands studied? Is there evidence that episodic large-scale disturbance events need to be involved to generate the species composition present? Runkle (1981) and Barden (1981) both found very good matches between the species composition of saplings in gaps and the species composition of the canopy in several different cove forests. Therefore, small-scale disturbance does seem adequate to perpetuate these forests. The distribution of gap sizes results in forests dominated by tolerant species, with intolerants persisting at low densities.

This analysis suggests that the relative abundance of tuliptree may be a good indicator of the disturbance regime present in a stand. Its importance should be related to the frequency of gaps >400 m² or so. Support for this suggestion comes from the fact that of the sites studied by Runkle (1981), Joyce Kilmer had both the largest gaps and the highest importance of tuliptree. Lorimer (1980), in a more intensive study of Joyce Kilmer, found average rates of disturbance (3.8–14.0% of total land area/decade) similar to those of other cove forests but concluded that tuliptree originated primarily after occasional large windthrows. The widespread distribution of tuliptree in climax forests of the Piedmont (Skeen et al., 1980) implies that such intermediate-size disturbances (say, 400 m² to 1 ha) are fairly common there. The virtual absence of tuliptree in most cove forests studied in the GSMNP implies that disturbances >400 m² are relatively rare there.

B. Forests of the Allegheny Plateau, Pennsylvania

The forests of the Allegheny Plateau in northwestern Pennsylvania differ from the cove forests of the southern Appalachians in their disturbance regime. The Allegheny forests are affected more often by large-scale disturbances. However, small-scale disturbances also occur and are important. Thus, the Allegheny forests represent a disturbance regime intermediate between the cove forests and forests whose dynamics are dominated almost completely by occasional large disturbances, such as the pine-dominated forests of northern Minnesota. Also, the literature on the Allegheny forests relates more clearly how topographic position and soil structure influence the disturbance regime and the vegetation.

The Allegheny Plateau contains broad, level uplands interspersed with narrow river valleys (Hough and Forbes, 1943). The uplands are 600–750 m above sea level south of the glacial border and held up by the hard sandstones of the Pottsville and Pocono series. The valleys are V-shaped, narrow, and winding, with a relief of ≥120–240 m. Slopes are usually steep and rocky. A mantle of surficial materials of varying thickness completely covers the bedrock of almost the whole region, becoming generally thicker on lower slopes (Goodlett, 1954).

Differences in soils and topography are reflected by differences in vegetation. Of several types of presettlement forests that occurred in this area, two will be examined here. Stands dominated by white pine occurred on sandy river flats and terraces and on lower slopes where the soil was loose and sandy, particularly on south-facing slopes (Hough, 1936; Hough and Forbes, 1943; Marquis, 1975b). American chestnut, red maple, northern red oak (*Quercus rubra*), and white oak (*Q. alba*) were confined mainly to these stands. A second major vegetation type was dominated by eastern hemlock and American beech. This vegetation type was the most widespread climax type, occupying most north-facing slopes and poorly drained upland sites (Hough, 1936; Hough and Forbes, 1943; Marquis, 1975b; Bjorkbom and Larson, 1977). Common associates of hemlock and beech in these stands were sugar maple, black cherry (*Prunus serotina*), and yellow birch.

These two different vegetation types were characterized by substantially different disturbance regimes, which interacted with the soils and topographic positions to determine the vegetation. White pine was associated with disturbances such as fires and windthrows large enough to allow light to reach the forest floor and severe enough to expose mineral soil. Fire frequency in the region is greater than in the Appalachian mountains, although less than in the forests of northern Minnesota (Bormann and Likens, 1979). In sites near Heart's Content, Lutz (1930b) found fire scars on 86 trees, accounting for 41 different years in the interval 1727–1927. Five fire years were noted on six or more trees each. Such fires are thought to have given rise to white pine stands in Heart's Content and Cook Forest, two of the only extant pine stands in the region (Lutz, 1930b; Hough and Forbes, 1943). In other places, white pine originated primarily after windthrows uprooted trees and exposed mineral soil on treefall mounds (Goodlett, 1954). Large white pine stumps are still abundant on treefall mounds in various stages of settling (Goodlett, 1954). This mechanism of establishment also helps explain the existence of white pine in several areas as scattered individuals rather than as a pure stand originating after one large-scale disturbance (Lutz and McComb, 1935; Goodlett, 1954).

The disturbance regime of the moister uplands varied considerably from that of the pine-dominated stands. Fires were rare or absent due to the moist forest floor and lack of inflammable undergrowth (Lutz, 1930a; Hough, 1936; Goodlett, 1954; Bjorkbom and Larson, 1977). Even at Heart's Content, the section without pines, which was cooler and moister than the section with pines, contained no evidence of fires (Lutz, 1930b). Occasional large-scale windthrows do occur (Hough, 1936; Goodlett, 1954). For instance, large storms in 1808 and 1870 uprooted trees in areas many hectares in extent in the Tionesta tracts (Bjorkbom and Larson, 1977). Areas affected by such large-scale disturbances regenerate into stands dominated by species with intermediate tolerance and long-lasting dormant seeds, such as red maple, black cherry, black birch (*Betula lenta*), and yellow birch. Surviving saplings of hemlock, beech, and sugar maple may also be present.

Despite the existence of these large-scale disturbances, "a widespread blow-down during a single intense storm is probably less common than the loss of a single tree here and there throughout the stand over a long period" (Hough 1936, p. 19). Many of the major sources of regional disturbances affect trees singly or in small clumps. Prolonged periodic droughts occur and result in heavy mortality of shallow-rooted trees species such as hemlock and yellow birch (Hough and Forbes, 1943; Bjorkbom and Larson, 1977). However, the effects of such droughts might be expected to be restricted to scattered individuals that are already weakened or located on unfavorable microsites. Similarly, ice or glaze storms occur but cause loss primarily of branches or scattered trees, particularly because the dominant species are fairly resistant to ice damage (Bjorkbom and Larson, 1977).

The regime of small-scale disturbances or gaps presently occurring in protected old-growth hemlock–beech stands is very similar to the one described earlier for the southern Appalachian cove forests (Runkle, 1981, 1982). Gaps were smaller on average than in the cove forests and rates of disturbance were only 0.5% of land surface area per year, near the low end for eastern forests although close to measurements from some parts of the southern Appalachians (Runkle, 1982). This low rate of disturbance is perhaps related to the more complete dominance in the Tionesta sites of two of the longest-lived species, beech and hemlock. Also, occasional cutting or large-scale disturbances may have decreased the number of old trees likely to form gaps. Beech dominated the gap regeneration for all gap sizes, but there was some tendency for hemlock to reach its maximum abundance in small gaps and sugar maple to reach its maximum abundance in intermediate-sized gaps. Overall, the species composition of saplings in gaps was very similar to the species composition of the canopy (Runkle, 1981). Therefore, a disturbance regime characterized by small-scale disturbances seems sufficient to maintain the beech–hemlock forests.

The effect of non-Indian settlement on the area was to disrupt the natural disturbance regime, with major direct and indirect consequences for the vegetation. These disruptions were not uniform, but affected some areas and some species much more than others.

The white pine-dominated areas were the most severely affected. White pine was the most prized timber species and was eliminated from the canopy almost completely by 1900 (Goodlett, 1954; Marquis, 1975b). Extensive fires from the logging slash eliminated the seedling pines (Marquis, 1975b). As a result, white pine is virtually absent from the forests today and is unlikely to return in the foreseeable future. A second important species, American chestnut, has been almost completely eliminated due to an introduced disease. As a result of these two species eliminations, the drier sites today are dominated by various species of oak (Goodlett, 1954).

The effects of human settlement on the upland forests have been less striking and more indirect, but still important. Hemlock remains important but less so than in the primeval forests, due partly to extensive logging for its tannin-rich bark. Hardwoods have increased in relative density due to their ability to sprout or survive as buried seeds following logging and fires (Marquis, 1975b). A large deer herd has become established due both to increased protection from hunting and to abundant forest growth following cutting, resulting in much available browse (Marquis, 1975b). Deer populations are now high enough to impede the growth of seedlings and saplings following natural or human-caused disturbances (Hough, 1965; Jordan, 1967; Marquis, 1974, 1981; Bjorkbom and Larson, 1977; Marquis and Brenneman, 1981). The net impact of deer browsing has been to favor beech at the expense of hemlock and other hardwoods. Because beech is one of the dominant species, the effect on the vegetation overall may be small. However, the elimination of many small hemlock stems is a concern and may result in sharp decreases in hemlock density in the future. On the other hand, hemlock regeneration in much of the region occurs irregularly, so the species may be able to survive a prolonged period of very little regeneration (Hough and Forbes, 1943).

Another change in disturbance regime affected by human use has been the elimination of large stems and therefore a decrease in the rate of gap formation. Forests characterized by small-scale disturbances have a sizeable fraction of their total area in or near gaps. Repeat disturbances are common. Therefore, saplings of many species are able to become established and be ready to respond to new openings. A second growth stand does not possess as many opportunities for saplings to become established. Unfortunately, especially given high deer-browsing pressure, the success of all cutting methods in establishing a favorable new stand requires that an abundance of seedlings already be established beneath the canopy of the existing overstory (Marquis and Brenneman, 1981). The most effective response of foresters is to mimic the primeval disturbance regime through shelterwood cutting, in which the canopy is removed in stages, gradually increasing light to the understory and increasing the number of saplings available to grow when the last of the old canopy is removed (Marquis and Brenneman, 1981).

In summary, the forest composition of the Allegheny Plateau is determined by the interaction of the natural disturbance regime, topography, and soils. South-facing slopes and sandy soils are affected by fires and blowdowns that uproot trees, both of which disturbances favor white pine and associated relatively shade-intolerant species. Upland moist sites are affected primarily by small-scale disturbances that favor shade-tolerant species. Large-scale blowdowns on these sites favor species of intermediate shade tolerance. Human influences on the area have disrupted the natural disturbance regimes, producing several changes in the species composition of the area.

IV. ARTIFICIAL DISTURBANCE REGIMES

In the preceding section, forest type and disturbance regime were found to be somewhat correlated. The causal relationship is not clear. Do the species otherwise adapted to an area (due to soils, climate, etc.) determine the disturbance regime, or does the potential disturbance regime in an area determine the vegetation? Both factors may interact simultaneously and reciprocally, so that simple causation is impossible to detect. To identify the chain of causation, it would be useful to conduct field studies, varying the disturbance regime to determine whether the pattern of disturbance by itself can affect species composition and the forest as a whole. Fortunately, such studies have been done many times at many different locations by foresters concerned with maximizing the harvest while selecting for a certain species composition in the new growth following disturbance. In the forestry literature, artificial disturbance regimes are referred to as "silvicultural systems."

Many such systems have been proposed and tested for particular locations and particular species (see, e.g., Smith, 1962; U.S. Forest Service, 1973, 1978; Tubbs, 1977). Two examples follow.

Trimble (1965) compared the effects of two different cutting regimes on cove hardwood forests in West Virginia. Uncontrolled clear-cutting on good sites had produced stands that included a high proportion of shade-intolerant species. Tuliptree, northern red oak, and black cherry made up more than half of the stems in the overstory. In contrast, Trimble (1965) used selection cuttings on 40- to 50-year-old stands to harvest individual trees. The trees removed were either large and salable or of poor quality (culls). The result of this cutting regime after 10–15 years was to favor sugar maple, which eventually seemed likely to make up over half of the stand. American beech would also greatly increase, except that it is heavily culled by foresters. The three relatively intolerant species listed above would shrink in importance to ≤20% of the future stand.

Leak and Filip (1977) obtained similar results from a stand of northern hardwoods in New Hampshire subjected to group selection. Groups of trees were removed, leaving openings averaging about 2000 m². This disturbance regime was sufficient to allow intermediate and intolerant species to maintain their relative importance in the stand at 25–35%. In contrast, under single-tree selection cuts, tolerant species came to represent 92% of canopy individuals.

One of the general conclusions of these and similar studies is that to reproduce the original species composition of the northern hardwood forest region, it is necessary to use a mixture of selection cuts (of one or a few trees at a time) and larger patch cuts or clearcuts. Selection cuts favor tolerant species such as American beech, sugar maple, and eastern hemlock. Larger cuts favor relatively intolerant species such as yellow birch and tuliptree. This mixture of gap sizes is precisely the one that characterized much of the primeval forest. Another useful silvicultural system for this forest type is the shelterwood system, in which scattered trees remain after the first cut and are removed only when the sapling layer is established. The scattered trees help shade and protect the young saplings. This system seems similar to damage by mildly severe natural windstorms or to glaze storms in which scattered trees are left standing.

The responses of individual species to different silvicultural systems can also be used to estimate the natural disturbance regime of forests originally dominated by those species. For instance, because beech and sugar maple are favored by selection cutting, it seems reasonable to hypothesize that the beech–maple forest region (Braun, 1950) was characterized by a prevalence of small-scale gap disturbance. Also, because beech is very susceptible to damage by fire (Fowells, 1965), the small-scale disturbances most common to this forest region must have been due to wind or glaze storms.

V. SUMMARY

Temperate zone forests such as those discussed here differ in species composition and structure. However, some broad similarities in their disturbance regimes exist.

Usually, these forests are affected by both large-scale and small-scale disturbance, with the relative importance and spatial distributions of each having great consequences for the regional vegetation (Whittaker and Levin, 1977; Pickett, 1980). The interplay of disturbances of different sizes is probably more important than the existence of a single intermediate type of disturbance (to oversimplify Connell, 1978) in determining species diversity and other community properties. Some forest types fit the generalization of Horn (1981a) and Oliver (1981) that large-scale (clumped in time and space and often severe) disturbances occur frequently enough so that most canopy individuals originate following such disturbances. However, small-scale gap disturbance is of primary importance for many areas and forest types. Most forests probably follow the pattern of cyclic development (succession) and steady state (climax) described by Loucks (1970), with great variations in the time and number of canopy tree generations between cycles, ranging from decades to millennia.

Further study is needed to clarify several aspects of the disturbance regimes of temperate zone forests. The primeval and therefore evolutionarily important disturbance regimes for many forest types need to be described in more detail. In North America, oak-dominated forests in particular require more study to determine the relative importance of fire, large- or small-scale windthrows, insect defoliation, and other factors. The effects of disturbance regimes greatly modified by human activity also need to be documented. The disturbance regimes associated with particular successional stages need to be clarified to determine at which successional stage a treefall provides sufficient resources for a sapling to reach the canopy instead of allowing solely for the crown expansion of its neighbors. It is also important to know the tolerance of different species of plants and animals to deviations in their primeval disturbance regimes so as to manage best their continued existence.

In summary, the concept of a disturbance regime has proved a useful way to summarize much information on the natural dynamics and regeneration of temperate zone forests. It also lends itself well to the continued development of a management theory for those forests.

RECOMMENDED READINGS

Bray, J. R. (1956). Gap phase replacement in a maple–basswood forest. *Ecology* **37**, 598–600.
Loucks, O. L. (1970). Evolution of diversity, efficiency, and community stability. *Am. Zool.* **10**, 17–25.
Watt, A. S. (1947). Pattern and process in the plant community. *J. Ecol.* **35**, 1–22.
White, P. S. (1979). Pattern, process, and natural disturbance in vegetation. *Bot. Rev.* **45**, 229–299.

Chapter 3

Stand Dynamics in Chilean *Nothofagus* Forests

THOMAS T. VEBLEN

Department of Geography
University of Colorado
Boulder, Colorado

I. INTRODUCTION

The problems resulting from a lack of appreciation of the importance of natural disturbance in forest stands have been reviewed by Oliver (1982). These problems include an uncritical acceptance of the all-aged type of structure for most old-growth stands and erroneous predictions of silvicultural responses to stand treatments due to inadequate understanding of natural stand development. One of the major objectives of stand dynamics research is the elucidation of patterns of stand development, which may, in turn, be related to successional processes in different types of ecosystems.

In simplest terms, the two general patterns of forest stand development are the autogenic and allogenic patterns (Tansley, 1935; Spurr and Barnes, 1980). The

THE ECOLOGY
OF NATURAL DISTURBANCE
AND PATCH DYNAMICS

importance attributed to "reaction" in traditional successional theory (Clements, 1916) and to community control in contemporary successional theory (Odum, 1975; Whittaker, 1975) reflects the autogenic successional pattern. Over the past decade or so, there has been a marked shift toward emphasis on allogenic influences on stand development (see, e.g., White, 1979; Oliver, 1981). According to this view, most trees reach main canopy stature (either by release or by new establishment) following disturbance of the previous canopy. While the importance of periodic disturbance in patterns of stand development is well established for forests of many regions, the allogenic and autogenic patterns are not mutually exclusive. Although one pattern may predominate in particular habitats, stand development in general is the result of the interaction of both types of processes (White, 1979).

To a certain degree, Connell and Slatyer (1977) have obviated the traditional autogenic–allogenic dichotomy by postulating three general mechanisms of successional change: facilitation, tolerance, and inhibition. They suggest that the facilitation mechanism applies well to some examples of primary succession and that evidence from a broad range of situations supports the inhibition mechanism. But Connell and Slatyer (1977) could find little empirical evidence in support of the tolerance mechanism. Oliver (1981) has postulated a general developmental pattern of forest stands following large-scale human-created or natural disturbances, which corresponds best to Connell and Slatyer's (1977) inhibition mechanism. This developmental pattern is divided into four broad stages: stand initiation, stem exclusion, understory reinitiation, and old growth. The final stage may be rare in many regions because the susceptibility of a stand to lethal disturbance increases with age and the probability of a natural catastrophe occurring increases with time.

Attempts to generalize about forest succession in the past depended greatly on the literature on North American forests. The existence of extensive areas of old-growth forests in the Chilean Lake District (c.38–42°S latitude) provides the opportunity to explore the applicability of several of these concepts to some southern temperate forests. Specifically, the questions to be addressed for the forests of the Chilean Lake District are as follows: (a) Are autogenic or allogenic influences more important in accounting for changes in species composition? (b) Do successional patterns conform best to facilitation, tolerance, or inhibition mechanisms? and (c) Does Oliver's four-stage scheme of forest stand development provide a satisfactory framework for understanding forest succession?

II. THE CHILEAN LAKE DISTRICT

This chapter describes the forests of the Andean Cordillera (Fig. 1). The Andean summits average 2000 m in this region and were extensively glaciated during the Pleistocene. Most glaciated surfaces have been covered by extensive andesitic volcanic deposits of Recent origin, underlain by Cretaceous–Tertiary sedimentary, metamorphic, and granitoid rocks (Hervé et al., 1974).

This area has a west coast maritime climate characterized by a mild temperature

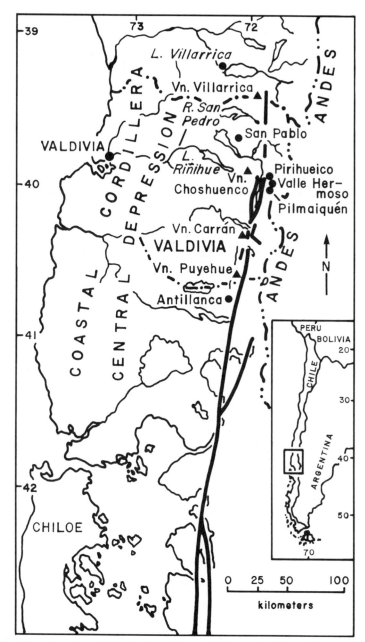

Fig. 1. Map of south-central Chile. The Reloncaví fault zone is indicated by the heavy lines, trending approximately north–south.

regime and high annual precipitation (Almeyda and Saez, 1958). A mild Mediterranean influence is reflected by the winter maximum distribution of precipitation; in the northern part of the region, a pronounced summer drought occurs, but the southern area is more continuously humid (Almeyda and Saez, 1958).

The forests of the Chilean Lake District have been described by Reiche (1934), Schmithüsen (1956, 1960), Oberdorfer (1960), Thomasson (1963), Heusser (1966, 1974), Donoso (1981), and Veblen *et al.* (1983). The lowland and lower montane forests (i.e., from sea level to c. 600 m) are mainly broadleaved evergreen with a few conifers admixed and a single major deciduous species (i.e., *Nothofagus obliqua*).[1] They are relatively species rich, with 15–20 tree species (>10 cm in diameter at breast height, dbh) per hectare and a luxuriant development of lianes and epiphytes. The most characteristic tree species of the emergent stratum are *Nothofagus dombeyi, N. obliqua,* and *Eucryphia cordifolia,* all exceeding 40 m in height and 1 m in diameter. Where canopy gaps permit light penetration to the forest floor, dense patches of the climbing bamboos, *Chusquea quila* and *C. macrostachya,* occur. Most of the forests below c. 600 m in elevation have been burned and/or logged since European settlement began in the mid-nineteenth century.

In the upper montane zone (600–1000 m), species richness decreases, but the abundance of lianes and epiphytes remains high and the total forest biomass appears to increase. The forests of this elevational belt in the Andes are characterized by emergent *N. dombeyi* and the deciduous *N. alpina,* often exceeding 45 m in height and 2 m in diameter. The common trees of the main canopy (typically 25–35 m in height) are the conifer, *Saxegothaea conspicua,* and the evergreen broadleaved *Laurelia philippiana,* and *Dasyphyllum diacanthoides.* Although the ground layer is relatively sparse, the 4- to 9-m-tall bamboo, *Chusquea culeou,* is abundant, particularly beneath canopy gaps, forming dense clumps of erect culms. The forests of the subalpine zone consist of mixtures of *N. dombeyi* and the deciduous *N. pumilio* at elevations of c. 1000–1200 m, and pure stands of *N. pumilio* above 1200 m to the upper forest limit at 1700 m. The 1.0- to 2.5-m-tall dwarf bamboo, *Chusquea tenuiflora,* forms a homogeneously dense understory in the lower subalpine forests but is absent from the upper 100–200 m of this elevational zone.

Where *Nothofagus* spp. occur in association with other tree genera in old-growth stands, they are confined almost entirely to the emergent and main canopy strata and are represented mainly by trees >50 cm dbh. In contrast, most of the associated tree species, such as *Aextoxicon punctatum* and *L. philippiana,* are represented by abundant individuals in all size classes from <5 cm dbh up to their maximum size of ~100 cm dbh. Thus, the size structure of these forests suggests that in old-growth stands, the *Nothofagus* spp. would be gradually replaced by the latter more shade-tolerant tree species. Nevertheless, extensive old-growth stands lacking *Nothofagus* spp. are not present, suggesting that this apparent successional trend is not progressing to a predicted steady-state or climatic climax forest.

In 1975 these circumstances led to the formulation of a general hypothesis of the

---

[1]Botanical nomenclature follows Muñoz (1966).

dynamics of *Nothofagus* forests (Veblen and Ashton, 1978). It was suggested that the regeneration of *Nothofagus* spp. in this region is largely dependent on massive natural disturbances (i.e., disturbances destroying forest patches measuring >0.25 ha) and that the frequency of these disturbances prevents the successional replacement of the *Nothofagus* spp. by the shade-tolerant tree species.

III. PATCH DYNAMICS

A. Large Patch Initiation

The hypothesis that the abundance of *Nothofagus* in the mid- and low-elevation Andean forests is largely a consequence of periodic massive disturbances was developed as a direct result of the 1960 earthquake that devastated southern Chile. From May 21 to 25, the Lake District was rocked by 11 earthquakes, each measuring at least 6.0 and one 8.5 on the Richter scale (Wright and Mella, 1963; Saint-Amand, 1962). Associated with this seismic activity were thousands of debris avalanches, landslides, and mudflows, which occurred in a 300- by 1000-km area of the Andes totaling at least 2.8% (25,000 ha) of the land surface of the province of Valdivia (Veblen and Ashton, 1978).

Although winter storms have triggered a few small mass movements since 1960, nearly all of the large mass movements still conspicuous in this landscape were triggered by the 1960 earthquake (Wright and Mella, 1963). Three conditions make this region especially susceptible to large, frequent mass movements. First, many slopes have been oversteepened by glacial erosion. Second, the volcanic lithology consists of alternating layers of very porous pumiceous lapilli (scoriaceous gravel-sized ash) and fine ash weathered to a stage high in allophane, which serves as a lubricant when moist, creating mudflows more than 1 km in length (Wright and Mella, 1963). Third, southern Chile is one of the most seismically active parts of the world (Weischet, 1960).

In 1975–1976 the colonization of the surface 350–850 m in elevation exposed by the 1960 slides was investigated (Veblen and Ashton, 1978). Toward the lower end of this elvational range, *Eucryphia cordifolia, Nothofagus obliqua,* and *Weinmannia trichosperma* were the most common tree species on the exposed surfaces, while above c. 500 m *Nothofagus dombeyi* was the most abundant colonizing species. On the most favorable sites (i.e., on the more finely textured substrates), dense thickets of the colonizing tree species with average heights in excess of 3 m had developed by 1975. Although shade-tolerant species are common in the adjacent forests, their seedlings were either absent or extremely rare on the surfaces exposed in 1960.

The survey of vegetation on the surfaces exposed in 1960 demonstrated the ability of *N. dombeyi* and *N. obliqua* (and of *E. cordifolia* and *W. trichosperma* at the lower elevations) to establish and become dominant relative to shrubs and herbs. *Nothofagus alpina* is by far the most valuable timber species in this region, and logging had removed nearly all reproductively mature trees from accessible areas. However, in more remote areas where seed trees remained, *N. alpina* was also a common colonist of the exposed sites.

While the 1975–1976 observations established that the tree species that occur mainly as large, emergent trees in the surrounding old-growth forests are capable of establishing and growing on bare surfaces, it was not immediately clear whether the events of 1960 represented a frequent disturbance or not. Inspection of soil profiles indicated that many old-growth stands had developed on old landslide scars or debris deposits (Veblen and Ashton, 1978). Historical sources showed that 47 notable earthquakes occurred between Santiago (33°30'S) and Chiloé Island (c. 42°S) between 1520 and 1946; 7 of these were roughly comparable in magnitude to the main shock of May 22, 1960 (Saint-Amand, 1962; Guarda, 1953; Weischet, 1960). In association with these major earthquakes, large mass movements were reported in the Andes of the Lake District (Fonck, 1896; Davis and Karzulovíc, 1963). Thus, the historical evidence indicates that events similar to those of 1960 have occurred several times during the potential life spans of the *Nothofagus* spp., which commonly exceed 400 years and occasionally 500 years.

Volcanism is another regionally extensive disturbance in the Lake District that creates bare surfaces suitable, in some cases, for the establishment of *Nothofagus* spp. The entire region has been devastated by catastrophic volcanism repeatedly during the Holocene, as indicated by regionally distributed volcanic ash deposits several meters thick overlying fossil soils and glacial deposits. Numerous volcanoes in the Lake District have been active during the past 2 centuries of historical observations (Table 1). Although these eruptions have been relatively minor, the lava flows and ash deposits have destroyed thousands of hectares of forest on the flanks of the volcanoes. The regional importance of ash falls from even minor eruptions, such as the 1971–1972 eruption of Volcán Villarrica, is shown by the deposition of large quantities of sand-sized ash in an experimental forest plot 25 km southeast of the source. Collections from 96 seed and litter traps indicated that over a period of a few days, 13.6 tons/ha of ash were deposited. Similarly, during the 1960 eruption of Volcán Puyehue, a 22,000-ha area was covered by pumice sand and gravel to a depth of 5 cm (Wright and Mella, 1963).

The three *Nothofagus* spp. of low and middle elevations (*N. obliqua, N. alpina,* and *N. dombeyi*) all colonize volcanic deposits (Veblen and Ashton, 1978; Donoso, 1981). Their abundance and growth rates vary greatly, depending on the properties of the volcanic deposits. Thick deposits of sand-sized ash allow the most rapid stand development, but stands can even develop on relatively little-weathered lava flows covered with a sparse layer of scoria. Colonization of volcanic deposits often leads to the development of relatively even-aged stands of *Nothofagus*. One even-aged (c. 150- to 175-year-old) forest of *N. dombeyi* covers c. 20 ha and occurs on sand- and small gravel-sized ash extending to a depth of at least 3 m. Buried stumps and logs of various sizes up to 1 m in diameter indicate that the forest destroyed by the ash deposit was an old-growth stand.

While volcanic eruptions occasionally ignite forests in the Lake District (e.g.,

TABLE 1

Volcanic Eruptions Recorded during Historic Times in the Andes of South-Central Chile[a]

Volcano	Location		Dates of eruptions
	Latitude	Longitude	
Antuco	37°24′S	71°22′W	1752, 1820, 1829, 1852–1853, 1861
Lonquimay	38°22′S	71°35′W	1853, 1887–1889
Llaima	38°42′S	71°42′W	1640, 1852, 1862–1866, 1876, 1887, 1889, 1895, 1896, 1903, 1912, 1917, 1927, 1941, 1955, 1957
Villarrica	39°25′S	71°57′W	1822, 1869, 1874, 1883, 1908, 1920, 1958–1959
Chos huenco	39°55′S	72°03′W	1864, 1869, 1893, 1907
Carrán	40°21′S	72°06′W	1955, 1979
Riñinahue	40°22′S	72°06′W	1907
Puyehue	40°35′S	72°08′W	1907, 1921–1922, 1926, 1960
Osorno	41°06′S	72°30′W	1719, 1790, 1835
Calbuco	41°20′S	72°37′W	1837–1838, 1917, 1929, 1961
Hequi	42°20′S	72°40′W	1890, 1893, 1896, 1906, 1907, 1920
Michinmávida	42°46′S	72°27′W	1742, 1835
Corcovado	43°11′S	72°48′W	1835

[a] From Martin (1923) and Casertano (1963).

during the eruption of Volcán Carrán in 1979), the frequency of lightning-ignited fires is essentially unknown. Noting the very humid climate of the region and the rare occurrence of lightning storms, Chilean foresters generally believe that only anthropogenic fire is significant in the Lake District. However, during some years when summer drought is intense (e.g., 1979) or when large populations of the native bamboo flower and die synchronously, a huge quantity of dry fuel becomes available (Veblen, 1982).

Although lightning-ignited fires may be infrequent relative to the human life span, they have probably occurred frequently enough during the 400- to 500-year potential life spans of *Nothofagus* spp. to have had a significant impact on the development of forest composition and structure. Human-set fires, on the other hand, have had a definite impact on the Lake District forests. Prior to European colonization in the mid-nineteenth century, the Lake District was occupied by an aboriginal population that used fire to clear forests in order to practice small-scale agriculture (Encina, 1954; Lauer, 1961). It is likely that pre-European burning had a major effect on forest patterns, particularly at the lower elevations (i.e., below c. 600 m) where agriculture was possible. Since European colonization, fire has had a great impact on the forests. In the late nineteenth century, many fires were set at low elevations to clear land for agriculture. Fires, both intentional and accidental, were also commonly associated with logging activities. Today, unburned and unlogged stands are rare at elevations below c. 600 m. Nevertheless, extensive areas of upper montane (c. 700–1000 m) and subalpine forests that show no signs of a fire history are still found in the Lake District.

Fire clearly favors *Nothofagus* spp. at the expense of most of the associated tree species. In the lower montane zone (300–600 m), extensive secondary stands of *N. obliqua* and *N. alpina* have developed following the intensive period of logging and burning of 50–20 years ago; both species stump-sprout vigorously after cutting or burning. Although *N. dombeyi* does not regenerate vegetatively, it is often established in abundance following fire in the upper montane zone. However, where the *Chusquea* bamboos are initially abundant in the understory, following burning or logging, dense bamboo thickets may develop, totally impeding *Nothofagus* regeneration and severely inhibiting the regeneration of the shade-tolerant trees; sparse, slow-growing seedlings of the shade-tolerant species, such as *Laurelia philippiana* and *Aextoxicon punctatum*, may occur beneath the bamboo canopy. Some *Chusquea* species also significantly influence patterns of forest stand development because of their habit of flowering, seeding, and dying synchronously at supra-annual intervals (Hosseus, 1915; Gunkel, 1948). Following a massive die-off of the bamboo, the advance regeneration of the most shade-tolerant tree species as well as the new establishment of less shade-tolerant tree species eventually take over the site.

In the subalpine forest zone, snow avalanches are another potentially stand-devastating disturbance. The common tree species at sites most frequently affected by avalanches is *N. antarctica*, which characteristically has a krummholz form that reduces the probability of uprooting or trunk breakage. *Nothofagus pumilio*, the dominant of the subalpine forests, may also assume a krummholz form in the timberline zone but occurs more commonly as erect, well-formed trees. Strips of uprooted and snapped-off trees are common at the upper forest limit. These avalanche paths are often over 50 m wide and may extend downslope for several hundred meters. Following the disturbance of the subalpine forests by an avalanche, dense thickets of *N. pumilio* seedlings establish, eventually developing into the even-aged *N. pumilio* stands that are so common near the upper forest limit (Veblen *et al.*, 1981).

B. Developmental Patterns of Large Patches

Sudden availability of at least some types of resources for plant growth is expected following the destruction of an existing forest by a large-scale disturbance. This sudden availability may permit the establishment of numerous tree seedlings over a relatively short period of time, resulting eventually in the development of a relatively even-aged population of mature trees. The span of tree ages in such stands would, of course, vary with the length of the colonization phase, which, in turn, would vary with such factors as dispersal capacities, favorability of the habitat, and the size of the disturbed area. Thus, in stands described as "even-aged," trees may have become established in a 2- to 3-year period or over a period of many decades.

Competition among the trees of a narrowly even-aged population may result in some characteristic patterns of stand development. For example, several studies have shown that soon after a stand of even-aged woody plants becomes established, the size–frequency distribution is a negatively skewed, bell-shaped curve; the distribution subsequently becomes positively skewed and eventually approaches normality after substantial thinning (Mohler *et al.*, 1978). Ford (1975) suggests that in relatively old even-aged populations (i.e., after self-thinning has begun), a slightly bimodal size–frequency distribution is likely to develop; detection of the bimodality depends upon selection of approximately 12 or more equal-sized classes for the frequency analysis. The bimodality is apparently due to the formation of a distinct subpopulation of larger trees, which have been more successful under the intense competition of an even-aged stand and have outgrown the majority of the population. These may be trees that established slightly earlier or grew under more favorable conditions due to microsite differences or to initial spacing of colonists. Intense competition through the mechanism of density-dependent mortality also favors the development of a regular rather than a contagious spatial distribution. These trends in the development of even-aged tree populations have been investigated mostly in tree plantations less than c. 50 years of age or in natural stands of relatively young (i.e., c. <100 years old) trees (Ford, 1975; Cooper, 1961; Laessle, 1965; Mohler *et al.*, 1978). Structural data on relatively even-aged *Nothofagus* stands, which originated following a variety of stand-devastating disturbances, suggest that their developmental patterns are similar to these general patterns.

TABLE 2

Characteristics of Even-Aged *Nothofagus* Stands Sampled

Stand	Dominant	Plot size (m²)	Elevation (m)	Age of oldest tree cored (years)	Mean age (± SD) of n trees (years)	Mean height of dominants (m)
A-2	*N. alpina*	500	680	27	21.3 ± 2.9 n = 12	14
A-1	*N. alpina*	500	680	28	23.6 ± 3.5 n = 12	14
B	*N. alpina*	500	680	25	20.6 ± 4.7 n = 13	16
O	*N. alpina*	500	680	26	22.9 ± 2.2 n = 14	14
E-3	*N. alpina*	500	620	35	31.5 ± 3.6 n = 6	20
E-1	*N. alpina*	500	620	60	51.7 ± 7.7 n = 7	28
W	*N. pumilio*	2754	1000	139	99.3 ± 30.3 n = 13	16
S	*N. pumilio*	3888	1120		110[a]	20
Q-1	*N. dombeyi*	3888	810	150	145.0 ± 5.0 n = 3	34
H-1	*N. dombeyi*	3888	940		250[a]	40

[a] Estimated ages.

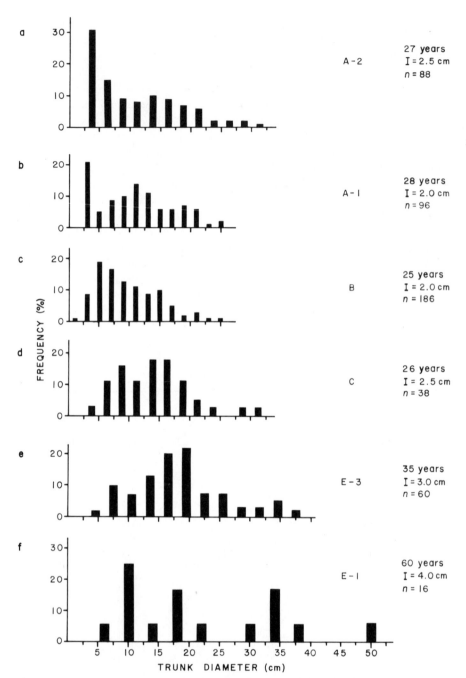

Fig. 2. Frequency distributions of trunk diameters (at breast height) for *Nothofagus* spp. in approximately even-aged stands. I, the interval of trunk diameters between two successive bars in the histogram, was determined by dividing the range of diameters observed into 12 or 13 equal intervals.

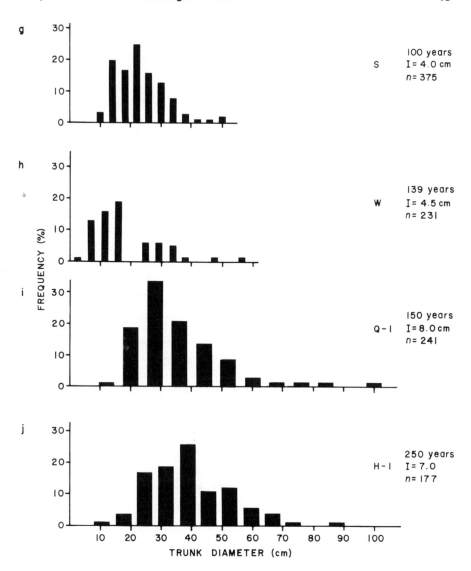

Six stands of relatively young (25–60 years old) *N. alpina*, which developed after old-growth forests were clear-cut, were sampled to illustrate early stages of stand development. Older stands of *N. dombeyi* or *N. pumilio* that originated after snow avalanches or burial by volcanic ash or flood deposits are used to illustrate the characteristics of older, even-aged stands (Table 2).

In the young *N. alpina* stands, 6–13 trees over the entire diameter size range were cored at a height of 30 cm for age determination. Trees were cored in only two of the four older stands; in the other two older stands, an estimate of stand age was derived from ring counts of tree stumps on adjacent, similar sites. In each stand a considerable range of ages was encountered; in fact, most natural "even-aged" stands consist of trees that establish over several years. In the case of *N. alpina* stands, new stems continued to originate (from both seedling establishment and sprouting) for approximately 10 years after the disturbance. Where a disturbance created large open areas (e.g., more than a few hectares), as in the cases of burial by ash or flood deposits, it is likely that colonization of the site continued for 2 or 3 decades. Given this variation in age, it is unlikely that the development of the *Nothofagus* stands will precisely follow the pattern postulated for truly even-aged tree populations such as plantations. Three of the four 25- to 28-year-old *N. alpina* stands have size-class distributions that are highly negatively skewed and roughly bell-shaped (Fig. 2a–c). The fourth stand (Fig. 2d), however, deviates markedly and has a broadly bell-shaped distribution; it is characterized by a dense understory of *Chusquea culeou* and *C. quila* that appears to have significantly diminished tree establishment. For all four stands, very slight tendencies toward bimodal distributions are evident. In the 35-year-old stand (Fig. 2e), the skewness of the bell-shaped distribution has shifted to the right and the tendency toward bimodality is slightly greater. In the 60-year-old stand (Fig. 2f), a strongly bimodal distribution has developed, and at this stage a definite subpopulation of more rapidly growing trees is evident.

In the two c. 100-year-old *N. pumilio* stands that were initiated after snow avalanches, the size class frequencies approach normal distributions, with a slight bimodal tendency due to the presence of a few larger trees (Fig. 2g–h). In the two stands of *N. dombeyi* that were initiated after the deposition of ash or flood deposits, similar patterns have developed (Fig. 2i, j). In general, the size-class distributions of these 10 successively older stands correspond approximately to the patterns found in other even-aged tree populations (Ford, 1975; Mohler *et al.*, 1978). At a scale of 9 m^2, none of the tree populations are significantly clustered ($P < .05$; F test applied to Morisita's index [Morisita, 1959]); however, in three of the four stands, there was a marked tendency toward a regular distribution, as expected in the development of even-aged stands (Cooper, 1961; Laessle, 1965).

C. Gap Dynamics in Old-Growth Forests

The degree to which *Nothofagus* spp. depend on stand-devastating disturbances for their regeneration was investigated by considering regeneration patterns in old-growth stands in the lowland, montane, and subalpine zones of the Lake District. Old-growth stands that showed no signs of large-scale (i.e., influencing an area >0.25 ha) disturbance for at least the past 200 years were selected to test the hypothesis that *Nothofagus* spp. may regenerate beneath canopy gaps created by the fall of the large-crowned emergent trees.

At c. 250 m near Lake Villarrica, emergent 40- to 42-m-tall *N. obliqua* project c. 10 m above the main canopy formed by the evergreen *Aextoxicon punctatum*,

Persea lingue, and *Laurelia sempervirens.* The other common emergent tree, *Eucryphia cordifolia,* attains heights of c. 35–40 m. The structure of this forest was studied in four 0.25-ha plots and one 10 × 320-m belt transect (Veblen *et al.,* 1979). It had one of the densest canopies of any old-growth forest in the Lake District and, consequently, had an extremely sparse understory. *Nothofagus obliqua* was represented only by trees >50 cm dbh and was most abundant in the 80- to 100-cm-dbh size class (Fig. 3a). Even in five treefall gaps ranging in size from 29 to 54 m², no saplings (i.e., trees >2 m tall but <5 cm dbh) of *N. obliqua* were present. Instead, the gaps were vigorously filled by saplings of shade-tolerant *A. punctatum,* *P. lingue,* and *L. sempervirens,* as well as by root suckers of *E. cordifolia. Aextoxicon punctatum* is sufficiently shade tolerant so that even beneath a continuous canopy it is represented by abundant saplings and small-diameter stems, while the other three species are mostly dependent on gaps for their regeneration. In the

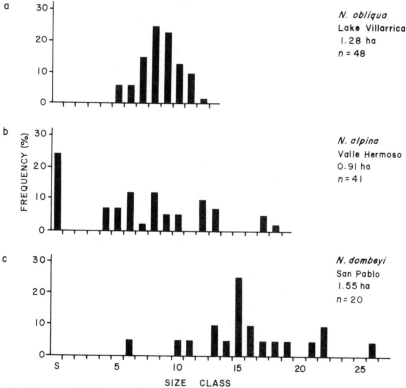

Fig. 3. Frequency distributions of *Nothofagus* spp. in size classes in old-growth stands. The size classes used are S (saplings) for trees <5 cm dbh but at least 2 m tall, and 1–26 for trees in 10-cm-dbh classes from 5 to 265 cm.

absence of stand-devastating disturbances, *N. obliqua* must eventually disappear from this forest. Since the Villarrica forest had been an area of aboriginal settlement in pre-European times, human-set fire was probably the disturbance that allowed the initial establishment of *N. obliqua* (Lauer, 1961). However, remnant, partially logged stands of similar structure elsewhere in the Lake District may owe their origin to ash deposition, mass movements, and flood deposition in addition to fire.

Stands representative of old-growth forests of the upper montane zone were studied at elevations of 800–950 m (Veblen *et al.*, 1980, 1981). Although seedlings of *N. dombeyi* often occur at densities of several thousand per hectare in old-growth stands, they are restricted almost entirely to heights <20 cm. Observations in permanent plots indicate that beneath closed canopies the *Nothofagus* seedlings fail to survive for more than 2 years or to reach heights greater than a few centimeters (Burschel *et al.*, 1976). In stands not recently affected by large-scale disturbances, *N. dombeyi* was represented mainly by trees >100 cm dbh (Fig. 3c). In stands that had been periodically disturbed by small debris flows and rockfalls, such as along the lower slopes at Valle Hermoso, a few *N. dombeyi* occurred in nearly all dbh classes from c. 10 to >150 cm. In the most undisturbed old-growth stands, *N. alpina* was also represented mainly by trees >50 cm dbh (Fig. 3b); however, it was also represented by numerous saplings that were clustered beneath canopy gaps of up to c. 150 m². *Nothofagus alpina* is more shade tolerant than *N. dombeyi* (Müller-Using, 1973) and, consequently, has a greater potential for regeneration beneath small treefall gaps. Nevertheless, the absence of small-diameter (i.e., 5–40 cm) stems of *N. alpina* in the most undisturbed stands indicates that regeneration beneath treefall gaps must be rare; the saplings that establish beneath gaps apparently fail to develop unless the stand is affected by larger-scale disturbances.

The failure, or at least the infrequency, of *N. dombeyi* and *N. alpina* to attain main canopy stature by developing beneath treefall gaps in otherwise undisturbed stands contrasts markedly to the regenerative behavior of the more shade-tolerant trees of the upper montane zone. In most old-growth stands, *L. philippiana* is represented by abundant stems <1.4 m tall (e.g., 5000–10,000/ha) and saplings (e.g., 500–2000/ha), most of which originated as root suckers. Beneath canopy gaps these root suckers proliferate, forming dense patches. Although less abundant than *L. philippiana*, *D. diacanthoides* and *S. conspicua* also regenerate beneath relatively small (i.e., 20–150 m²) canopy gaps. The lack of gap-phase regeneration of *Nothofagus* in the upper montane forests is due to the preemption of gaps by these three tree species, which are usually already present as seedlings or root suckers at the time of gap creation, and, more importantly, by the 4- to 9-m-tall bamboo, *Chusquea culeou*. In most old-growth stands, *C. culeou* attains a density of at least 5000 culms/ha and occurs in dense patches of hundreds of culms beneath canopy gaps. Given its ability to grow up to 9 m in length in a single vegetative season (Veblen, 1982), *C. culeou* poses a formidable obstacle to tree regeneration. The regeneration of the more shade-tolerant tree species is less seriously affected, than that of *Nothofagus*, but the vegetational response to treefall gaps is strongly dominated by *C. culeou* in the upper montane zone.

Above the upper montane zone, a narrow transitional zone of forest is dominated by *N. alpina* and *N. dombeyi* from the montane zone and *N. pumilio* from the subalpine zone. In the absence of the dense intermediate tree stratum of shade-tolerant trees characteristic of the montane forests, the canopies of these forests are relatively open. *Chusquea culeou* is replaced by the much smaller *C. tenuiflora*, which has mean and maximum culm lengths of c. 1.5 and 2.5 m, respectively. In contrast to the patchy distribution of *C. culeou* beneath gaps, *C. tenuiflora* is homogeneously and densely distributed over areas of several hectares of forest (Veblen, 1982); this is probably due to the more open canopies of the subalpine, old-growth forests. In these transitional forests, clusters of 6–12 saplings or three to five 5- to 10-cm-dbh stems of *N. alpina* often occur beneath canopy openings; among the three *Nothofagus* spp. of this zone, *N. alpina* is the most effective in regenerating in a gap-phase mode (Veblen *et al.*, 1980). Its success here may be attributed both to the more open tree overstory and to the much shorter bamboo layer; seedlings need only attain heights of c. 1.5 m to project above the bamboo layer, in contrast to heights three to five times those required in the montane forests.

At slightly greater elevations (c. 1100–1250 m at 40°S), the subalpine forests are characterized by *N. dombeyi* and *N. pumilio* and, above this zone, consist of pure stands of *N. pumilio*. In old-growth stands, the presence of dense clusters of small trees of both species beneath canopy gaps suggests a gap-phase mode of regeneration. Given the shade intolerance of these two *Nothofagus* spp. and the relatively continuous bamboo layer, the occurrence of any regeneration of *Nothofagus* is surprising. However, where treefalls have created canopy gaps (commonly c. 12–25 m in diameter), the microenvironment appears less favorable to the bamboos than to *Nothofagus* seedlings. Beneath canopy gaps, *C. tenuiflora* is shorter, has fewer culms, and provides less cover than sites beneath the adjacent forest canopy (Veblen *et al.*, 1981). Although *Nothofagus* seedlings <25 cm are usually abundant beneath the bamboo layer, 1- to 2-m-tall plants are found only in areas of reduced *C. tenuiflora* cover. In mixed stands, *N. pumilio*, which is often represented by >100,000 seedlings <25 cm tall/ha beneath the bamboo, responds more vigorously to the reduced bamboo cover. Where a stump or log is present in a gap, *N. dombeyi*, which establishes and grows vigorously on elevated woody substrates, is favored. Observations during late spring indicate that snow persists longer beneath canopy gaps in the subalpine forests, presumably due to greater accumulation. Thus, sites beneath canopy gaps have an effectively shorter vegetative period. Furthermore, during summer the frequency of temperatures below 0°C is greater beneath canopy gaps than in the adjacent forests (Veblen *et al.*, 1981). These microenvironmental conditions beneath gaps may account for the reduced importance of the bamboos in gaps, which, in turn, permits the regeneration of the *Nothofagus* spp.

IV. CONCLUSIONS

In the Andes of the Chilean Lake District, allogenic processes appear to be more important in accounting for changes in forest composition than do autogenic processes. The failure to recognize the importance of periodic natural disturbance impeded the attempts of earlier workers to arrive at a comprehensive interpretation of *Nothofagus* stand dynamics. For example, Brun (1975) and Burschel *et al.* (1976) suggested that in old-growth, mixed stands of the upper montane zone, as old senescent trees gradually died, the resulting canopy gaps would be filled by newly established *N. dombeyi* and *N. alpina*. However, neither could provide any descriptions of this process, and both neglected to consider the possibility that old-growth stands originated following massive disturbances. Although it is likely that *Nothofagus* spp. occasionally regenerate following relatively small treefalls, establishment following stand-devastating disturbances is far more important in accounting for the abundance of these trees. Earlier attempts to interpret these old-growth stands as self-maintaining steady states may be another example of the oppressive influence of the climax notion (Sprugel, 1976).

The developmental patterns of *Nothofagus*-dominated forests following massive disturbance correspond reasonably well to both Oliver's (1981) general model of stand development and the more detailed pattern of structural changes predicted for even-aged, single-species stands (Mohler *et al.,* 1978; Ford, 1975). Following logging and burning of an old-growth stand in the Lake District, both shade-intolerant and shade-tolerant trees are present at the initiation of stand development. The shade-tolerant species initiated growth (both from sprouts and from new establishment) simultaneously with *N. alpina* but, under the open conditions of the site, had grown more slowly in height and diameter. However, where a more severe disturbance, such as a mass movement or volcanic ash deposition, has eliminated the possibility of resprouting from stumps or roots as well as any buried viable seeds, the pattern of stand development is quite distinct. In primary successional development, *Nothofagus* seedlings establish in abundance, whereas seedlings of the shade-tolerant species are absent despite the availability of seed from adjacent stands. After development of a closed canopy of *Nothofagus* spp., seedlings of the shade-tolerant species establish and gradually become more abundant as the stand ages. Their success appears to depend on modification of the site by the colonizing *Nothofagus* spp. Thus, primary succession resulting in old-growth, mixed-species stands conforms to Connell and Slatyer's (1977) facilitation mechanism. Once the old-growth stage is attained, however, further changes in stand composition (assuming no massive exogenous disturbance) correspond best to the inhibition mechanism. Following a treefall in an old-growth stand, the species that benefits the most is already present on the site.

There are interesting parallels between the forest dynamics described for the Chilean Lake District and those in *Nothofagus*-dominated forests in New Zealand. On both the North and South Islands of New Zealand, large-scale mass movements have been triggered by earthquakes, and in some habitats *Nothofagus* spp. are the initial colonizers (Robbins, 1958). In general, the regeneration of *Nothofagus* spp. in New Zealand follows a whole-stand replacement process in which even-aged populations develop following disturbances such as mass movements, extensive windthrow, lethal insect attack, and severe snow damage (Wardle, 1970; Wardle and Guest, 1977; June and Ogden, 1978). Periodic large-scale disturbances, including volcanism and drought, in addition to those already noted, are also highly important influences on the composition and structure of the New Zealand conifer–hardwood forests lacking *Nothofagus* (Molloy, 1969; Veblen and Stewart, 1982; Jane and Green, 1983). In southeastern Australia, the other major region of temperate *Nothofagus* forests, catastrophic fire has a pervasive influence on forest dynamics. Here, *N. cunninghamii* sometimes develops into even-aged stands following fire but, in general, appears to be more an equilibrium-type species than any of the New Zealand or South American *Nothofagus* spp. (Howard and Ashton, 1973; Mount, 1979).

Drury and Nisbet (1971) have proposed that equilibrium models of succession be abandoned in favor of what they have termed a "kinetic" view of vegetation change. Such a view accepts the fundamental instability of the physical site, emphasizes continuous change, and does not require the existence of stable endpoints. The major objective of a kinetic theory of vegetation change would be to predict the physical and biological disturbances that are so important in determining changes in forest composition. This is clearly a challenging goal that is most likely to be achieved by the interdisciplinary efforts of scientists concerned with both the physical and biological processes that create a forested landscape. While there is yet no kinetic *theory* of vegetation change that can totally replace successional theory, a kinetic *viewpoint* is certainly reflected in the current research focus on disturbance (White, 1979; Oliver, 1981). Adoption of a kinetic viewpoint is essential to understanding the dynamics of the *Nothofagus*-dominated forests of the Chilean Lake District and also appears appropriate for many other forested landscapes.

RECOMMENDED READINGS

Burschel N., P., Gallegos G., C., Martinez M., O., and Moll, W. (1976). Composición y dinámica regenerativa de un bosque virgen mixto de raulí y coigue. *Bosque* **1**, 55–74.

Donoso Z., C. (1981). "Tipos Forestales de Los Bosques Nativos de Chile." Corporación Nacional Forestal, Santiago, Chile.

Heusser, C. J. (1974). Vegetation and climate of the southern Chilean Lake District during and since the last interglaciation. *Quat. Res.* **4**, 290–315.

Veblen, T. T. (1982). Growth patterns of *Chusquea* bamboos in the understory of Chilean *Nothofagus* forests and their influences in forest dynamics. *Bull. Torrey Bot. Cl.* **109**, 474–487.

Veblen, T. T., Donoso Z., C., Schlegel, F. M., and Escobar R., B. (1981). Forest dynamics in south-central Chile. *J. Biogeogr.* **8**, 211–247.

Chapter 4

Treefalls, Regrowth, and Community Structure in Tropical Forests

NICHOLAS V. L. BROKAW[1]

Department of Biology
Kenyon College
Gambier, Ohio
and
Smithsonian Tropical Research Institute
Balboa, Panama

I. INTRODUCTION

The idea that tropical forest is a structural or floristic mosaic dates at least from Aubréville's 1937 ''mosaic theory of regeneration'' (cited in Richards, 1952). Since then, authors have depicted this community as a mosaic of treefall-created patches differing in age, size, and species composition (Jones, 1950; Richards,

[1]Present address: Manomet Bird Observatory, P.O. Box 936, Manomet, Massachusetts 02345.

THE ECOLOGY
OF NATURAL DISTURBANCE
AND PATCH DYNAMICS

1952; Baur, 1964; Whitmore, 1975; Oldeman, 1978; Hladik, 1982). Treefall gaps are one phase in a forest regeneration cycle; colonization and growth in the gap phase lead to building and then mature phases, in which treefalls renew the cycle (Whitmore, 1975; Hallé et al., 1978; Oldeman, 1978). In many tropical forests, the majority of canopy tree species may depend on growth in a gap to reach maturity (see, e.g., Hartshorn, 1980). Thus, endogenous disturbance by treefalls is central to the function and structure of this community. Indeed, Strong (1977) attributes much of the high tree species richness of some tropical forests to falling trees. This disturbance may prevent competitive exclusion and reduced richness among ecologically equivalent species (Connell, 1978; Huston, 1979). At the same time, treefall gaps could provide a variety of "regeneration niches" (Grubb, 1977), meeting the requirements of ecologically distinct species (Ricklefs, 1977; Denslow, 1980a; Pickett, 1983).

Section II of this chapter deals with treefall gap regimes: parameters of frequency, size, and related phenomena that determine patch character and dimensions in the vegetation mosaic. Section III surveys plant regeneration behavior in relation to gaps. Section IV describes regrowth in gaps and further explains plant adaptations for gap regeneration. Finally, in Section V, all perspectives are brought together in a discussion of tropical forest community structure.

Except where noted, this chapter focuses on lowland tropical wet (evergreen) or moist (semideciduous) forest (Holdridge et al., 1971) and concerns only small-scale disturbance. Garwood et al. (1979) and Foster (1980) treat large-scale disturbance.

II. TREEFALL REGIMES

A. Gap Creation

Falling trees do not always create clearly delimited holes in the forest. The complex configurations resulting from a treefall can be seen in the idealized dumbbell-shaped "chablis" (Oldeman, 1978). Chablis is a medieval French word, without an English equivalent, denoting the fall of a tree, the resulting damage, and the fallen tree itself. In the ideal chablis, the fallen trunk forms the axis of the damaged area, outlined as a dumbbell on the ground. At one end, above the former base of the tree, is a gap in the upper canopy, while lower forest layers remain intact. Along the axis, the trunk may have sliced down among its neighbors, disturbing the subcanopy but otherwise causing little change in forest structure. At the other end, the tree crown obliterates the understory, while the upper canopy layers may be completely or partly damaged. Although the ideal chablis may be infrequent, treefalls creating a structure primarily like the axis region or either end of the dumbbell are common. Often, a gap will include regions corresponding to all those of the dumbbell, but not in that exact spatial arrangement or proportions.

Structural complexity makes gap area difficult to define. Depending on the definition used, the measured area of any gap can easily vary by a factor of 2. This frustrates interpretation of data on treefall disturbance regimes not accompanied by a gap definition. Nonetheless, data on gap size, turnover rate (mean time between successive creations of gap area at any one point in the forest), and related parameters of structurally mature tropical forests are presented in Table 1. The most reliable turnover rates reported are based on actual records of treefall gaps created over a period of time (Hartshorn, 1978; Brokaw, 1982b; Foster and Brokaw, 1982; Uhl, 1982b). Other estimates are based on surveys of gaps extant at one time, with turnover rates derived from the apparent age–class distribution of those gaps or estimated gap closure time (both of which are difficult to assess). The wide range of estimates in Table 1 results from disparities in method, real differences among forests, and year-to-year and place-to-place variations within forests (Putz and Milton, 1982).

Because gap size is related to regenerator composition, the size distribution of gaps is of interest. Most gaps are small (Fig. 1), although a disproportionate amount of total gap area is contributed by large gaps. The smallest gaps result from fallen liana tangles, limbs, or the gradual disintegration of a standing dead tree; the largest gaps result from multiple treefalls. In the MPassa forest, Gabon, Florence (1981) found that falling tree parts created 23% of all gaps and 10% of total gap area in the forest; single falling trees created 51% of all gaps and 38% of gap area; synchronous treefalls in domino fashion created 14% of all gaps and 16% of gap area; and temporally separated abutting or overlapping treefalls created 13% of gaps and 36% of gap area. On Barro Colorado Island (BCI), Panama, liana connections raise the incidence of multiple treefalls (Putz, 1984). Naturally, falling tree size is correlated with gap area created (Brokaw, 1982b). So, as tree size increases during secondary succession, large gaps become more frequent (Brokaw, 1982b).

Species-specific modes of death affect gap size. For example, in the Ituri forest, Zaire, *Gilbertiodendron dewevrai* typically dies standing, gradually drops limbs, and forms only small gaps (Hart, personal communication). By contrast, *Brachystegia laurentii* is generally uprooted, producing large gaps. On BCI, the emergent *Ceiba pentandra* sheds its limbs, failing to produce a gap commensurate with tree size (Brokaw, personal observation).

The shape of gaps, a determinant of gap microclimate, varies within and perhaps consistently among some forests. In the tall forest at Pasoh, Malaysia, canopy trees have relatively small crowns (Putz, personal communication) and create gaps significantly more linear than those on BCI (Foster and Brokaw, unpublished data). In heath forests (white sands) in the Sierra Madre, Luzon, falling trees generally cut gaps too narrow for light penetration to the soil (Ashton *et al.*, 1984).

Periodicity of gap creation is of interest for three reasons: (a) as a possible selective factor in the timing of dispersal or germination of species depending on gaps for early establishment, (b) implications for gap regenerator composition, considering the intermittent availability of some species' seeds, and (c) as insight into causes of treefalls. Monthly records over a period of 4 years indicate that treefalls peak every year at about the middle of the rainy season on BCI (Rand, 1976; Brokaw, 1982b), about 2 months after peak germination of many gap-depen-

TABLE 1

Various Parameters of Gap Creation Regimes in Lowland Tropical Forests[a]

Location	Mean gap size (m²)	Gaps/ha	% gap[b]	Turnover rate (years) (see text)	Critical size for pioneers (m²)	Source
Mexico	200	12.8	25		200–250	Torquebiau (1981)
Costa Rica	87–125	3.4–7.25	4–8	80–135	500	Hartshorn (1978)
Costa Rica[c]					200–300	Barton (1984)
Costa Rica			4.43		150	Acevedo and Marquis (1978)
Panama[d]	86	3.2[g]	3[g]	62–114	150–200	Brokaw (1982b); Foster and Brokaw (1982)
Venezuela[e]	160		4.8	104		Brandani (1984); Uhl and Murphy (1982)
Surinam					1000	Schulz (1960)
French Guiana	628	6–13	7			Mutoji-a-Kazadi (1977, cited in Torquebiau, 1981)
French Guiana			3–5			Hallé et al. (1978)
French Guiana		0.85[h]				Riera (1984)
Ecuador		15 (on level) 18 (on slopes)				Oldeman (1978)
Ivory Coast	140 (on level) 208 (on slopes)			75–416		Bonnis (1980, cited in Torquebiau, 1981)
Ivory Coast		5	15			Nierstrasz (1975, cited in Florence, 1981)
Gabon[f]		3	8	60		Florence (1981)
Malaysia		6	13			Whitmore (1975)
Malaysia			8			Poore (1968)

[a] See Brokaw (1982b) and Lang and Knight (1983) for data on a tropical forest not structurally mature.

[b] % gap = total gap area as a percentage of forest area.

[c] Gap area defined as the disturbed area under a canopy opening extending to the bases of canopy trees (≥20 cm dbh) delimiting the opening.

[d] Gap area defined as a vertical "hole" in the forest vegetation, extending through all levels to within an average of 2 m from the ground.

[e] Canopy gap area defined as the area with no tree crown directly above, resulting from a treefall.

[f] Gap area defined as a hole in the canopy, with "some attention to understory aspect."

[g] Gaps ≤3 years old.

[h] Gaps ≤1 year old.

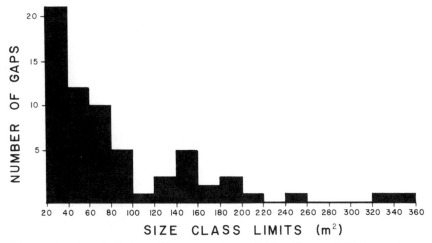

Fig. 1. Size-class distribution of 66 gaps occurring on 13.4 ha of forest on BCI, Panama, between August 1975 and April 1980. Gaps were defined as vertical holes in the forest vegetation extending through all layers down to within an average of 2 m from the ground (Brokaw, 1982a). [Reproduced by permission of *Biotropica*.]

dent species in this forest (Garwood, 1983). Trees fall most often at the beginning of the rainy season in the Tai forest, Ivory Coast (Alexandre, 1982a). At Los Tuxtlas, Mexico, treefalls peak in the season of windy, rainy "northers" (Sarukhán, 1978), and exceptionally rainy years produce many treefalls there (Torquebiau, 1981). In French Guiana, trees come down most often in the rainy season (Oldeman, 1972; Riera, 1984), but dry season winds fell the most trees in montane forest at Monteverde, Costa Rica (Lawton, personal communication). A seasonal pattern may occur in areas subject to seasonal cyclonic storms (Webb, 1958; Doyle, 1981).

Rain and wind, separately or together, cause treefalls. Rain causes trees to fall for three reasons: (a) loosened soil and root grip, (b) increased weight of water on tree surfaces during showers, and (c) reduced ability of a wet tree to assume streamlining (Gloyne, 1968). In lowland forests, gusty wind seems more effective than steady, albeit strong, wind. Thus, treefalls peak during the season of rain and gusty wind on BCI and are fewer during the steady trade winds of the dry season (Brokaw, 1982b). Trees along the periphery of gaps and exposed to turbulent air are likely to fall (Brunig, 1964; Lawton and Dryer, 1980; Putz and Milton, 1982).

We can partly distinguish the effect of rainload and wind from that of high soil moisture by considering how trees actually fall. About 75% of falling trees are snapped off at MPassa and BCI (Florence, 1981; Putz et al., 1983). Putz et al. (1983) report 63–85% snapped off trees at other sites. In these forests, it appears that aboveground agents are more closely associated with treefalls than soil loosening. By contrast, 90% of the gaps in wet forest at La Selva, Costa Rica, are caused by uprooted trees (Hartshorn, 1980), and in a big blowdown at San Carlos, Venezuela, 80% of the large fallen trees had uprooted (Uhl, 1982a). [But snap-offs predominate in single treefalls at San Carlos (Uhl, personal communication).] Of course, a tree may be uprooted by combined wind, rainload, and soil wetting; uprooting in Queensland, Australia, often occurs during gales following continuous rain (Webb, 1958). Large trees are more likely than small trees to uproot on BCI (Putz *et al.*, 1983) but not at MPassa (Florence, 1981).

Rain and wind are often simply proximate causes of treefalls, bringing down trees already weakened by fungus or disease (Hubert, 1918). In wet forest on New Britain, 25 Americans were killed by falling trees during a World War II campaign; while bombardment had reduced the stability of many trees, subsequent very heavy rains and wind were the proximate agents of mortality (Morison, 1950; Hough and Crown, 1952). Treefall gaps are also created by elephants (Leigh, 1975; Whitmore, 1975; Hladik, 1982) and by lightning (Anderson, 1964; Brunig, 1964).

Treefall rates are linked to topography. In Ecuador (Oldeman, 1978), French Guiana, and the Ivory Coast, treefalls are more common on slopes and in places exposed to wind than on level ground (Mutoji-a-Kazadi, 1977, and Bonnis, 1980, both cited in Torquebiau, 1981). Tree life expectancy is relatively short on BCI slopes (Putz and Milton, 1982). In the montane forests at Monteverde, Costa Rica, local variation in topography is closely associated with different treefall regimes and forest types. For instance, as wind stress increases with proximity to ridge crests, canopy continuity decreases (Lawton and Dryer, 1980; Lawton, 1982). Near ridge tops, forest stature is so low and gaps are so common that microclimatic and floristic distinctions between light gaps and shaded understory are blurred. Oldeman (1978) states that treefall frequency increases with altitude in Ecuador.

Soil characteristics account for variation in treefall regimes. For example, the turnover rate is higher in swampy than upland sections of La Selva, Costa Rica (Hartshorn, 1978). In some forests trees are shallowly rooted and unstable on fertile soils; in contrast, they are deeply rooted and stable on infertile, freely draining soils (Ashton *et al.*, 1984).

B. Gap Environment

Microclimate and environment in gaps are determined by many factors. For example, the duration and intensity of light received in a gap depend on its size, shape, slope, orientation, height of surrounding mature-phase forest, and characteristics of posttreefall debris and surviving vegetation. Denslow (1980a; see also Whitmore, 1975; Alexandre, 1982b) assembled data from several studies comparing the microclimate of newly opened gaps and the understory of intact forest. Light intensity and duration are greater in gaps. Soil and air temperature can be much higher in gaps and fluctuate over a greater range. Humidity in gaps is comparatively low, while evaporation from the soil surface is high. But at a few centimeters in depth, soil moisture may be higher in gaps than beneath mature-phase forest, where root uptake is greater (Lee, 1978). As a rule, the degree of microclimatological contrast produced by a treefall increases with gap size (Lee, 1978).

Treefall debris undoubtedly provides a nutrient pulse in gaps, as observed in newly cut-over areas at San Carlos, Venezuela (Uhl *et al.*, 1982). Initially, reduction in root competition is probably positively related to gap size, but the roots of peripheral trees and surviving plants do provide adequate mycorrhizal inoculum for regenerators (Janos, 1980).

Of course, uniform conditions do not prevail throughout a gap. Florence (1981) measured light intensity in gaps at MPassa. Starting at the area beneath the hole in the upper canopy, mean light intensity declines centrifugally. But recalling the structural variety of the dumbbell-shaped chablis, it is clear that environmental factors within gaps do not vary in regular fashion. Figure 2 is a crude map of a gap

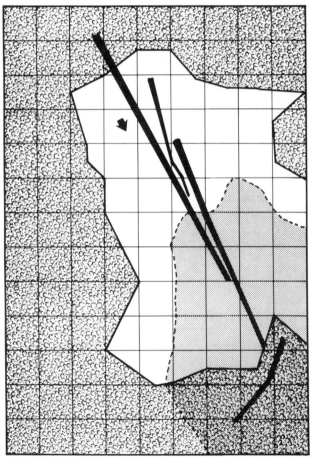

Fig. 2. Diagram of treefall gap components on a 2 × 2-m grid, Gigante forest, Panama. The canopy hole (————) extends, free of vegetation, to within an average of 2 m from the ground. The debris area (- - - -) and direction in which trees fell (arrow) are shown. The gap was created by felling trees, with care taken to simulate a natural treefall (Smith *et al.*, unpublished data).

on which the canopy opening, the area of branch and foliage debris, and fallen trunks are shown. Taking the gap as the area of the canopy opening, we see that the distribution of nutrients from decomposing trunks and debris will be quite local. Debris piles maintain dark, cool, moist regions below, while other areas of the gap are potentially bright, hot, and dry, depending on the arrangement of saplings and understory plants. The patchy occurrence of smothering lianas, irregularities in vertical structure, discrete areas of soil disturbance, possible allelopathic influences (Anaya, 1976), and the variety of surrounding tree species, each with its separate influence, could further complicate this picture. Newly formed gaps are thus composed of abutting yet highly contrasting microenvironments, and the types of microenvironments and their spatial arrangement in gaps vary greatly from site to site.

III. REGENERATION BEHAVIOR

Trees regenerate in gaps via plants established prior to the treefall and via seeds germinating after gap creation. These two modes characterize two principal regeneration types: primary species and pioneers. Although a continuum of regeneration behaviors exists among tropical forest trees, workers have found it convenient to assign species to broad classes (see, e.g., Hartshorn, 1980).

A. Primary Tree Species

In any tropical forest a large group of tree species—variously termed "shade tolerant," "persistent," "climax," or "primary forest species"—germinate in the closed forest understory and exist as suppressed juveniles until a gap opens above them, permitting accelerated growth. In this group dispersal is sometimes quite limited, resulting in clumped seedlings (Jones, 1956; Poore, 1964; Liew and Wong, 1973) and consequently a clumped distribution of species among small gaps and within large gaps (Schulz, 1960). It is frequently stated that primary species show little seed dormancy (Schulz, 1960; Whitmore, 1975; Moreno, 1976). But in a thorough study on BCI, Garwood (1983) observed wide interspecific variation in periods of seed dormancy. For a given species, germination may occur only in shade or in both light and shade (Baur, 1964; Whitmore, 1975). For example, *Dacryodes excelsa,* a dominant canopy species at El Verde, Puerto Rico, germinates better in the shade (Quarterman, 1970).

Primary species are shade tolerant in a limited sense only; they ultimately depend on growth in a gap to reach maturity. In the understory, seedling mortality is high (Richards, 1952; Garwood, 1983), but some seedlings persist. The degree of tolerance and the length of the bearable suppression period vary among species; thus, the potential composition of juvenile primary species in the gap phase changes in time as well as in space.

Understory tolerance requires a low compensation point, photosynthetic rate, and growth rate (Lugo, 1970; del Amo and Gómez-Pompa, 1976; see also review in

Bazzaz and Pickett, 1980). Growth of understory saplings can be nil (Liew and Wong, 1973; Hladik, 1982) and dieback is common, often due to mechanical damage (Enright and Hartshorn, 1981; Hladik, 1982; Uhl, 1982b). The required light increment for growth varies among species (Schulz, 1960; Baur, 1964; Whitmore, 1974), but it is generally believed that canopy opening occasions accelerated seedling and sapling growth in perhaps all middle- and upper-story tree species, as shown by comparative and experimental studies (Brown and Mathews, 1914; Schulz, 1960; Whitmore, 1975; Baur, 1964; Nicholson, 1965; Bannister, 1970; Edmisten, 1970; Lugo, 1970; Liew and Wong, 1973; Hartshorn, 1980). Repeated episodes of suppression and release may precede canopy attainment. Though saplings are often bent and broken by treefalls, sprouting and recovery are common (Uhl, 1982b; Alexandre, 1982a). Small trees recover better than large ones (Putz *et al.*, 1983), but sprouting of large trees may be more frequent in dry forests (Ewel, 1980).

Richards (1952) and Whitmore (1966) emphasize the broad range of conditions a canopy tree must tolerate during its development, necessitating wide physiological tolerance or substantial shifts in resource allocation. Hence, within a species, size-specific tolerance and gap response may vary. Also, among forests with different physiognomies and thus different light regimes, we may expect contrast in typical adaptations to understory conditions. Huber (1978) examined light compensation points of plants growing in the understory of a tropical montane forest. He concluded that most plants were not highly shade adapted, but instead were capable of widely varying responses to the variable light conditions in that forest.

B. Pioneer Trees Species

A large group of tree species are loosely grouped together as "light-demanding," "shade-intolerant," "secondary," "nomadic," or "pioneer species" (Vázquez-Yanes, 1980b). These are chiefly distinguished from primary species in that juvenile pioneers, as a rule, are found only in gaps. Thus, pioneers typically colonize gaps from seed. Among pioneers there are two commonly recognized classes: short lived and long lived (Dawkins, 1966). The short-lived class are the most deserving of the name "pioneer," as most are quick to invade both large natural gaps and areas cleared by humans. They belong at the opposite end of the regeneration behavior continuum from primary species. The long-lived pioneers (late secondary species) are less distinct from primary species.

Pioneers produce abundant small seeds, in some cases throughout the year (Pinto, 1970; Vázquez-Yanes *et al.*, 1975; Trejo, 1976; Alexandre, 1978). Most of the conspicuous gap colonizers are broadly dispersed by birds or bats (Richards, 1952; Guevara and Gómez-Pompa, 1972; Hartshorn, 1978). Seventy-six bird species are known to feed on *Cecropia* spp. (Holthuijzen and Boerboom, 1982). Bats deposit seeds in clumps, producing clumps of pioneer trees (Fleming and Heithaus, 1981).

Seed dormancy is more prevalent among pioneer than primary species (Schulz, 1960; Whitmore, 1975; Moreno, 1976), and pioneer germination typically is cued to a disturbance indicator. Exposure to light stimulates germination of *Cecropia* spp., *Didymopanax morototoni*, *Trema* spp., and probably other pioneer species (Blum, 1968; C. R. Bell, 1970; Lóez and Vázquez-Yanes, 1976; Vázquez-Yanes, 1976a,b, 1977, 1980a). Light intensity is not as important as light quality. Vázquez-Yanes and Smith (1982) showed that germination of *C. obtusifolia* is triggered by a high ratio of red to far red light. Beneath a dense canopy, far red light predominates and germination is inhibited. With canopy removal, red light predominates and germination occurs. Experiments with alternating red and far red light show the need for long periods of exposure to red and demonstrate a quick reversability of the red stimulation by far red. Thus, *C. obtusifolia* seeds can distinguish the long exposure afforded by a large gap from the short exposure in small gaps or sunflecks. As a corollary, *C. obtusifolia* seeds show a marked ability to distinguish between light regimes in central versus peripheral parts of a gap.

Seeds of *Heliocarpus donnell-smithii* detect canopy opening on the basis of widened soil temperature fluctuation (Vázquez-Yanes and Orozco-Segovia, 1982b), and the germination response showed that its seeds respond differentially to within-gap position. Germination of some pioneers is cued to soil disturbance (Liew, 1973; Alexandre, 1978; Putz, 1983).

Adaptive dormancy of pioneers is less characteristic of the long-lived type. In the Tai forest, Ivory Coast, seeds of long-lived *Khaya ivorensis* germinate shortly after dispersal and the seedlings survive only in gaps (Alexandre, 1978; see also Garwood, 1983).

Abundant dispersal of pioneer seeds with disturbance-cued germination creates a soil seed bank in mature forests. Numerous experiments, typically consisting of soil transfer from beneath to outside the forest canopy, have demonstrated the prevalence of viable pioneer seeds in the soil (Symington, 1933, cited in Richards, 1952; Keay, 1960; Blum, 1968; C. R. Bell, 1970; Guevara and Gómez-Pompa, 1972; Liew, 1973; Cheke *et al.*, 1979; Hall and Swaine, 1980; Holthuijzen and Boerboom, 1982; Uhl, 1982a; Putz, 1983). Densities can be quite high but vary greatly from point to point (Ewel *et al.*, 1981). Some species occur all year long due to frequent dispersal or long viability. The latter is known for several species (Castro and Guevara, 1976; Vázquez-Yanes and Orozco-Segovia, 1982b). Though a seed bank must be important, the relative contribution to regrowth of pioneer seeds arriving before rather than after treefalls has not been clarified for any forest.

Growth of pioneers is rapid due to high photosynthetic rates (Lugo, 1970; Kira, 1978) and investment in rapid internode and leaf expansion (Coombe, 1960; Coombe and Hadfield, 1962; Bazzaz and Pickett, 1980; Jordan and Farnsworth, 1980; Coley, 1982). Pioneers tend to grow continually (Ashton, 1978; Coley, 1982) and to conform strictly to simple architectural models (Oldeman, 1978). Short-lived pioneers reach canopy stature, while the emergent tree class is largely composed of long-lived pioneers. Short-lived pioneers are thought to attain 20–30 years of age, but the ''short-lived'' *Cecropia peltata* may reach 50 years on Puerto Rico (Doyle, 1981). Leaf life span is comparatively short (Givnish, 1978; Coley, 1982).

C. Understory Plants, Palms, Lianas, and Herbs

Understory trees and shrubs are thought to be less gap dependent than canopy trees (Hartshorn, 1978; Whitmore, 1978; Denslow, 1980a). Many of these plants survive in closed forest conditions and less growth is needed to reach mature stature, but for some species life cycle completion is enhanced in gaps. For instance, though the shrub *Piper hispidum* germinates in both gaps and mature-phase forests at Los Tuxtlas, Mexico, it probably establishes in gaps. Then it tolerates canopy closure; however, gap populations show greater growth and reproductive capacity (Vázquez-Yanes and Orozco-Segovia, 1982a; see also Lebron, 1979). In BCI gaps the subcanopy tree *Protium panamensis* is one of the faster-growing primary species (Brokaw, 1985). And some shrubs are true pioneers; on BCI the tall shrubs *Acalypha macrostachya* and *Croton billbergianus* are mostly found in gaps (Brokaw, personal observation).

Other than the climbing varieties, palms are generally not gap pioneers (Whitmore, 1978), though some colonize large disturbances (Foster and Brokaw, 1982). Growth of *Euterpe globosa* on Puerto Rico (Bannister, 1970) and leaf and fruit production of *Astrocaryum mexicanum* in Mexico (Pinero and Sarukhán, 1982) are enhanced in gaps.

The first significant growth of perhaps most lianas occurs in gaps (Whitmore, 1978). Beneath a closed canopy, liana growth is limited not only by low resource levels but also by lack of suitable natural trellises for climbing (Putz, 1980). The sides of gaps and the small-diameter trees within them provide such trellises. Also, shoots from mature lianas, having fallen with trees from the canopy, reascend via gaps, leaving great climber coils in the understory. Hence, the distribution and abundance of lianas are linked to the availability of trellises furnished by treefalls (Putz, 1984), and lianas are particularly characteristic of heavily disturbed forest (Wyatt-Smith, 1954; Webb, 1958; Leigh, 1975). The term "chablis" itself may be derived from the frequent association of chablis grapevines with gaps in medieval France (Oldeman, personal communication).

Lianas interfere strongly with tree regeneration (Jones, 1950; Nicholson. 1965). Comparatively little investment in self-support permits vigorous climbing, demonstrated by Putz (1983) to retard the growth and increase the mortality of host trees. Liana removal enhances tree survival (Schulz, 1960; Nicholson, 1965). Fast-growing trees and those with large compound leaves or deciduous branches, such as many pioneers, can avoid or rid themselves of lianas (Putz, 1980).

Herb distribution and reproduction in tropical forest are quite patchy and in many cases strongly related to gaps (Richards, 1952; Hall and Swaine, 1981). At La Selva, Costa Rica, *Heliconia* spp. are conspicuous in gaps, while patch size and reproductive activity increase with gap area (Stiles, 1975). Marantaceae colonize the more open parts of MPassa gaps (Florence, 1981), and herbaceous vines are found in openings on BCI (Croat, 1975). In the most definitive study, Mulkey (unpublished data) compared BCI gaps with adjacent intact forest. The number of species, frequency of reproductives, and percentage of cover for ground herbs were all markedly greater in gaps.

IV. REGROWTH

Regrowth in gaps comes from three sources: previously established plants (juveniles and damaged plants), seeds arriving before or after gap formation, and ingrowth by trees along the gap margin. Ingrowth by margin trees undoubtedly occurs around all gaps. Small gaps may thus close entirely. The amount of ingrowth depends on the developmental stage of margin individuals, species-specific traits, and variation in margin exposure and other environmental factors, including the distribution of inhibitory lianas.

Brokaw (1980, 1985) studied gap-phase regeneration of canopy trees on BCI. Beginning near the time of gap opening, he marked and measured all saplings >1 m tall at intervals over 6 years in 30 gaps ranging from 20 to 705 m^2. After gap formation, species and stem densities rose sharply and then leveled off or declined, reflecting reduced recruitment and increased mortality, especially of pioneers. These patterns were most apparent in large gaps (>150–200 m^2). By contrast, regeneration was slow in small gaps and in gaps filled with debris or dense understory vegetation surviving the treefall. There is no evidence that seed input is greater to large gaps. Undergrowth bird species that preferentially frequent gaps in central Panama are not likely seed dispersers (Schemske and Brokaw, 1981), and seed-dispersing bats in the neotropics tend to avoid large openings (Heithaus, personal communication).

Pioneer mortality was notably severe in dense patches, with shorter trees suffering most. This mortality was more likely due to intense competition among colonizers (Garwood, 1983) than to differential herbivory. Isolated pioneer and primary individuals suffer no less herbivory than those in conspecific clumps on BCI (Coley, 1983). Per unit leaf area, pioneers are eaten more than primary species (Coley, 1983).

In the 30 sites studied by Brokaw, a few species of both regeneration types were relatively common and were usually found in large gaps, but most were uncommon and unpredictable in occurrence. Large gaps were more similar to each other in terms of primary rather than pioneer species composition, as observed by Florence (1981) at MPassa, due to the local, dense distribution of pioneers (Whitmore, 1978; Fleming and Heithaus, 1981). Hartshorn (1978) discussed the many variables potentially affecting gap composition.

Pioneers colonized gaps of all sizes but as a group were more abundant in large gaps, where they established in areas not occupied by other plants or treefall debris. Pioneer prominence in big gaps has been observed elsewhere (Kramer, 1933; Richards, 1952; Beard, 1945; Schulz, 1960; Baur, 1964; Whitmore, 1975; Hartshorn, 1978; Acevedo and Marquis, 1978; Barton, 1984; Riera, 1984), but reported threshold gap size for pioneers varies greatly (Table 1). This is expected, because variation among workers in their definitions of pioneer, gap area, and threshold size are surely added to real differences in pioneer behaviors. Pioneer prominence in large gaps is most likely due to the availability of light, reduced root competition, and frequent occurrence of disturbed soil there (Putz, 1983). At MPassa, pioneers colonize gaps in which light intensity is at least 20% of full light (Florence, 1981).

In the BCI study, stem density of primary individuals varied from site to site but was not related to gap size or pioneer density in this gap size range. Barton (1984) observed the same at La Selva, Costa Rica. In very large gaps in other forests, primary species regeneration is hindered by rigorous microclimate and profusion of pioneers (Schulz, 1960; Baur, 1964; Whitmore, 1982). For example, in the Ivory Coast, *Turreanthus africana* benefits from minor canopy opening but suffers fatal desiccation in large gaps (Alexandre, 1977).

Pioneers grew faster than primary species in large gaps, while primary species grew in gaps too small for pioneers. The data suggested that pioneer growth increases within a certain range of gap size, paralleling resource release (Schulz, 1960; Baur, 1964; Whitmore, 1974). The growth of primary species as a group was faster in gaps than in shaded understory but did not equal the growth of pioneers and did not increase with gap size. Liew and Wong (1973), however, observed a growth increase with the degree of overstory opening among dipterocarp seedlings.

At several of the BCI sites, early recruits to the 1-m-tall class grew faster than later ones, and growth of the first recruits of primary and pioneer species declined with time (see Liew and Wong, 1973). Uhl *et al.* (1982) reported decreasing growth of *C. obtusifolia* in a cut-over area at San Carlos, Venezuela, and related it to an observed depletion of the debris-derived nutrient flush. Additional factors in growth decline are encroachment of bordering trees and lianas, increasing competition among regenerators, and the rising ratio of respiring to photosynthesizing tissue in a growing plant.

These events in the gap phase bring regeneration behavior into focus. Diminishing recruitment and the apparent decline in growth among successive cohorts of both regeneration types, and the mortality of shorter pioneers, suggest that early presence affords the best chance of reaching maturity in a gap. Both pioneer and primary species depend on growth in gaps to reach maturity, but they differ fundamentally with respect to the stage in which they persist until gap formation occurs (Alexandre, 1982a). For pioneers, which colonize gaps from seed, the necessity of early arrival would select for effective and abundant dispersal, seed persistence until germination is cued by gap microclimate, and rapid growth. Presumably, rapid growth is possible only at sites of major resource release; hence, pioneers are found mainly in large gaps. An extreme case is *Trema micrantha,* which is the fastest-growing pioneer on BCI and is strictly limited to big gaps (Brokaw, 1985). Coley's (1983) work suggests that pioneers do not escape herbivores via this spot-wise establishment.

For primary species, which generally regenerate in gaps from the seedling or sapling stage, the necessity of early presence would select for the ability to persist under a closed canopy while awaiting gap formation. This capacity allows primary species to capitalize on the resources of small as well as large gaps [to this extent, they are not small gap specialists (Barton, 1984)], but precludes exceptionally rapid growth in large gaps. This adaptive trade-off could account for their even distribution and the lack of correlation of growth rate with gap size in the BCI study. The advantages of the primary species' strategy are juvenile presence in the comparatively noncompetitive understory (Garwood, 1983), the likely occurrence of growth opportunities in relatively numerous small gaps (Fig. 1), the ability to persist through successive gap closures, and the capacity to invest in herbivore defenses (Coley, 1983).

Decline in growth and recruitment signals a transition from the gap phase to the building and mature phases of the forest growth cycle. As in Oldeman's (1978) model of "sylvigenesis," the "dynamic" gap phase produces a dense canopy of early recruits that suppresses further colonization and brings on a "homeostatic" phase of reduced growth. Oldeman also proposes that further alternating and distinct functional phases precede stand maturity. These may occur during regrowth in huge gaps or secondary succession, but it is not clear if multiple phases take place in the endogenous cycle considered here (Whitmore, 1982). In most cases, the building phase probably consists of gradual thinning, continued growth of early recruits, and the slow entry of new age classes of primary species.

Gap, building, and mature phases in large gaps display distinctive architectures (Ashton, 1978; Hallé et al., 1978; Oldeman, 1978; Torquebiau, 1981). In the first years of the gap phase, it is difficult to walk through the dense tangle of saplings, lianas, and debris. But thinning and the decomposition of debris clear the understory within 3–4 years. In the building phase differential growth produces a range in height distribution (Crow, 1980), height growth dominates diameter increment (Baur, 1964; Whitmore, 1975; del Amo and Nieto de Pascual, 1981), and trees develop deep conical crowns (Whitmore, 1978). As trees attain the canopy, diameter growth predominates, lower branches die, and tree crowns become wider than deep.

V. COMMUNITY STRUCTURE

Endogenous disturbance by treefalls is essential for the regeneration of many tropical forest plants, and the degree of canopy opening influences species composition. The examples below illustrate connections among treefall regimes, species requirements, and community structure.

On Kolombangara, in the Solomon Islands, Whitmore (1974) recognized four distinctive tree regeneration types on the basis of age-specific requirements and gap size preference. Due to the interaction of topography and prevailing winds, treefall frequency differs consistently over parts of the island, resulting in predictable differences in the relative abundance of the four regeneration types.

Hart (unpublished data) studied two forest types in the Ituri region, Zaire. In the "mbau forest" a fairly continuous canopy is heavily dominated by *Gilbertio-*

dendron dewevrai, which usually dies standing and creates small gaps suitable for its own, but for few other species', regeneration. The mbau forest is species poor. The nearby "mixed forest" has a broken, open canopy, and there are 3.5 times as many saplings per unit area as in the mbau forest. In the mixed forest the dominants usually uproot, creating large gaps in which many species regenerate. The mixed forest is species rich. Between the mbau and the mixed forest there is no apparent difference in history, soil, or climate.

At Pasoh, Malaysia, short-lived pioneers are rare (Ashton, personal communication). Pasoh trees often die standing (Ashton, personal communication; Putz, personal communication) or create large but linear, functionally small gaps. That tendency, plus an exceptionally dense understory, common in Southeast Asia, may limit pioneer success (Ashton *et al.,* 1984). As a result or a cause, there are few buried seeds at Pasoh (Putz, personal communication).

On the infertile soils of Southeast Asian heath (white sands) forest, trees are deep-rooted, snap off rather than uproot, and sucker profusely (P. S. Ashton, 1981). Lack of disturbed soil and space, plus infertility, inhibit pioneer colonization. The reverse occurs in nearby forest on fertile soils.

Treefall regime and regeneration opportunities can depend on overall forest development (Oldeman, 1978). For instance, about half of BCI is structurally mature forest, 300–400 years old, while the remainder dates from clearings abandoned around 1880 (Foster and Brokaw, 1982). Large trees and consequently large gaps and pioneer saplings are more common in the old forest (Brokaw, 1982a). Of course, as the young forest matures, large trees, large gaps, and pioneer regeneration will become more common there.

Denslow (1980b) reasoned that a large group of species in a community will be those whose regeneration is adapted to the most common type of disturbance-derived regeneration site. That notion links the predominance of primary species (in terms of species richness) in the BCI old forest with the predominance of small gaps there (Fig. 1). Some authors have devised mathematical models of gap creation and regrowth to explain tropical forest composition (Acevedo, 1981) or physical structure (Torquebiau, 1981). Doyle (1981), with the advantage of good autecological data for a species-poor community, closely simulated the dynamics of the montane Luquillo Forest, Puerto Rico. His model projections showed that maximum species richness is provided by an intermediate disturbance regime, including periodic hurricanes to create large gaps.

The link between gap size and regenerator composition suggests that environmental heterogeneity within large gaps would also provide different regeneration niches (Ricklefs, 1977). A degree of site preference within gaps has been observed. The understory palm *Cryosophila guagara* grows best at the root end of treefalls in the Osa forest, Costa Rica (Richards and Williamson, 1975). On BCI, pioneers are most common on disturbed soil within gaps (Putz, 1983). Orians (personal communication) reports that root, bole, and crown zones are more like the same zone in another gap, in terms of species present, than they are like another zone in the same gap at La Selva, Costa Rica. Also at La Selva, Barton (1984) showed that among a selected group of species, pioneers tend to be found in the middle of large gaps, while for primary species systematic patterns within gaps are negligible. Baur (1964) and Florence (1981) observed middle-to-edge transitions from pioneer to understory-tolerant individuals in large gaps, but transects across gaps at MPassa did not reveal consistent species-specific distributions (Florence, 1981). Oldeman (1978) attributes the crescent-shaped local distributions of some species in Ecuador to regeneration on gap edges.

Site preferences, however, may not be fine-tuned or varied enough to explain the larger part of tropical forest species richness (Connell, 1978). Short-lived pioneers, for example, are "notoriously site tolerant" (Ewel, 1980), and some authors believe that large groups of tropical forest trees are equivalent with respect to regeneration requirements (Richards, 1952; van Steenis, 1958; Poore, 1964; Ashton, 1969; Aubréville, 1971; Whitmore, 1982). Barton (1984) points out that even with regard to the major regeneration types, namely, pioneer and primary species, distributional overlap on the gap size gradient weakens the concept of gap partitioning. Hubbell's (1979) close simulation of dominance–diversity curves for a tropical dry forest, using a model based on gap-phase regeneration of ecologically equivalent tree species and stochastic immigration and extinction, suggests that we need not invoke niche differentiation to explain a major aspect of community structure. Further, the characteristic clumped distributions of many species are better explained by dispersal patterns and episodic gap-phase regeneration of equivalent species than by site specialization (Poore, 1964; Paijmans, 1970; Hubbell, 1979). Indeed, many regard gap composition as largely a probabilistic mix of interchangeable species (Richards, 1952; van Steenis, 1958; Schulz, 1960; Baur, 1964; Poore, 1964; Aubréville, 1971; Webb *et al.*, 1972; Hall and Swaine, 1976; Whitmore, 1982). Still, the examples above show that the abundance of pioneers as a group can depend on gap environment, and the extent of niche partitioning within gaps remains untested.

VI. CONCLUSIONS

The treefall and regrowth cycle connects three kinds of structure inherent to the forest: physiognomic structure, population structure, and community structure. The irregular topography of a tropical forest canopy and the widely varying stem densities on the ground are properties of physiognomic structure, reflecting the mix of gap, building, and mature phases in the forest mosaic. The flux of phases is driven by the cycle of treefalls and subsequent regrowth. At the same time, physiognomic structure feeds back on the forest cycle: the size of trees on treefall regime and the size of gaps on the rate and composition of regrowth.

The population or size class structure of many tropical forest trees is related to the persistent stage in which they await gap opening. Some species persist as seeds in the soil or by repeated abundant dispersal; others persist as suppressed juveniles in the understory. The type of persistence involves associated traits that enhance a species' ability to reach maturity in certain situations but restrict its ability to capitalize on all situations. This trade-off produces a degree of regeneration niche differentiation regarding gap size and sites within gaps that explains a portion of tropical forest species richness, an aspect of community structure. In terms of the relative abundance of regeneration types, if not of particular species, tropical forest community structure can depend on the frequency of large gaps, a factor in pioneer abundance. Also, the spatial arrangement of large gaps describes a coarse floristic mosaic of pioneer stands. A finer floristic mosaic exists to the extent that individual species are clumped, often as a·result of local regeneration in gaps.

The diversity of seemingly equivalent species, their patchy distributions, and the wide microenvironmental variation among gaps make it impossible to predict exactly the species composition at sites in a tropical forest. Hence we recognize a stochastic component in gap composition. But many systematic patterns, emphasized in this chapter, also influence tropical forest dynamics and composition. These patterns include seasonal gap creation, numerical predominance of small gaps, increasing number of large gaps with secondary succession, existence of alternative regeneration behaviors, pioneer prominence in large gaps, and the predictable structure and function of phases in the forest cycle.

ACKNOWLEDGMENTS

I thank Carlos Vázquez-Yanes, Daniel-Yves Alexandre, Emmanuel Torquebiau, and Terese Hart for exceptional aid with source material.

Chapter **5**

Gap Processes and Large-Scale Disturbances in Sand Prairies

ORIE L. LOUCKS

Butler University
Indianapolis, Indiana

MARY L. PLUMB-MENTJES

U.S. Army Corps of Engineers
New Orleans, Louisiana

DEBORAH ROGERS

Technical Information Project
Pierre, South Dakota

I. INTRODUCTION

Since the first monographic studies of the vegetation of North America, the grasslands have been described as a stable or mature community in which the principal species are viewed as climax dominants. The most comprehensive treatment of grassland stability is that of Clements and Shelford (1939). These authors differentiate several types of prairie, ranging from Tallgrass Prairie in Illinois to True Prairie in the tier of states west of the Mississippi to Shortgrass Prairie nearer the Rocky Mountains. However, they note that physical disturbance is infrequent and that fires, although possibly frequent, have little influence on species composition or succession in the usual sense. This view is continued, although in different form, in the monograph by Risser *et al.* (1981) on the True Prairie ecosystem: characterization of grassland types is in relation to a climatic type, implicitly a climax community. The section on succession in grasslands discusses mainly the outcome of interaction among the plants themselves and is presented in terms of compositional replacement following major human disturbances.

A different perspective emerges from the work of Platt (1975), Giesel (1976), and Wiens (1976). These authors, in conjunction with Levin and Paine (1974), suggest that patchy environments can be essential in maintaining the biological diversity characteristic of certain communities. An extension of this view leads to the hypothesis that the populations of animals in the arid grasslands could have produced annually a sufficient disturbance habitat to assure the maintenance of certain species requiring openings for seedling establishment. In Platt's study, the disturbance agents were populations of badgers and pocket gophers. In the patchy environments reported on by Baxter and Hole (1967), the disturbance agents were several grassland species of mound-building ants. In both examples, an alternate view of the grassland community emerges, one in which a large number of small disturbances produces a dynamic mosaic of small, short-lived patches utilized by opportunistic species. A type of secondary succession may take place in the larger disturbances, or the patches may simply disappear by edge closure. Thus, the prairie, as sampled at the relatively coarse level of most community measurements, may need to be examined as a composite of both pioneer and K-selected or climax species, not necessarily mixed in a single community but as a mosaic of two or more communities (or age-states) in a dynamic, fine-textured mosaic.

Earlier literature notes two types of larger-scale phenomena in the grasslands with properties possibly qualifying as disturbances: prolonged drought and prairie fires. The work reported by Weaver and Albertson (1943) shows the changes in species composition in grasslands during recovery from a prolonged drought, giving them certain characteristics of a disturbance. However, the effects of fire documented by Curtis (1959) and his students are essentially physiological rather than compositional, at least for the mesic Tallgrass Prairie, and do not qualify as a disturbance in terms of an immediate compositional impact on the grassland community.

In this chapter, we consider compositional and life history evidence indicating both species and community responses to a spectrum of environmental disturbances ranging from short-term and small-scale patches (produced by several animal species) to long-term and large-scale disturbances (due to prolonged drought and wind or water erosion). We will consider life histories and the specializations evident among species utilizing disturbances in grasslands to evaluate possible community responses to the dynamics of disturbances at different temporal and spatial scales.

II. PERSPECTIVE ON GRASSLAND DISTURBANCES

One of the more comprehensive studies of the eastern grasslands is the work summarized for Wisconsin by Curtis (1959). Although he presents a static view of species composition (as gradients), Curtis devotes considerable attention to the general question of the evolutionary origin of the prairie and its adaptation to fire. Until the studies by Curtis, there had been little appreciation of periodic burning as a stimulus for the annual growth of grassland species in the mesic prairies, enhancing flower and fruit production while producing no measurable effect on composition. Writings by early explorers suggested to Curtis that they had recognized the role of fire and the potential influence of animals in the grassland, but following the decline of native animal populations and widespread fire protection by the mid-nineteenth century, little further reference to fire was made. The effects of biennial burning on the mesic prairie, however, have not come to be viewed as a demographic or successional influence on composition, but rather as a sustained constraint like the annual return of a cold period. Frequent burning of sand prairies, however, can alter species composition (Zedler and Loucks, 1969) and, in this sense, fire can be a large-scale, immediate disturbance in certain types of grassland.

Contrast these viewpoints to those in the extensive recent literature on compositional responses to disturbance in a wide variety of forest ecosystem types. In "Forest Succession: Concepts and Application" (West et al., 1981), species changes in a number of biome types are examined, but grasslands are not. A decade earlier, a simple succession model had been fitted to postcultivation grassland recolonization data by Bledsoe and Van Dyne (1971). A major reason for the dearth of data and models for grassland patch dynamics, as opposed to forest patch processes, appears to lie in the problems of hierarchical phenomena as they influence our perception of communities (Allen and Starr, 1982). Although the vegetation recovery on the animal disturbance sites described by Platt (1975) can be viewed as microsuccession, the utility of conceptualizing the prairie as a mosaic of small patches at different successional age-states has not yet been worked out as well as it has for the intertidal zone (Levin and Paine, 1974).

Still another perspective, also from the forest, is the view that recurring disturbances (Loucks, 1970) create initial conditions for opportunistic species at intervals such that a relationship can be recognized between the species' life history and the average return time of the disturbance. The work of Canham and Loucks (1984) indicates that the average return time for severe wind disturbance in Wisconsin forests is a little more than 1000 years, or three times the life span of the first-generation recovery forest. The return time for fire in these northern forests varies with soil conditions but is closer to the maximum life span of the species. Swain (1973) has shown that in an area of frequent fires in northern Minnesota, seed production of the dominant trees begins at about the shortest return time of fire (15 years), and the life span of the dominants (150 years) is about twice the average return time of fire (75 years). Such studies suggest that similar relationships on a much smaller scale could exist in the grasslands, where frequent fire, drought, and animals produce a number of disturbances with a relatively short return time. The small-scale disturbance agents could maintain opportunities for many short-lived species (annuals or biennials), while larger-scale agents (e.g., drought, irregularly timed fires, and wind erosion) could produce larger and longer-term disturbances of a magnitude more closely related to the succession observed in forest systems.

A major difficulty in examining the significance of disturbance agents in grasslands is the profound dearth of natural systems where the disturbance regime even approaches that of historical conditions. Most of the former grassland area is now under some form of management, ranging from the intensive cultivation of much of the Tallgrass and True Prairie to management for livestock production in most of the Shortgrass Prairie. In both, of course, the native animal populations have been much modified and, in the case of prairie dogs and buffalo, virtually eliminated. Even the abundance of ant species can be altered by changes in the density of aboveground herbage production, a factor of some concern where the grasslands have only been grazed. Although it is probable that we will never really know the details of the grassland as a dynamic system because we cannot fully reconstruct the role of animals in it, we should formulate and examine the simplest hypotheses as to their potential role in conjunction with drought, fire, and wind.

Over the past 20 years, the preserve systems established by The Nature Conservancy and other organizations have had some success in redeveloping the animal components. Many of these systems are now being managed to remove infringing hedgerows, to maintain relatively realistic burning frequencies, and to reestablish normal population densities of small mammals and their predators. The hypothesis that grassland communities should be viewed as a system with a high frequency of small-patch, pioneer habitats, as in other arid-region biomes, is sufficiently credible that it should be examined wherever possible. Studies of even partial disturbance processes in restored and protected sites, as reported here, appear to be worth careful scrutiny.

III. A SOUTHWESTERN WISCONSIN CASE STUDY: THE SPRING GREEN SAND PRAIRIE

In the late 1960s, the Wisconsin Scientific Areas Preservation Council and The Nature Conservancy began taking steps to establish the Spring Green Cactus and Reptile Reserve in southwestern Sauk County, Wisconsin. The site is part of a valley bottom sand plain, a mile or more in width, lying between the Wisconsin

River on the south and calcareous bluffs to the north. The area is particularly unusual because it supports a dozen or more species of reptiles, many of which are uncommon in Wisconsin. In addition, the vegetation of the area is dominated by a sand prairie community relatively rich in annual and biennial species and locally dominated by the prickly pear cactus (*Opuntia compressa* [Salisb.] Macbr.). Approximately 600 acres (2.5 km^2) were protected by a combination of private and Nature Conservancy ownership. Portions of the site appear to have been unplowed, but a substantial area has experienced cultivation at least sometime in the previous century; the remainder was intensively grazed as recently as the 1940s. Only the presence of a large number of uncommon species (both plant and animal) suggested that the site had important values representative of a natural system.

Research on the area during the 1970s focused first on the stability of the prickly pear cactus population (Carosella, 1978), a dominant on about 1 km^2. The patterns of herbaceous colonization following intensive pocket gopher disturbance on the one hand, and cultivation in the sand prairie on the other, were investigated by Rogers (1979), and the phenology and colonization dynamics of short-lived species were studied by Plumb (1979). This discussion is a synthesis of the contributions from these studies within a framework that asks whether the multiple scales of disturbance (large and small, spatial and temporal) in this grassland could be conceptually equivalent to the similar phenomena widely documented in forest systems.

A. Scale in the Physical Environment at Spring Green

Sand prairies are communities with relatively little physical resilience during periods of drought or erosion. The soil is derived from glacial outwash, which, at Spring Green, is a terrace system up to 30 m above the modern level of the Wisconsin River. The coarse, predominantly quartz sands have little water-holding capacity, and heavy rains percolate quickly to a water table considerably below the reach of most herbaceous species. With little storage of water in the system, drought conditions develop quickly during the growing season and are usually extreme by the middle of each summer (Plumb, 1979).

In addition to the aridity of the site due to its physical characteristics, this portion of southwestern Wisconsin, like the remainder of the grassland region, is subject to extreme droughts that may persist for several years. The annual, thin snow layer and subsequent spring rains bring some moisture even during drought years, but the infrequent wetting during the summer and the high temperatures that accompany drought periods produce a sustained alteration of the competitive relationships among the sand prairie dominants. By the end of two growing seasons, such a drought can affect short-lived species to produce the equivalent of a disturbance (Weaver and Albertson, 1943). Our studies have not yet followed plant populations at Spring Green Prairie through a complete drought cycle, but the species are sufficiently short-lived and so closely coupled to their environment that changes in composition would be an inevitable result of a drought lasting for 2 years or more.

Any given year is influenced to some degree by the prior year's rainfall and temperature (Plumb, 1979).

Other types of large-scale influences must be considered within the framework of weather fluctuations: fire, and soil erosion by either wind or water. The stimulation of grassland species by fire, as reported by Curtis (1959), has been recognized as applicable primarily to the mesic prairie and its relatively large moisture-storage capacity. The same fire treatment on dry prairies weakens some of the climax dominants and enhances colonization by short-lived opportunistic species (Zedler and Loucks, 1969). We do not have many details about fire frequency in sand prairies prior to settlement, and even less about fire frequency during drought. However, we do know that most grasslands, dry and mesic, burn at intervals of 1–3 years, far more frequently than the once in 5–10 years that is a common management practice for dry-prairie systems today. Thus, burning as frequently as every second year, which seems probable, must be viewed as a large-scale disturbance capable of reducing the vigor of perennial species and enhancing the populations of annual, biennial, and opportunistic perennials. This effect would be expressed more strongly during the drought phase of a climatic cycle when moisture is even more limiting. The temporal and spatial patterns in those relationships are summarized in Table 1.

In addition, few windbreaks occur in a sand prairie under natural conditions, and with the combination of the drought cycles and burning, the light sands have tended to form sandblows and sand dunes, apparently since the recession of high water levels after the melting of glaciers from northern Wisconsin. Certainly, the extent of sandblows increased considerably in the Spring Green area during the drought of the 1930s under a combination of cultivation and moderate grazing pressure (equivalent in some respects to frequent burning).

TABLE 1

Spatially and Temporally Determined Components of Small-Scale and Large-Scale Disturbances in the Spring Green Sand Prairie[a]

System component	Observed property
Spatial scale (habitat perturbation)	Sand dunes, blowouts, and alluvial fans (tens of meters)
	Gopher mounds and ant mounds (tens of decimeters)
	Insect burrows, ant hills, and interplant wind erosion (decimeters)
Temporal scale (habitat availability)	Drought cycles (portion of a decade)
	Annual drying cycle (year)
	Within-season drought (month)
	Rain/sun drying cycle (days–week)

[a] After Plumb (1979).

B. Biological Setting

Preliminary knowledge of the life histories of the biota in the area indicates an unusual distribution of reproductive and colonization strategies. The studies by both Plumb (1979) and Rogers (1979) focused first on the simplest division of life history strategies: annuals, biennials, and perennials. Second, these studies recognized several common categories of life form: grass, forb, and shrub. Third, Plumb adopted terms for three occupancy strategies, one of which incorporates the concept of "fugitive" species (Platt, 1975):

"Persistent": Species that determine the structuring of biotic utilization of spatial resources in the sand prairie by their continued presence, numbers, size, and spacing.

"Ephemeral": Either annual or perennial species that respond to the availability of unutilized resources at various times during the growing season; populations tend to be sparsely distributed and the species are independent of changes in spatial occupancy by other plants.

"Peripheral": Species that utilize, for a year or two, small, physically determined spatial openings such as gopher mounds or small-scale erosion sites; includes what is understood as "fugitive species."

All three of these sets of descriptors (life form, life history, and occupancy strategy) can be applied simultaneously to define functionally distinct guilds of the species utilizing the sand prairie. These will be referred to as "spatial/temporal life form/occupancy guilds."

Plumb's study also sought to understand the effects of within-season climatic changes that appear to influence demographic outcomes during competition between aggressive perennial species and the annual and biennial species. To examine these questions, four periods of phenological activity in the growing season were distinguished:

Phenoperiod I: April through mid-May, characterized by cool temperatures, low light, high wind, and low but predictable rainfall.

Phenoperiod II: Mid-May through late June, characterized by optimal growing conditions, warm temperatures, adequate light, low wind, and a relatively dependable moisture supply.

Phenoperiod III: Late June through July, characterized by higher temperatures, high light, and erratic rainfall.

Phenoperiod IV: August through mid-September, characterized by very high temperatures, erratic rainfall, and the greatest probability of severe seasonal drought.

The utilization of each phenoperiod by species in the various life form occupancy guilds is summarized in Table 2. The term "prominent" is used to describe those species completing the majority of their growth (and making most of their demands

TABLE 2

The Number of Prominent Species Observed in Each of 10 Life Form/Occupancy Guilds across the Four Phenoperiods of the Growing Season, Spring Green Sand Prairie

Life form/occupancy guild	Number of prominent species, by phenoperiod			
	Phenoperiod I	Phenoperiod II	Phenoperiod III	Phenoperiod IV
Ephemeral perennials	5	3		
Ephemeral annuals	1	5		
Persistent perennials		6	5	4
Persistent perennial grasses		4	7	
Persistent shrublike perennials		3	2	
Persistent biennials			2	
Persistent annual grasses			1	
Persistent annuals			2	4
Peripheral biennials		2		
Peripheral annuals		3	4	
Total species	6	26	23	8

on resources) during the period shown. Although moisture resources are more abundant in the spring than in the summer and fall, a simple examination shows that the annual and biennial life-history strategy peaks numerically in June and July, while the largest percentage occurs in August. If the short life span and adoption of annual reproduction are frequent adaptations of opportunistic species for occupying spatially dispersed short-term habitats, then an appreciable portion of the prominent species at Spring Green are opportunistic within a small dispersal range.

C. Spatial/Temporal Utilization within the Phenoperiods

The results in Table 2 can be explained more fully as follows:

Phenoperiod I is characterized by ephemeral species (100%). The occupancy strategy appears to be one of avoiding competition by adapting to a short growth period and the cool season. Birdfoot-violet (*Viola pedata* L.) and pasque-flower (*Anemone patens* L.) are representative perennials. Whitlow-grass (*Draba verna* L.) is the annual.

Phenoperiod II shows 50% persistent, 30% ephemeral, and 20% peripheral species. The overall occupancy strategy appears to be primarily one of rapid growth and early completion of flowering and fruiting prior to active growth by warm-season grasses. Dwarf dandelion (*Krigia virginica* [L.] Willd.) and lyre-shaped rock-cress (*Arabis lyrata* L.) are representative of ephemeral annuals in seasonally open habitats, while pepperwort (*Lepidium densiflorum* Schrad.) and carpetweed (*Mollugo verticillata* L.) are indicative of the peripheral annuals on the small patch disturbances.

Phenoperiod III shows 83% persistent and 17% peripheral species; 43% of all the prominent species are grasses. The occupancy strategy is primarily persistence, indicating strong competition from plants in undisturbed sites. The peripheral annuals are mostly of arid-region origin and thus are able to exploit the xeric conditions present at this time in small-scale openings. Winged pigweed (*Cycloloma atriplicifolium* [Spreng.] Coult.), lamb's-quarters (*Chenopodium album* L.), and Russian thistle (*Salsola Kali,* var. *tenuifolia* Tausch) are representative and are frequent colonizers of the patch disturbances.

Phenoperiod IV has fewer prominent species, and although 50% are annuals, they meet the criteria for being persistent rather than ephemeral or peripheral. The dominant strategy for this period appears to be physiological adaptation to hot, droughty conditions without regard to disturbance patterns. Chamaesyce spurge (*Euphorbia glyptosperma* Engelm.), cottonweed (*Froehlichia floridana* Nutt. Mog.), jointweed (*Polygonella articulata* [L.] Meisn.), and slender knotwood (*Polygonum tenue* Michx.) are the annuals showing their greatest growth at this time of year without any evident influence from the smaller-scale disturbances.

One might like to examine directly the relationship between seedling establishment (by annuals and biennials) and subsequent succession on the disturbance patches, but the year-to-year and month-to-month variations in moisture conditions and in demographic survivorship patterns make data collection and interpretation very difficult. The 2 years of data for a few species at Spring Green (Plumb, 1979) indicate that greater extremes in growing conditions occur from early spring to midsummer than between the disturbed habitat of a gopher mound and adjacent undisturbed prairie at any time during the season. Thus, the importance of the annual array of disturbed environments may not be as significant as is suggested simply by the large number of annual and biennial species.

What is also evident from these studies is that during phenoperiod II, when moisture is initially optimum, opportunism can be expressed by both ephemeral and peripheral species, which complete their life history among the few perennials active in this period by taking advantage of both the more open environment of the spring season and the disturbance patches. During phenoperiod III, competition from surrounding perennials is sufficient so that only a few species (the peripheral annuals) utilize the openings provided by gopher mounds; other annual species simply occupy space among the perennials, possibly fostered by earlier mounds, ant hills, or the heavy grasshopper herbivory.

To the extent that the mounds are used in both phenoperiods III and IV, pocket gopher activity serves to provide a colonizable habitat for fewer than a dozen species in the sand prairie system. By phenoperiod IV, patch disturbances appear to play no role at all. Open habitat remains, but it is used by species able either to root deeply or to tolerate the limited amount of moisture received in rainfall and stored briefly near the surface.

A related phenomenon at Spring Green is the remarkable assemblage of animals associated with the disturbance regime and the type of herbaceous production.

Thirteen mammals are known for the site, but two, the pocket gopher (*Geomys* sp. Rafinesque) and the badger (*Taxidea taxus* Waterhouse), are the principal animal agents of disturbance in the system. In addition, however, studies of the numerous reptile species show that the burrows of the pocket gopher are probably essential as cover for the Bull snake (*Pituophis melanoleucus sayi* [Schlegel] Smith and Kennedy) and possibly others during both the hot and cold periods of the year. The ornate box turtle (*Terrapene ornata ornata* [Smith and Ramsey]), blue racer (*Coluber constrictor foxi* [Schmidt] Conant), western fox snake (*Elaphe vulpina vulpina* [Baird and Girard] Conant), and sixlined racerunner (*Cnemidophorus sexlineatus* [L.] Dumeril and Bibron) are among the more unusual species whose populations must be managed by The Nature Conservancy, if possible. The widespread occurrence of prickly pear cactus on the site appears to be important as cover and food for several of the rodent species and as cover for the lizards. In turn, the abundance of the prickly pear is influenced by the frequency and scale of the animal disturbance agents (Carosella, 1978). The midsummer hoards of grasshoppers cause a considerable reduction in the plant canopy and serve as the food source for some of the fauna.

D. Macroresponse to Pocket Gopher Populations

Because of qualitative similarities in the occurrence of annual species between areas of high-density pocket gopher populations at Spring Green and areas of old-field recolonization following cultivation, Rogers (1979) investigated the differences and similarities between these communities in more detail. Her goal was to determine whether these two very different types of disturbance could be functionally similar. Gopher digging and cultivation can both be viewed as large-scale treatments in that 2–5 ha undergo the treatment at any given time, and within these treatment areas extensive populations of disturbance-adapted species can develop. An important difference, however, is that in cultivation the entire range of perennial species is removed, whereas in the presence of a high gopher population a broad spectrum of perennial species serves as a matrix within which the disturbance occurs. Thus, in the latter case, the opportunistic species might be viewed as colonizing patches in a mosaic, whereas in the former they are viewed as colonizing a relatively uniform, large-scale disturbance.

The differences and similarities between the two systems are evident from data comparing undisturbed sand prairie with pocket gopher–rich sites and old-field sites 1–3, 4–10, 11–24, and over 25 years of age (Rogers, 1979). The proportion of bare surface habitat found in the gopher sites is similar to that of the old-field recovery when it is between 4 and 10 years from abandonment and is about four times the 8% bare area of the undisturbed sand prairie. The relative dominance by perennials on the gopher-rich sites is similar to the abundance of perennials on old fields of 11–20 years following cultivation, 5–10% less than in the undisturbed prairie. Nineteen species found to be more abundant in pocket gopher-dominated sites than in the undisturbed prairie were also colonizers of the old field. Ten others more common in the gopher sites were species that persisted in abundance on old fields.

IV. DISCUSSION: DISTURBANCE SCALE, RETURN TIMES, AND PREDICTABILITY IN GRASSLANDS

For years our understanding of succession, even in forest systems, was limited greatly by the failure to deal precisely with the processes that create the initial conditions for succession. The importance of considering disturbances from the viewpoint of size and frequency distribution was noted by Johnson (1977), Loucks et al. (1980), and others. In this view, the aggregate of successional processes in a forest landscape produces an equilibrium between events that return advanced successional states to earlier states and the process of succession itself. As succession and the processes controlling the frequency and types of initial conditions have been modeled (Loucks et al., 1980), we have become increasingly aware of the attributes of forest species that allow the success of some species in the early stages, while the characteristics of others foster the replacement pattern we know as succession.

There are analogous processes in grassland systems. We have considered only one site, and we know from its history and the absence of balanced animal populations that it is functioning in less than pristine conditions, but certain principles are evident. We can differentiate among the several types of disturbance processes that influence opportunities for species to become established in grasslands. Certain phenomena in the forest system (e.g., droughts) are not ordinarily thought of as disturbance agents, but disastrous windstorms with a return time of a thousand years or so (Canham and Loucks, 1984) are disturbances. The reverse may be true in grasslands: individual windstorms are usually not a disturbance, but a drought may be. The difference, apparently, is the extent to which a qualitatively different community emerges following the event. By such a criterion, a drought lasting for 2 years in a community characterized by many annuals and biennials can produce considerable change in the community and must qualify as a disturbance.

Similarly, the occurrence of fire in a system with relatively little moisture retention and minimal hydrocarbon storage must be viewed as a disturbance if it fosters a community the following year with a higher number of disturbance-adapted species (Zedler and Loucks, 1969). This occurs more often in sand prairie than in mesic prairie.

With the sand prairie, the spatial scale of disturbances also raises some issues that may modify our concept of disturbance. While an area of badger digging is clearly recognizable as a disturbance in a grassland, and can support a population of opportunistic species, a small area such as a single pocket gopher mound, serving a much smaller number of individuals and species, might not be recognized immediately as a disturbance. However, the aggregate of such areas can be equivalent functionally to badger diggings. The gopher mound seems to meet the criteria of a disturbance in much the same way that windfalls of a few large trees in the forest create a patch disturbance. On the other hand, areas of ant hills, no more than 1 dm in diameter individually, also represent a fresh disturbance area with an open habitat capable of colonization. However, no such colonization has been detectable. Because there are thousands of 1-dm anthills per hectare, with only tens to hundreds of gopher mounds or a single badger disturbance per hectare, all three processes must be viewed as contributing a colonizable habitat for short-lived species in the grassland community. Examination of hierarchical relationships in the ecological system, as discussed by Allen and Starr (1982), appears to be essential for understanding the interaction between organisms of varying size and life span, and environments of varying size and duration.

While the microdynamics of gaps in the sand prairie described here are superficially similar to those quantified by Levin and Paine (1974) for an intertidal zone, the degree of these similarities has not yet been examined fully. If such studies can be done, we may find that for certain species in the sand prairie, models will need to incorporate the demographics of small-scale disturbances and to monitor the seasonal changes as well as the aging and extinction of the patches. Other species and the larger-scale disturbance agents on the sand prairie, however, produce phenomena that are more like the large disturbances in forests and may be followed by a succession over tens to hundreds of years. The model of succession by Bledsoe and Van Dyne (1971) is one example, applied to a human disturbance area, but it has the potential for describing recovery in arid grasslands following severe drought or following wind or massive water erosion and deposition.

Finally, the qualitative results summarized here indicate that the sand prairie grassland, as much as the forest, is a community of plants and animals bound closely to multiple-scale disturbances. The potential for influence by small and short-term patch disturbances is much greater in a system of 100 or more species, each capable of contributing at different seasons and at different times in a recovery transient. Thus, unlike the forests, both the processes of disturbance and the responses of organisms to these processes have to be evaluated within a framework of appropriate hierarchical constructs. More precise data will also be needed to quantify the interactions among both physical and biotic disturbance agents and the short- and long-lived species. Perhaps through the grassland preservation programs already underway at various research sites in the plains states, we can look forward to such data and a possible validation of links between adaptations and the disturbance regimes described here.

RECOMMENDED READINGS

Allen, T. F. H., and Starr, T. B. (1982). "Hierarchy: Perspectives for Ecological Complexity." Univ. of Chicago Press, Chicago, Illinois.

Levin, S. A., and Paine, R. T. (1974). Disturbance, patch formation, and community structure. *Proc. Natl. Acad. Sci. U.S.A.* **71**, 2744–2747.

Loucks, O. L., Ek, A. R., Johnson, W. C., and Monserud, R. A. (1980). Growth, aging and succession. *In* "Dynamic Properties of Forest Ecosystems" (D. E. Reichle, ed.), pp. 37–85. Cambridge Univ. Press, London and New York.

Platt, W. J. (1975). The colonization and formation of equilibrium plant species associations on badger disturbances in a tall-grass prairie. *Ecol. Monogr.* **45,** 285–305.

Weaver, J. E., and Albertson, F. W. (1943). Resurvey of grasses, forbs, and underground plant parts at the end of the great drought. *Ecol. Monogr.* **13,** 63–117.

Chapter 6

Shrubland Fire Regimes and Their Evolutionary Consequences

NORMAN L. CHRISTENSEN

Department of Botany
Duke University
Durham, North Carolina

THE ECOLOGY
OF NATURAL DISTURBANCE
AND PATCH DYNAMICS

I. INTRODUCTION

Shrub-dominated ecosystems extend from arctic to tropical regions and occur over the entire moisture gradient from deserts to wetland (Specht, 1979; DiCastri, 1981; West, 1983). Shrublands are defined as ecosystems dominated by plants with persistent woody stems but no central trunk (Stebbins, 1972). The specific reasons for the success of shrubby species in particular shrubland types vary from region to region, but nearly all shrublands, save perhaps warm deserts, are to some degree oligotrophic (especially with respect to nitrogen and potassium), and adaptations to nutrient limitation are common. These include evergreenness, sclerophylly, nutrient reabsorption, and comparatively high root-to-shoot ratios (Chapin, 1981).

In the majority of shrublands, fires play a significant role in the maintenance of community structure and function, and have been important selective forces in the evolution of plant form and life history. The literature on fire is replete with generalizations regarding fire effects and vegetation response. Like the politician who searches the scriptures for phrases to match an ideology, the asciduous shrubland ecologist can find data to support any plausible and some implausible generalizations. This situation arises not as a consequence of poor techniques or shoddy research but because of the high variance in shrubland fire regimes. Fires are patchy phenomena and contribute significantly to temporal and spatial heterogeneity among and within shrublands. Furthermore, the occurrence of shrublands in such a wide range of environments provides an opportunity to compare the effects of and responses to a variety of fire regimes within the context of a single physiognomic type. Thus, I shall focus especially on variability with respect to (a) variations among fire regimes in shrublands and their causes, (b) variation in environmental consequences of shrubland fires, (c) variation in the evolution of fire-related adaptations, and (d) variation in the nature and mechanisms of community response to fire.

II. FIRE REGIMES AND THEIR CONTROL

Many of the variations in effects and response to fire result from variations in fire regime. The fire regime is determined by fire type and intensity, fire frequency, season of burning, and predictability (Gill and Groves, 1981). Although each of these factors is discussed separately, they are clearly not independent of one another.

A. Fire Type and Intensity

Because of the low stature and single vegetational stratum of shrublands, comparatively intense crown fires are characteristic of these areas. "Intensity" is used here to refer to energy released during burning, which may or may not have a relationship to the intensity of disturbance or perturbation [*sensu* Connell (1978)].

Byram (1959) suggested that fire intensity was a consequence of the rate of spread of the fire and its fuel mass and energy content, and would best be expressed as energy release per unit length of burning front per unit time. Such measures have been incorporated into computer models to predict shrubland fire properties (Lindenmuth and Davis, 1973; Rothermal, 1972; Rothermal and Philpot, 1973; Aston and Gill, 1976), in particular the rate of fire spread.

In addition to affecting the rate and likelihood of spread, variations in fire intensity have other ecological consequences. Within-fuel temperatures in shrublands vary from 200°C (Kayll, 1966) to well over 1000°C (Martin *et al.*, 1969; Traubaud, 1974, 1975, 1977), although intermediate values are more common (Bentley and Fenner, 1958; Christensen and Wilbur, 1984). The proportion of available fuel consumed, as well as nutrient losses due to volatilization and smoke, are largely determined by fire temperature (Kentworthy, 1963; Allen, 1971; Evans and Allen, 1971; Raison, 1979). The extent of these losses clearly influences the postfire nutrient regime. Although temperatures are attenuated quickly down the soil profile, the temperatures at the high end of this range may greatly influence shrub and seed survivorship (Sampson, 1944; Hadley, 1961). Low temperatures may be insufficient to break dormancy in seeds with a heat requirement.

Fire intensity is determined by abiotic factors such as weather and topography, as well as characteristics of the fuel such as quantity, structure, and flammability (Rundel, 1981b). The potential importance of fuel quantity and structure is obvious (Countryman and Philpot, 1970). Flammability, or likelihood of ignition, is affected by several factors. Fuel moisture content is often the factor determining the season of burning in addition to the intensity of the fire. For example, in mediterranean shrublands, fuel moisture is sufficiently high in winter, spring, and early summer that fires are unlikely (Green, 1982). In climates where rain occurs year-round, high fuel flammability is correlated with drought during periods of high evapotranspiration (Christensen *et al.*, 1981).

Fire intensity increases in proportion to the ratio of dead-to-live fuel, which varies considerably among and within shrublands (Rundel, 1981b; Green, 1982). This ratio is determined by the relationship between litter production (particularly branch mortality) and decomposition. In southern California the dead-to-live ratio in coastal sage scrub (soft chaparral) after 20 years may exceed 60% (Green, 1982; Gray, 1982), whereas this value is between 25 and 50% in hard chaparral at the same age (Green, 1982; Gray and Schlesinger, 1981). Green (1982) proposes that this difference accounts in part for the higher frequency of fire in coastal sage communities. In nearly all shrub communities, the dead-to-live ratio increases with the age since the last fire (Blaisdell, 1953; Gimingham *et al.*, 1979; Christensen *et al.*, 1981; Gray and Schlesinger, 1981; Specht, 1981). The rate at which this ratio increases may greatly affect the fire return interval.

Inorganic and organic chemical properties of the fuel are known to affect flammability. In general, flammability is inversely related to mineral nutrient content (Shafizadeh, 1968; Philpot, 1970), and tissue nutrient concentrations, particularly in sclerophyllous shrub species, are in general quite low (see, e.g., Groves, 1981b;

Rapp and Lossaint, 1981; Gray and Schlesinger, 1981). Low tissue nutrient content also slows decomposition and thus increases the dead-to-live ratio. Variations in the content of ether-extractable organic chemicals have attracted even more interest (Shafizadeh *et al.*, 1977; Traubaud, 1977; Rundel, 1981b). In excess of 30% of the caloric content of the leaves of many shrub species, including species from such disparate shrublands as southeastern shrub bogs and southwestern chaparral, may be due to such ether extractives. The ecological role of these extractives is not certain, but herbivory and allelopathy have been most frequently implicated (Mooney and Dunn, 1970; Muller 1966; Muller *et al.*, 1968). Several authors have suggested that plant characteristics that increase flammability may have been selected for in order to guarantee a favorable fire regime (Mutch, 1970; Rundel, 1981b). This possibility is explored in more detail below.

A number of shrublands occur on peat or organic soils (shrub bogs, arctic heath, and various heathlands) and are visited by "ground fires." Wilbur and Christensen (1983) found that a relatively light fire in a pocosin shrubland in North Carolina consumed 0.5–6.0 cm of peat, accounting for nearly 75% of the carbon oxidized during the fire. Similar effects have been noted for heathlands on organic soils (Gimingham *et al.*, 1979) and arctic shrublands (Wein, 1974). Fires burning in these organic layers may move very slowly and burn for months.

Fire intensity may vary considerably within a fire as a consequence of small-scale variations in weather and topography. Fires moving with the wind or uphill (head fires) move rapidly, burn less fuel, and release less heat over a shorter time period than do fires moving downhill or against the wind (Albini and Anderson, 1982). Thus, considerable local variation in fire intensity may be unrelated to fuel characteristics.

B. Fire Frequency

Fire frequency can be defined in terms of the return time at a particular location (see, e.g., Heinselman, 1981b) or the number of fires per unit area over a landscape (regional frequency). From an evolutionary standpoint, both definitions are important.

Fire return interval is determined by the frequency and availability of ignition sources, factors influencing the spread of fires, the frequency of weather favorable for ignition, and successional changes in the amount and flammability of fuels. The frequency of natural ignition events (particularly lightning) varies widely among shrub ecosystems (Komarek, 1968) and may limit the fire return interval and the regional frequency in very arid shrublands (Wright, 1972) and in arctic heathlands (Bliss, 1979). As a result of human activity, ignition sources do not appear to be limiting in most shrublands. However, the regional fire frequency will depend on the frequency of ignition events.

Since the probability of any particular location being ignited directly (e.g., struck by lightning) is low, the fire return interval at a particular location depends more on the likelihood of fire spreading from adjacent areas (Heinselmann, 1981b). Such factors as fuel continuity, topography, and flammability of adjacent ecosystems will obviously affect this probability. Thus, in areas of extensive, continuous shrubland (e.g., the chaparral of southern California or desert grass–shrublands), fires of large extent were the norm prior to human intervention (Humphrey, 1962; Hanes, 1971; Byrne et al., 1977; Minnich, 1983). Fire return times in arctic heathland are highest near the transition to more easily ignited boreal forests (Rowe et al., 1974). Hard chaparral stands near more flammable coastal sage shrub communities have shorter return intervals (Radtke et al., 1982). Fire return intervals appear to have been altered in several shrubland systems due to dissection of the environment by roads and agricultural development (Radtke et al., 1982; Forman and Boener, 1981).

In many shrublands, intrinsic factors appear to set lower limits on fire return intervals. Rates of primary production, patterns of stem and whole plant mortality, and rates of decomposition determine the rate at which flammable fuel accumulates in a shrubland. In extreme situations (e.g., very arid deserts), production is so low that there is rarely sufficient fuel to carry a fire despite otherwise favorable conditions (Humphrey, 1962). In shrub bog ecosystems, the standing crop may return to prefire levels within 3 years following fire (Christensen and Wilbur, 1984), but flammability remains low until the dead-to-live ratio increases. It is reasonable to expect that any factor that increases flammability will decrease the fire return interval.

C. Season of Burning

Season of burning is largely determined by the availability of ignition sources and seasonal variation in fuel flammability. A major factor affecting flammability is seasonal variation in moisture. However, seasonal variation in the dead-to-live ratio is important in shrub bogs (Christensen et al., 1981) and California coastal sage scrub (Westman, 1981a; Gray, 1982), where a major component of the shrub community is deciduous. Philpot (1969) has documented seasonal changes in ether extractives in the leaves of chamise (*Adenostoma fasciculatum*) that render it more flammable in late summer.

Season of burning may alter the effects of and response to burning considerably. For example, fires that occur just prior to a period when plant growth is limited by cold may cause release of nutrients to a plant community unable to make efficient use of them. This factor remains largely unstudied.

III. ENVIRONMENTAL RESPONSES TO BURNING

A. Heat Release during Burning

The aboveground portions of most shrub species are killed during most shrubland fires, although only 60–80% of aboveground biomass may be consumed (Countryman and Philpot, 1970; Wilbur and Christensen, 1983). Surface soil temperatures during fires in British heath (Gimingham, 1971), chaparral (Craddock, 1929; Sweeney, 1956), and southeastern shrub bogs (Christensen and Wilbur, 1984) are normally between 200° and 300°C; however, temperatures exceeding 500°C have been measured. Heat transfer down the soil profile is quickly attenuated, so that temperatures rarely rise above 100°C below 3 cm (Craddock, 1929; Priestley, 1959). The duration of these elevated soil temperatures depends on the rate of fire spread and the burning of residual material. Thus, while temperature increase in the soil is reduced by a litter or humus layer (Priestley, 1959), the ignition of this layer can result in elevated temperatures over a longer time period.

B. Postfire Microclimate

Differences between the microclimates of recently burned and unburned shrublands are profound. Increased insolation coupled with reduced soil albedo may result in considerable soil heating (Sweeney, 1956). Christensen and Muller (1975a) measured surface soil temperatures in excess of 70°C on south-facing slopes of recently burned chamise chaparral. Such temperatures were sufficient to stimulate germination of seed of species with a heat requirement to break dormancy. Wilbur (unpublished data) found that peat temperatures in a recently burned shrub bog had both higher maxima and lower minima than those of unburned areas. Air temperatures 10 cm above the peat also displayed considerably more variation in burned than unburned areas. Similar observations have been made by Sweeney (1956) for chaparral. Such changes could potentially alter plant survival and growth.

The direct importance of increased available light to seedling and sprout growth depends on the extent to which prefire light levels are limiting. Total shrub light levels are limiting. Total shrub leaf area indices vary from below 1 in desert scrub (Chew and Chew, 1965) and arctic heath (Bliss, 1979) to approximately 1 in *Adenostoma* chaparral (P. C. Miller, 1981) and to 2–4 in moister chaparral areas and temperate shrub bogs (Gray, 1982; P. C. Miller, 1981; Christensen and Wilbur, 1984). Thus, in moister temperate shrublands, increased light following fire may greatly influence seedling success.

C. Nutrient Changes

Considerable attention has been focused on nutrient release following fire. Despite potential nutrient losses due to volatilization and increased runoff, nutrient availability has in general been found to be higher immediately following fire (see, e.g., Christensen and Muller, 1975a; Wein and Bliss, 1974; Gimingham *et al.*, 1979; Westman *et al.*, 1981). The importance of such increased postfire availability is undoubtedly related to the extent of prefire limitations. Specht (1969) suggested that nutrients are not limiting in many chaparral locations. However, most shrubland ecosystems have been found to be deficient in nitrogen and/or phosphorus (Groves, 1981a,b). Indeed, Loveless (1961, 1962) proposed that the sclerophylly characteristic of many shrub species was a response to nutrient deficiencies.

Although not emphasized in most papers dealing with fire-caused nutrient changes, soil nutrient concentrations tend to be considerably more variable after than before fire (see, e.g., Christensen and Muller, 1975a; Wilbur and Christensen, 1983). This variation arises due to local variations in fire intensity as well as the uneven distribution of ash. Wilbur and Christensen (1983) found that peat phosphorus concentrations in some locations in a shrub bog were unaffected by fire, whereas other areas showed a tenfold increase. Similar variations in nutrient availability relative to site conditions and fire intensity have been documented by Westman et al. (1981). Thus, the postfire shrubland environment may be considerably more heterogeneous with respect to potentially limiting resources.

D. Water Relations

The effect of fire on water relations varies from shrubland to shrubland. In general, reduced leaf area following fire results in reduced evapotranspiration and more available water to deeply rooted species (Bauer, 1936; Adams et al., 1947). However, increased surface soil heating due to insolation often results in less available water in surface soil horizons in the first season following fire (Christensen and Muller, 1975a; Westman, 1979; Wilbur, unpublished data). Surface soil moisture availability may increase above prefire conditions once sufficient leaf area is available to provide shade. Thus, in some shrubland systems, conditions for seedling germination and establishment may be most favorable in the second or third year following burning.

E. Allelochemicals

Muller et al. (1968) proposed that germination of many shrubs and herb species was inhibited by water-soluble toxins produced by chaparral shrubs. Fire removed the source of those toxins and perhaps denatured residual toxins in the soil, resulting in a release from this extrinsically enforced dormancy. Thus, the pulse of germination observed in these shrublands following fire was considered to be a result of changes in the biochemical environment of the seed pool. Research by McPherson and Muller (1969) and Chou and Muller (1972) provided evidence supporting this hypothesis. In particular, several chaparral shrubs were shown to produce potentially toxic phenolic chemicals that were leached from the shrub foliage. Kaminsky (1981) has presented evidence that toxins are actually produced in decomposition of shrub litter by soil microbes. In many cases, removal of the shrub cover in the absence of fire (and presumably soil heating and nutrient pulse) results in a germination response similar to that observed in burned areas. Christensen and Muller (1975a) verified the production of toxins by chamise (A. fasciculatum), a chaparral dominant, but found that many of the alleged allelochemicals identified by McPherson et al. (1971) were actually more common in burned than unburned soil. Furthermore, they noted that soil temperatures in cleared areas were often sufficient to stimulate germination. Further work (Christensen and Muller, 1975b) indicated that dormancy for many chaparral herbs and shrubs was endogenously enforced, and that once dormancy was broken shrub toxins played little role in either germination or survival. Germination was not inhibited in any of these species by leaf leachate. Christensen and Muller (1975a) noted that germination of some ruderal species (with short-lived, wind-dispersed seeds) was inhibited by shrub toxins. The potential importance of changes in allelopathic influences in shrublands is still open to debate; however, the sweeping generalization of "allelopathic control of the fire cycle" (Muller et al., 1968) has not been widely supported.

Wicklow (1977) raised the possibility that biochemical changes following fire may actually stimulate germination in some species. He found that *Emmenanthe penduliflora*, a chaparral, fire-endemic herb, germinated best in the presence of charred *Adenostoma* wood. Keeley et al. (1985) and Nitzberg and Keeley (1984) have provided additional data for this mechanism of dormancy release in several other fire-response endemics. The potential importance of this phenomenon in other shrublands has not been investigated.

Biochemical changes associated with burning may also influence the physical and hydrological properties of soil. Best studied in this regard is the effect of fire on soil wettability in arid shrublands (see the review in Debano et al., 1967). Fire is thought to cause the downward migration of hydrophobic chemicals increasing surface soil wettability and the potential for surface erosion. S. Adams et al. (1970) found that the heterogeneous distribution of such hydrophobic layers following a wildfire in a California desert creosote shrubland played an important role in postfire vegetational patterns.

F. Animal Populations

Animals are known to influence plant establishment and growth in many shrubland ecosystems. The prevalence of secondary chemicals in the leaves of many shrub species is usually attributed to selective pressures from insect and mammal grazing (Whittaker and Feeny, 1971; Mabry and Difeo, 1973; Force, 1981, Fuentes et al., 1981). The prevalence of evergreenness in many shrublands may select for increased energy expenditure for herbivore defense (Mooney and Dunn, 1970; Mooney, 1972). The importance of mammal grazing in shrubland understories has been emphasized in several studies (Bartholomew, 1970; Christensen and Muller, 1975a,b; Halligan, 1973; Wirtz, 1982).

Fires profoundly affect shrubland herbivore populations and, thereby, the patterns of grazing. Force (1981, 1982) has documented increased insect abundance and diversity following chaparral fire and attributes this to increased floristic diversity. However, populations of phytophagous insects were reduced in the first postfire year. Small mammal populations are depleted in the immediate postfire years in chaparral (Lawrence, 1966; R. D. Quinn, 1979), although they may increase to a peak 5 years following burning before complete shrub recovery (Wirtz,

1982). Similar observations have been made for heathlands (Main, 1981). Christensen and Muller (1975a,b) considered that postfire release from herbivory may have been a significant selective force in the evolution of mechanisms that concentrate reproductive effort among shrubland plants in immediate postfire years.

IV. PLANT ADAPTATIONS TO FIRE IN SHRUBLANDS

The significance of fire in the history of shrubland ecosystems can best be appreciated by the range of adaptations to fire displayed among shrubland species. Several excellent reviews have considered the range of such adaptations (Gill, 1981; Gill and Groves, 1981; Keeley, 1981; Noble, 1981). In this section, I shall describe the most obvious of these "fire traits" and shall consider the basis for their enhancement of plant fitness and their occurrence among different shrublands.

A. Survivorship

Many shrubland species possess the ability to sprout from belowground buds following fire. As mentioned earlier, large temperature fluctuations occur only in the upper few centimeters of soil. In shrublands where intense fires return at regular intervals, many shrub species have basal bud burls or lignotubers. These are particularly common in shrubs from mediterranean climates (Gill, 1981; Keeley, 1981). Such burls may account for a considerable portion of total shrub biomass (Specht and Rayson, 1957; Kumerow, 1981) and appear to store considerable quantities of nutrients and carbohydrates (Mullette and Bamber, 1978). Buds on such burls are often protected by fire-resistant coverings.

Susceptibility to fire kill varies seasonally and in relation to fire intensity within a shrubland type (see, e.g., Sampson, 1944; Kologiski, 1977). Among shrublands, fire-caused shrub mortality is greatest in those areas where fires are least frequent and predictable, for example, deserts (Blaisdell, 1953; Humphrey, 1962, 1974; Wright, 1972; O'Leary and Minnich, 1981).

Shrub species in mediterranean climates vary along a gradient from obligate sprouters (genets rarely killed, regenerating primarily from sprouts) to facultative sprouters (genet mortality variable, reproduction from seeds and sprouts common) to obligate nonsprouters (genets always killed, reproduction entirely from seed) (Naveh, 1975; Keeley and Zedler, 1978; van Wilgen, 1981; van Wilgen and le Maitre, 1981; Kruger, 1982). Wells (1969) observed that obligately seeding species (nonsprouters) of *Arctostaphylos* and *Ceanothus* evolved from sprouting species. He noted that obligate seeding species produce considerably more seed and that their seeds often have stronger dormancy mechanisms than sprouters. Wells suggested that obligate seeding strategy would result in more frequent genetic recombination and higher potential rates of evolution. He argued that this would be adaptive given the intense selection due to recurrent fires (see also Raven, 1973). Keeley and Zedler (1978) pointed out that in order to have selection for sexual reproduction, there must be an immediate benefit to individual fitness. They contend that obligate seeding evolved in response to long fire return intervals (perhaps >100 years). Old chaparral stands have a high dead-to-live ratio, experience very intense fires, and have very high shrub mortality even among obligate sprouters (Hanes, 1971). Under these conditions, natural selection would favor increased energetic investment in long-lived seeds that would accumulate in the soil seed pool rather than in perenneating organs with a low probability of survival. They suggest that fire return intervals of <25 years would favor the sprouting over the seeding strategy. Furthermore, van Wilgen (1918) found that frequent burning in South African fynbos resulted in elimination of "seed regenerating shrubs," whereas longer return times favored vegetative reproduction. Based on these considerations, we should expect that the relative fitness of resource investment in seed production versus vegetative structures such as lignotubers will depend on the fire return interval, prefire resource limitations, fire intensity and associated shrub mortality, and postfire seedling demography.

B. Fire-Stimulated Germination

Seed dormancy and extensive germination following fire are characteristic of many shrubland species, particularly in mediterranean climates (see, e.g., Sampson, 1944; Sweeney, 1956; Naveh, 1975; Gill and Groves, 1981). In many cases, this dormancy is profound and germination is obligately tied to fire. As discussed earlier, these dormancy mechanisms appear to be enforced endogenously (Keeley *et al.*, 1985), as opposed to by exogenous allelochemicals (Muller *et al.*, 1968). Selection for dormancy suggests that the probabilities of successful seedling establishment in interfire years are very low and that they are improved in the immediate postfire year. Christensen and Muller (1975b) found that low light levels, nutrient limitations, and herbivory combined to cause near-zero seedling survivorship in unburned chaparral. Burning ameliorated each of these negative effects.

Several cues have been identified that "inform" these dormant seeds that a fire has passed. Many mediterranean shrub species are stimulated to germinate by a short burst of heat (Quick, 1935; Stone and Juhren, 1951, 1952; Sweeney, 1956; Gardner, 1957; Christensen and Muller, 1975a,b; Arianoutsou and Margaris, 1981a,b, 1982; Keeley *et al.*, 1985). In many cases, this effect is a consequence of the rupture of impermeable seed coats or the melting of seed coat waxes. As mentioned earlier, chemicals leached from shrub char stimulate germination in several fire-endemic species (Wicklow, 1977; Jones and Schlesinger, 1980; Keeley *et al.*, 1985). The wide soil temperature variations characteristic of recently burned areas appear to stimulate germination in many shrubland species (Gimingham, 1972; Christensen and Muller, 1975a).

The extent to which germination patterns depend on fire is variable among and within species. In the California chaparral there are several annual herbs that germinate only after heat treatment. They are seen only in recent burns and survive from one burn to the next in the seed pool (Sweeney, 1956). In *Adenostoma,* however, a portion of any year's seed crop will germinate after a brief after-ripening, whereas the remainder requires heat treatment (Stone and Juhren, 1952). Jones and Schlesinger (1980) found that seeds of *Emmenanthe penduliflora,* an annual herb, collected from recently burned chaparral sites required a char treatment to effect germination. *Emmenanthe penduliflora* seeds collected from desert sites where fires are uncommon did not.

It is difficult to find examples of fire-stimulated germination outside of the mediterranean shrublands. In shrublands with organic soils, such as wet heaths and shrubs bogs, any soil seed pool is likely to be consumed by the fire. Furthermore, this life history pattern concentrates germination in the first postfire year, when moisture and microclimate conditions may be harsh in some shrublands.

C. Seed Retention and Fire-Stimulated Seed Release

Emergent trees in several shrubland types have serotinous fruit or cones. This is particularly well documented in the Australian flora (Gill, 1975; Gill and Groves, 1981) and in 10 species of North American pines (see, e.g., Lotan, 1975; Vogl, 1973; Woodwell, 1958). In these species, a seed pool is maintained on the tree and the heat of the fire causes dehiscence. The size of this arboreal seed pool depends on the rate of fruit or cone production, the rate of prefire dehiscence, and the rate of fruit or cone predation. To the extent that potential nonfire or nearby fire establishment opportunities are missed or that unopened reproductive structures are more vulnerable to predation (Smith, 1970), total serotiny will reduce fitness. Thus, it is not surprising that levels of serotiny vary among and within species. For example, knobcone pine (*Pinus attenuata*) occurs in very dense chaparral in which nonfire reproductive opportunities are minmal and fires occur with some regularity (Vogl, 1973). It is intensely serotinous. Pond pine (*Pinus serotina*), on the other hand, occurs in shrub bogs that have a more patchy distribution and a more irregular and unpredictable fire cycle. Cones of this species remain closed for 2–8 years. Thus, a tree will have at any time a 2- to 8-year accumulation of cones but will still be releasing seeds, which may become established in other openings or nearby burns (Christensen and Wilbur, 1984).

D. Fire-Stimulated Flowering

Flowering is increased in a number of species in many shrublands following burning. In several cases, flowering of perennial species is obligately tied to burning (Stone, 1951; Horton and Kraebel, 1955; Levyns, 1966; Gill and Ingwerson, 1976). In such cases, heat or simply leaf removal may stimulate floral development (Gill and Groves, 1981). Flowering in other shrubland species appears to be tied to increased rates of production following fire (due to increased nutrients, light, etc.) and the availability of resources to be allocated to flowering. Seedling establishment in these species occurs 2 to several years after fire. Christensen and Wilbur (1984) found that conditions for successful seedling establishment for many species in a southeastern shrub bog were most favorable in the second postfire year.

Few generalizations have emerged regarding the relative importance of these different life history characteristics in different shrubland types. Although nearly all shrublands burn, the evolution of life histories that are obligately tied to fire is characteristic only of mediterranean climate shrublands, and even there the importance of such life history patterns is variable. Which life history characters will be the most fit depends not only on the fire's intensity and frequency but also on its predictability, limitations on establishment and growth between fires, and the nature of the postfire environment. All of these factors vary within as well as among shrublands.

E. Flammability

Mutch (1970) suggested that in areas where plant fitness, for whatever reason, is positively correlated with fire, natural selection might favor plant traits that increased flammability. This hypothesis has been very attractive to fire ecologists (see, e.g., Mooney *et al.,* 1981) because it implies that fire cycles may be under intrinsic genetic control. Shrubland species may "manipulate" their fire regime to optimize fitness. There is no question that fuels in areas that are prone to fire are more flammable than those in areas that are not (Rundel, 1981b). Implicit in the hypothesis that such increased flammability evolved in response to natural selection are several assumptions: (a) traits controlling flammability are, at least to some extent, under genetic control and are genetically variable; (b) a substantial portion of the variability in fire regime is determined by individual plant flammability; and (c) it is possible to imagine selection for flammability per se based on its improvement of individual plant fitness.

Plant traits contributing to flammability have been discussed previously and thoroughly reviewed by Rundel (1981b). There is, indeed, no reason to doubt that many of these traits are under genetic control. Furthermore, genetic variation in factors such as ether-extractive concentration and mineral nutrient content undoubtedly exists in many populations. However, it should also be recognized that there is no shortage of alternative selective forces to account for any of these flammability traits that are unrelated to fire (see, e.g., Mooney and Dunn, 1970).

Although traits of individual plants contribute to shrubland fire behavior, flammability is, in general, a collective community property that varies widely from time to time as a consequence of variations in macro- and microclimate. Most of the models of fire behavior and spread discussed earlier employ multiple regression techniques to predict flammability. The relative importance of plant characteristics (such as biochemistry or fine twig production) varies considerably depending on climatic variables. Such plant features often contribute least to these models when fires are more common (i.e., during periods of drought). It is difficult to imagine how increased flammability of an individual plant might influence significantly the fire regime in the community at large. The likelihood of *de novo* ignition (e.g., by lightning) of a particular shrub is infinitesimally small, and the likelihood of ignition from adjacent vegetation depends on a host of factors; one plant or a small group of closely related plants has little effect on overall fire behavior. It is possible that the flammability characteristics of an individual shrub may influence fire intensity in its direct vicinity, but given the physical structure of most shrubland communities, this effect must be small.

Is it possible that increased flammability of an individual or small group of related individuals in a sea of somewhat less flammable shrubs would result in those individuals contributing more offspring to the next generation than their competitors? It is doubtful that such an individual or group of individuals would measurably increase the likelihood of fire in such a community; however, their flammability might alter the local effects of fire when it occurred and thereby improve conditions for their offspring. Considering the variety of factors that influence local fire intensity on the scale of meters, the importance of plant-to-plant variations in flammability is, at best, unpredictable and, at worst very small. In situations where large clones dominate the landscape and the characteristics of a particular genet and the community are nearly synonymous, advantages to the individual can be more easily argued. Such situations are not common in shrublands and are least common in the most flammable types. In general, it seems unlikely that variations in plant traits affecting flammability could be selected because of their influence on the fire regime.

There is, however, no doubt that shrublands are quite flammable relative to other ecosystems (Mutch, 1970). I would argue that such flammability has arisen in response to selection unrelated to fire. Shrublands occur in areas in which production is to some extent limited, often as a consequence of water or nutrient shortages. Under such conditions, there is selection for sclerophylly and low litter nutrient content. These features diminish decomposition rates, which are additionally slowed by other features of the shrubland environment. Thus, high dead-to-live ratios are characteristic of most shrublands (even those where fire is infrequent). The presence of flammable secondary chemicals in many shrubland species may be accounted for by their role in herbivore defense or, perhaps, allelopathy. I do not wish to discount totally the notion that fire cycles in shrublands are controlled by intrinsic feedback processes (e.g., increasing flammability with age since burning), but I question whether flammability evolved in response to selection to optimize the fire cycle.

V. COMMUNITY RESPONSES TO BURNING

A. Successional Patterns

In most shrubland systems, postfire community composition closely resembles that of the prefire community, a situation referred to as "autosuccession" (Muller,

1952; Hanes, 1971). In general, the response would best fit the "tolerance" model of Connell and Slatyer (1977) in that species turnover is largely a consequence of innate species life history patterns rather than interspecies interactions. Shrubland succession following fire, however, fits very few of the generalizations regarding linear successional sequences described by Odum (1969). Rather, it is probably best viewed as a cyclic process in which changes within the system over time enhance the probability of disturbance (i.e., decreased stability *sensu* resistance). Clearly, succession in shrublands does not lead to self-perpetuating communities in which genet mortality and recruitment occur in association with small-scale disturbance.

It is worth noting that in some shrub systems the cyclic nature of this process is intensity dependent. In shrub bogs, for example, normal fires of low to moderate intensity result in rapid shrub regeneration, with reproduction from seed varying in direct proportion to fire intensity. Very intense fires, however, result in high shrub mortality and a postfire herb community that may persist for decades (Kologiski, 1977). Zedler *et al.* (1983) reported a similar response to an abrupt change in fire return time in California coastal sage scrub.

Given the apparent dependence of many shrublands on fire, the absence of fire might be considered a more severe "stress." In most such shrublands, the complete absence of fire is a moot issue. However, there is no doubt that fire suppression policies have greatly altered natural fire regimes in many shrublands (see, e.g., Pyne, 1982; Radtke *et al.*, 1982; Minnich, 1983). Such changes could presumably result in shifts in relative species importance. In at least one case, the cyclic change from herb to shrub dominance can proceed (albeit more slowly) in the absence of fire. The so-called *Calluna* cycle has been described by Watt (1947) and Barclay-Estrup and Gimingham (1969; Barclay-Estrup, 1970). In this case, the degeneration of the shrubs in the absence of fire opens space for herb (particularly bracken) invasion. Subsequently, shrubs reinvade. No similar autogenic cyclic process has been described in other shrublands. However, Hanes (1971) noted the reinvasion of early successional suffrutescent shrubs into decadent stands of chamise in the California chaparral.

B. Patterns of Production

In general, net aboveground ecosystem production (biomass accumulation) in shrubland ecosystems diminishes in old stands and may approach zero (Bliss, 1979; Specht *et al.*, 1958; Jones *et al.*, 1969; Rundel and Parsons, 1979; Gimingham *et al.*, 1979; Wilbur and Christensen, 1983). In some cases (e.g., heathland: Gimingham *et al.*, 1979; *Adenostoma* chaparral: Rundel and Parsons, 1979), this change comes about as a consequence of declines in net primary production, whereas in other shrublands (e.g., *Ceanothus* chaparral: Schlesinger *et al.*, 1982), net primary production becomes constant and decay processes begin to come to equilibrium. In either case, the years immediately following fire typically have the highest aboveground biomass accumulation rates.

Biomass accumulation rates in immediate postfire years depend to some extent on the reproductive mode. Following intense fires in which vegetative structures are killed or in communities dominated by obligate seeders, there is a lag in shrub biomass accumulation rates probably associated with the development of leaf area (Horton and Kraebel, 1955; Jones et al., 1969). Where vegetation recovery is predominantly due to sprouting, the rate of recovery initially is probably related to the amount of belowground translocatable resources. Thus, responses in these situations are highly variable. Recovery to prefire aboveground biomass levels in heathlands and mediterranean shrublands occurs in 7–20 years (see, e.g., Specht et al., 1958; Horton and Kraebel, 1955; Kruger, 1977; Specht, 1969; van Wilgen, 1982; Hanes, 1971; Pond and Cable, 1962). In warm desert (Humphrey, 1974) and arctic heathland (Wein, 1974) ecosystems, postfire shrub production may be much slower. At the other extreme, Wilbur and Christensen (1983) have found that shrub bog ecosystems return to approximately 80% of prefire biomass within two growing seasons. By the end of the first growing season, the shrub leaf area index (LAI) averaged 2.2 and was >3.0 in the second growing season (prefire LAI = 3.8). However, Christensen et al. (1981) point out that postfire production in these ecosystems is greatly affected by fire intensity which is highly variable. Postfire production in shrublands is also greatly affected by site conditions unrelated to fire, such as moisture availability (Jones et al., 1969).

Herbaceous species contribute most to shrubland production in immediate postfire years, but their importance varies from shrubland to shrubland. Herb production in immediate postfire years may equal or exceed shrub production at any stage in the fire cycle in some shrublands (Keeley and Keeley, 1983; Specht, 1969). Several workers have noted the potential importance of this herb response to postfire ecosystem nutrient retention (see, e.g., Christensen and Muller, 1975a; Debano and Conrad, 1976, 1978; Arianoutsou and Margaris, 1981b; Boerner, 1982). In shrub communities in which shrub production is rapid following fire (e.g., shrub bogs [Wilbur and Christensen, 1983]), herb production is considerably lower.

C. Patterns of Change in Community Composition and Diversity

Fire has been associated with the maintenance of species diversity in virtually all shrubland types. In general, species richness and equitability tend to be highest immediately following fire and decline more or less quickly thereafter. Dominance–diversity curves reflect an almost random distribution of resources among species in immediate postfire years, suggesting that competition for resources may play a comparatively small role in structuring communities at this stage. Subsequently, the distribution of dominance becomes log normal or, in some cases, almost geometric (Christensen et al., 1981).

When the seed pool is included, species richness changes very little during the fire cycle in many shrublands. Most regeneration following burns is vegetative or from a prefire seed pool, although seed dispersal from adjacent burns or communities may, over time, replenish seed pools (Westman, 1979). Ruderal species are relatively uncommon. Thus, genets of most constituents of shrubland communities are present throughout the fire cycle, if only as seeds. Shrubland niche space is partitioned such that only a fraction of the community is competing for resources at any time. Spatial and temporal variation in postfire environments must result in variation in the relative success of community constituents over the landscape and from fire to fire. This situation clearly exemplifies Grubb's (1977) contention that establishment opportunities represent a very important niche axis (the regeneration niche). This view of the relationship of the fire cycle to shrub community diversity differs from other recent disturbance–diversity hypotheses (see, e.g., Loucks, 1970; Connell, 1978; Huston, 1979) in that it does not necessarily invoke none-quilibrium conditions (see also Denslow, Chapter 17, this volume). Indeed, it speaks more to the *evolution* of richness than its *maintenance,* and may account for the high richness and endemism of many shrubland floras. There are many more ways to succeed in a shrubland than analysis of the mature community or mean disturbance conditions might suggest. To the extent that variability in shrubland species composition is related to variability in fire regimes and fire effects, it may be necessary to incorporate variation into shrubland management strategies. In such systems, the concept of an optimal fire regime may be meaningless.

ACKNOWLEDGMENTS

I thank Peter Vitousek, Steward Pickett, and Peter White for very helpful reviews of the manuscript. Several of the ideas expressed here matured during discussions with Janis Antonovics, Henry Wilbur, and Bob Peet. The University of North Carolina provided me office space during my sabbatical in order to complete this and other writing chores. I appreciate very much the financial support of the Duke University Research Council.

Chapter 7

Disturbance and Patch Dynamics on Rocky Intertidal Shores

WAYNE P. SOUSA

Department of Zoology
University of California
Berkeley, California

I. INTRODUCTION

The zonation of animal and plant populations with tidal height is one of the most cosmopolitan and often-studied features of rocky intertidal shores (Lewis, 1964; Stephenson and Stephenson, 1972). Physiological stress due to exposure and bio-

THE ECOLOGY
OF NATURAL DISTURBANCE
AND PATCH DYNAMICS

logical interactions such as predation (or grazing) and interspecific competition constrain organisms to live within characteristic, if somewhat fluctuating, vertical limits on the seashore (see, e.g., Connell, 1961a, 1972; Paine, 1974; Schonbeck and Norton, 1978, 1980; Lubchenco, 1980; Underwood, 1980; Robles and Cubit, 1981). Within these distributional limits, populations are subjected to a variety of natural disturbances. Selective pressures associated with different regimes of disturbance have probably been important in the evolution of the diverse life histories exhibited by intertidal organisms (see, e.g., Suchanek, 1981). That disturbance is common in this habitat does not negate the fact that biological processes such as larval settlement, interspecific competition, and predation are also important determinants of community structure (Connell, 1972; Paine, 1977; Underwood and Denley, 1984). In fact, the interplay between disturbance and these biological processes accounts for much of the organization and spatial patterning of intertidal assemblages.

This chapter deals primarily with the effects of natural disturbance on the assemblages of sessile algae and invertebrates that occupy the surfaces of intertidal rocks. Space for attachment and/or resources associated with open substrates (e.g., light, access to suspended food) are usually the prime limiting requisites. Not surprisingly, species that are poor competitors for primary space [colonizable rocky substratum including that encrusted by some coralline algae; *sensu* Dayton, (1971) and Paine, (1974)] commonly exhibit facultative epizoism or epiphytism. Here, I consider only the dynamics of assemblages occupying primary space. Organisms secure space by overgrowing or laterally crushing their neighbors, by spreading vegetatively into open space, or by growing from settled propagules (zygotes, spores, or larvae). Propagules are often unable to invade occupied surfaces (see, e.g., Sousa, 1979a; Dayton, 1973; Denley and Underwood, 1979; Paine, 1979) and depend on disturbances to create the open space required for colonization. Since the damage caused by most disturbances is localized, primary space is often provided as more or less discrete patches. Various attributes of a newly created patch—for example, its size, shape, and location—can affect its colonization and ensuing biological interactions. Thus, localized disturbances transform an assemblage of sessile organisms into a mosaic of patches varying in such characteristics as size and age (time since last disturbed) and, consequently, in species composition.

While the majority of the examples in this chapter are drawn from studies of temperate rocky shores, this does not imply that disturbance is uncommon on seashores at other latitudes. This bias merely reflects the geographical distribution of research activity.

The term "disturbance" has been used rather loosely in the ecological literature. Some authors (see, e.g., Dayton, 1971) apply it to any mechanism that renews the limiting space resource but distinguish between disturbances of physical versus biological origin. The latter are referred to as "biological disturbances" and encompass everything from singular acts of predation that free space occupied by the prey individuals killed (or displaced via behavioral escape from the vicinity of the predator) to nonpredatory acts (e.g., hauling-out behavior of pinnipeds: Boal, 1980) that inadvertently kill or displace other organisms. Similarly, the foraging or territorial defense behavior of motile consumers sometimes causes mortality of nonprey species (e.g., bulldozing of sessile species by limpets: Dayton, 1971; Stimson, 1973; Choat, 1977). This chapter will focus primarily on the effects of physical disturbance.

Physical disturbances to intertidal assemblages include burial under sand or terrigenous sediments and associated scouring (Markham, 1973; Daly and Mathieson, 1977; Robles, 1982; Seapy and Littler, 1982; Taylor and Littler, 1982; Littler *et al.,* 1983), overturning of unstable substrata by wave action (Sousa, 1979b), exfoliation of the rock surface (Frank, 1965), impact and abrasion by large water-borne objects (see, e.g., logs: Dayton, 1971; Dethier, 1984; Sousa, 1984; J. Cubit, on intertidal reef flats in Panama: personal communication; cobble: Lubchenco and Menge, 1978; Sousa, 1979b; Wethey, 1979; Robles, 1982; Dethier, 1984; D. Lindberg, personal communication; ice: Stephenson and Stephenson, 1954, 1972; Dayton *et al.,* 1970; Schwenke, 1971; Wethey, 1979), and stresses associated with wave action that detach organisms from the rock surface (Jones and Demetropoulos, 1968; Seed, 1969; Harger and Landenberger, 1971; Paine, 1974; Grant, 1977; Lubchenco and Menge, 1978; Suchanek, 1978; Menge, 1979; Underwood, 1980; Paine and Levin, 1981; Paine and Suchanek, 1983; Dethier, 1984; Sousa, 1984). Climatic extremes experienced during prolonged low tides can also act as physical disturbances. For example, low (Crisp, 1964; Connell, 1970; Dayton, 1971; Paine, 1974) or high air temperatures with associated desiccation stress (Lewis, 1954; Hodgkin, 1960; Connell, 1961b; Frank, 1965; Glynn, 1968, 1976; Sutherland, 1970; Branch, 1975; Emerson and Zedler, 1978; Suchanek, 1978; Sousa, 1979b; Underwood, 1980; Hay, 1981; Seapy and Littler, 1982; Taylor and Littler, 1982; Tsuchiya, 1983) also kill or defoliate algae and cause mortality of sessile and mobile invertebrates. With notable exceptions (see, e.g., Seapy and Littler, 1982), few studies have quantified the amount of colonizable space generated by such severe climatic events or documented subsequent patterns of recolonization.

Very extensive areas of open space are occasionally generated by large-scale earth movements such as landslides (Garwood *et al., 1979*), lava flows (Townsley *et al., 1962*; see also Grigg and Maragos 1974, Fig. 1), and uplifting by earthquakes (Haven, 1971; Johansen, 1971). For obvious reasons, studies of the effects of such phenomena on intertidal communities are rare. J. Cubit and S. Garrity (personal communication) are presently documenting patterns of succession along the Pacific coast of Panama on shorelines newly created by landslides resulting from an earthquake in 1976 (Garwood *et al., 1979*). Haven (1971) and Johansen (1971) studied the effects of uplifting by the Alaskan earthquake of 1964 on intertidal invertebrate and algal assemblages, respectively. They examined patterns of mortality and changes in zonation caused by vertical displacement of portions of the shore. Lebednik (1973) conducted a similar study of changes in algal zonation on an intertidal rocky bench uplifted by underground nuclear testing on Amchitka Island, Alaska. The remainder of this chapter explores the effects of the more common, smaller-scale disturbances described above.

II. KINDS OF OPEN SPACE PRODUCED BY DISTURBANCE

Disturbances can generate at least two very different kinds of open space: (a) patches within occupied sites, and (b) patches isolated from occupied sites (hereafter referred to as types 1 and 2, respectively). The extent of the damage and the physical dimensions of the rock surface will determine whether a disturbance generates type 1 or type 2 patches.

Patches of bare space cleared in mussel or algal beds are examples of the first kind of patch. The space opened up by the disturbance (e.g., wave impact or log battering) is smaller than the area of continuous occupied substratum so that the open space is bounded by living organisms (this is, in fact, how the patch is discerned). The overturning of a boulder by wave action also frees space for colonization. When this event clears the entire surface of the boulder, the area of the damage is identical to that of the substratum. The resulting patch of open space is not contiguous with the occupied substratum (except when boulders come into direct contact) and is an example of the second kind of patch. Type 2 patches are more characteristic of systems in which areas of continuous substrata are relatively small (e.g., boulder fields). However, even a relatively large section of shoreline bounded by surge channels or sand beaches may become a large isolated patch if it is subjected to a large disturbance over its entire surface. Paine and Levin (1981, p. 161) described such a situation on shores dominated by mussel beds. Conversely, small type 1 patches may be generated in the cover of sessile organisms on a boulder if, for example, the boulder is overturned only briefly or is struck by wave-borne cobbles or logs (Sousa, 1979a, 1980).

A similar dichotomy in patch types applies to subtidal assemblages of sessile invertebrates (Connell and Keough, Chapter 8, this volume). As discussed below, this distinction has important implications for the mode of patch colonization and for regional population dynamics.

III. CHARACTERISTICS OF THE DISTURBANCE REGIME

While physical disturbances have been recognized as having an important influence on intertidal community structure for more than a decade (Dayton, 1971), surprisingly few studies have gathered quantitative information concerning regimes of natural disturbance in rocky intertidal habitats. The effects of disturbances vary in areal extent, intensity, and frequency. These features of the disturbance regime often influence subsequent patterns of secondary succession within disturbance-generated patches of open space.

A. Areal Extent

The patches of open space created by physical disturbances vary in size. Some are quite small, as when a single plant or animal is removed from the substratum. At the opposite extreme, some disturbances (e.g., sand inundation) may denude large sections of seashore. To my knowledge, only two studies have documented the size distributions of such patches. Paine and Levin (1981) have shown that the size distribution of storm-generated clearings in mussel beds on the outer coast of Washington State is well approximated by a lognormal distribution. The mean patch area was found to vary with the season, year, and site, reflecting the temporal and spatial variation in wave stress. Patches born in the relatively calm summer months tended to be smaller (range of means at four sites: 103–652 cm^2) than those produced in the winter (162–6374 cm^2), when wave energy was greater. Patches at relatively protected mainland sites on the Olympic Peninsula were about an order of magnitude smaller than those on nearby Tatoosh Island, which receives heavy wave action.

The precise mechanism by which patches are created in mussel beds and the factors that determine the size of a patch at birth are unknown (see Paine and Levin, 1981, pp. 150–151). There is also some disagreement as to whether the exposed edges of extant clearings are more likely to experience subsequent disruption (resulting in an expansion of the patch area) than are intact portions of the mussel bed. Dayton (1971) inferred from occasional observations made at three outer coast sites that some patches expanded in area by an average of 24–4884% (low to high exposure sites, respectively) over a 24-month period following initial disruption of the bed. Such a situation would be analogous to the phenomenon of wave-regenerated balsam fir (*Abies balsamea*) forests at high altitudes in the northeastern United States (Sprugel, 1976; Sprugel and Bormann, 1981). However, Paine and Levin (1981) found that patch size in mussel beds is generally fixed at birth and suggested that coalescence of several small neighboring patches into one large one may give the false impression (particularly if observations are only occasional) that patches grow in size by erosion along their edges. No other studies have measured the size distributions of type 1 patches on the seashore.

The size distributions of type 2 patches are determined by the dimensions of the discrete substrata. For example, when a boulder is overturned by waves and remains so for some months (Sousa, 1980), the algae and sessile invertebrates on what was formerly its top surface are killed. If subsequent wave action rights the boulder, a discrete patch of open space becomes available for colonization. The size distribution of such patches is identical to that of the population of boulders that has experienced disturbance of this intensity.

I have shown (Sousa, 1979b) that the size of boulders overturned by waves varies with season and site. The percentage of large boulders moved by waves was greater in winter months and at the more exposed of my two study sites. Casual observations over the past 8 years indicate tremendous annual variation in the size of boulders moved by storm waves, a reflection of annual variation in the intensity of storms. On the mainland near Santa Barbara and on San Nicholas Island in southern California, S. Swarbrick, R. Dean, and D. Lindberg observed (personal communication) the horizontal displacement of entire boulder fields following an unusually severe storm in the winter of 1983.

B. Intensity

Landslides, lava flows, geological uplifting, and smaller-scale exfoliation or fracture of a rock face produce surfaces that are biologically sterile, but such events are rare. The more common forces of disturbance in the intertidal zone (listed earlier) cause variable amounts of damage to attached organisms. At maximal intensity, they remove all macroscopic organisms, leaving only microorganisms on the rock surface. The existence of surviving dormant propagules of macroorganisms (e.g., algal spores) analogous to buried dormant seeds in terrestrial soils (Kivilaan and Bandurski, 1973; Marks, 1974; Harper, 1977; Gross, 1980; Gross and Werner, 1982) remains an interesting possibility that has yet to be rigorously verified in the field. Less intense disturbances often leave survivors with reparable injuries. This is especially true for some species of macroalgae that possess a remarkable ability to regenerate from small surviving fragments or crusts persisting within cracks or crevices in the rock surface (Dayton, 1975; Lubchenco and Menge, 1978; Lubchenco, 1980; Sousa, 1980; Dethier, 1984). In contrast, the effects of disturbance on the predominantly solitary intertidal invertebrate fauna (as opposed to colonial ones: Jackson, 1977a; Connell and Keough, Chapter 8, this volume) tend to be all or none. That is, either an individual is removed from the rock surface or it is unaffected altogether (except possibly indirectly due to changes in the composition and abundance of its neighbors or natural enemies). As a result, patches created in monospecific stands of mussels or barnacles are usually devoid of living macroorganisms at birth (Paine and Levin, 1981; W. Sousa, personal observation).

The degree of damage caused by a disturbing force depends on (a) the magnitude of the force, (b) physiological and morphological characteristics of the organisms in question, and (c) the nature of the substratum to which the organisms are attached. A stronger disturbing force is more likely to cause intensive damage; however, an organism's size and shape affect the magnitude of the force it encounters. Intertidal organisms on wave-swept shores experience drag, lift, and accelerational forces due to the moving water (Koehl, 1977, 1982; Denny et al., 1985). As an organism grows larger, the drag and lift forces it experiences (due to instantaneous nonaccelerating water flow) increase. If, however, the organism grows isometrically (i.e., with constant shape), these increased forces associated with large size are distributed over a proportionately larger area (that projected in the direction of flow), and the stress (force/area) resisted by the attachment or support structures remains constant. If this were the case, the organism would be no more likely to be detached by drag and lift forces when it becomes large than when it was small. However, this situation is sometimes unrealistic. Not all intertidal organisms maintain a constant shape as they grow. It is my subjective impression that the basal holdfast area (or cross-sectional area of the stipe) of many algae becomes proportionately smaller (compared to the area exposed to flow) as the plant grows. Morphological measurements made by Santelices et al. (1980) indicate that this is true of the brown algae *Lessonia nigrescens* and *Durvillaea antarctica,* which inhabit the low intertidal zone of central Chile. Thus, the stress on the holdfast (or stipe) due to drag and lift probably increases with size. More importantly, water flow in the intertidal is not steady but alternately moves toward and away from shore and thus is periodically accelerating. Breaking waves and associated eddies also induce large accelerations in the flow. Therefore, in addition to drag and lift, an attached intertidal organism experiences a force termed the "acceleration reaction." In accelerating flow, the magnitude of this force increases with the organism's volume (rather than the projected area), so that the organism experiences increased stress as it grows larger even if growth is isometric. Stress increases with size to some maximum level, the breaking stress, that the organism's tissues or adhesive material can sustain, at which point the organism is partially or completely removed from the rock surface. Wounds caused by grazing or boring organisms are likely sites of breakage (Black, 1976; Koehl and Wainwright, 1977; Santelices *et al.*, 1980) since the stress is imposed on a smaller cross-sectional area. In short, the larger an organism is, the more likely it will be damaged or killed by wave action of a given magnitude. However, the influence of neighboring organisms on the wave forces an individual plant or animal experiences must also be considered. The forces impinging on an isolated individual may be very different from those acting on an individual that is surrounded by organisms of equal or greater stature. To my knowledge, the latter phenomenon has yet to be investigated in detail.

It is not surprising that those organisms that do attain large sizes on the seashore, such as some species of macroalgae, exhibit morphological traits that serve to reduce the hydrodynamic forces acting on them (Koehl, 1979, 1982; Denny *et al.*, 1985). Regardless, overgrowth by epizoic or epiphytic organisms may increase the forces acting on a host and thus increase the likelihood that it will be dislodged during heavy wave action (Menge, 1975; Sousa, 1979a).

Similar considerations may apply to groups of sessile organisms. Assemblages can become more vulnerable to disruption as their individual members increase in size and number. Two examples illustrate this point. Following an initial dense settlement in an area with few predators, newly recruited individuals of the barnacle *Balanus balanoides* often grow to fill the available space. The ensuing competition for space forces neighboring barnacles to coalesce and form hummocks of weakly attached, elongate individuals (Barnes and Powell, 1950; Connell, 1961b; Menge, 1976; Grant, 1977). Such hummocks of barnacles are more easily torn loose by wave action than are individuals of the conical or columnar form that develops under less crowded conditions. Similarly, as beds of the mussel *Mytilus californianus* exhaust the available primary space, continued recruitment from the plankton causes them to become multilayered. Multilayered beds have a higher profile than single-layered ones, and a higher proportion of individuals in the former are attached by byssal threads to neighboring mussels rather than to the rock surface. This makes multilayered beds less stable and more subject to disruption by wave forces than single-layered beds (Paine, 1974; Paine and Levin, 1981).

In some situations, the opposite relationship between individual size and the likelihood of disturbance may pertain. Small individuals may succumb more rapidly than large individuals to physiologically stressful conditions such as low or high temperatures, desiccation, and so on, presumably due to their greater surface-to-volume ratio (see, e.g., Connell, 1961b; Wolcott, 1973). However, in many instances, the microclimate that a small individual experiences may be moderated by the presence of neighboring organisms that block the flow of air, cast shade, or retain moisture. Small organisms can also more easily seek refuge from extremes of the physical environment within cracks and crevices in the rock surface than can large organisms.

There are likely to be species-specific differences in vulnerability to particular kinds and magnitudes of disturbing forces. My own experiments (Sousa, 1980) demonstrated that macroalgal species differ in the degree of damage they suffer when the substrate to which they are attached is overturned by waves. Similarly, there are marked differences in the responses (i.e., amount of cover lost) of intertidal plants and animals to aerial exposure and burial under sand or terrigenous sediments (Seapy and Littler, 1982; Taylor and Littler, 1982; Littler *et al.*, 1983). Littler and Littler (1980) found that under the same regime of wave exposure, some species of macroalgae suffer greater loss of tissue than others.

The nature of the substratum to which an organism is attached may also influence the vulnerability of the organism to wave forces. Barnes and Topinka (1969) showed that the force required to detach the alga *Fucus* is less if the plant is growing on the test of a barnacle than if it is attached directly to the rock surface. In some cases, the adhesive strength of an organism may exceed that of other organisms to which it is attached or the breaking stress of the rock surface itself. When the rock or host organism breaks, the attached individual is carried away (Dayton, 1973; Denny *et al.*, 1985).

C. Frequency and Seasonality

The frequency of disturbance determines the interval of time over which recolonization can occur. Some kinds of disturbance such as landslides are rare, while others such as wave shear occur frequently. The precise frequency of a particular kind of disturbance is determined (a) by temporal variation in the strength of the force that creates the disturbance and (b) by the "intrinsic vulnerability" of the organisms or substratum (if movable, as in the case of boulders) to which a force of given magnitude is applied.

The magnitudes of forces capable of disturbing intertidal assemblages vary in time; in most cases, they are highly seasonal. Mortality due to heat and/or desiccation stress at low tide occurs during periods of calm, clear weather when there is little wave splash to ameliorate the harsh conditions of exposure. The seasonal timing of such events will depend on the local tidal and climatic regime. For example, in southern California, these conditions occur in winter, when extreme low tides shift into daylight hours and frequently coincide with hot and dry Santa Ana wind conditions (Sousa, 1979a; Seapy and Littler, 1982). At more northern latitudes on the same coast (northern California, Oregon, and Washington), daytime low tides and long exposures occur in summer, as does the mortality caused by desiccation stress (Frank, 1965; Sutherland, 1970; Dayton, 1971; Cubit, 1984).

Wave energy is maximal in winter along most temperate coasts, causing marked seasonality in related disturbances. The mean rate at which patches are cleared in beds of the mussel *M. californianus* on the outer coast of Washington State (measured as the percentage of the bed cleared per month) is more than an order of magnitude greater in winter than in summer (Paine and Levin, 1981). Similarly, the probability that a boulder will be disturbed by wave action is much higher in winter (Sousa, 1979b). In any season, not unexpectedly, this probability decreases with the size (i.e., mass) of a boulder, an intrinsic property of the substrate. As mentioned earlier, the frequency of disturbance varies among years and sites in both of the above systems.

Wethey (1979) documented marked seasonality in the physical disturbances of *B. balanoides* populations on the shore at Nahant, Massachusetts. Scour by floating ice and loose cobbles causes extensive mortality in winter. Ice scour was most damaging to barnacle populations at wave-exposed sites in the mid-intertidal zone, while the impact of cobbles was greatest at wave-protected areas during periods of high winds. Mortality from both sources varied considerably among the three winters of the study. Wethey estimated from weather records that cobble-related mortality in wave-protected areas on the Massachusetts coast has been important every 3–5 years since 1630. Heavy mortality from ice damage at wave-exposed sites has probably occurred once every 10 years in the high intertidal zone and once every 4 years in the mid-intertidal zone over the past 360 years.

Assemblages of organisms living in tide pools may also experience seasonal disturbances; the seasonal likelihood of disturbance varies with the height of the pool (Dethier, 1984). The occupants of high pools tend to be disturbed by heat stress in summer, while those of low pools are disturbed primarily in winter by strong wave forces and the impact of wave-driven logs or cobbles. On average, the inhabitants of a tide pool on the coast of Washington State experience a disturbance (loss of >10% cover within 6 months due to an exogenous factor) every 1.5 years.

Some forms of disturbance such as sand inundation occur exclusively in one season. Most commonly, sand is deposited in summer, when wave energy and current velocities are low (Hedgpeth, 1957; Markham, 1973; Daly and Mathieson, 1977; Robles, 1982), but at certain sites, unique aspects of the local topography and current regime combine to cause deposition only in the winter (Taylor and Littler, 1982; Littler *et al.*, 1983). During the major period of sand intrusion, the degree of deposition may vary in a regular fashion with the tide as well as irregularly with storm activity. Local topographic features (e.g., channels and outcrops) can induce substantial small-scale spatial variation in both the depth and the frequency of sand burial. The deposition of terrigenous sediments carried to the sea by flooded streams and rivers is also highly seasonal and varies in degree from year to year. Seapy and Littler (1982) studied the effects of such an inundation by fine sediments following heavy rains on Santa Cruz Island off the coast of southern California.

In some regions of the world, annual patterns of storm-related mortality are unlikely to be strongly seasonal. For example, on the coast of New South Wales, Australia, rough weather and large storm waves can occur at any time of the year with no obvious seasonality (Underwood, 1981).

The frequency of disturbance, like its intensity, may be influenced by temporal changes in the intrinsic vulnerability of the target organisms. As discussed earlier, such changes occur during ontogeny. As an organism grows, its vulnerability to some disturbing forces (e.g., the drag and accelerational forces associated with flowing water) may increase, while its susceptibility to other stresses (e.g., desiccation) may decrease because of a smaller surface-to-volume ratio. The frequency of disturbance may therefore be linked to the growth rate of an organism. When wave forces are the major disturbing agent, a high rate of individual growth could increase the frequency of disturbance, all else being equal. The opposite might be true in the case of desiccation stress.

The phenomenon of variability in intrinsic vulnerability may operate at the community level as well. As succession proceeds, the vulnerability of the assemblage to disturbance may change, due in some cases to changes in individual as well as group morphology. For example, recolonization of patches cleared in mussel beds on the outer coast of Washington State proceeds through a series of stages: diatoms → macroalgae → barnacles → mussels (Dayton, 1971; Paine, 1974; Paine and Levin, 1981). Beds of mussels, particularly older multilayered ones, are more likely to be dislodged by wave action than are stands of macroalgae or barnacles. Thus, the probability that a patch of clear space will be created depends on the successional age of the assemblage occupying the space. The rate of successional species replacements will therefore control, to some degree, the frequency of disturbance. After a patch of open space has been generated in a bed of *M. californianus,* a minimum of 7–8 years elapse before the same area can be subjected to another major disturbance (Paine and Levin, 1981). This was the shortest rotation period observed; intervals between successive disturbances were longer at less exposed sites. The apparent cyclic vulnerability of some intertidal assemblages to disruption by waves seems analogous to cycles of inflammability controlled by the rate of accumulation of combustible plant material, characteristic of some terrestrial plant communities (White, 1979; Horn, 1981b; Minnich, 1983). It should be noted as well that in some instances the vulnerability of an assemblage to a particular kind of disturbing force may decrease with successional age (Sousa, 1980) if the species characteristic of later successional stages are more resistant to the force than are those of early stages.

The frequency of disturbance need not always be influenced by the successional state of the assemblage. In the boulder fields I studied, the short sessile cover (mostly turflike macroalgae) probably had little influence on the probability that a rock would be disturbed by wave action. On the other hand, in similar habitats supporting larger organisms, the forces associated with waves that act on such individuals may increase the probability that the underlying mobile substratum will be displaced (see, e.g., Schwenke, 1971, p. 1116). If such large species are typical of later successional stages, the frequency of disturbance will, as discussed above, be influenced by the rate at which succession proceeds.

D. Correlations among Characteristics of the Disturbance Regime

The preceding discussion alludes to several important correlations among the individual characteristics of the areal extent, intensity, and frequency of disturbance. In boulder field habitats, all three are linked. When a boulder is overturned by wave action, the organisms on what was formerly its top surface begin to suffer damage and may eventually be killed by a combination of sea urchin grazing, anoxia, light levels below compensation intensity, crushing, and abrasion (Sousa, 1980). The longer a boulder remains overturned, the greater is the mortality of the attached organisms. Since a greater force is required to overturn a large boulder than a small one, and since large winter storms that generate such forces occur less frequently than small storms, large boulders are less likely to be overturned than small boulders. For the same reason, when a large boulder is overturned, it remains turned for a longer period of time than a small boulder. When righted, the former top surface of a large boulder is usually devoid of living macroorganisms. Therefore, in the boulder field system of type 2 patches, large clearings (overturned large boulders) are the least frequent and contain the fewest survivors (they are subject to the most intense disturbances).

There are also interactive effects among some characteristics of the disturbance regimes in more continuous intertidal habitats where type 1 patches predominate. The largest clearings in beds of *M. californianus* are created by forceful winter storm waves (Paine and Levin, 1981) and are therefore less frequent than small clearings that can be produced by waves of lesser magnitude in any season, though most commonly in winter. In mussel beds, unlike boulder fields, the intensity of disturbance does not vary among patches. This is a consequence of both the solitary habit of mussels and the way in which a disturbed patch is operationally identified. A patch exists only when a sufficient number of mussels has been removed to reveal the primary substratum, so by definition there are no "survivors" in a newly created patch regardless of its size or how frequently it is redisturbed. The same is true of patches in assemblages of other solitary species such as barnacles. If, however, one measured the intensity of disturbance as the number of resident organisms removed per unit area, rather than the number that survived within a patch, a three-way interaction among the disturbance characteristics would be expected. Infrequent large storms would affect large areas and cause intense damage. As discussed earlier, the frequency of disturbance might be mediated by temporal changes in the intrinsic vulnerability of the assemblage.

Unlike mussels or barnacles, many species of marine algae are capable of regenerating from small portions of thalli that survive within a patch. Rather than the all-or-none response to disturbance exhibited by solitary invertebrates, the intensity of disturbance to stands of such algae can vary from slight tissue damage to complete mortality. This being the case, I would expect that in assemblages of algae capable of vegetative regeneration, all three characteristics of disturbance would interact in the manner described earlier.

IV. MODES OF PATCH COLONIZATION

A patch of open space is colonized (a) by dispersed propagules such as spores or larvae and/or (b) by vegetatively propagating macroalgae and attached but semi-mobile solitary invertebrates (e.g., mussels and sea anemones) that encroach inward from its perimeter. Drifting fragments of macroalgae may sometimes become established within a patch (Sousa, 1979a), but this method of colonization seems far more significant for corals (Highsmith, 1982; Connell and Keough, Chapter 8, this volume). One reason is that coral fragments are much denser in water than are pieces of algae. The former sink to the bottom and can become lodged in a position suitable for attachment and growth. It is much more difficult for a drifting algal fragment to attach firmly to the newly exposed surface of a rock. Colonization by drifting algal fragments will not be considered further.

Rates of colonization by dispersed propagules and vegetative encroachment will depend on a variety of patch characteristics as briefly described in the following sections.

A. Patch Type

Earlier, I distinguished between patches that are formed within occupied areas (type 1) and those that are isolated from occupied areas (type 2). Type 1 patches can potentially be colonized either by water-borne propagules or by vegetative encroachment. The relative importance of these modes of colonization will depend on other characteristics of the patch including its location, size, and shape. Type 2 patches can be colonized only by water-borne propagules. Space within both kinds of patches can be filled by vegetative regrowth of surviving algal fragments if the disturbance is not too intense.

B. Patch Location

The location of a patch can greatly influence the mode and rate of colonization. Colonization by vegetative ingrowth from the perimeter (or the analogous inward spreading of attached but mobile individuals) is unimportant for most type 1 patches cleared in the midst of nonmobile solitary species such as barnacles or fucoid algae. Only the smallest such patches can be filled by the lateral growth of neighboring solitary organisms. The contribution of water-borne propagules to patch colonization may, in some cases, depend on the proximity of the patch to a source of dispersing propagules (i.e., reproductively mature individuals). Several studies

(Dayton, 1973; Paine, 1979; Sousa, 1984) provide evidence of relatively short-range dispersal (i.e., a few meters or less) by the propagules (spores or zygotes) of some species of intertidal macroalgae.

A patch's position with respect to various environmental gradients (tidal height, wave exposure, current velocity, degree of insolation) can directly affect colonization rates (see, e.g., Denley and Underwood, 1979). Those species constrained to live within a particular range of tidal heights will, quite obviously, become important occupants only of patches created in that zone.

C. Patch Surface Characteristics

Variation in the composition and rugosity of the rock surface among patches may also cause spatial variation in patterns and rates of patch colonization. A number of field (see, e.g., Connell, 1961b; Harlin and Lindbergh, 1977) and laboratory studies (see, e.g., Crisp, 1974; Norton and Fetter, 1981) have demonstrated the influence of relatively small-scale surface texture on the settlement and/or recruitment of algae and sessile invertebrates. Larger-scale heterogeneity in surface texture such as cracks and crevices may affect settlement as well as enhance recruitment by providing refuge from consumers (see, e.g., Lubchenco, 1983, also references cited earlier). On the other hand, at upper tidal levels, consumers themselves may be constrained by physical exposure to forage only over the substratum immediately adjacent to such crevices in the rock surface (see, e.g., Menge, 1976, 1978a; Levings and Garrity, 1983). The type of rock on which the patch occurs may influence the settlement and/or recruitment of some species (Nienhuis, 1969; den Hartog, 1972) but not others (Caffey, 1982). Variation in surface characteristics is likely to be greater among small patches than large ones because the former "sample" a smaller portion of the substratum.

D. Patch Size and Shape

The size of a disturbance-generated clearing may affect colonization in a variety of ways. Simply as a consequence of greater area alone, large patches would be expected to sample a greater proportion of the pool of available spores and larvae, and might therefore support a greater number of species. In my own study of algal succession in the low intertidal in southern California (Sousa, 1979a), at least two of the algal species that became abundant on 225-cm^2 concrete settling surfaces were nearly absent from smaller 100-cm^2 clearings that were created in monocultures of the dominant alga. However, this result could be attributed to differences other than size between the two kinds of patches. The smaller patches were more spatially isolated from adult plants of these species than were the large patches. In addition, the small patches were surrounded by a dense algal turf that may have harbored small grazers (e.g., amphipods or polychaetes) that preferentially fed on recruits of the two species. The turf surrounding the small patches may have also altered the internal physical environment of these patches in such a way as to discourage their recruitment. Finally, the type of substrata, one being of concrete and the other of natural sandstone, may have caused different patterns of colonization. This explanation seems unlikely, however, since patterns of recruitment on small (163 cm^2), experimentally stabilized sandstone boulders placed closer to the concrete blocks, were very similar to those seen on the larger artificial surfaces.

More recently (Sousa, 1984), I documented the patterns of succession within clearings of two sizes in mussel beds. The abundance of only one out of the eight common species of macroalgae that colonized the patches was influenced by the size of the patch per se. This species, the red alga *Endocladia,* colonized patches of both sizes but became more abundant in the larger ones. Again, it is impossible to say whether this resulted from the effect of sampling area alone or from some difference in the internal physical environments of the patches (e.g., humidity, insolation, patterns of water flow) to which this particular species was sensitive. To my knowledge, neither of these effects has been explicitly investigated in intertidal habitats.

Few studies have investigated whether the larvae of intertidal invertebrates can actively select a patch of bare space in which to settle based on its size alone. In the study just described, the size of a patch had no direct effect on recruitment by the mussel *Mytilus edulis.* In contrast, Jernakoff (1983) found that the barnacle *Tesseropora rosea* settled more densely in small (25 cm^2) clearings in beds of algae than in large (200 cm^2) ones. The mechanism causing this pattern was not identified.

Differences in the biological environments of patches varying in size have been examined to a limited degree. As discussed in Section V, several studies have demonstrated differences in the density of consumers, especially herbivores, in patches of different size, as well as the effects of this interaction on within-patch dynamics.

Perhaps the most obvious influence of patch size on colonization derives not from the size of a patch per se but from the manner in which the ratio of patch perimeter length to area changes with patch size. This effect is probably most important for type 1 patches cleared in assemblages of species that are able to invade by lateral movement of individuals or by vegetative ingrowth of clones. The rate at which clearings in mussel beds are filled by the lateral encroachment of adults is affected by a variety of features of the patch environment [e.g., depth of the surrounding bed, angle of the substratum (Paine and Levin, 1981)]. All else being equal (including patch shape), however, smaller patches have a greater ratio of edge to area and are therefore closed by the lateral encroachment of mussels at a correspondingly higher rate (Paine and Levin, 1981; Sousa, 1984). Very large clearings in mussel beds are recolonized primarily by settlement of larvae from the plankton. Vegetative encroachment is also an important means of securing space for some turf-forming species of macroalgae (Sousa, 1979a; Sousa et al., 1981), and I would expect the result described above for mussel beds to obtain in patches of different sizes cleared in stands of these species. Connell and Keough (Chapter 8, this volume) and Miller (1982) describe examples of similar phenomena in assemblages of subtidal colonial invertebrates and terrestrial plants, respectively.

The rate of colonization by propagules may also vary with patch size. The number of nearby adults per unit patch area may be greater for small patches than for large patches, and may result in a greater density of settling propagules and more rapid colonization of small patches. This would be important only if these propagules are not dispersed very far from the parent organisms. On the other hand, the recruitment and growth of plants and animals in clearings within stands of erect macroalgae are more likely to be negatively influenced (e.g., through shading, whiplash, competition for nutrients) by the surrounding adult plants when clearings are small than when they are large (Sousa, 1979a). Also, as Connell and Keough (Chapter 8, this volume) suggest, colonization of small patches will be more negatively influenced by the presence of neighboring assemblages of suspension-feeding invertebrates that prey on settling propagules.

The relationship between perimeter length and patch area may also influence the pattern of colonization of isolated discrete patches (type 2). The roughly laminar flow of water over the surface of such patches will often be disrupted at their edges by the formation of turbulent eddies (Hoerner, 1965; Evans, 1968; Foster, 1975; Munteanu and Maly, 1981). Water velocity is slower in these eddies than in the region of more laminar flow near the center of the patch. This slower-moving water may trap spores and larvae and thus enhance their settlement along patch edges (Foster, 1975; Munteanu and Maly, 1981; W. Sousa, personal observation). Once settled, these individuals could enjoy higher rates of growth and/or survival than organisms living in more central areas of the patch because the eddies might concentrate the particulate foods of suspension feeders or make food capture easier. Eddies also reduce the thickness of the boundary layer. The increased water circulation close to the substratum due to the reduction of the boundary layer may enhance rates of nutrient renewal and waste removal for small, newly settled individuals on the edge of the patch. Since a greater percentage of the internal area of a small patch is likely to be influenced by such turbulent patterns of water flow, small patches may be disproportionately colonized by species that benefit from turbulence created at the edges. If the species whose recruitment is enhanced by this "edge effect" is also capable of vegetatively spreading into the center of a patch, it may dominate small patches at an even faster rate (Schoener and Schoener, 1981).

The influence of patch shape on colonization is likely a simple extension of the above arguments. The more circular a type 1 patch of given area, the lower its ratio of perimeter length to area will be, and the more slowly it will be filled by invasion along its perimeter. Predictions concerning rates of colonization of isolated discrete patches (type 2) that differ in shape will be dictated by the extent to which edge-induced turbulence disproportionately enhances colonization of patches with a high perimeter length-to-area ratio.

E. Time of Patch Creation

The time at which a patch is created by disturbance may determine, to some degree, the course of succession because the propagules of certain species are only seasonally available to colonize (see, e.g., Connell, 1961a,b; Dayton, 1971; Paine, 1977, 1979; Emerson and Zedler, 1978; Denley and Underwood, 1979; Sousa, 1979a; Hawkins, 1981; Suchanek, 1981). Several studies (Paine, 1977; Emerson and Zedler, 1978; Hawkins, 1981) have found that the early successional algal colonists were very different in experimental clearings made in different seasons. In theory, such initial differences might influence subsequent colonization so as to cause large differences in the composition of later successional stages between patches created in different seasons. There is, however, little evidence that this occurs in intertidal communities. Patches created in different seasons eventually come to be dominated by the same long-lived dominant animal or plant species (Paine, 1977; Sousa, 1979a; Hawkins, 1981), although the rate of successional species replacement varies because recruitment and growth are seasonal.

The abundance of a species will be most enhanced by disturbances that create space during periods of time when its propagules are available for settlement. However, any temporal or spatial variation in the distribution and abundance of propagules in the water column will enhance variation in rates of settlement and recruitment among similar patches (Connell, 1961a; Lewis and Bowman, 1975; Bowman and Lewis, 1977; Hruby and Norton, 1979; Caffey, 1982; Hawkins and Hartnoll, 1982; Salman, 1982).

V. RESPONSES OF MOBILE CONSUMERS TO PATCH CHARACTERISTICS

Within-patch dynamics are influenced by a variety of biotic interactions (Thompson, Chapter 14, this volume). In some systems, the abundance of consumer species within patches and their influence on the composition of the patch assemblage vary predictably with the size of the patch. This phenomenon has been observed in both type 1 and type 2 patches.

Several investigations have found that small patches in beds of the mussel *M. californianus* support higher densities of grazers (especially limpets) than do large patches (Suchanek, 1978, 1979; Paine and Levin, 1979; Sousa, 1984). It has been hypothesized that this pattern occurs because the bed of mussels surrounding a patch serves as a refuge for small grazers from wave shock, desiccation stress, and predation. The observation that grazing by these organisms is largely restricted to a 10- to 20-cm browse zone around the perimeter of the patch (see photos in Dayton, 1973; Suchanek, 1978, 1979; Sousa, 1984) is circumstantial evidence that the beds provide some sort of protection. The consequence of this spatially restricted foraging is that the total area of small patches is subject to relatively intense grazing, while the centers of large patches experience little grazing and therefore develop a more extensive cover of algae. The assemblages of algae that developed within small and large experimental patches differed markedly in composition (Sousa, 1984). The assemblage in small patches consisted of grazer-resistant but apparently competitively inferior species, whereas that in large patches included grazer-vulnerable but competitively superior species. The abundance of the mussel *M. edulis* in the experimental clearings was also affected by the interaction between grazing and patch size (Sousa, 1984). The density of *M. edulis* was lowest in small patches from which the green alga *Ulva,* a preferred substratum for the settlement of the mussel, was rapidly removed by grazing. Small mussels may also have been bulldozed off the rock surface by limpets. Similarly, Hawkins (1981) observed that green algae grew only in the centers of cleared areas in stands of barnacles, and attributed this pattern to the grazing forays of small snails (*Littorina*) that apparently refuged among the barnacle tests. Presumably, larger clearings in stands of barnacles would develop a more extensive algal cover than small clearings. Menge (1978b) has shown that predation by the gastropod *Thais* is more intense under a canopy of algae where physical conditions are benign than in the open where conditions are harsher. One could infer that snails consume more prey in small clearings in a stand of algae than in large clearings. This phenomenon is likely to be observed in any system of type 1 patches where natural enemies prefer, or are forced, to live largely within the phase(s) of the community mosaic that surrounds the patch, while their prey occupy the interior of the patch.

The opposite pattern of natural enemies preferring the conditions of the patch interior to those of the neighboring assemblage was demonstrated by Underwood *et al.* (1983). In natural populations, most individuals of the limpet *Cellana transoserica* are found in clearings free of barnacles. When confined with fences to areas with varying degrees of barnacle cover, limpets suffered greater weight loss and mortality as the density of barnacles increased. The negative influence of barnacle cover was stronger when limpets were enclosed in areas dominated by the larger of the two species of barnacles tested. In a separate experiment, adult limpets were transplanted into unfenced plots in which clearings of different sizes had been created in the barnacle cover. The total cleared area in each plot was equal. Within 2 days, 67–69% of the limpets had emigrated from the plots containing small clearings (mean area = 37–39 cm²), while only 3–27% had left the plots with large clearings (mean area = 177–200 cm²). Almost all limpets (88–94%) emigrated from control plots in which barnacle cover was not manipulated. The rate of emigration was higher from plots dominated by the larger species of barnacle. In other words, limpets exhibited a behavioral avoidance of areas in which only small patches of bare substratum were available. This avoidance was presumably related to the increased cost of foraging in such areas. The net effect might be that small clearings in stands of barnacles are less intensively grazed by *Cellana* than are large clearings. Interestingly, another species of limpet living on the same shore, *Patelloida latistrigata,* recruited equally well at all barnacle densities, and its adults survived better in the presence of barnacles (Creese, 1982; Underwood *et al.,* 1983). The impact of grazing by this species might be greater in small clearings than in large ones, since its density increases with barnacle density (Creese, 1982).

I documented (Sousa, 1977) an interaction between the size of type 2 patches and the density of consumer species in intertidal boulder fields. The density of limpets was considerably higher on small boulders than on large ones (Fig. 1). The cause of

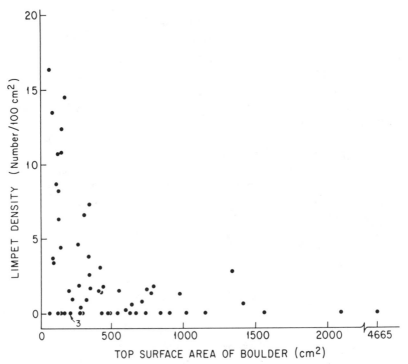

Fig. 1. Density of limpets (*Notoacmea fenestrata, Collisella strigatella*, and *Collisella scabra*) on 60 randomly selected boulders of different sizes sampled at Ellwood, California, on July 16, 1974. (See Sousa, 1979a,b, for details.)

this pattern was not investigated but is probably explained by the fact that small boulders are more frequently disturbed by wave action and therefore have more open space for limpet attachment and grazing than do large boulders (see Table 2 in Sousa, 1979b). In addition, limpets prefer to graze on microscopic algae or immature thalli of macroalgae (Nicotri, 1977). These would be more available on more frequently disturbed substrata. The upper surfaces of rarely disturbed large boulders are covered with larger perennial macroalgae. These plants, which are not grazed by limpets, trap sediment or otherwise fill the space, making the rock surface uninhabitable to limpets. A number of studies present evidence that dense stands of macroalgae, particularly in the middle and low intertidal zones, outcompete limpets for space (Dixon, 1978; Sousa, 1979a; Underwood and Jernakoff, 1981; Sousa, 1984). The net effect of this pattern of limpet distribution is that small boulders are more intensively grazed by limpets than are large boulders.

The impact of consumers on prey populations within a patch will also vary with the location of the patch. Herbivores and predators often forage less effectively and/or are least abundant in areas exposed to strong wave action (Menge 1976, 1978a,b; Lubchenco and Menge, 1978; J. F. Quinn, 1979) or shifting sediments

(Robles, 1982). Consequently, their influence on patch colonization by sessile species is likely to be less in such areas. The demography and abundance of populations of mobile consumers may also vary with the composition of sessile organisms among which they live. For example, the density, rate of recruitment, and size distribution of limpet (*Patella vulgata*) populations differ in areas dominated by mussels versus barnacles at the same tidal height (Lewis and Bowman, 1975). Therefore, if a significant proportion of the consumers that forage within a patch come from surrounding occupied areas, identical patches created within different background assemblages may experience different levels of consumer pressure.

VI. WITHIN-PATCH DYNAMICS

A sequence of species replacements follows the initial colonization of a newly created patch of open space. In its general features, the sequence will be characteristic of the particular tidal height, regime of wave exposure, and geographical locale, all of which will determine the pool of species available to colonize the patch. Patterns of species replacement in various regions of the world are documented in many of the references cited earlier. Any of several mechanisms of replacement may operate at each step in the successional sequence (Connell and Slatyer, 1977; Sousa, 1979a). However, regardless of the mechanisms operating, the sequence of successional replacement tends to be deterministic and usually leads to the monopolization of space within the patch by one or a few competitively dominant and/or long-lived species (see e.g., Paine, 1966, 1974; Dayton, 1971, 1975; Lubchenco and Menge, 1978; Sousa, 1979a; Paine and Levin, 1981). Unless some mechanism (see below) acts to prevent local dominance, populations of species characteristic of earlier stages of succession become locally extinct.

The time to extinction from a patch of those species characteristic of earlier stages of succession will depend on (a) the mode of colonization by the late successional, dominant species, (b) the kind of patch (type 1 versus 2), and (c) the size and shape of the patch. If the late successional species can invade open space both by vegetative encroachment from the borders of the patch and by recruitment of planktonic propagules, dominance will be attained more rapidly than if space becomes occupied primarily through recruitment from the plankton. However, in type 2 patches, even if the late successional dominant is capable of spreading vegetatively or by lateral movement of individuals into open space, the initial colonization must be by dispersed propagules. Consequently, in this situation, dominance of patches will occur more slowly and populations of early successional species will persist longer in type 2 patches than in comparable type 1 patches.

Along a continuous rocky shore, the local extinction and successional replacement of early colonists will be most rapid in small type 1 patches shaped irregularly and created within stands of late successional dominants that are able to fill the patch rapidly by lateral encroachment. Populations of early successional species will be more persistent in larger, more circular patches (with a smaller ratio of perimeter length to area) since the rate of encroachment by the dominant species will be slower.

Small patches will also be dominated more quickly than large ones when the late successional dominant disperses its propagules only a short distance. In large patches whose radii exceed the dispersal distance of the late successional species, dominance will develop only gradually as colonists spread slowly inward from the border over several generations. No such spatial gradient in settlement will occur in small patches whose radii are short relative to the dispersal distance of the dominant. Propagules will be deposited relatively uniformly and densely throughout such patches, and successional replacement will be relatively rapid.

The opposite relationship between patch size and rates of successional replacement has been hypothesized for systems of type 2 patches in subtidal epifaunal assemblages (Jackson, 1977b; Karlson, 1978; Kay and Keough, 1981; Keough, 1984b; Connell and Keough, Chapter 8, this volume). Early colonists may persist longer in small rather than in large type 2 patches. This could occur if, as often appears to be the case (see, e.g., Sousa, 1979a), the late successional dominant produces fewer propagules and/or disperses them more seasonally than do the species characteristic of earlier stages. Since small patches "sample" less of the available pool of spores and larvae than do large patches, this alone will cause late successional species to be underrepresented in some small patches. Successional replacement of earlier colonists in such patches will be slow and may not occur at all if late successional species have very limited powers of dispersal. In systems of discrete substrata, one might expect a strong selective advantage to accrue to individuals of an early successional species whose propagules selectively settle in smaller patches (Keough, 1984b; Connell and Keough, Chapter 8, this volume). To my knowledge, this phenomenon has not been studied on rocky intertidal shores.

The replacement of a species within a patch may not always be gradual. It may disappear precipitously as the density of its population falls below some lower threshold level necessary for local replacement. For example, populations of the annual brown alga *Postelsia palmaeformis* within open patches in mussel beds rarely persist at densities below 20–30 plants per square meter (Paine, 1979). Apparently, this density of plants produces insufficient spores to compensate for losses to grazers and to competition with the coralline algae-dominated understory and/or adjacent mussels.

Several mechanisms may act to prevent local extinction and thus maintain a diverse assemblage within a patch. The mechanisms most commonly documented to maintain local diversity in intertidal assemblages are compensatory mortality and intermediate disturbance (Connell, 1978). The first of these mechanisms refers to the situation in which the potential late successional dominant suffers a disproportionate rate of mortality from causes largely unrelated to its interaction with earlier successional species. Selective predation by the starfish *Pisaster* on the competitively dominant mussel *Mytilus californianus* is a classic example (Paine, 1966, 1974). Local diversity will also be maintained if populations within a patch are repeatedly disturbed at intermediate frequencies and intensities. These intermediate scales of disturbance allow species to accumulate within the patch but prevent domination by a few competitively dominant and/or long-lived species. This relationship between frequency of disturbance and diversity within a patch can be readily observed in those boulder field habitats where wave forces are the predominant source of disturbance. Boulders of intermediate size (or mass) are overturned by wave action more frequently than large boulders and less frequently than small boulders. As a result, boulders of intermediate size support, on average, the most diverse assemblage of sessile organisms (Sousa, 1979b). Intermediate frequencies and intensities of biological and physical disturbance also maintain diversity in tidepool assemblages (Paine and Vadas, 1969; Lubchenco, 1978; Dethier, 1984).

At mid-intertidal levels on the outer coast of Washington State, an assemblage of organisms occupying a particular area of substratum is disturbed at most once every 7–8 years (Paine and Levin, 1981). This interval is sufficiently long that a large number of species is able to colonize a newly created patch of open substratum. For a time, the diversity of the assemblage within the patch increases as these species accumulate. However, the frequency of disturbance is too low to maintain this state of high diversity. The interval between disturbances is long enough that all but the very largest of clearings will be closed by lateral encroachment of adult mussels or by larval recruitment of mussels from the plankton. The diversity of the assemblage within the patch will gradually decline as species are competitively excluded. Eventually all species that occupy primary space other than mussels will be driven locally extinct. Therefore, while localized disturbance enhances the regional diversity of sessile organisms by creating opportunities for colonization in an area that mussels can potentially dominate, rates of disturbance at the sites studied by Paine and Levin (1981) are too low to maintain a persistently diverse assemblage on any one area of substratum.

VII. REGIONAL PERSISTENCE OF FUGITIVE SPECIES

In situations where no mechanism is effective in maintaining local diversity, species other than the late successional dominant(s) that are restricted to live within a particular range of tidal heights can only persist in that zone as true fugitive species (Hutchinson, 1951). Fugitive species persist as members of the assemblage by dispersing their offspring into space newly cleared by disturbance with suitable conditions for recruitment, growth, and reproduction. Other species may have broader distributional boundaries that include relatively exclusive refugia outside the zone in question. Although disturbance is not strictly required for the persistence of the latter species in the community, the exploitation of the additional space created by disturbance will enhance the size of their populations within the region.

As suggested by earlier comments, the regional dynamics and abundance of a species on the seashore reflects the interplay of that species' life history and the regime of disturbance. For a fugitive species to persist, disturbances must generate colonizable space within the dispersal range of extant populations and within the period of time it takes for those populations to go extinct. Dispersal distance and the time to local extinction are, to some degree, functions of the life history of the particular species as well as local population density. Paine's (1979) long-term study of the sea palm *Postelsia* provides one of the clearest examples of an intertidal species that requires a specific frequency and quantity (percentage of space cleared) of disturbance to persist. During a 10-year period, populations of the plant were monitored at 26 sites on Tatoosh Island off the coast of Washington State. The regimes of wave-induced disturbance of mussel beds varied among the sites. Populations of sea palms, which recruit primarily to clearings within the beds (Dayton, 1973), were present in all 10 years at the 7 sites that had high levels of regular disturbance. Sea palms were absent from the other 19 sites where either the frequency of disturbance or the quantity of open space generated by disturbance was lower. The experimental establishment and persistence (for at least two generations) of populations of sea palms at the latter sites ruled out the possibility that sites lacking natural populations were simply unsuitable for their establishment. At sites where populations were naturally persistent, the subpopulation within any particular clearing was doomed to local extinction due primarily to competition with a variety of sessile organisms, including mussels and coralline algae. The regional persistence of populations at such sites depended upon adequate dispersal of spores to new clearings within about 1.5 m (the approximate dispersal distance) of extant stands. At other locations along the Washington coast, individual stands of sea palms seem to persist longer than they do on Tatoosh Island. P. Dayton (personal communication) monitored 15–20 stands of *Postelsia* on Waadah Island from 1965–1974. All of the stands, with the exception of 3 that had been experimentally manipulated, persisted through the 9 years of observation. Reasons for the greater persistence of stands at the latter site have not been investigated. Competition for space with other sessile organisms may have been less intense on Waadah than on Tatoosh. In addition, the stands Dayton observed may have been sufficiently dense that enough open space was generated each fall and winter, when detached adult plants carried away overgrown competitors (Dayton, 1973), to ensure repopulation of the patches of substratum for many generations. According to Paine's (1979) calculations, stands of sea palms on Tatoosh rarely reach densities at which this mechanism could maintain a local population for more than a few years. A detailed comparative study of *Postelsia* populations at these and other sites might reveal the causes of variation in local dynamics.

The observation that many species of algae and invertebrates on the Pacific coast of North America disperse their propagules in winter provides circumstantial evidence that the open space generated by disturbance is important to the persistence of some intertidal populations. Winter is the season in which disturbance is most prevalent and open space for recruitment is most available (Sousa, 1979a; Paine and Levin, 1981; Suchanek, 1981, and references therein).

Not all cleared space is of equal quality in promoting the persistence of, or enhancing the abundance of, a particular species. For example, as discussed earlier, small clearings in mussel beds are closed more quickly by lateral encroachment of adult mussels and are also subjected to more intense grazing than are large patches. Therefore, species that are vulnerable to grazing will have a low probability of becoming established in small patches (Suchanek, 1979; Sousa, 1984). Even if some individuals are successful in recruiting to such patches and grow to an invulnerable size, competitive exclusion by mussels will be relatively rapid. Regional populations of species that are vulnerable to grazing will be smaller, in proportion to the total area cleared, on shores where the patches are small than on shores where disturbances generate large clearings. On the other hand, populations of grazer-resistant species that lose in competition with grazer-vulnerable species may become established more successfully and persist longer in relatively small patches where the abundance of grazer-vulnerable species is low. This seems to occur in spite of the faster rate at which small patches are closed by mussels (Sousa, 1984). The expectation is that grazer-resistant species will be disproportionately more abundant on shores characterized by small-scale disturbances.

Similar considerations regarding the effect of patch quality on the regional abundance and persistence of populations apply to intertidal boulder fields. Boulders smaller than a certain size are disturbed so frequently that populations of most perennial species cannot become established on their surfaces (Sousa, 1979b). The sparse cover on these boulders is dominated by highly vagile and fast-growing early successional species. All else equal, the regional abundance of perennial species should be less in fields where the mean size of boulders is small than in fields where the mean size is large. The opposite pattern would be predicted for populations of early successional species.

VIII. CONCLUDING REMARKS

Nearly a half-century of experimentation and quantitative observation has yielded considerable information concerning the role of biological interactions in structuring rocky intertidal communities. Only recently, however, have the effects of disturbance on the local and regional dynamics of intertidal assemblages and on the evolution of life histories been recognized as potentially important. Consequently, much of the available data gathered from this habitat cannot be used to test the predictions made in this chapter. It is my hope that the ideas presented herein will serve as a useful guide for future studies as to the kinds of data needed to elucidate the effects of natural disturbance, including its interplay with biological processes. By viewing intertidal assemblages as mosaics of patches generated by disturbance, we can better understand their dynamics and comparison with communities of terrestrial plants discussed in other chapters of this volume will be facilitated.

ACKNOWLEDGMENTS

For comments on the manuscript I thank J. Connell, P. Dayton, P. Fairweather, M. Koehl, K. McGuinness, B. Okamura, R. Paine, G. Roderick, A. Underwood, and the editors, S. Pickett and P.

White. My research on rocky intertidal communities has been generously supported by NSF grants OCE 75-23635 and OCE 80-08530.

RECOMMENDED READINGS

Dayton, P, K. (1971). Competition, disturbance, and community organization: the provision and subsequent utilization of space in a rocky intertidal community. *Ecol. Monogr.* **41**, 351–389.

Dethier, M. N. (1984). Disturbance and recovery in intertidal pools: maintenance of mosaic patterns. *Ecol. Monogr.* **54**, 99–118.

Koehl, M. A. R. (1982). The interaction of moving water and sessile organisms. *Sci. Amer.* **247**, 124–134.

Paine, R. T. (1979). Disaster, catastrophe and local persistence of the sea palm *Postelsia palmaeformis*. *Science* **205**, 685–687.

Paine, R. T., and Levin, S. A. (1981). Intertidal landscapes: disturbance and the dynamics of pattern. *Ecol. Monogr.* **51**, 145–178.

Sousa, W. P. (1979). Disturbance in marine intertidal boulder fields: the nonequilibrium maintenance of species diversity. *Ecology* **60**, 1225–1239.

Taylor, P. R., and Littler, M. M. (1982). The roles of compensatory mortality, physical disturbance, and substrate retention in the development and organization of a sand-influenced, rocky-intertidal community. *Ecology* **63**, 135–146.

Chapter **8**

Disturbance and Patch Dynamics of Subtidal Marine Animals on Hard Substrata

JOSEPH H. CONNELL AND MICHAEL J. KEOUGH[1]

Department of Biological Sciences
University of California
Santa Barbara, California

I. INTRODUCTION

On marine subtidal hard substrata, the community consists of attached plants and animals, together with the mobile animals that prey upon them. This chapter will consider the mechanisms that determine the species composition, relative abundance, and diversity of the attached animals (including some, such as anemones, crinoids, and holothurians, which are capable of slight movements). We will concentrate on two faunas that have received much attention: the scleractinian reef-

[1]Present address: Department of Biological Science, Florida State University, Tallahassee, Florida 32306.

THE ECOLOGY
OF NATURAL DISTURBANCE
AND PATCH DYNAMICS

building corals on tropical reefs, and the invertebrates that live attached to hard substrata (hereafter referred to as "epifauna") in temperate regions or in cryptic tropical habitats. There are many studies of the organisms that comprise "fouling communities," but in the majority of these, the relation between the fouling community on artificial substrata and the same assemblages on natural substrata has not been demonstrated. For the most part, we have not included studies of fouling communities in this review.

These organisms attach to and spread over the substratum, gaining energy by photosynthesis or by capturing food from the water column. In either case, space is a primary requirement. Species may gain space in three ways: (a) invasion of open patches of substratum, e.g., "primary space"; (b) expansion by killing neighbors, either indirectly by overgrowth that cuts off the supply of resources or by direct means, e.g., crushing, smothering, undercutting, extracoelenteric digestion, etc.; (c) invasion of "secondary space," i.e., the nonliving surfaces of other organisms attached to hard substrata, e.g., mollusc shells or polychaete tubes. There are also a few examples of larvae of subtidal invertebrates attaching themselves to surfaces covered with living tissues of some sponges and bryozoans. However, once a substratum is occupied, most occupants are able to prevent propagules from attaching to their living surfaces, as has been observed on fouling panels (Sutherland, 1974, 1976, 1981; Sutherland and Karlson, 1977).

Open patches occur at different scales of time and space, ranging from frequent small openings to rare events when vast areas are cleared. Organisms invade such clearings in two ways: (a) survivors at the edge of the patch spread into the opening by vegetative growth; (b) propagules arrive from outside. The propagules can be of either sexual or asexual origin. The relative importance of these two sorts of colonization depends on the size of the opening and its isolation from other occupied sites (Jackson, 1977b; Keough, 1984b).

A. Characteristics of Open Patches in Marine Subtidal Habitats

Most substrata in marine subtidal habitats are soft sediments; hard substrata are discontinous pieces of habitat that are separated from other such pieces by a "sea" of mud or sand. These substrata vary greatly in size, from small stones or the shells of sand-dwelling molluscs to coral reefs and rocky extensions of terrestrial land masses that may be many square kilometers in area. Primary space is made available in two sorts of patches. The first (hereafter referred to as type 1) are created by death of some of the residents, leaving a patch bounded at least in part by the survivors; the second (type 2) are patches isolated from existing assemblages. Type 2 patches can arise either by the creation of new hard substrata, e.g., from recruitment and growth of organisms with hard skeletons, e.g., coralline algae, corals, molluscs, etc., and from landslides or submarine lava flows, or on discrete existing substrata, e.g., cobbles overturned to expose either bare surface or organisms that are then removed by predators. These are the same two types referred to in Sousa (Chapter 7, this volume). The agents of patch formation can be biological (e.g., predators, grazers, pathogens, or senescence) or physical (e.g., heavy wave action, deposition of sediments, floods of fresh water, aerial exposure during extremely low tides, etc.). These mechanisms create open patches over a wide range of spatial scales.

We know of few studies of corals or epifauna in which patches have been experimentally produced (Bak *et al.*, 1977; Vance, 1978, 1979; Karlson, 1978, 1983; Kay and Keough, 1981; A. M. Ayling, 1981; Palumbi and Jackson, 1982; A. L. Ayling, 1983a,b; Keough, 1984a,b), although there are many observational studies of events following a disturbance, especially on coral reefs (Connell, 1978; Pearson, 1981; Colgan, 1983).

B. Characteristics of Subtidal Epifauna

While the fauna of the intertidal zone is composed mainly of solitary forms, the greatest proportion of species occupying subtidal primary space on natural hard substrata appears to be modular colonial forms, especially when coral communities are included (Jackson, 1977a). Solitary organisms may still be common in many subtidal habitats, and in some they may occupy most of the space (Wells *et al.*, 1964; Paine, 1976; Sutherland, 1980; Young, 1982). Among the colonial species,

TABLE 1

Summary of Major Growth Forms of Sessile Invertebrates
Attached to Hard Substrata Subtidally[a]

Growth form	Characteristics
Runners	Linear or branching organisms parallel to the substratum; either completely attached or with a few points of attachment
Vines	Linear or irregularly branching forms growing vertically; either growing free or attached to other organisms
Trees	Upright, extensively branching forms, with a restricted area of attachment to the substratum
Bushes	Upright forms branching from the base, with a restricted area of attachment
Plates	Upright forms with a flat or upwardly concave plate, supported by a central column
Mounds	
Determinate	Encrusting forms with extensive vertical as well as horizontal growth, completely attached to the substratum. Either with a regular shape or with horizontal spread restricted
Indeterminate	As for determinate mounds, but no limit on horizontal spread
Sheets	
Determinate	Thinly encrusting forms, completely attached to the substratum, but with limited horizontal spread
Indeterminate	As for determinate sheets, but with extensive capacity to expand by vegetation propagation

[a]The classification scheme is modified from that suggested by Jackson (1979a).

TABLE 2

Distribution of Growth Forms among Major Taxonomic Groups of Attached Organisms, with Examples of Each Growth Form

Taxonomic group	Growth form	Example
Porifera	Determinate mounds	*Sycon, Tethya, Ircinia*
	Indeterminate mounds	*Dysidea, Ancorina, Spongia, Halichondria*
	Indeterminate sheets	*Dendrocia,* some *Mycale, Tedania, Dysidea*
	Runners	Some *Callyspongia*
	Bushes	*Callyspongia, Carteriospongia*
	Trees	*Raspailia, Axinella*
Cnidaria	Runners	*Culicia*
	Trees	*Acropora*
	Plates	*Acropora, Agaricia*
	Bushes	*Acropora, Stylophora, Pocillopora, Porites*
	Determinate mounds	Most anemones and solitary corals
	Indeterminate mounds	*Porites, Fungia,* most massive species
	Indeterminate sheets	*Montipora, Acropora*
	Trees	Gorgonians, some *Acropora* ("staghorns")
Mollusca	Determinate mounds	Scallops, oysters
Annelida	Determinate mounds	Most solitary serpulids
	Sheets	Colonial species, e.g., *Filograna*
Crustacea	Determinate mounds	Acorn barnacles; tube-dwelling amphipods
	Bushes	Stalked species; *Lepas*
Brachiopoda	Determinate mounds	*Argyrotheca*
Entoprocta	Determinate mounds	Solitary forms: Loxosomatidae
	Runners	Pedicellinidae
	Bushes	Some *Barentsia*
Bryozoa	Vines	Some *Scruparia*
	Runners	*Aetea, Bowerbankia*
	Bushes	*Bugula,* Crusiidae, *Scrupocellaria*
	Determinate mounds	*Lichenopora, Heteropora* (some)
	Indeterminate mounds	*Celleporaria, Diaperoecia,* Reteporidae
	Determinate sheets	Some *Membranipora, Schizoporella*
	Indeterminate sheets	Formed by fusion of adjacent colonies, e.g., *Membranipora*
Echinodermata	Determinate mounds	Dendrochirotid holothurians, crinoids
Chordata	Runners	*Metandrocarpa*
Ascidiacea	Bushes	*Podoclavella, Clavelina, Sycozoa*
	Determinate mounds	Unstalked solitary species
	Indeterminate mounds	*Aplidium*
	Indeterminate sheets	*Botrylloides, Cystodytes,* Didemnids
	Determinate sheets	*Trididemnum* (if limited by predators)

TABLE 3

Major Occupiers of Space in Subtidal Epifaunal Communities Attached to Natural Substrata or Where the Artificial Substrata Are Known to Be Similar to Natural Ones

Locality	Substratum	Organisms[a]	Source
Tropical			
Australia			
Great Barrier Reef	Caves, crevices	CE	Day (1977)
Caribbean			
Jamaica	Beneath corals	CE	Jackson (1977b)
	Crevices, walls	CE	Jackson and Winston (1982)
	Vertical walls	CE	Lang (1974)
	Back reef, rubble	CE	Karlson (1980)
Curacao	Rocky reef	CE	Bak *et al.* (1981)
Venezuela	Mangrove roots	S, CE	Sutherland (1980)
India	Oyster shells	S, C	Alagarswarmi and Chellam (1976)
Temperate			
New Zealand	Rocky reef	CE	Ayling (1981)
	Vertical wall	CE	Ayling (1984a)
	Boulders	CE	Gordon (1972)
Australia			
Victoria	Pier pilings	C	Harris (1978)
South Australia	Pier pilings	CE	Kay and Butler (1983)
	Pier pilings, rocky reefs	CE	Kay and Keough (1981) Kay (1980)
	Chlamys asperrimus	CE	Pitcher (1981)
	Chlamys bifrons	CE, S	Keough (1981)
	Pinna bicolor in sand	CE, S	Keough (1981, 1984a)
	Pinna bicolor in sand	CE	Butler and Brewster (1979)
	Pinna bicolor in mud	S	Butler and Keough (1981)
	Pinna bicolor in fine sand/mud	S, CE	Butler and Keough (1981)
	Rocky reef	C	Shepherd and Womersley (1976)
Europe			
Mediterranean	Cave	CE	Sara (1970)
Adriatic Sea	Boulders	CE	Rutzler (1970)
France	Rocks	C	Castric-Fey *et al.* (1978)
Great Britain	*Fucus serratus*	CE	Wood and Seed (1980); Boaden *et al.* (1976); Stebbing (1973); Hayward (1973)
	Rocky reefs, wrecks	CE	Forster (1958)

(*continued*)

TABLE 3   *(Continued)*

Locality	Substratum	Organisms[a]	Source
	Undersides of rocks	CE	Rubin (1982)
	Shells, stones, organisms	CE	Eggleston (1972)
Norway	Lava grounds	C	Gulliksen *et al.* (1980)
North America— east coast			
Gulf of Mexico	Rocky reefs	CE	Storr (1976)
Florida	Scallop shells	CE	Bloom (1975)
North Carolina	Scallop shells	S	Wells *et al.* (1964)
	Pier pilings	CE	Karlson (1978)
New York	Shells	CE	Abbott (1973)
Massachusetts	Shells	S,C	Driscoll (1968)
	Underneath boulders	CE	Osman (1977)
Bay of Fundy	Shells	CE	Powell (1968)
North America— west coast			
California	*Macrocystis pyrifera*	CE	Bernstein and Jung (1979)
	Macrocystis pyrifera	CE	Dixon *et al.* (1982)
	Rock walls	C	Vance (1979)
	Chama shells	CE	Vance (1978)
	Vertical rock walls	C	Keough and Downes (unpublished observations)
	Subtidal algae	C	Benson (1983)
	Rock reef	S, CE	Breitburg (1984)
	Rocky reef	C	McLean (1962)
Washington	Vertical rock walls	CE	Young (1982)
	Vertical and horizontal rock walls	S	Young (1982)
British Columbia	Underneath boulders	S, C	Keen and Neill (1980)
Alaska	Rock walls	CE	Young (1982)

[a]Growth forms are denoted as follows: S, solitary, moundlike; CA, colonial, arborescent (trees, bushes, vines); CE, colonial, encrusting (sheets, mounds); C, both colonial types.

there is a diversity of growth forms, and Jackson (1979a) proposed a classification of these that is particularly relevant to the way in which they occupy space. He suggested the following forms as models for colonial marine invertebrates, although they may also be applied to solitary species: runners, vines, sheets, plates, trees, and mounds. These, with some additions, are summarized in Table 1. He divided mounds into two categories: those exhibiting either indeterminate or determinate growth. We will also divide sheets into these two categories because some species, e.g., certain cheilostome bryozoans, may be determinate, senescing after reaching a particular size (Keough, 1984a). Each major sessile invertebrate group and geographic region has some of these forms (Tables 2 and 3), and each form may be viewed as an adaptation for colonizing and persisting in different sorts of habitats.

Subtidal epifauna are dispersed by propagules produced either sexually or asexually. The latter are formed either as larvae (e.g., sponge gemmules), as buds that detach from the parent, e.g., in some sponges (Ayling, 1983b) and corals (Scheer, 1959), or as fragments broken off and moved by water currents, gravity, or animals.

Jackson (1979a, Table 13) predicted that particular growth forms vary in abundance as a function of the strength of disturbances (both physical and biological in origin) and of the supply of food or light. It is likely, however, that additional factors need to be taken into account, e.g., areal size and frequency of disturbances, as described above. We will consider the types of clearings that the various growth forms are best suited to invade, and use these patterns of occupation of space to predict the kinds of community change that are likely to result from disturbances of a given intensity. These arguments will be illustrated with the results of our observations and experiments on coral reefs and open coast epifaunal assemblages.

II. EXAMPLES OF THE PATCH DYNAMICS OF SUBTIDAL COMMUNITIES

A. Reef-Building Corals

1. Characteristics of Reef-Building (Hermatypic) Corals

Hermatypic corals get a large proportion of their energy from photosynthesis of their zooxanthellae, and thus require light as well as space. Corals can survive for some period beneath other colonies in shallow water, providing the upper colony is high enough above them to allow sufficient light to reach them (Sheppard, 1981), but they cannot survive overgrowth at close range. Corals secure new space by three means: (a) vegetative growth of existing colonies, (b) propagules produced vegetatively as detached buds or fragments broken off existing colonies, and (c) planula larvae. The different growth forms of corals are given in Table 2, based upon our modifications of Jackson's (1979a) categories. (In this section on corals, we will not discuss the other clonal organisms that occupy exposed hard substrata on coral reefs, e.g., plants, sponges, soft corals and other cnidaria, bryozoans, and ascidians. Interactions between scleractinian corals and these organisms may be quite important, but because of limitations of space, they were not included.)

2. The Disturbance Regime on Coral Reefs

a. Amount of Damage and Patch Size Many descriptions of damage from disturbances state that "all corals were destroyed." However, when small areas are examined closely, it is often found that some colonies or parts of them have survived (Connell, 1973; Highsmith *et al.*, 1980; Knowlton *et al.*, 1981). These survivors are often capable of regenerating and reoccupying vacant patches by growing in from the edge. Therefore, the diameter of an open patch is defined by the minimum distance between survivors capable of regenerating. From this definition, it follows that the greater the degree of mortality caused by disturbance, the larger the patch size. Open patches range in size from small gaps created by the death of one or a few polyps to larger areas created by disease, aggregations of predators, and physical extremes that may result in the death of whole sections of reefs. The main physical causes of mortality are storms, cold water episodes, sedimentation, and unusually low tides. Biotic causes include grazing and predation by fishes, echinoids, asteroids, molluscs, polychaetes, and microorganisms [see Pearson (1981) for a review of all of these].

Quantitative data on patch sizes are few. It may be that severe physical extremes or aggregations of predators have sometimes destroyed all corals over fairly large tracts, but in no case has such complete destruction been demonstrated. To do so would require a detailed search for any surviving portions or fragments and then a study of their survival and regeneration. Such a study has been done for at least one species. Knowlton *et al.* (1981) found that fragments of *Acropora cervicornis* that had survived a very intense cyclone (hurricane) in Jamaica suffered over 99% mortality during the following 5 months. Predation, and perhaps disease, caused much of this mortality. The number of predators had been reduced by only half by the storm, whereas the corals were reduced by 99%, so predator pressure on the survivors was much more intense after the storm.

Estimates of the amount of damage and patch sizes caused by severe storms are available from several detailed studies. On the reef of Heron Island, Queensland, Connell (1973, 1978, 1984) found considerable spatial and temporal heterogeneity in the effects of four cyclones in 1967, 1972, 1976, and 1980. On permanently marked quadrats totaling 16 m^2 at four sites, the hurricanes produced significant damage at only two of the four. These two are 1150 m apart, one on the northern reef crest, the other in shallow outer pools just beyond the crest. The damage occurred only in the first two of the cyclones; the latter two produced no significant damage.

The 1967 cyclone caused only minor damage on the north reef crest site, producing small patches up to 0.5 m^2 in diameter, whereas in and near the other (outer pool) site, all corals were killed. In contrast, the 1972 hurricane caused a major physical change to the north reef crest by removing a short section of the crest. As a result, at low tide the water on the reef flat drained through this gap instead of flowing over the reef crest. At low tide the crest then dried out completely for several hundred meters on each side of this gap, killing all but a few scattered massive mounds that happened to be located in small tide pools. However, at the same time in the two quadrats in the outer pool site, the corals were badly damaged, but 18 and 25% of the colonies survived the storm and were all regenerating 2 years later.

Thus, in each of these two cyclones, very large open patches were cleared in one of the two study sites, while only small patches were formed in the other. Also, the location of greater damage was reversed in the second storm as compared to the first. Little damage occurred at either of these sites in the subsequent two hurricanes.

The impression of extreme spatial and temporal heterogeneity of storm effects is reinforced when we consider other localities on the reef. Three of the four cyclones produced no significant damage in the other two study sites on the south crest and the south inner reef flat, whereas the last cyclone in 1980 caused considerable damage at one site (south crest) but not on the other (inner flat). Also, the degree of damage suffered by corals at any particular site was not related in any simple way to the intensity of these hurricanes, in all of which extreme wind velocities were recorded at Heron Island. Censuses made in the intervals between these hurricanes showed that small patches up to about 50 cm in diameter occurred frequently.

Similar heterogeneity in patch formation has been found on other reefs. Davis (1982) compared two maps of Dry Tortugas reef in the Caribbean made 95 years apart. One stretch of *Acropora palmata* measuring 44 ha had been reduced to a small stand 0.06 ha in extent at that time, the rest being dead coral rubble. A large new stand of *A. cervicornis* measuring 200 ha had appeared between 1881 and 1976, but in the next winter 90% of it was killed at all depths when the water temperature fell to 14°C. *Acropora palmata* suffered only 60–70% mortality from this cold stress. Porter *et al.* (1982) found that at the same time at another site on this same reef, 96% of the corals died in shallow water during this cold winter, whereas in deeper water mortality was much less. Again, damage from cold stress varied among species and sites on this reef. The study of Porter *et al.* (1982) indicates that the diameter of the patches cleared by the low temperatures may have been less than 1 m.

Dollar (1982) found that a moderately severe storm on Hawaii reduced coral cover only 9 to 11% between 0- and 40-m depths, whereas a very severe storm reduced cover by 59% in shallow depths, increasing to 89% at the deepest sites. The greater cover reduction in the deepest zone may have resulted in larger open patches than in shallower zones. Live fragments from these sites were carried down to an even deeper rubble zone. Woodley *et al.* (1981) described marked spatial patchiness of damage from a severe hurricane on Jamaican coral reefs.

Two quantitative studies of the effect of exposure to air during unusually depressed sea levels have been made. Loya (1976) found that mortality in the nine most common species on the reef flat of an undisturbed reef near Eilat, Israel, ranged from 63 to 93% during an unusual low tide that exposed them to air for 2 hr per day for 4 days. These data are based upon numbers of colonies; the reduction in the percentage of cover was probably greater since many colonies were damaged but not killed. Corals with larger polyps were less damaged than those with smaller polyps (Fishelson, 1973). Yamaguchi (1975) studied the effects of a period of unusually low sea levels on Guam. All but one species on the reef flat were killed completely; a few colonies of *Porites lutea* survived and regenerated. In neither of these studies is it possible to estimate the patch sizes created, although the high mortality at Guam may have produced large patches.

Patches formed on coral reefs by biological agents have been described in several instances. The phenomenon most intensively studied is the effect of the crown-of-thorns starfish, *Acanthaster planci* [see reviews in Pearson (1981), Potts (1981), and Birkeland (1982)]. Large outbreaks of this starfish have caused extensive damage since 1966 on many Pacific reefs. Mortality varied among habitats and species. In Guam, 99% of the corals in deeper habitats were eaten, and 57% of those in shallow habitats, although no patch sizes were recorded. Corals that survived were either in inaccessible places, were low in *Acanthaster*'s hierarchy of preferences, or were defended by commensal crustaceans. Although no patch sizes were recorded, the series of outbreaks on the Great Barrier Reef probably constitute the most extensive creation of open patches from any source in recent times.

Other predators and diseases produce lower mortality and, presumably, smaller patches. Aggregations of the carnivorous snail *Drupella* have been described by Moyer *et al.* (1982) in Japan and the Philippines. They fed preferentially on fast-growing species of coral; over 2 years, mortality of about 35% of one reef was attributed to this predator. Most reports of other predators indicate that they produce more minor effects, often killing only parts of colonies, and so result in much smaller patch sizes than those made by aggregations of *Acanthaster* or *Drupella*. Competition for space creates open patches in some intertidal species; barnacles in dense aggregations often grow into unstable "hummocks," which are then washed away in storms, whereas those at lower densities survive (Barnes and Powell, 1950). However, we know of no similar effect of competition on coral reefs.

b. Frequency of Disturbances

Small-sized, less intense disturbances occur frequently. In contrast, large, intense disturbances occur less frequently, and may also occur only at certain seasons. Hurricanes, thermal stress, and low sea levels tend to occur at certain seasons, and aggregations of predators may also feed more heavily at certain seasons. The occurrence and magnitude of large, infrequent disturbances are unpredictable from year to year even though their seasonal pattern may be known.

Frequency of disturbance is also affected by variations in the "intrinsic vulnerability" of different types of corals. Small colonies are more likely to be completely killed by individual predators or by sediments smothering them than are large colonies. In contrast, larger trees or bushes (e.g., *Acropora* spp.) are probably more likely to be broken by wave stress than either smaller ones of the same growth form or mound or sheet forms. For this reason, and also because boring organisms weaken their skeletons, trees or bushes are more likely to be disturbed by wave action as they grow and age. In this way they resemble intertidal mussels, or large algae, whereas mound or sheet corals that do not become more vulnerable with age resemble uncrowded intertidal barnacles or short turf algae.

c. The Production of Coral Fragment Propagules by Disturbances

Disturbances may create new clearings and the fragments that reoccupy them (Connell, 1973; Highsmith *et al.*, 1980; Tunnicliffe, 1981; Highsmith, 1982; Bothwell, 1983). Large waves either break corals directly or carry fragments or rubble that act as projectiles that break corals (Highsmith, 1982). Also, some fish turn over and/or break up corals as they feed either on the corals or on the boring animals within the skeleton (Al-Hussaini, 1947; Randall, 1967, 1974; Glynn *et al.*, 1972). Borers weaken the skeleton and increase the likelihood of breakage. Although several sorts of disturbances produce fragments, the survival of these propagules depends upon the intensity of the disturbance. Very severe storms break fragments into smaller pieces and abrade off the living tissue to a much greater extent than do smaller storms. In addition, when the area of coral surviving a disturbance is quite small, the surviving predator population may concentrate its attacks more intensely. Thus, larger disturbances probably result in few surviving fragment propagules (Knowlton *et al.*, 1981). Some causes of disturbances do not produce fragments; groups of starfish (*A. planci*) predators kill large areas of coral but produce few or no new fragments (Pearson, 1981); likewise, reduction of the water level may cause large areas of reef flats or crests to dry out at low tide, killing but not breaking the corals. Partial death of a colony may divide it into separate parts but not into propagules.

3. Modes of Patch Colonization

a. Invasion of Type 1 Patches: Clearings within Existing Assemblages

i. Smaller patches. When small clearings are made in coral cover by the death of whole small colonies or parts of larger ones, vegetative growth of the bordering colonies into the patch is probably the main way that such space is reoccupied. Attachment of propagules, either fragments or larvae, is also possible. However, for such propagules to survive, they must be able to grow up quickly to the level of the bordering colonies and then be able to hold off these colonies that are growing into the space from the border. Because they are much larger at initial colonization, fragments satisfy these criteria better than larvae. Therefore, species with growth forms that readily break into fragments (trees, bushes, plates) should probably be superior invaders of small patches. In contrast, massive mound species are probably poor invaders of small clearings, since they seldom break into fragments and grow slowly. In summary, we suggest that vegetative growth into small clearings by sheets, plates, or trees, and invasion by fragments, are the main methods by which small clearings are filled by corals. The relative importance of vegetative growth versus propagules depends upon the nature of the colonies bordering the patch. If they are slow-growing (e.g., mounds), then one would expect propagules to be the major form of recolonization. If fast-growing sheets or plates border the edge, the patch would be filled mainly by vegetative growth.

ii. Larger patches. Severe disturbances occur relatively infrequently. Because they are intense, there are few survivors, so patches are larger. Thus, larger patches are created less frequently than smaller ones. Vegetative growth of survivors at the edges of the patch occurs, but must be proportionately much less important in these wide clearings than in small ones. As a result, invasion by propagules will be more important than vegetative growth. Although fragments are often produced in great abundance by the intense storms that cause large clearings, these fragments are often greatly scoured and abraded in the process. Therefore, unless the coral has a robust skeleton, fragments may be so small and damaged that they seldom survive long (Knowlton *et al.,* 1981; Highsmith, 1982). Fragments of robust forms such as certain trees (e.g., staghorn corals), as well as detached whole colonies of massive mounds, are likely to be the only vegetatively produced propagules available soon after a large patch is created. For these reasons, we predict that most of the re-colonization soon after large disturbances will be by larval attachment.

However, it is likely that recolonization of large patches will not occur rapidly unless propagules are quickly available in large numbers and unless survival after settlement is high. Both of the studies of large patches at Heron Island indicate slow recolonization. Three years after the 1967 cyclone had killed all corals in the outer pools beyond the crest, live coral cover had reached only 3 and 4%. In contrast, although the 1972 hurricane immediately reduced the cover to 4 and 5% in these same quadrats, portions of many colonies survived, so patch sizes were never larger than 0.3 m in diameter. Within 2 years the cover of live corals had reached 20 and 35%, mainly by vegetative growth.

The other large patch was created when the reef crust was dried out after the 1972 cyclone, as described earlier. A series of censuses in the next 11 years has shown that recolonization is very slow. Live coral cover in 1983 was still only 4% in each of the two 34-m^2 transects. Thus, in contrast to the rapid recolonization in small patches, large amounts of free space remained available for several years in both of these large patches.

Sexual reproduction is highly seasonal for some species of corals. On the Great Barrier Reef, many species release gametes within the same short period once a year, usually in early summer (Kojis and Quinn, 1981, 1983; Harrison *et al.,* 1984). In contrast, other species, usually those that brood larvae, release these planula larvae over an extended period of several months (Marshall and Stephenson, 1933; Stimson, 1978; Kojis and Quinn, 1983). If space for larval settlement is in short supply and is available only for short periods at various times of the year, the observed mass spawning of many species once a year would probably be a poor strategy, as compared to each species spawning at a different time of year. Howev-er, open space for settlement in single large patches remained available for many years in the situation studied at Heron Island, as well as in Guam, in large patches cleared by predatory *Acanthaster* (Colgan, 1983). As a consequence, seasonal or annual variability in the supply of propagules (Wallace and Bull, 1983) will have less influence on recolonization of larger patches than on smaller ones that are open to colonization for only short periods.

b. Invasion of Type 2 Patches   Coral larvae do not settle directly on soft sediments, requiring some stable hard substrata, e.g., stones, dead corals, or mol-lusc shells. However, living buds and fragments of coral can, if sufficiently large, survive and grow when carried into an area of sandy substratum (Highsmith, 1980).

By producing durable skeletons, they can invade an area of soft sediments and change it to hard substratum, literally paving the way for later species that require such substratum. Corals also invade large patches of new hard substrata such as lava flows (Grigg and Maragos, 1974).

4. Within-Patch Dynamics: Acquisition of Space without Clearings from Disturbances

Although patches often do not become completely occupied, individual colonies sometimes become surrounded by corals or other sessile organisms. Such colonies can gain space on the substrate by several means. Some species can attack and kill the tissue of a neighboring coral using mesenterial filaments or defend against such attacks by sweeper tentacles (Lang, 1971, 1973; Glynn et al., 1972; Porter, 1974; Dustan, 1975; Glynn, 1976; Sheppard, 1979; Richardson et al., 1979; G. M. Wellington, 1980). Branching species may overgrow neighbors, blocking the supply of light and/or reducing water flow, which increases the rate of deposition of sediment (Connell, 1979). In addition to such vegetative growth, it may be possible for propagules to invade occupied patches. A large fragment might fall onto a living colony, kill the underlying tissue, and attach to the skeleton (G. Wellington, personal communication). Also, it may be possible for larvae to settle on a living coral colony, but in our opinion this is extremely unlikely and we know of no evidence for it. The first method, vegetative growth of existing colonies, is undoubtedly the main way space is gained within occupied patches.

B. Subtidal Epifauna

1. Characteristics of Attached Epifauna That Affect Their Colonizing Ability

In contrast to hermatypic corals, certain other cnidarians, some tropical sponges, and didemnid ascidians that require light, most epifauna gain energy exclusively through suspension feeding, and so require access only to space and the water column. The acquisition of space by epifaunal organisms varies with the growth form. Solitary epifauna have an extremely limited ability to invade space by vegetative growth, and most species do not regenerate readily from damaged fragments. Therefore, they invade newly created patches almost exclusively by means of larvae. In contrast, colonial epifauna may invade space by all of the mechanisms detailed for corals: as propagules (sexually produced planktonic larvae and asexually produced larvae, buds, and fragments) and by vegetative extension of existing colonies, including regrowth of surviving fragments. Some taxa, such as sponges, tunicates, and bryozoans, are able to regenerate rapidly from pieces left behind after disturbances (Kay and Keough, 1981; Knowlton et al., 1981; Palumbi and Jackson, 1982; Ayling, 1983a,b), so this means of reoccupation is probably common. Asexually produced fragments may act as propagules in some sponges (Fell, 1974; Sivaramakrishnan, 1951; Woodley et al., 1981).

Of the many studies of natural epifaunal assemblages, few have measured their responses to events that produce open primary space. Those studies known to us are listed in Table 4. The sizes of open patches usually range up to about 1 m², about the same order of magnitude as in the intertidal zone (Sousa, Chapter 7, this volume). Occasionally, much larger patches are cleared by exceptional storms. One

TABLE 4

Sizes of Patches and the Events That Produce Them on Subtidal Natural or Artificial Substrata

Locality, substrate	Maximum sizes of clearings (cm²)	Events producing free space	Source
Australia			
Pier pilings, rock reef	50	Fish, asteroids	Kay (1980)
Pier pilings	10,000	Waves, senescence	Kay and Keough (1981)
Pier pilings	1,000	Senescence	Kay (1980)
Mollusc shells	100	*Pinna* recruitment	Keough (1983a)
New Zealand			
Rock reef	1,000	Grazers, disease	Ayling (1981)
Rock reef	10,000	Waves, senescence	Ayling (1984a)
North America—west coast			
California, rock wall	100	Urchins	Vance (1979)
California, shells	50	*Chama* mortality	Vance (1978)
California, rock	250	Urchins, fish	M. J. Keough and B. J. Downes (unpublished data)
California, rock	10,000	Rock falls	M. J. Keough and B.J. Downes (unpublished data)
California, rock	40,000	Severe storm	A. Ebeling (personal communication)
Washington, rock reef	400	Asteroids	C. M. Young (personal communication)
Washington, rock reef	10,000	Senescence	Young (1982)
North America—east coast			
North Carolina:			
pier pilings	1,000	Urchin grazing	Karlson (1978)
Caribbean			
Jamaica:			
coral skeletons	200	Coral recruitment and growth	Jackson (1977b)
Jamaica:			
coral skeletons	2	Lesions in bryozoans	Palumbi and Jackson (1982)
Venezuela:			
mangrove roots	200	Root growth	Sutherland (1980)

such storm in the winter of 1983 exposed subtidal bare rock in patches up to 4 m across when ledges were broken off at Coal Oil Point, near Santa Barbara, California (A. Ebeling, personal communication). Another example occurred in Australia in intertidal pools near the low tide line that were essentially subtidal. During extreme low tides during calm summer weather, the water temperatures in such pools may rise above the lethal point of the organisms in them. This occurred at Rottnest Island, Western Australia, in 1959 (Hodgkin, 1959) and 1982 (R. Black, personal communication); in 1982 all organisms were killed over an area of 1200 m^2.

The studies have been concentrated in a few geographical areas (see Tables 3 and 4), so that it may be premature to generalize from them. Therefore, rather than attempt to summarize this work, we will use the attributes of the various growth forms to predict the kinds of recolonization that may occur, and will illustrate these patterns with the results of experimental studies. Invasion into open patches of type 1 (cleared within existing assemblages) is likely to be affected to some degree by the surviving epifauna bordering the patch. In contrast, type 2 patches, being isolated, will be less affected by existing epifauna. We will consider these two categories in turn.

2. Modes of Patch Colonization

a. Invasion of Type 1 Patches

i. Smaller patches. Small clearings are made by the death of small individuals or parts of colonies caused by senescence, physical forces or individual predators such as urchins, sea stars, or fish, which eat a small number of organisms or a part of a colony and then move on. These events create a mosaic of small clearings that are surrounded by resident organisms. The effect of the surviving residents on recolonization will vary depending on their growth forms. Solitary organisms will not affect directly the recolonization of any but the smallest clearings, where they may slowly shift or grow into the clearing (e.g., intertidal organisms: Sousa, Chapter 7, this volume). Small clearings among solitary organisms will therefore be occupied mainly by larval recruitment from the plankton, and residents may affect this process by preying upon larvae as they settle from the plankton (see Young and Chia, 1984, for a review of this phenomenon). Many solitary organisms produce large numbers of long-lived larvae, which may disperse long distances (Thorson, 1950; Scheltema, 1971), so the species that colonize from the plankton may bear very little resemblance to the residents that surround it.

With colonial forms, the survivors bordering the patch may exert a strong influence on its recolonization, either by vegetative regrowth, production of larvae, predation on settling larvae, or modification of the physical conditions in the patch. With regard to vegetative growth, colonies with indeterminate growth (certain sheets and mounds) are the forms most likely to invade by vegetative extension into small openings. Some mounds and sheetlike sponges have the ability to modify their growth form in response to disturbance. Ayling (1983a) has shown that when space is cleared alongside a sheetlike sponge, an extremely thin extension of the colony is produced. The clearing is rapidly filled by this sheetlike extension, and the gain of space is later consolidated by thickening with additional tissue. In contrast, upright, bushy growth forms such as arborescent bryozoans, gorgonians, and clavelinid ascidians will not reoccupy by vegetative extension, because they have a small area of attachment to the substratum and grow mostly at the tips of branches. Similarly, mounds with determinate growth will not be able to expand rapidly laterally, and vines and runners cannot completely cover large areas of space quickly.

Kay and Keough (1981) examined the reoccupation of two sorts of type 1 clearings on pier pilings in South Australia: randomly positioned experimental clearings made four times a year, and clearings of various sizes that were deliberately positioned so that they were surrounded by one of a pair of common sheetlike sponges. These clearings were reoccupied almost exclusively by the vegetative extension of adjacent colonies, with recruits from the plankton being overgrown quickly. Small clearings were reoccupied much more rapidly than large ones, although in the case of the fast-growing sponges, even a large (0.5-m diameter) clearing offered little chance for recruits to become established. The two sponges, *Mycale* sp. and *Dendrocia* sp. [referred to in Kay and Keough (1981) as *Crella* sp.], differed in growth rate, and consequently patches that were cleared alongside *Mycale* were reoccupied much more quickly than comparable clearings made alongside *Dendrocia*. The actual location of a clearing was more important than its size; a small clearing alongside the slow-growing coral *Culicia tenella* would provide more opportunities for establishment than a large clearing alongside the faster-growing sponges. Keough (1984b) examined the composition of clearings of various sizes to compare the abundance of sheetlike sponges and colonial tunicates to that of species that were relatively poor competitors for space, such as bryozoans and serpulid polychaetes. The species composition of clearings was not strongly influenced by their size, and after 1 year, most were dominated by sponges and tunicates.

With regard to larval production, those species released from local colonies have a greater chance of invading small openings than those from distant colonies only if their larvae crawl on the surface or spend only a short time in the plankton when local currents are relatively weak. Many colonial species produce only a few short-lived larvae (Thorson, 1950; Ryland, 1974, 1976), so if currents are weak, the openings may be invaded mainly by locally produced larvae (Ostarello, 1976; Gerrodette, 1981). In regard to predation, all forms, particularly those with extensive vertical growth, may reduce colonization by predation on larvae as they arrive from plankton. Lastly, colonies bordering a patch may modify the local environment and as a consequence affect larval settlement, e.g., by changing water flow patterns or by influencing the behavior of larvae. If skeletons of former occupants remain in the patch, they may modify the surface texture and so influence the invasion of propagules. Since, for all of these reasons, the species surrounding a small clearing influence its colonization, the spatial position of the clearing is very important (Harris, 1978; Kay and Keough, 1981; Palumbi and Jackson, 1982; Ayling, 1983a).

ii. Larger patches. Larger cleared patches are usually created by intense storms, by the senescence of either large colonies (Kay, 1980) or aggregations of solitary organisms (Karlson, 1978), or from attacks by large aggregations of predators (Russ, 1980). Again, in areas where most forms are either solitary or colonial with determinate growth, recruitment by larvae will be the main means of invasion, so that the reoccupation of a clearing will be relatively unaffected by surviving residents. In contrast, where there are many sheets or mounds with indeterminate growth, larval recruitment will be of somewhat lesser importance, since the surrounding colonies will grow in from the edges of the clearing. Near the center of very large clearings, recruits will be able to grow for some time before encountering those growing in from the edge. Since larger colonies tend to win over smaller ones in competition (Day, 1977; Buss, 1980; Russ, 1982), this initial period of growth without competition will confer an advantage on colonies recruited to the centers of large clearings, enabling them to reach a size large enough to compete successfully with colonies growing in from the edges.

Aggregations of predators or intense wave action do not always clear an open space completely. Small depressions in the surface may provide protection for whole organisms (see, e.g., Connell, 1961b; Keough and Downes, 1982) or small remnants of colonies (see, e.g., Kay, 1980; Ayling, 1983b). Alternatively, certain taxa may be particularly resistant to the events that clear space. Karlson (1980) showed that *Hydractinia* and *Xestospongia* are very resistant to grazing by the urchin *Arbacia punctata,* whereas other species were removed completely from the substratum. These species were able to become very abundant because they were the first species to recover from intense grazing; they quickly regrew into the bare space before other species could recruit. Thus, any circumstance that increases the chance of survival of residents will reduce both the effective size of the clearing and the relative importance of larval recruitment in its recolonization (Keough, 1984b).

b. Invasion of Type 2 Patches

i. Small, discrete pieces of substratum. Patches of primary substratum such as stones, small cobbles, large algae, worm tubes, and mollusc shells are usually occupied only by recruitment from the plankton, since they are ordinarily separated physically from each other. In such cases, reoccupation will mainly reflect the relative abundance of larvae in the plankton, their behavior in choosing settlement sites, and their survival after settlement. These small, discrete patches offer little scope for expansion to encrusting sheets with indeterminate growth.

Kay and Keough (1981) observed the epifauna of small, discrete type 2 patches represented by the large bivalve *Pinna,* which protrude from soft sediments at their study site in South Australia. Their epifauna differs markedly from that of the nearby pilings; small determinate bryozoan sheets and solitary organisms were much more common, while indeterminate sheets were relatively uncommon on the shells (Kay and Keough, 1981). Keough (1984a) showed that the species composition on individual *Pinna* shells is determined primarily by patterns of recruitment and is influenced relatively little by interference competition. Juvenile monacanthid fish prey upon newly metamorphosed colonial tunicates, further reducing the abundance of indeterminate sheets (Keough, 1984a).

The recruitment rate was very low (Keough, 1983), but the abundance of all attached species combined remained fairly constant. Over a long period, there was a considerable turnover of colonies on individual shells and a consequent large fluctuation in the number of epifaunal species through time (Keough and Butler, 1983). This situation does not hold for all *Pinna* populations; in northern parts of the Gulf St. Vincent and Spencer Gulf, southern Australia, where sediments are much finer, solitary eipfaunal species are relatively more abundant. Jackson (1979a) suggested that sheets are relatively unsuited to habitats where rates of sediment deposition are high. In habitats with low water movement and much suspended matter, there is usually an abundance of solitary species (Wells *et al.*, 1964; Alagarswami and Chellam, 1976; Gulliksen, 1980).

ii. Large, discrete Type 2 patches.   Such patches are created by landslides or submarine lava flows (Grigg and Maragos, 1974; Gulliksen *et al.*, 1980), but there are no detailed studies of the recolonization of such substrata by epifaunal organisms other than corals. Most studies of attached epifaunas have been done in temperate waters, where catastrophic events such as hurricanes occur less often than in the tropics. Most fouling panel studies have been done in very protected waters, and even the studies on natural epifaunas in temperate open waters have not been done in the very exposed localities in which such large-scale disturbances might be expected to occur.

In contrast to small type 2 patches, it might be expected that large ones would be more likely to be colonized by clonal forms with indeterminate growth, such as sheets and mounds in regions where these are common. Since these forms are generally superior in competition to vines, runners, bushes, etc. (Jackson, 1977a, 1979a; Buss, 1981; Kay and Keough, 1981; Russ, 1982), indeterminately growing sheets and mounds should eventually eliminate the latter and come to dominate. In sites where the competitively inferior forms (solitary or clonal runners, vines, bushes, or determinate sheets or mounds) predominate, there should be little or no difference in the pattern of occupation of large versus small type 2 patches, both being more influenced by recruitment than by interference competitive ability. Such would also be the case if colonies with indeterminate growth were restricted in size by continual heavy grazing, so that their growth became effectively determinate.

c. Direct Comparisons between the Dynamics of Type 1 and Type 2 Patches   The difference between the faunas of type 1 and type 2 patches results from differences in the relative importance of various biological processes, particularly recruitment versus interference competition (Kay and Keough, 1981). Keough (1984b) suggested that such a difference should be expected when there are tradeoffs between vegetative growth and the number of dispersive propagules produced. The situation in which colonies with indeterminate growth are more abundant on large type 1 substrata, while poor competitors for space are more abundant on small type 2 patches, has been reported a number of times from both natural and artificial substrata (Jackson, 1977b; Karlson, 1978; Kay and Keough, 1981).

The above comparisons of large and small, discrete substrata were mainly observational. Keough (1984b) made an experimental test of the effect of patch type and size on the abundance of various taxa, and showed that indeterminate sheets (colonial tunicates) were more common in type 1 than type 2 patches and also on larger than smaller type 2 patches. In contrast, bryozoans and serpulids showed the opposite trends. Since rocky reefs and *Pinna* populations have been present for thousands of years at this locality, this is ample time for any evolutionary change to have occurred, given the life spans of these attached organisms. One would predict a selective advantage for a bryozoan or serpulid larva that chooses to settle on a small type 2 patch, since mortality declines with substrate size. Likewise, it is advantageous for an indeterminate sheet species to settle on a large type 2 patch or in a large type 1 clearing, since expected reproductive output increases with colony size and so with substratum size. This pattern of recruitment has been observed (Keough, 1984b).

d. Occupation of Previously Unsuitable Habitats

Although attached species all require hard surfaces to which they can attach, some species require very little area. Once established, they may be able to expand over the neighboring soft sediments. If, like corals, these species have hard skeletons, the remnants after the colony has died may be colonized by other species and eventually consolidated. This appears to be the case for some bryozoans, such as *Schizoporella errata* in northern Florida, where large, moundlike colonies may form around small shell fragments, sea grass blades, or worm tubes; these colonies may be colonized subsequently by a variety of other species (M. J. Keough, personal observation). Some annelids and tunicates may also invade soft sediments in the same way (C. M. Young, personal communication). The general importance of this phenomenon is unknown.

e. Within-Patch Dynamics: Effects on Diversity

Once an open patch is occupied, or in sites where disturbances are uncommon, interactions between neighbors become more important. Examples in the literature come from Great Britain (Stebbing, 1973; Rubin, 1982), Canada (Keen and Neill, 1980), United States (Osman, 1977), Australia and New Zealand (Day, 1977; Russ, 1980, 1982; Kay, 1980; Kay and Keough, 1981; Ayling, 1983a; Keough, 1984a), and the Caribbean and Panama (Jackson and Buss, 1975; Jackson, 1977a, 1979b; Buss and Jackson, 1979; Buss, 1980; Jackson and Winston, 1982). These studies observed contacts between neighbors either at one time or in a series of censuses over time. In the latter case, several outcomes have been observed. A common observation in most studies is that one species overgrows and kills the other. However, several alternative outcomes may occur: (a) the overgrown colony may survive for a long period (Rutzler, 1970; Sara, 1970; Harris, 1978; Kay, 1980; Ayling, 1983a); (b) the overgrowing colony may subsequently retreat (Karlson, 1978; Kay and Keough,

TABLE 5

Competitive Symmetry in Subtidal Colonial Organisms[a]

| | Number of pairs in which the two species are in: | | | | | | | |
| | Same phylum | | | Different phyla | | | Total % symm. | Source |
Organism	Symmetrical	Asymmetrical	% Symm.	Symmetrical	Asymmetrical	% Symm.		
Temperate								
Sponges	4	3		4	20	17	29	Kay and Keough (1981)[b]
Bryozoa	1	—	60		(All plus Cnidaria)			
Tunicates	1	1						
Sponges	3	0		13	33	28	36	Russ (1982)
Bryozoa	9	13	48		(All three phyla)			
Tunicates	3	3						
Bryozoa	7	18	28				28	Rubin (1982)
Bryozoa	3	2	60	1	4	20	40	Keough (1984a)[b]
				(Bryozoa, sponge, tunicates)				
Sponges	41	4	91				91	Ayling (1984a)[b]

Tropical								
Bryozoa	4	2	67	0		0	67	Jackson (1979b)
Bryozoa	0	1	0		2		0	Buss (1980)
Corals	11	2	85		2 (2 Bryozoa, 1 alga)		85	Connell (1976, 1984)[b]
Corals, algae, foraminifera				4	2	67	67	Sammarco and Carleton (1983)
All studies combined								
All sponges	48	7	87	22	61 (Total, different phyla)	27	48	
All Bryozoa	24	36	40		57			
All tunicates	4	4	50		4			
All temperate	72	44	62	18	57	24	47	
All tropical	15	5	75	4	4	50	68	

[a] All known studies that presented sufficient data to test statistically were included. All interactions between pairs of species having five or more observations were included. Analyses used the Binomial test, two-tailed, $p < .10$, with ties equally divided between the two species.

[b] Studies in which the outcome of contact was followed over a period of time; in all other studies, observations were made at a single time.

145

1981; Ayling, 1983a); (c) the overgrown colony may advance, gaining space from the colony lying above it (Harris, 1978). Thus, a colony that has its edge above that of its neighbor is not necessarily the winner of the encounter, as has been assumed in all of the studies that have used single-time observations. Because of this problem, Russ (1982) scored a win only when one colony had overgrown at least 25% of the area of its neighbor; this figure was based upon his observation that once this amount of overgrowth had occurred, it was probable that it would continue until the subordinate colony was completely overgrown.

Sometimes two colonies remain in stationary contact for a period of time. Connell (1976) termed this a "standoff" and Russ (1982) a "delay/tie." Two studies indicate that such standoffs are sometimes simply delays in the process of competition between the colonies. First, Connell (1976, 1984) found that corals on a reef crest often had standoffs that lasted a year or more, after which one overgrew the other (although one standoff persisted for the entire period, almost 9 years). Second, Ayling (1984a), studying the growth of sponges in contact, compared undisturbed ones with others in which open space was created at the juncture of the two sponges. She found that when space was provided among 17 species pairs that maintained standoffs for 9 months if undisturbed, in 13 pairs one species gained space from the other, growing beyond the original boundary line. The other four pairs grew back only to the original standoff line. Both of these studies suggest that standoffs may sometimes be only temporary, and in other instances may last a very long time.

We have used these studies to answer the following question: To what degree are competitive interactions in epifauna asymmetrical, i.e., with one of a pair of species consistently winning over the other? To test the null hypothesis that the interaction is symmetrical, i.e., that one species did not win significantly more often than the other, we have used a modification of the method of Kay and Keough (1981). In studies where standoffs or delay/ties occurred, we have assigned these as wins divided equally between the two species. This procedure seems reasonable, based upon the observation described above that standoffs often change to wins for one or other species in the pair. Table 5 gives the results of analyses of all studies known to us in which sufficient data were published to allow statistical analysis. We applied the Binomial test (Siegel, 1956) to all pairs with at least five observations, using a level of significance of $p < .10$. Since the direction of the outcome was not predicted, two-tailed probabilities were used.

Table 5 indicates that when both species in a competing pair were in the same phylum, competition was usually more symmetrical than when each was in a different phylum. Within a phylum, sponges and corals showed more symmetry than did bryozoans and tunicates. The numbers in some studies are very small, so that these tendencies should not be regarded as definitive.

The studies of interactions cited above give many possible reasons for symmetry in competition. In a homogeneous space, larger colonies usually win over smaller ones, as discussed earlier. The angle at which bryozoan colonies happen to meet sometimes determines the winner (Jackson, 1979b). Spatial heterogeneity may also affect the outcome; if one colony is on slightly higher ground, this may give it an initial advantage in overgrowing its neighbor. Other types of environmental variation in space or time may shift the advantage from one to the other. While some reasons for competitive superiority are fairly well documented (e.g., relative size or angle of contact), most mechanisms of competition remain unknown.

Although none of these studies followed changes in diversity, in the absence of disturbance symmetrical competition should reduce the probability of a species becoming locally extinct. However, with no disturbance, the diversity in a patch should eventually decline to a monoculture as species become extinct by chance. Therefore, even when disturbances are rare, they may be extremely important for the long-term maintenance of species richness.

III. DISCUSSION

A. Variations in the Disturbance Regime

The various characteristics of the disturbance regime—areal extent, intensity, frequency, and seasonality—are highly correlated. Small, less intense disturbances occur frequently; also, intensity and frequency of storms vary seasonally. Different organisms have different intrinsic vulnerabilities to disturbance. This vulnerability may increase with size, age, and population density in certain growth forms (bushes and trees). Some growth forms are intrinsically more vulnerable to certain types of disturbance: encrusting forms are more readily smothered by sediments than upright forms, whereas wave stress acts in the opposite way. The maximum sizes of patches cleared completely of all organisms vary widely. In epifaunal assemblages these seldom reach 1 m^2 (Table 4); in corals they can reach two or more orders of magnitude larger than this. However as discussed above, much larger patches may very occasionally be cleared of epifauna.

B. Predicted Responses to the Disturbance Regime

1. Among-Patch Dynamics: Variations in Colonization and Species Composition

We predict that the main determinant of the variation in species composition of epifaunal and coral communities is the degree to which surviving residents influence the invasion of the patch. Their influence can take several forms: by preempting the space as they grow into the patch, by contributing propagules (fragments or larvae), by capturing and eating larvae as they invade from the plankton, and by affecting the local physical environment. All of these influences will be maximal in small type 1 patches cleared in assemblages composed mainly of forms with indeterminate growth. In corals these are sheets, plates, trees, and bushes; in epifauna they are certain sheet and mound sponges that are able to extend quickly as thin sheets.

The influence of surviving residents should decline with increasing size of type 1 patches; in contrast, the relative importance of invasion by propagules should increase. (The latter is the only way that type 2 patches are colonized.) The influence of local residents near large type 1 patches is greater if they produce many propagules that do not travel far; this situation, in turn, is determined by the nature of the propagules (both fragments and larvae) and the strength of local water movements.

The degree to which the species composition of a community varies in space and time is therefore determined in great part by such among-patch dynamics. Patches that are colonized by planktonic larvae will probably vary most in space and time. These are type 2 patches and larger type 1 patches, particularly those that occur in assemblages of epifauna of determinate growth form. Keough and Butler (1983) found that the species composition of small type 2 patches (*Pinna* shells) varied greatly over time. They adopted the definition by Chesson (1978) of "stochastic boundedness," and proposed a method by which the fluctuations in either population or community attributes could be measured independently of scale or sample size. The species richness of epifauna on *Pinna* shells fluctuated a great deal; in their terminology, it was seldom "narrowly bounded." Likewise, Sutherland and Karlson (1977) found extensive fluctuations in species composition over time on artificial type 2 patches (the underside of small panels suspended beneath a pier), although the data were not sufficient to analyze by the method of Keough and Butler (1983).

In contrast, species composition should be less variable in sites where colonization by planktonic larvae is less important. We would expect this situation to occur where local residents strongly affect recolonization, i.e., in small type 1 patches in assemblages with high proportions of species with indeterminate growth forms. Kay and Butler (1983) found that with small type 1 patches among such species on pier pilings in South Australia, the overall species richness, cover, and individual species abundance tended to be narrowly bounded, using the method proposed by Keough and Butler (1983).

If species composition is measured by the number of species or by the relative numbers of individuals or colonies of the different species, it can change only by recruitment or mortality of individuals or whole colonies. If, on the other hand, it is expressed as the percentage of cover of the different species, it can change by differential growth, without recruitment or mortality of whole colonies or individuals. Keough and Butler (1983) used the number of species, whereas Kay and Butler (1983) used both numbers and percentage of cover of species. Kay and Butler (1983) point out that one of the reasons for the observed low degree of fluctuation in species numbers and percentage of cover was that some colonies had survived the whole period of study. However, there was also a substantial turnover in space, so that about half of the space on their quadrats had "changed hands" at least once. The species on the type 2 patches of Keough and Butler (1983) were mainly short-lived solitary and colonial forms. When judging the degree of stability or persistence, it is necessary to compare the abundance of a population before and after at least one complete turnover of individuals (Connell and Sousa, 1983).

These predictions, concerning variations in population and community characteristics relative to both the disturbance regime and the growth forms of the assemblage, were developed using the existing studies described above, and so have not yet been tested with data collected independently. They are based upon only a few studies in a few different sites, and so need to be tested more broadly.

2. Within-Patch Dynamics: Competitive Symmetries, Hierarchies, and Networks

Competitive interactions between species in subtidal assemblages of plants, corals, and epifauna have been considered in several ways. Within a single pair of competing species, one species may win all or most of the contests; competition is asymmetrical in this case. Alternatively, neither wins significantly more often than the other, or there are many standoffs when both stop growing after contact; competition is then symmetrical, with neither species dominant (Connell, 1976, 1983; Kay and Keough, 1981; Russ, 1982).

The proportion of symmetry versus asymmetry in competing pairs of species has only recently begun to be assessed. Lawton and Hassell (1981) found that 34% of the 35 pairs of insects tested showed symmetrical competition. Connell (1983) found that among 54 pairs (including terrestrial and aquatic plants and animals), competition was symmetrical in 39%. Table 5 indicates a considerable degree of competitive symmetry in epifauna, particularly between members of the same phylum, in which percentages ranged up to 91% symmetrical.

When all of the interactions among more than one pair of species are considered, at least three arrangements exist. If all pairs show asymmetrical competition, the arrangement can be either a transitive hierarchy (e.g., A > B, B > C, A > C) or an intransitive network (e.g., A > B, B >, C, C > A). Buss and Jackson (1979) defined the latter as "the occurrence of a loop in an otherwise hierarchical sequence of interference competitive abilities." Third, if some pairs in the assemblage exhibit symmetrical competitive interactions, the community can be regarded as a network intermediate in the degree of intransitivity (Buss, 1980; Rubin, 1982; Russ, 1982).

In only one of the assemblages listed in Table 5 did all pairs show asymmetrical competition. The three species in this study (Buss, 1980) formed a perfectly intransitive network, each winning over one and losing to another. Among the other eight studies cited in Table 5, five had between 4 and 49 asymmetrical pairs, yet there were no perfectly intransitive networks among any of the species in these assemblages. However, there were many pairs showing symmetrical competition, and these pairs impart an intermediate degree of intransitivity to these communities.

3. Within-Patch Dynamics: Effects on Species Diversity and Succession

The species richness of communities in relation to disturbance has been hypothesized to be highest at intermediate levels of disturbance and lowest under conditions of both high and low disturbance (Connell, 1978; Lubchenco, 1978). With frequent disturbances, the diversity is low either because few species have evolved with the capability of very rapid recolonization and/or tolerance of the physical environment in a new patch, or because there has not been sufficient time for many species to invade between successive disturbances. With low levels of disturbance, diversity is reduced because, in the long intervals between perturbations, a few species have gradually eliminated all others in competition for space. This hypothesis assumes a transitive hierarchical ranking of competitive abilities among the species, with competitive outcomes being consistent and asymmetrical, i.e., one of a pair of competitors always winning over the other. The net result of competition for space in the absence of disturbance would be monopolization of space by one or a few species.

However, as discussed above, subtidal assemblages are usually not completely hierarchical, because some of the competing pairs are symmetrical. In such cases, the position of any peak of species richness will be difficult to predict, even with low rates of disturbance. Beginning with an empty patch, diversity will rise with invasion, but the degree to which it will decline later depends upon the abundance of the invaders, how much they interact, and the degree of competitive asymmetry of the interactions. Among corals, certain trees (large branching "staghorn" *Acropora* sp.) appear to be capable of overgrowing almost all other species. However, there are other situations in which no such clear dominance exists. For example, where interactions are constrained to two dimensions, as with assemblages composed of encrusting sheets and mounds, or in very shallow water on reef crests or flats, interactions are so symmetrical that some pairs of species appear to be almost equivalent in competitive ability. Here we predict that, particularly when the interactions are between members of the same phylum, (Table 5), diversity may not decline to any degree even after a long period without further disturbance.

The number of species that will coexist without further disturbance will then depend upon the position in the overall competitive hierarchy of the symmetrical pairs. If such a pair were low in the hierarchy, both species could be eliminated by higher-ranking species. If, on the other hand frequent reversals or standoffs occurred between high-ranking species, they might continue to coexist. Buss and Jackson (1979), Karlson and Jackson (1981), and Quinn (1982) have reasoned that such symmetries should slow down the rate of elimination of competitively inferior species. These suggestions and some theoretical models incorporating them (Horn, 1975; Usher, 1979; Hastings, 1980) have not yet been tested against data from natural communities.

Following a disturbance, a bare patch is almost always invaded by a succession of species. However, this succession may or may not converge toward a particular "climax" species composition. We would not expect convergence in a situation in which those species with the greatest competitive abilities interact symmetrically with each other in an intransitive network instead of in a transitive hierarchy. With no clear dominant, the species composition would continually vary. In such cases, although a succession of species might occur, the mechanisms causing it should not produce a predictable equilibrium species composition but rather a continuously changing one. Succession might follow one or more of the pathways described in the models of Connell and Slatyer (1977), but the last stage would not be a single dominant.

At present there is little information from subtidal assemblages on the proportions of asymmetrical hierarchies versus symmetrical networks. When corals competed with each other on the reef crest at Heron Island (Table 5), they showed a high degree of symmetry, but at moderate depths they competed asymmetrically, a few species dominating the rest (Connell, 1978, 1984). No subtidal epifaunal assemblage has, to our knowledge, been followed long enough without disturbance to see whether or not a few dominants will emerge.

What we have sought to emphasize here is that there is no uniform way in which disturbances influence communities of subtidal sessile organisms. Natural disturbances occur on a wide range of spatial and temporal scales, and the responses of natural communities to them depend on their size, intensity, and frequency, and also on the characteristics and life histories of the resident species. We have attempted to provide a conceptual framework that recognizes this variability and uses it to predict the dynamics of natural communities exposed to a given disturbance regime. The results from our own observations and experiments are consistent with these predictions, but are not a test of them.

ACKNOWLEDGMENTS

We thank the following persons for comments on the manuscript: N. Andrew, A. Ayling, C. Battershill, R. Black, A. Bothwell, A. Butler, J. Choat, R. Creese, R. Day, R. Dean, Z. Dinesen, B. Downes, S. Holbrook, T. Hughes, J. Jackson, M. Kingsford, P. Moran, S. Osborn, G. Russ, R. Schmitt, K. Sebens, D. Shiel, W. Sousa, L. Stocker, D. Strong, S. Swarbrick, D. Thistle, J. Thompson, R. Vance, C. Wallace, G. Wellington, and C. Young. Also, we thank R. Black, A. Ebeling, L. Harris, and D. Laur, who gave us unpublished data on patch sizes created after severe disturbances. The research was supported in part by NSF grant DEB80-11123.

Chapter 9

Disturbance and Vertebrates: An Integrative Perspective

JAMES R. KARR[1]

Department of Ecology, Ethology, and Evolution
University of Illinois
Champaign, Illinois

KATHRYN E. FREEMARK

Department of Biology
Carleton University
Ottawa, Ontario, Canada

[1]Present address: Smithsonian Tropical Research Institute, P.O. Box 2072, Balboa, Panama.

THE ECOLOGY
OF NATURAL DISTURBANCE
AND PATCH DYNAMICS

I. INTRODUCTION

Although ecologists often assume that homogeneous ecosystems are a reality, virtually all naturally occurring and man-disturbed ecosystems are mosaics of environmental conditions. A major share of that heterogeneity arises from the consequences of disturbances operating at various temporal and spatial scales. Because of the primary role of disturbance and the inherent complexity of disturbance types and impacts, a systematic approach to the study of disturbance, and heterogeneity is essential.

Disturbance may stimulate vertebrate individuals to change physiologically (e.g., acclimatization), behaviorally (adopt a different resting posture), or ecologically (feed in a different place or alter reproductive output). In addition, a disturbance may change relative fitnesses of phenotypes. Individual responses or modified relative fitnesses collectively produce changes recognizable at population, community, or ecosystem levels. At the population level, attributes such as density, spatial distribution, mortality rate, reproductive rate, genetics, and extinction probabilities may change. At the community level, changes may occur in species composition, relative abundance, or trophic structure because of differences in the responses of species to disturbances. For ecosystems, functional rate processes such as energy flow and nutrient cycling may also be altered as a result of the responses of component species.

II. THE COMPLEXITY OF DISTURBANCES IN VERTEBRATE COMMUNITIES: AN EXAMPLE FROM THE EVERGLADES

Difficulties associated with the definition of disturbance can be illustrated by the responses of vertebrates of the Everglades National Park to changing water levels (Kushlan, 1976, 1979). On average, rainfall and thus water levels in the Everglades are high from May to October, with little rain and low water levels during the spring. Many herons, ibises, and storks reproduce during the relatively dry period when fishes, their primary food, are concentrated. Reproduction by wading birds may be unsuccessful in years without seasonal water shortage. During prolonged periods of water shortage, colonies may shift to areas where food is more available. Without other suitable sites, regional persistence of species and the probability of recolonization of old sites are low.

When fishes are concentrated in deeper pools during periods of seasonally low water and high bird populations are present, birds remove 76% of fish. As a result, fish abundances are low but species richness is high and dominated by small omnivores and herbivores. In the absence of predation by birds, fish suffer up to 96% mortality from overpopulation and depressed oxygen concentrations in deeper pools. Local extinctions are common and recolonization varies with the availability of fish colonists from surrounding areas. In this case, a biological disturbance (predation by birds), when present, reduces the overall mortality from a physical disturbance (drought).

In years without spring drought, predatory fish increase in abundance and in species richness, while the omnivorous and herbivorous groups decline because of increased pressure from abundant predatory fish. Mean size of fish increases.

Thus, temporal instability in water levels due to spring drought represents a disturbance that ensures successful reproduction in the resident wading birds and high populations of small omnivorous fishes. During periods of drought but without the disturbance of avian predation, fish populations are reduced and species richness is low. The alternative disturbance, stabilized water levels, reduces populations of birds and omnivorous fishes but results in major increases in predatory fishes. The seasonal change in water availability, regardless of the magnitude in any year, is a disturbance to varying segments of the resident vertebrate community and emphasizes the importance of defining disturbances from the relative perspective of the receiving organism.

Further, the cascade of disturbance impacts among trophic levels from producers to top carnivores, although not discussed here in detail, should be kept in mind. Rather than arbitrarily define only one of these as a disturbance, we adopt the perspective that all are disturbances. In our view, that perspective is required for the development of an orderly theoretical framework that considers disturbances and the heterogeneity they generate in biological systems.

III. DISTURBANCE AS CONSIDERED BY ECOLOGISTS

Historically, ecologists have been slow to recognize the importance of disturbances and the heterogeneity they generate. Several factors are responsible for slow progress in this area. First and perhaps foremost is the tendency to concentrate on the presumed equilibrium nature of biological systems (Wiens, 1977; Connell, 1978). Notions about successional development of equilibrium (climax) communities dominated ecological thought early in this century. More recently, mathematical approaches to ecology required equilibrium assumptions to be tractable. Exceptions to these generalities clearly exist in the work of Andrewartha and Birch (1954) and more recently in theoretical work (Levins, 1968), in symposia (Anderson et al., 1979), and in review papers (Levin, 1976; Stenseth and Hansson, 1981). But the majority of both theoretical and empirical work, especially that dealing with vertebrates, has been dominated by an equilibrium perspective. When large-scale disturbances were examined, they were often considered as exceptional circumstances rather than as an extreme on a gradient of disturbance intensities.

Second, ecologists have interpreted averages derived from short-term studies (e.g., a single breeding season in vertebrates) as deterministic values. Further, they may combine data from several study plots and/or the same study plot over several years, and thus overlook dynamics in space and time. Variability is often assumed to reflect noise and sampling inefficiencies rather than real dynamics of biological systems.

Third, pressure to publish, the short duration of research grants, and the focus on problems that are tractable over the short term have often imposed planning horizons involving time scales inappropriate for study of the interaction between environmental heterogeneity and disturbance. Longer periods of study require methods of analysis that have only recently become widely available with the expansion of computer technologies.

Fourth, the large size, relatively long life cycle, and high mobility of vertebrates make them difficult to study in space and time.

The fact that adaptations of vertebrates and vertebrate assemblages exhibit recurrent patterns leads us to attempt a synthesis of disturbance impacts on vertebrates. We also emphasize the need to consider differences among vertebrates that develop due to selective pressures at least as strong as those that produce recurrent patterns. We are guided in this effort by a philosophy that lies "between the devil of oversimplification [generality] on the one hand and a deep blue sea of endless unrelated facts on the other. The important thing is to ensure that neither gets the upper hand" (Lawton and McNeill, 1979).

IV. DISTURBANCE AND VERTEBRATE ADAPTATIONS

The existence of order in vertebrate adaptations is clear from the convergence of attributes among individuals, populations, and communities in similar environments throughout the world (Cody, 1975; Karr and James, 1975; Mooney, 1977). Insight into this complexity may be increased if it is interpreted as a response to disturbance. In this section, we examine the interplay of disturbances and vertebrate adaptation in three distinct environments: tropical forest, freshwater streams, and freshwater marshes.

A. Tropical Forest

Biologists have long considered tropical forest areas to be stable environments. However, detailed studies have clearly documented the variability inherent in both physical and biological processes in the tropical forest (Leigh et al., 1982). Of special significance to organisms is seasonal and between-year variation in moisture availability. Reproductive responses to rainfall variation have been documented for amphibians (Toft et al., 1982), reptiles (Sexton et al., 1971; Andrews et al., 1983), mammals (Fleming et al., 1972; Russell, 1982; Glanz, 1982), and birds (Fogden, 1972; Worthington, 1982). In addition, effects of variation in habitat distribution on seasonal cycles are known for frogs (Toft, 1980; Toft et al., 1982) and in considerable detail for birds (Karr, 1981a, 1985; Karr and Freemark, 1983). Further, when data from more than 1 year are available, responses by organisms to both seasonal and between-year variation in rainfall are clear.

We netted birds in lowland forest (Karr and Freemark, 1983) for 4 years at about 70 net sites that differed in vegetation structure (primarily shrub density) and undergrowth microclimate (relative humidity). Variation in vegetation structure at a site was subtle over time (except for sites experiencing treefalls; Schemske and Brokaw, 1981), in contrast to marked diurnal, seasonal, and between-year variability in microclimate. Spatial distribution of birds along a moisture gradient varied in the forest on these three time scales. Avian activity is generally highest in the early morning, with a lower peak late in the afternoon and a trough at midday (Okia, 1976; Karr, 1979). However, the relative change during the day varies among sites as a function of moisture conditions. Avian activity at dry sites decreases more from morning to midday than at wet sites (Fig. 1A). Similarly, some species shift their pattern of distribution between wet and dry season (Fig. 1B). Further, activity varies among years; more birds are active at the dry end of a moisture gradient in those dry seasons with high rainfall than in dry seasons with low rainfall (Fig. 1C). During the extraordinarily dry "El Niño" year of 1983, many species moved closer to the ground in the forest, presumably to find a more equitable (less dry) environment. During 2-week sample periods from 1979 to 1982, an average of 6.9 individuals of canopy species (range 5–11, SE = 1.0, N = 8) were included in samples of 355–416 captures as compared to 14 individuals in 1983 among 409 captures (t = 7.07, p < .001).

The distribution of frogs among stream-edge, midslope and ridgetop sites also varies with the season (Toft *et al.*, 1982; J. R. Karr, personal observation). In the extraordinarily dry season of 1983, snakes and frogs (especially *Colostethus* sp.) were concentrated in stream-edge environments in central Panama (J. R. Karr, personal observation).

Further, the impact of the local mosaic of microhabitats caused by physical disturbances and biotic responses seems to be important in determining population attributes (Karr, 1982a). That is, as local populations of a variety of bird species track appropriate microclimatic conditions and vegetation structure, local abundances shift on a variety of temporal and spatial scales. Areas without habitat mosaics suitable to allow species to adjust their distribution and behavior in response to disturbance may be subject to high extinction probabilities (Karr, 1982b).

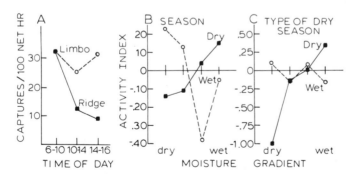

Fig. 1. (A) Capture rate (captures per 100 net-hours) as a function of time of day at dry (ridge) and relatively wet (limbo) study areas in central Panama. (B, C) Activity patterns of the ochre-bellied flycatcher (*Pipromorpha oleaginea*). (B) During wet and dry seasons, 1979–1982. (C) During the dry (1979) and wet (1981) dry seasons. The activity index displayed here is an assessment of capture rates in mist nets at a gradient position relative to the sample effort (in mist-net-hours at that gradient position). (After Karr and Freemark, 1983.)

These dynamics are implicated in a number of extinctions on Barro Colorado Island (BCI), Panama (Karr, 1982a).

In addition, a biotic disturbance (predation) may also be important in the extinction of some species. Predation rates on artificial nests placed in the leaf litter were higher on BCI than on the nearby mainland (Loiselle and Hoppes, 1983). The cryptic behavior of adult birds at the nest and the well-concealed nests (or false nests) of many tropical forest birds also suggest that predation is a major disturbance to birds of tropical forests. Disease as a biological disturbance affecting populations of tropical forest vertebrates is known for yellow fever in howler monkey (*Alouatta palliata*) (Milton, 1982). In contrast, malaria in the lizard, *Anolis limifrons* (Rand *et al.*, 1983) seems to be a well-adapted parasite in a closely co-evolved host–parasite relationship with its host. At least in its chronic phase, malaria in *Anolis* behaves as a prudent parasite. However, Rand *et al.* (1983) were not able to rule out the possibility that initial stages of infection may be fatal to young animals.

Seasonal changes in available moisture affect food availability for many tropical vertebrates. Food shortage created by unusual rainfall patterns lowers recruitment in coatis (*Nasua narica*), agoutis (*Dasyprocta punctata*), pacas (*Cuniculus paca*), and squirrels (*Sciurus granatensis*), resulting in changes in the size and age structure of populations (Leigh *et al.*, 1982). Reproductive success may also be determined by year-to-year variability in climatic conditions. For *A. limifrons,* at one site on BCI the survival rate of adults is constant (74% per 28 days) regardless of sex or time of year (Andrews and Rand, 1982). The seasonal pattern of rainfall affects the time when eggs are laid, as well as their number and probability of survival. Population fluctuations among years are high, apparently because of variation in recruitment caused by dry seasons of varying length. In contrast, at another site on BCI, Andrews *et al.* (1983) found lower food intake and slower growth rates, suggesting at least occasional periods of food limitation in some habitats.

Earlier, Sexton *et al.* (1972) argued in favor of the hypothesis that populations were limited in the dry season by insufficient soil moisture for eggs to complete development successfully. Similarity in food in the stomachs of lizards in both wet and dry seasons led them to that conclusion.

In contrast, Wright (1979) argued in favor of food limitation for the population during dry periods. He cited seasonal changes in arthropod abundance and the fact that the results of Sexton *et al.* (1972) came from a dry season that was unusually wet. Wright felt that food was likely to be limiting in drier years.

The long-term studies of Andrews *et al.* (1983) suggest Wright's hypothesis that food is limiting during the dry season. However, year-to-year variation in food intake by females was not closely associated with annual recruitment variation. Comparisons between their two sites on BCI showed that populations were more food limited at one area than another. Overall, selection pressures in the egg environment (soil–leaf litter interface) and the lizard environment were quite different in space and time. This heterogeneity produced both spatial and temporal variation in lizard densities that defy simple univariate hypotheses. Disturbances in space and time seem to be instrumental in generating this complex pattern.

In another example, the timing and duration of recruitment for two anurans on BCI seem to be most influenced by suitable water levels for their aquatic tadpoles (Toft *et al.*, 1982).

Historically, studies in tropical forest ecosystems concentrated on problems that could be investigated in short periods. More recently, studies on seasonal and annual cycles have expanded knowledge of the role of disturbance in tropical forests. Both the scale of disturbance and the scale of investigation are important as determinants of results in a field study. When data from the bird studies of Karr and Freemark (1983) are examined by lumping data from all sites for each sample period, variability in the number of species in standard samples is low (mean number of species = 53.9; CV = 4%). However, analysis by study plot yields much higher variation ($\bar{X}$ = 29.5 − 32.1; CV = 10 − 14%). In addition, among the study plots, the relative abundances and even identities of species change as a function of climatic conditions (Karr, 1981a). Many of these dynamics would be overlooked in a study focusing on only one spatial or temporal scale.

We expect further clarification of the role of many kinds of disturbance at various spatial and temporal scales as field biologists consider observations in the context of disturbance.

B. Freshwater Streams

Like the vertebrates of the tropical forest, fishes of freshwater streams illustrate the influence of disturbance on the biota. Life history attributes of fishes that are tied to disturbance are myriad: seasonal movement to escape drought (Lowe-Mc-Connell, 1975); migration to spawning areas with abundant food supplies and/or lower predation rates (Schlosser, 1982); differential growth rates and return of young fish to areas occupied by adults (Karr and Gorman, 1975; Schlosser, 1982); air-breathing capabilities in many tropical species (Graham *et al.*, 1977, 1978); and local extinction in temporally inhospitable habitats and recolonization through migration (Kushlan, 1974). Seasonal changes in temperature and flow volume triggered by variable rainfall seem to be the main physical disturbances stimulating these patterns.

Biological events that affect stream vertebrates include the seasonal pulse of leaves in autumn (McDowell and Fisher, 1976) that are subsequently colonized by bacteria and fungi (Kaushik and Hynes, 1971), and provide food for invertebrates and ultimately fish. This cascade of influences through trophic levels is common. However, the relation between food supply and vertebrates is not always clear-cut. Panama stream fishes often avoid foraging in areas where they are susceptible to predation even when food is abundant there (Power, 1984). Overall, terrestrial predators seemed to be more important than food availability in determining distributions of fish among habitats (Angermeier and Karr, 1983). However, trophic diversity of fish assemblages may be related to the reliability of available food resources (Angermeier and Karr, 1983).

The importance of interactions of several disturbances in affecting the fishes of headwater streams is illustrated by a 10-year study in Black Creek, Allen County, Indiana (Toth *et al.*, 1982). Local extinction of a small cyprinid, *Ericymba buccata*, refects the synergistic interplay between natural and man-induced disturbances. Prior to 1976, rainfall in the Black Creek watershed was sufficient to keep stream channels relatively free of silt, and *Ericymba* populations flourished despite extensive channel modifications. Thereafter, periods of climatic severity (low stream discharge due to summer drought and consecutive harsh winters) combined with man-induced habitat alterations (increased sediment loads following removal of riparian vegetation and habitat disturbance of newly channelized streams) resulted in heavy siltation and limited reproductive success for *Ericymba*. *Ericymba* populations declined from 25–30% of fish samples in 1973–1976 to less than 1% in 1978–1980 (Fig. 2). Thus, *Ericymba* was able to withstand man-induced habitat disturbances alone, but when these were coupled with severe climatic events, the species was incapable of maintaining a viable population.

But fishes of these streams have evolved with periodic disturbance. As a result, they recolonize streams rapidly following local extinction. Given the dynamics of these systems, a key problem facing ecologists is to distinguish between shifts in species composition attributable to alteration in biological interactions (e.g., competition) mediated by temporal variability in physical environment and those attributable to individualistic responses to environmental changes (Schlosser, 1985). This question is central to the ongoing controversy in ecology about the relative roles of physical disturbances versus biological interactions as the primary organizing forces in communities. Individualistic responses by species to a diversity of physical, chemical, and biological disturbances may be responsible for assemblage attributes. When sets of species are subjected to a disturbance regime, the historical baggage that each carries (physiology, morphology, behavior, etc.) produces nearly

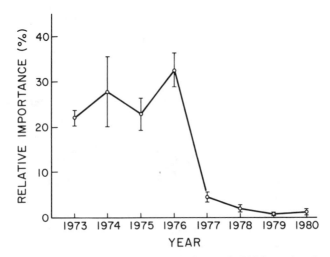

Fig. 2.   Mean proportional representation ($\pm$ 1 SE) of *Ericymba* in fish samples taken at stations on the main channel of Black Creek. (From Toth *et al.*, 1982.)

as many responses as there are species. Thus, competitive interactions may be an important determinant of niche structure in relatively few of the interactions in multispecies assemblages (Hutchinson, 1961).

The pervasive influence of disturbance and the complexity of disturbance-related attributes in biological systems argue for an individualistic perspective in many species' distributions.

A practical application of the study of the effects of disturbance on stream fishes is an Index of Biotic Integrity (Karr, 1981b) developed for use in the assessment of water resource quality. This ecologically based index, with 12 metrics in three major classes (species composition, trophic structure, and fish condition), provides a mechanism to evaluate man-induced and natural disturbance in running-water ecosystems (Fausch et al., 1984).

C. Freshwater Wetlands

Rainfall variability and associated surface and groundwater flows are also important in determining attributes of wetland ecosystems (Good et al., 1978). In freshwater marshes of the prairie pothole region of north-central North America, a "marsh habitat cycle" (Weller and Spatcher, 1965) is tied to water level fluctuations on time scales of 4–35 years. During low water years, muskrat populations decline as a consequence of emigration, increased vulnerability to predation (especially by mink), and lowered reproductive success (Weller, 1981). As water levels increase in later years, recolonization is often nonrandom, proceeding from long-established population centers within marshes (Errington, 1963). Muskrat populations increase as water levels rise and their primary foods (e.g., bulrushes, cattails) spread through the marsh basin. Muskrats open areas of marsh as they feed on emergent plants, producing an interspersion of open water and emergent plants. This interspersion of cover and water is ideal for nesting by many marsh birds (Weller, 1981). Salamanders and fish that require several years to mature are associated only with more permanent marshes and ponds. They may expand the number of marshes occupied during prolonged wet periods, but their distribution shrinks during prolonged droughts. During regional droughts in the prairie pothole region, birds that normally breed in that area move farther north into the tundra (Sanderson and Bellrose, 1969).

On a time scale of 100 years, observations would show 3 to 10 cycles of drought and high water levels with appropriate changes in vegetation, birds, muskrats, and mink. This short-term cycle is repeated numerous times during the general successional process in which marshes are considered a transient stage in the progression from lake to terrestrial environments. Yet most freshwater marshes in north-central North America are of glacial origin and are thus thousands of years old; their rate of filling may be very slow, and some processes (wave and ice action) may even tend to deepen them (Weller and Spatcher, 1965). The ecology of wetlands can be examined carefully by considering the cyclical periodicities of the annual cycle, the erratic periodicity of rainfall among years, and perhaps the directional pattern as marsh basins silt in and emerge as terrestrial habitats.

TABLE 1

Factors to Be Evaluated in the Study of Disturbance, Examples Involving Vertebrates, and Characteristic Type of Biotic Response

Factor	Example	Biotic response result
Type of disturbance		
Physical factor	Drought impact on distribution of birds and frogs in tropical forest	Behavior and ecology of habitat selection
Biological factor	Yellow fever decimates howler monkey population	Evolution of disease resistance, extinction of host, or cyclic abundance changes in monkey
Interaction of physical and biological factors	Spring rain and bird distribution impact fishes in Everglades	Growth rates, survivorship, and reproduction of fish
Regime of disturbance		
Spatial dimension (i.e., scale)	Treefall vs. hurricane distribution of large area of forest	Time to recolonization and assemblage of vertebrates will vary
Temporal dimension		
Frequency	Annual vs. irregular, severe dry season	Physiological adaptation vs. behavioral ecological and evolutionary responses
Time of occurrence	Food and sedimentation during spawning period of fish	Destruction of cohort
Type of biological system		
Individual–species–population	Birds vs. lizards vary in mobility, longevity, and ability to turn off reproduction on land bridge island	Bird extinction rates high, whereas lizard extinction rates low on Barro Colorado Island, Panama
Assemblage/ecosystem environment type—biotic and abiotic	Fish more constrained by the chemistry of their aqueous environment than birds are by chemical toxicity of air	Different suite of physiological and behavioral adaptations. But air pollution from human society alters this balance.
Regional context		
within area	Size of area and nature of the internal habitat mosaic	Survivorship, colonization, extinction pattern varies among islands
between area and adjacent regions	Forest habitat islands in ocean vs. in agricultural area	Varying extinction and colonization dynamics among vertebrate groups. Differing effects of colonists from adjacent patch of grass vs. water

V. STUDY OF VERTEBRATE ADAPTATIONS IN THE CONTEXT OF DISTURBANCE

Ecologists normally view irregular, erratic, or unusual events that force vertebrates away from a static, near-equilibrium condition as disturbances. Floods, windstorms, and epidemics are easily conceived examples of single-event disturbances, as are very severe wet or dry periods. Similarly, cyclical biological events such as the seasonal influx of large mammals on the Serengeti Plains of Africa (McNaughton, 1983) are disturbances to natural vegetation and, presumably, to the many other organisms associated with the habitat.

Vertebrates regularly experience a continuum of major and minor disturbances that alter them, and the objective definition of a threshold at which a periodicity becomes a disturbance is difficult at best. Indeed, disturbance thresholds certainly vary among organisms in the same environment (e.g., the everglades environment already discussed). Thus, to gain insight about the role of disturbance in the evolution of vertebrates, one should perceive disturbance, as generally conceived, as the extreme on a continuum of perturbations that affect vertebrates.

With this perspective, major disturbance can be integrated with more minor (less extensive or less intensive) perturbations. Our goal is to see the study of disturbance, in the classical sense of large disturbance, evolve into the study of the full range of perturbations that influence vertebrates. (We favor discussion and study of the ecology of ''perturbations'' rather than ''disturbance.'' The latter has a connotation of extreme, which we feel should be avoided. However, we have used disturbance throughout this chapter to conform with more conventional usage in this volume.)

Studies of disturbances and their consequences for vertebrates can be approached with two fundamentally different approaches. First, one might measure and evaluate the effect of an observed disturbance. Alternatively, one might examine the adaptations of vertebrates and interpret them in the context of past disturbance regimes. In either case, to obtain useful data from a study, disturbances of consequence to vertebrates must be measured on a scale that reflects the study organism's perception or response to the perturbation.

Knowledge about four sets of factors are required to effectively evaluate the influence of disturbance: type of disturbance; regime of disturbance; type of biological system; and regional context (Table 1). All should be evaluated in efforts to interpret the type of response (Table 1) stimulated by a disturbance.

A. Type of Disturbance

Both physical and biological events may have an impact on vertebrates. Further, physical and biological events may interact to alter biological systems. For example, leaf morphology of arrowleaf changes with water depth (Wallace and Srb, 1964), and the resulting change in structure of the physical habitat could affect fish use of the area.

B. Regime of Disturbance

Effective evaluation of the impact of a disturbance requires careful documentation of temporal and spatial scale of that disturbance. Size of area impacted should be evaluated as should frequency and time of occurrence of the disturbance. A disturbance may be a single event or may occur cyclically, erratically, or with directional change through time. Daily, monthly, seasonal, and annual variations in environmental conditions are examples of cyclical patterns. However, superimposed on these periodic temporal patterns are erratic changes associated with variability from one time period to the next. In Panama, two periods (dry season and early wet season) have predictable rainfall (Rand and Rand, 1982), whereas rainfall is least predictable at other times (start and end of wet season). This presents a challenge to frogs evaluating whether late dry-season rains actually signal the onset of the rainy season and, thus, a high chance of successful reproduction.

Long-term, directional change also occurs, such as the temperature increase associated with the recession of glaciers in the Pliestocene. But when considered over hundreds of thousands of years, this phenomenon appears to be cyclical, with successive ice ages punctuated by interglacials. Impacts of such events on vertebrates include progressive movement of geographic ranges and even rapid bouts of speciation such as in warblers (Parulidae) in the Pleistocene.

In contrast to frequency of disturbance, time of occurrence relates to the specific time during which a disturbance occurs. The impact on fish of a flood carrying heavy sediment concentrations may vary, depending on the time of the flood's occurrence relative to the spawning cycle. Winds at night may have less direct impact on a vertebrate that spends the night in a burrow than on one active at night.

Intensity of disturbance must also be evaluated. Intensity designates magnitude as measured in some standard unit of measure (e.g., temperature change of 2 versus 10°C). In addition, it is important to distinguish between severity and intensity. We noted already that intensity relates to the magnitude of the change caused by the disturbance, but changes of similar magnitude (e.g., 3°C) may have different impacts on different fish assemblages. A change of 3°C from 2–5°C might have little effect on cold water fishes with a tolerance range of 2–8°C, but a shift from 7 to 10°C would be catastrophic. Similarly, a desert lizard may easily contend with a decrease in humidity of 25% from 50%, but a 20% decrease in a tropical forest lizard from 90 to 70% might be a great deal more severe (Janzen, 1967). Thus, the severity of a disturbance of a certain intensity is a function of the magnitude of the stress placed on the organism by that disturbance.

C. Type of Biological System

Attributes of the biological system under study also affect the specific results of a disturbance. At the species level, taxon and associated life history attributes (e.g., size, mobility, generation time, sex, and age) affect responses to a specific disturbance.

At community and ecosystem levels, responses vary with the type of environment: terrestrial versus aquatic, marine versus fresh water, forest versus grassland, and many others. Thresholds of response, lag times to response, and many other system attributes may change among vertebrates of different environments subjected to the same disturbance.

D. Regional Context

Beginning with the pioneering work of MacArthur and Wilson (1967) and continuing with numerous recent advances (see, e.g., Pickett and Thompson, 1978; Foster, 1980; Forman and Godron, 1981) in regional or landscape ecology, biologists have begun to explore the role of regional pattern in determining both static and dynamic aspects of ecology (Forman, 1983). Availability of colonists and stepping-stone effects are obvious examples. The mosaic of physical and biotic conditions within a study region, as well as conditions in adjacent areas or regions, are important in evaluating the regional context of a disturbance.

E. Type of Response

Only after accumulating information on these four aspects of disturbance can one begin to understand and to identify the nature of the response by vertebrates. In response to relatively minor disturbances, organisms may alter their physiology or behavior for short periods; physiological responses may include altered enzyme activity, and behavioral responses may involve changes in posture or sociality or movement to another area. As a disturbance increases in severity, behavior may be altered (e.g., through postural changes at the same site or movement to another area). In addition, behavioral changes may result in alteration of social systems (e.g., clumping of animals to avoid the effects of very low temperatures). These alterations may be short-term and short-distance or longer on either spatial or temporal scales, depending on the organisms' capabilities and the magnitude and duration of the disturbance. An excellent example of an evolutionary response to a climatological disturbance is short-term shifts in morphology of Darwin's finches as a result of changing rainfall pattern (Boag and Grant, 1981). Eventually, if disturbance is frequent enough, responses may evolve from facultative to obligate, with alteration of the genetic structure of the population.

F. Overview

A systematic and comprehensive approach to the study of disturbance is essential, given its primary role and the inherent complexity of disturbance types and impacts. Without such an approach, ecologists will continue to catalog disturbance patterns and biological responses, with little hope of comprehension emerging from the chaos.

VI. SYNTHESIS

Environmental shifts, whether natural or man-created, involve disturbances to which biological systems may respond. The disturbance caused by seed-caching behavior in agoutis (Smythe, 1970) is no less a disturbance than is the digging behavior of a badger as it constructs a burrow or searches for food, or the response of stream fish to channel alterations by man. Because of the arbitrary way of defining thresholds that are unusual, and thus a disturbance, we believe that all factors that cause biological systems to respond should be considered as a portion of the disturbance continuum.

Rainfall (and the runoff that it produces) appears to be the principal extrinsic variable driving all of the biological systems that we have examined. The same conclusion obtains in the grassland and shrub steppe communities of western North America (Wiens, Chapter 10, this volume). The most obvious exceptions to this might be oceans and large lakes, where the buffering capacity of large water volumes may dampen seasonal and year-to-year variation in rainfall. This generality may apply only in regions with extended growth periods in the medium temperature range (a broadly defined biological zone of thermoneutrality). At higher elevations or latitudes, the relative impact of temperature may increase.

The importance of rainfall and temperature regimes to biological systems is clear from concepts such as biomes and life zones. Historically, pattern analysis on broad biogeographic scales resulted in useful systems of classification and insights into the specific physical factors responsible for attributes of biological systems. Analyses of discontinuities at this scale circumstantially implicate a disturbance (e.g., seasonality) in creating patterns.

To some extent, the development of generalities at this scale has slowed. Although the need to expand studies at finer (smaller) scales is clear (Karr and Freemark, 1983; Wiens, 1983; Rice *et al.*, 1983), it would be an error for studies at finer scales to dominate future research (Karr, 1983). Disturbance clearly has impacts at different spatial and temporal scales, and the greatest insight should develop following the study and integration of results at a hierarchy of levels (Allen and Starr, 1982). With that approach, the potential for understanding pattern and process in biological systems seems high.

Several examples illustrate the need for a hierarchical approach and integration at a variety of spatial and temporal scales. Although Tennessee warblers (*Vermivora peregrina*) feed primarily on insects in their temperate breeding areas, they feed to a considerable extent on nectar and fruit in their tropical wintering grounds (Morton, 1980). A focus on fine-scale pattern in their temperate breeding grounds would fail to identify the adaptive nature of their tongue morphology. Analogs of this spatial scale pattern can be seen in the impact of drought on BCI (Leigh *et al.*, 1982) or the lack of a dry period in the Everglades (Kushlan, 1979). Finally, the changing migration patterns of East African elephants with more intensive land use by human society requires integration at several spatial and temporal scales for comprehension of the biological pattern and conservation of biological resources.

These dynamics often involve a dichotomy between breeding and nonbreeding areas. Historically, biologists have emphasized ecological and evolutionary processes during the breeding season (but see Fretwell, 1969, on winter mortality), especially in the context of limiting resources. Some biologists suggest that in reality resources are hardly ever more than locally limiting (Taylor and Taylor, 1979). Rather, they feel that the first premise of natural selection is not maximized fecundity, as classical theory assumes, but the survival of the individual. Surely this view would not deny the importance of maximized fecundity integrated over the life of the organism. Rather, we believe that the comment should be directed toward the concentration of most studies during the relatively short breeding season. Too frequently, reproductive efforts are considered without evaluation of an organism's ability to do those things that ensure its survival until it can reproduce. That is, an important product of evolution is the ability to behave in ways that avoid premature mortality. For vertebrates, a key component of that must be a decision on how to move among the patches in their world in ways that maximize survival potential. That survival potential must be enhanced by an ability to react appropriately (i.e., in regard to survival and reproduction) to a wide variety of disturbances.

As human populations and their demands for resources increase, the consequence of natural and man-induced disturbances will be of increasing concern to human society. For numerous reasons (Ehrlich and Ehrlich, 1981), protection of the world's biota is essential. Such actions must also protect the processes that resulted in the evolution and persistence of that biota. Disturbance regimes, such as the water-level dynamics of the Everglades, must be protected to preserve associated genetic (Frankel and Soulé, 1981), population (Franklin, 1980), and assemblage (Kushlan, 1979; Karr, 1982a,b) dynamics.

In this chapter, we have dealt explicitly with some of the conceptual issues central to consideration of the role of disturbance as ecological and evolutionary phenomena. Such a discussion is essential to an adequate description of disturbance impact and, more importantly, to the formulation of testable hypotheses and the design of appropriate sampling and analysis programs to study disturbance. We emphasize that the task is not complete with this chapter. We offer it as an early guide in defining more clearly the path ahead.

Finally, we want to note our agreement with McNaughton's (1983) view that the proximate mechanism regulating the abundance of species results from many weak forces acting probabilistically. The cumulative effects are large but the individual effects are minor, interactive, and uncertain. Spatial and temporal heterogeneity in species attributes and population characteristics (distribution, etc.), although a bane to early ecologists as they searched for communities that were homogeneous in space and time, may be particularly important contributors to the integrity and continuity of ecological systems. Study of the ecology of disturbance may well tie together many of these more problematic areas of ecology.

ACKNOWLEDGMENTS

This chapter was developed during periods of research support from the U.S. Environmental Protection Agency (G005103, R806391, CR807677, and CR810745) and the National Science Foundation (DEB 82-06672) to JRK and a Natural Sciences and Engineering Research Council of Canada Scholarship to KEF. Robin Andrews, Thomas Martin, Roland Roth, Isaac Schlosser, Jared Verner, and Philip Yant commented constructively on an earlier draft of the chapter.

RECOMMENDED READINGS

Karr, J. R., and Freemark, K. E. (1983). Habitat selection and environmental gradients: Dynamics in the "stable" tropics. *Ecology* **64,** 1481–1494.
Leigh, E. G., Jr., Rand, A. S., and Windsor, D. M. (1982). "The Ecology of a Tropical Forest: Seasonal Rhythms and Long-Term Changes." Smithsonian Inst. Press, Washington, D.C.

Chapter **10**

Vertebrate Responses to Environmental Patchiness in Arid and Semiarid Ecosystems

JOHN A. WIENS

Department of Biology
University of New Mexico
Albuquerque, New Mexico

I. INTRODUCTION

Natural systems at any scale of resolution are mosaics of patches. Each patch follows its own pattern of development and dynamics, but this in turn is influenced by the nature of the mosaic itself—how the patches are arrayed in landscape patterns. Individual organisms and populations ebb and flow through such patch mosaics in varying manners, some spending entire lives or generations within single

THE ECOLOGY
OF NATURAL DISTURBANCE
AND PATCH DYNAMICS

169

patches, others using some patches for reproduction and others for feeding, and still others drifting over a mosaic in a random fashion but distinguishing between different mosaics. Communities and ecosystems contain populations exhibiting a diversity of such patch-use strategies.

Much of the theoretical foundation of ecology and evolutionary biology that has been built over the past several decades, however, has ignored this complexity. To simplify the theory (indeed, to make its development possible), nature has often been assumed to be spatially homogeneous. Such theory often fails when taken into nature. Schluter (1981), for example, demonstrated how a neglect of heterogeneity in optimal foraging models severely compromises their ability to predict actual diets of organisms, and the detailed graphic models of resource use developed by Tilman (1982) were extremely sensitive to spatial patchiness. To be sure, some recent efforts have incorporated the effects of heterogeneity into population or community theory (see, e.g., Roughgarden, 1978; Roff, 1978; McMurtie, 1978; Levin, 1978; Paine and Levin, 1981; de Jong, 1979; Weiner and Conte, 1981; May, 1981; Nisbet and Gurney, 1982). [Haggett et al. (1977) provide a valuable review of the use of spatial models in human geography.] Often, however, separate submodels are simply generated for individual patches, within and between which relationships are deterministic. As Chesson (1981) has shown, incorporation of local stochastic effects into spatially structured models may produce quite different results than are obtained from their deterministic analogs. There is a clear, continuing need for models that realistically consider the spatial dynamics of systems and predict the consequences of these dynamics (Noy-Meir, 1981). As den Boer (1981) has observed, "heterogeneity and variability should not be considered as just drawbacks of field situations, that can best be circumvented by retreating into the laboratory or even into deterministical mathematics. On the contrary, heterogeneity and changeability must be recognized as fundamental features, not only of the natural environment of a population but also of life itself."

The development of such theory, however, cannot be based on abstract, armchair notions of heterogeneity, but must incorporate basic knowledge of how patch dynamics are expressed in nature and how organisms and populations respond to them. Indeed, this is a major focus of this volume. My treatment in this chapter emphasizes selected aspects of patch effects on individuals, populations, and communities, with special reference to relatively large, mobile vertebrates living in arid and semiarid environments (deserts, shrubsteppe, and grasslands).

II. PATCHINESS IN ARID ENVIRONMENTS

Arid and semiarid environments are characterized by extreme variability in climate, topography, and soil, which acts to produce a clear but complex patch structure. The details of vegetation patch dynamics in these ecosystems have been discussed elsewhere in this volume (Loucks et al., Chapter 5; Christensen, Chapter 6), but a brief review is in order here. Because rainfall in these systems is scarce and generally variable (Wiens, 1974a), basic ecosystem processes such as production, consumption, and decomposition are closely linked to "pulses" of precipitation (Noy-Meir, 1973, 1974; Crawford and Gosz, 1982). Temporally, rainfall varies across a spectrum of scales, over centuries, periods of a few decades or years, annually, seasonally, and so on (see, e.g., Tyson, 1980). Coupled with this temporal heterogeneity is spatial variance. In Africa, for example, general movements of pressure systems back and forth across the equator produce distinct wet and dry seasons in the Serengeti. The timing and intensity of these seasons, however, are further modified by topography (Norton-Griffiths et al., 1975). At a more local scale, thunderstorms or topographic features may produce great differences in rainfall over short distances (see, e.g., Rotenberry et al., 1976). Because such local rainfall pulses occur at different times, an otherwise homogeneous landscape may become a mosaic of patches differing from one another in the time since the last rainfall or in the magnitude of that rainfall.

Topography and soil also vary tremendously in arid and semiarid environments, perhaps more so than in any other life zone (Crawford and Gosz, 1982). At a local scale, these landform variations may be of greater importance to the biota than rainfall variations through their influences on microclimatic characteristics, salinity, soil stability, and moisture penetration and runoff patterns (Ayyad, 1981).

Collectively, such complexes of climatic, edaphic, and topographic factors may have major influences upon the spatial distribution, composition, and productivity of the vegetation in an area. In Wyoming shrubsteppe, for example, variation in annual herbage production is associated with variations in fall precipitation, spring precipitation, summer potential evapotranspiration, percentage of coarse soil fragments, percentage of very fine sands, and percentage of grass and forb composition (Ries and Fisser, 1979). To the degree that these factors vary spatially, production will assume a patchy distribution as well. Moreover, patchiness in the distribution of vegetation in an environment may enhance spatial variance in the distribution of other environmental factors. Stark (1973), for example, demonstrated that water infiltration is 8–10 times faster under shrubs than in the open in desert soils that are high in clay, and Charley and West (1975) documented clear patterns in the distributions of chemicals in desert soils that are related to the occurrence of shrubs. To the degree that the shrubs in such environments exhibit a complex demographic age structure (see, e.g., Crisp, 1978), spatial variance in these feedback relationships will be further enhanced.

Superimposed upon these basic environment–vegetation patch dynamics are the effects of disturbances. In arid and semiarid systems there is no counterpart of the treefall disturbance that is so important in generating patchiness in forested habitats (see Runkle, Chapter 2, and Brokaw, Chapter 3, this volume). Instead, most disturbance of an existing environment–vegetation mosaic is through erosion by water or wind, fire, or the actions of animals in disturbing the soil [e.g., burrowing mammals (Platt, 1975; Loucks et al., Chapter 5, this volume)], clearing areas of vegetation and harvesting nutrients to a central place [e.g., ants (Wiens, 1976; Culver and Beattie, 1983)], or grazing by domestic herbivores (Crisp, 1978; Risser et al.,

1981). Grazing effects may vary greatly as a consequence of local or regional cultural practices or regulations. In general, however, they are likely to be relatively more severe in less arid environments (e.g., tallgrass or montane grasslands; Grant *et al.,* 1982), where grass and forage production are greater and there is thus more material to be removed by the grazers. In more arid zones, habitats dominated by woody shrubs are less likely to be altered by domestic grazing than are habitats containing less robust growth forms (Jones, 1981). The potential for grazing to produce greater localized between-patch variance is thus also greater in such arid shrublands.

Fire may also be a major disturbance, especially in grasslands, savannas, and shrubsteppe. There, fire often suppresses woody vegetation while enhancing grass production (Walker, 1981). Fires tend to increase vegetational homogeneity within the burned areas, but if they are localized (e.g., many lightning-caused range fires), the mosaic structure at a larger landscape scale may be quite heterogeneous, a consequence of the varied past history of fires in the area. Norton-Griffiths (1979) has reviewed the effects of fire on the Serengeti ecosystem, and Bock and Bock (1978) have noted some of the consequences of such disturbances on bird and mammal populations occupying desert grassland in Arizona.

Because production rates are low in most arid and semiarid systems, the recovery of vegetation from fire, grazing, or other disturbances is likely to be quite slow. Areas in the shrubsteppe that have been burned may not have regenerated shrub cover several decades later, and heavily grazed desert grasslands may show the effects long after cattle or sheep have been removed. Just as recovery is slow in these systems, so is any sort of community change. Many shrubs and cacti, once established, are long-lived and grow quite slowly, giving habitats considerable structural stability over time. Patch structure persists.

The overall result of this variation in environmental factors, vegetation responses, and disturbances is a complex spatial pattern of resources available to vertebrates and other consumers. Shrubs provide shelter, breeding sites, or display locations for birds in many deserts, and the availability of these resources varies as a consequence of patchiness in shrub dispersion. Production of seeds may be quite heterogeneous as a function of the distribution, age structure, and reproductive output of individual plants, and the effects of wind, water movement, topography, and animal vectors on seed dispersal may increase spatial variance even further. The outcome of these influences is the sort of extreme patchiness in seed distributions documented by Reichman and Oberstein (1977) for desert soils. Invertebrates also respond to the distribution of plants, often in a more specific manner than vertebrates (since a greater proportion of them are monophagous herbivores, often of limited mobility). The invertebrate food resources of vertebrates may thus also exhibit substantial spatial heterogeneity.

III. ENVIRONMENTAL PATCHINESS AND INDIVIDUALS

The existence of patchiness in an environment offers organisms the option of responding to the patch structure in a nonrandom fashion. Much of the evidence that

TABLE 1

Patterns of Patch Use by Foraging Sage Sparrows and Brewer's Sparrows in a Sagebrush-Dominated Shrubsteppe in Oregon[a]

Plot	Component	Random		Sage sparrow			Brewer's sparrow			Difference between species (P)
		$\bar{x}$	SD	$\bar{x}$	SD	p^b	$\bar{x}$	SD	p^b	
East	I	0.11	1.22	0.05	0.67	.64	0.02	0.59	.41	.61
	II	−0.03	0.87	−0.34	0.55	<.001***	−0.43	0.52	<.001***	.14
	III	−0.08	0.02	0.08	0.93	.15	0.05	0.81	.23	.78
	IV	−0.01	0.85	0.18	1.36	.14	−0.06	1.00	.72	.10
West	I	−0.11	0.70	0.27	0.84	<.001***	0.17	0.77	<.001***	.31
	II	0.03	1.12	−0.46	0.38	<.001***	−0.38	0.68	<.001***	.23
	III	0.08	0.97	0.23	0.69	.16	0.22	0.74	.17	.91
	IV	0.01	1.14	−0.17	0.70	.13	−0.13	0.67	.18	.67

[a]Means, standards deviations, and significance levels for factor scores for randomly sampled patches and patches used by sage sparrows and Brewer's sparrows for foraging in two Oregon shrubsteppe plots, as related to principal components of variation in patch attributes (see text).

[b]Test of difference between bird patches and random patches, using the t'-test.

patches are in fact used nonrandomly is intuitive or anecdotal, stemming from observations of apparent patch preferences or avoidances by organisms. To document such patterns more rigorously, patch use must be related to patch availability in the environment, as John Rotenberry and I have done (unpublished data) for birds in sagebrush (*Artemisia tridentata*)-dominated shrubsteppe in Oregon. During one summer we measured 11 attributes of vegetation patches sampled at random in two survey plots, and then used principal components analysis (PCA) to summarize the patterns of covariation among these variables. Overall, four components had eigenvalues greater than 1, cumulatively accounting for 73% of the total variation in the original data set. The first (PC I, 30% of the variation) was a shrub size component; PC II (20%) was a "compositional" component, contrasting large sagebrush to small green rabbitbrush (*Chrysothamnus viscidiflorus*); PC III (10%) was a patch shape component; and PC IV (10%) was another "compositional" component, contrasting sagebrush to gray rabbitbrush (*C. nauseousus*). We also documented the patterns of patch use by foraging sage sparrows (*Amphispiza belli*) and Brewer's sparrows (*Spizella breweri*) on these plots. The attributes of the patches used by the birds could then be related statistically to the randomly sampled patches, using the PC axes of variation.

Both of the bird species clearly differed from random in their use of patches for foraging (Table 1), showing high negative factor scores on PC II and, therefore, apparently strong preferences for using sagebrush-dominated patches rather than those composed of green rabbitbrush. We observed no significant discrimination between sagebrush and gray rabbitbrush patches (PC IV) on either plot, however, which suggests that the strong response on PC II is indeed quite real. The two plots were quite similar vegetationally, although the patches on the east plot were generally larger than those on the west plot (Table 1). The use of patches with respect to size (PC I) did not differ from random on the east plot, but on the west plot birds foraged in patches that were decidedly larger than expected on the basis of random encounters (Table 1). This suggests that in general the birds may select relatively large patches for foraging. Thus, in areas containing mostly large patches, their use is indiscriminate with respect to this patch attribute, but where smaller patches predominate, overall patch use is shifted toward the larger patches.

The patterns of nonrandom patch use that we observe in shrubsteppe birds may be related to food distribution, as sagebrush may support a quantitatively and qualitatively different arthropod fauna than green rabbitbrush by virtue of differences in structure or secondary chemical compounds contained in leaf tissues, or patches of different sizes may contain substantially different quantities of prey.

More generally, nonrandom patterns of patch use by individuals are likely to be related to (a) patch microclimate, (b) predation pressures, or (c) food availability or foraging efficiency.

A. Microclimate

To an individual, environmental patchiness may be important not merely in the physical configuration that is expressed but also in the profiles of microclimatic variation that it produces. Summer midday temperatures in the center of a desert shrub, for example, may be substantially lower than those on the exposed edges of the shrub or on the open ground a meter away, and humidity levels may exhibit similar variation in relation to patch structure. Such microclimatic heterogeneity provides individuals in arid and semiarid environments with options to regulate their thermal or water balance behaviorally by changing habitat positions. By doing so, overall energetic or water requirements may be reduced through a reduction in wind velocity, the radiation of thermal energy to the environment, the incoming short-wave solar radiation, or other environmental exchange pathways (Gates, 1980). Many desert lizards practice such tactics to perfection (see Huey and Slatkin, 1976). *Uta stansburiana* in Colorado adjusts its daily and seasonal use of microhabitats so as to maintain body temperatures within a limited range in which sprint speed (and thus predator avoidance) is maximal (Waldschmidt and Tracy, 1983), and Galapagos land iguanas (*Conolophus pallidus*) alter their use of hot cliff-face micro-habitats daily and seasonally in such a way that they are able to maintain a constant body temperature for a maximum length of time (Christian *et al.*, 1983; see also Porter *et al.*, 1973; Porter and James, 1979; Bowker and Johnson, 1980).

Birds and mammals exhibit similar responses to the thermal profiles of their

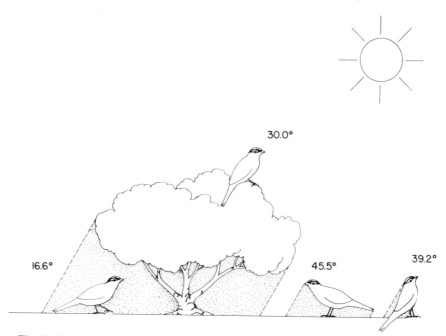

Fig. 1. Representative equivalent temperatures (T_e, °C) for a white-crowned sparrow in alpine breeding habitat in Colorado in early June. Time: 1400; $T_A = 20$°C. The bird in the shade on the ground experiences the lowest T_e; the other two birds on the ground are in full sun but experience different T_e's due only to their orientation to the sun. The bird on the right is oriented so that minimum effective area is exposed, while the bird to its left demonstrates maximum exposed area. The bird perched at 1 m demonstrates the effects of the greater wind speed at that height in lowering T_e; in all other respects, this bird is comparable to the ground bird on the right. (From Mahoney, 1976.)

habitats (see Lustick, 1983; Mugaas and King, 1981; Martindale, 1983). Mahoney's (1976) studies of white-crowned sparrows (*Zonotrichia leucophrys*) provide an especially clear example of the energetic consequences of patch use. Using estimates of equivalent black-body temperature, T_e (which integrates animal surface properties and the radiative and convective properties of the environment into a single temperature; Mahoney and King, 1977), she calculated the effective thermal environment of birds occupying various positions in the environment. Thus, a bird breeding in a subalpine habitat in Colorado in early June might have available the thermal options shown in Fig. 1. The upper boundary of this species' thermoneutral zone is 37°C; under these conditions, occupancy of either position on the ground in the open could impose heat stress. On the basis of the diurnal thermal regime for a bird breeding in an Oregon shrubsteppe habitat in early July (Fig. 2), one might expect that activity early in the morning would be concentrated on the ground and in exposed locations, while after 0800 individuals would increasingly confine their activities to locations in or beneath shrubs or other sheltering patches. Different species with different shapes, surface properties, or sizes would likely exhibit different T_e profiles in this habitat, and thus have the potential to respond to patch microclimates in different ways.

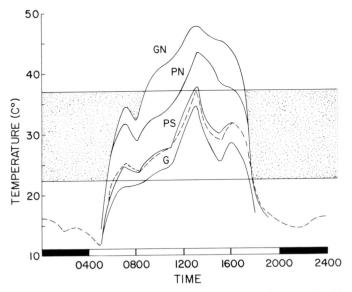

Fig. 2. Diurnal variation in air temperature (dashed line) and T_e for four typical positions occupied by a White-crowned Sparrow in its shrubsteppe habitat in Oregon in early July. GN = on the ground in full sun; PN = perched 1 m above the ground in full sun; PS = perched 1 m above the ground in shade but open to sky long-wave radiation; G = on the ground in shade open to sky long-wave radiation. The solid bar at the base of the graph indicates nighttime. The stippled area indicates the thermoneutral zone. (From Mahoney, 1976.)

B. Predation

There is less direct evidence of the role of predation pressures in determining patterns of nonrandom patch use by individuals, primarily because of the difficulty of documenting predation risk in nature. Predation on vertebrates is rarely witnessed in most systems, yet there is considerable intuitive appeal in the notion that risk should increase with habitat openness. Patches such as shrubs in deserts or tall grass in prairies should be expected to provide cover from predators, leading to an increased use of such patches in environments in which predation pressures are important. Certainly many lizards and birds flee into such cover when a potential predator (e.g., falcon, human) appears. Fuentes and Cancino (1979) and Pulliam and Mills (1977) have documented how different species use microhabitats differentially in relation to predation risk, often maintaining characteristic maximum distances from patches of protective cover while foraging in exposed situations. Group size and the amount of time spent by individuals in scanning for possible predators may also vary as functions of cover, although this relationship is likely to differ for organisms with differing food habits, body size, and/or speed or crypticity (as Jarman and Jarman, 1979, have elaborated for Serengeti herbivores). Features of the predators themselves may also influence prey use of cover. Bertram (1979), for example, noted the differences in prey distance thresholds at which several Serengeti predators initiate their chases (e.g., 5–20 m for leopards, 50–200 m for wild dogs). Herbivores subject to predation by species differing in their chase distances or search tactics might be expected to maintain different relationships to habitat patches. From observations such as these, one might predict variations in the degree of social aggregation, the patterns of habitat occupancy, or the budgeting of individual activity for various microhabitat components as either predation level or the degree of habitat patchiness varies. The development of such notions into more formal predictions, however, requires more extensive information on natural systems than is currently available for most systems (but see Jarman and Jarman, 1979).

C. Foraging

The importance of habitat patchiness to foraging individuals is much more apparent than its role in reducing predation risk, probably because foraging can be so much more easily observed. There is little doubt that foraging is generally patch related, primarily because food resources are almost always unevenly distributed, creating resource "hot" and "cold" spots (Pleasants and Zimmerman, 1979; Wiens, 1984). From his studies of diet choice in *Geospiza* finches, for example, Schluter (1982) concluded that the choice of patches in which to forage was probably of greater importance than the choice of specific food items. Such patch choice is often expressed as an area-restricted search in which an individual confines its searching and feeding behavior to a particular area after initially finding food there (Croze, 1970). This behavior, in turn, leads an individual to spend some time within a patch before giving up and continuing its broader foraging movements.

A large body of theory has developed over the last 2 decades dealing with aspects of such foraging patterns (for reviews, see Kamil and Sargent, 1981; Pyke *et al.*, 1977; Krebs *et al.*, 1983). The presumption of most models is not simply that foraging patch use is nonrandom, but that it is optimal with regard to some currency and constraints (usually maximization of energy intake per unit time). The models thus predict how an individual *should* forage optimally. Initial models, such as those of MacArthur and Pianka (1966) and Royama (1970), predicted that individuals should concentrate their activities in the most profitable patches and that the range of patch use should expand as overall food abundance declines. Charnov (1976; Krebs *et al.*, 1974) specifically considered the question of how long individuals should remain foraging once the patch has been selected. They concluded from theoretical arguments that a foraging individual should maximize its net energy gain by leaving a patch when its instantaneous rate of net energy intake falls to the average rate of net energy intake for the habitat as a whole (the "marginal value theorem"). The giving-up time for patches in a given habitat should thus be a constant. The theory predicts that initially all patches more productive than the habitat average should be visited, but because some patches are only slightly more productive than average and will rapidly be depleted, foraging should become progressively concentrated into fewer and fewer patches until all have been reduced to the average productivity of the habitat as a whole. Moreover, in habitats that are of generally poor quality (low average net energy gain), an individual should remain longer in a patch of a given quality than it should in a more productive habitat (since it will take longer to reduce this patch to the lower average value for the habitat). These are intriguing expectations, and initial tests (mostly on arthropods or on birds in laboratory settings) have provided some support (reviewed in McNair, 1982).

As is often the case, however, the real world is much more complex than portrayed in such models. For example, McNair (1982) has noted that the giving-up time is likely to vary for patches within a habitat. His model suggests that the average net energy gain per visit and the patch residence time will be greater in better patches, and that both of these variables will increase as the patches become more widely spaced. On our study plots in Oregon shrubsteppe habitat, sage sparrows (but not Brewer's sparrows) showed a significant relationship between time spent in patches and patch size (a possible indicator of quality) on the West plot but not on the East plot, where the patches were, on average, larger. Thus, increasing patch size was associated with the time sage sparrows spent foraging in the patch up to a point; beyond that point, visit time became largely independent of patch size. Patch residency time may not only vary between patches, but may do so in a distinctly nonlinear manner.

Optimal patch foraging theory generally presumes that the patches are discrete and arrayed against a neutral background matrix, and that they differ from one another only quantitatively, not qualitatively. This may be so for some organisms in some environments (e.g., cactus feeders in a desert), but often the matrix itself contains resources, and foraging individuals are therefore able to forage while traveling between patches: the "matrix," in a sense, becomes a patch itself. If patches are indeed discrete, the ability of organisms to discriminate between patches of differing quality may depend upon their movement patterns and the distance separating the patches. Relatively sedentary individuals or species might fail to discriminate patches of differing quality if the patches are very far apart (since direct comparisons would be difficult), while more mobile organisms might discriminate between the same patches with little difficulty but fail to differentiate still more widely separated patches (Fig. 3). Different species occupying the same habitat may thus have strikingly different patterns of patch use, as may individuals of the same species occupying different habitats. Kareiva's (1982) experiments with flea beetles feeding on collards provide some support for this view.

The resources contained in patches generally have their own dynamics, and the quality of a patch for a foraging individual is therefore a time-dependent variable. In particular, the pattern of patch use by foragers may depend upon the renewal rates of patch resources (e.g., floral nectar for nectarivores; Gill and Wolf, 1977). For example, experiments with hummingbirds indicate that the ability of individuals to learn feeding patch locations is influenced by renewal rates in a way that promotes the visitation of new feeding locations (Cole *et al.*, 1982). Whether or not resource renewal rates are important, of course, depends upon whether patch foraging appreciably depletes resources to the point at which resource limitation may occur (Wiens, 1984). Optimality-based theory presumes limitation (Calow and Townsend, 1981) and does not address situations in which resources may be abundant. Under such conditions, some of the constraints on patch use are presumably removed, and ''adequate'' rather than optimal foraging may be all that is required (Wiens, 1977, 1984; Schluter, 1982). To the extent that natural environments contain superabundant resources or that resource levels fluctuate between supera-

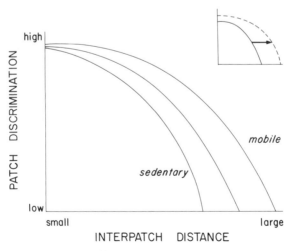

Fig. 3.   The relationship between the distance separating patches and the ability of organisms with differing degrees of mobility to discriminate among them. The inset shows the effect of increasing the degree of difference between patches for a given mobility level.

bundance and limitation, the predictions (indeed, the applicability) of optimal patch use theory may be limited.

Many of the problems with optimal patch foraging theory are consequences of the simplicity of the theory and, therefore, its neglect of the realities of nature. Some idea of the importance of the details of individual foraging tactics may be gleaned from the studies of Holmes and Robinson (1981; Robinson and Holmes, 1982) on hardwood forest birds. There, different species differed in their foraging movements and searching tactics, which in turn determined how large an area might be searched at a given time and how far a bird might move before searching again. Both foliage structure and the distribution and abundance of prey of various types influenced foraging options in important ways. Out studies of birds in shrubsteppe habitats have revealed similar variations in foraging movements and tactics of individuals, both between and within species, and point to the importance of such behaviors in determining the details of patch use (J. A. Wiens, J. T. Rotenberry, and B. Van Horne, unpublished data).

D. A Case Study: Heteromyid Rodents in Deserts

The various heteromyid rodents of the deserts of the American Southwest provide an instructive example of the multiplicity and complexity of factors that may affect patch use by individuals, especially inasmuch as they are a much-studied (and much-discussed) group. They feed primarily upon seeds, a resource whose dynamics and distribution can be quantified more readily than many (Harper, 1977; Rabinowitz, 1981; Brown et al., 1979; Reichman and Oberstein, 1977; Pulliam and Brand, 1975), albeit with difficulty. Moreover, rodents may at times significantly reduce the abundance of seeds in their habitats (Reichman, 1979) or may account for a major portion of the overall energy flow from this resource (Ludwig and Whitford, 1981). The microhabitat of these rodents is also characterized by a clear patch structure produced by the desert shrubs. As a consequence, patch response can occur with respect to either or both of the key resources: shrubs and seeds.

Brown and Lieberman (1973) initially suggested that several of the rodent species exhibited clear preferences for seeds of particular types or sizes, and Brown (1975) related these to the body sizes of the mice (but see Lemen, 1978). Both Mares and Williams (1977) and Lawhon and Hafner (1981) explored aspects of seed type selection in laboratory settings. More attention, however, has been focused on the responses of the animals to the dispersion patterns of seeds. Laboratory experiments (Reichman and Oberstein, 1977; Price, 1978; Hutto, 1978) and field observations (Bowers, 1982; Thompson, 1982a) suggest that some species of *Dipodomys* select and preferentially forage upon clumped seeds, while *Perognathus* species occurring in the same habitat may forage more efficiently upon scattered seeds (but see Frye and Rosenzweig, 1980; Trombulak and Kenagy, 1980).

These rodents also appear to respond strongly to microhabitat configuration (Rosenzweig et al., 1975; Brown, 1975; Reichman, 1975); *Perognathus* generally concentrates its activities in close proximity to shrubs, while *Dipodomys* forages more widely in open areas between shrubs. Rosenzweig and others have related these patterns to predation risk, which may be greater for the smaller *Perognathus* that are neither morphologically nor behaviorally adapted to the use of more risky open spaces, at least in comparison with *Dipodomys* (Thompson, 1982b). Working with several Namib Desert rodent species, however, Christian (1980) demonstrated that patterns of microhabitat patch use were influenced strongly by water availability, and M'Closkey (1981) related heteromyid patch use to population density. There may also be important interactions between seed selection and microhabitat use, as seed distributions are related to microhabitat (Reichman and Oberstein, 1977). M'Closkey (1980), for example, suggested that the use of seed types by several species was a consequence of their patterns of microhabitat selection, while Hay and Fuller (1981) related the use of seed types to both their preference ranking for the rodents and their distance from shrub cover. Clearly, several factors would seem to interact in determining features of microhabitat use in heteromyids.

Unfortunately, the observations on which these conclusions regarding seed- or microhabitat-patch use are based have generally been indirect, based upon tabulations of cheek pouch contents (seeds) or measures of habitat features in the vicinity of traps in which individuals were captured (microhabitat). Observations of individuals tagged with beta lights (Bowers, 1982; Thompson, 1982a) provide more direct evidence of patch use patterns, although the findings are contradictory. Bowers found that *Dipodomys* moved over greater distances and collected a smaller variety of seed types than *Perognathus,* which foraged closer to shrubs on scattered seeds. He suggested that the microhabitats differed in the extent to which they contained seed clumps, and that the basic preferences for clumped versus dispersed seeds determined the patterns of microhabitat use. While Bowers' observations of microhabitat use were direct, his information on seed use was derived from cheek pouch samples gathered some distance away over a 3-day period, and is thus of questionable value. Thompson's more intensive studies documented that all of the species concentrated their foraging efforts beneath or near shrub canopies, although *Dipodomys* traversed the open intershrub areas more often than *Perognathus.* Thompson related these patterns to predation pressures, concluding that "shrubs represent a critical resource for heteromyids since they offer relatively high concentrations of food and appear to be areas of relatively low predation risk." The findings of Kaufman and Kaufman (1982), that *Dipodomys* activity is reduced on moonlit nights, when the animals seem to express a greater preference for shaded microhabitats, lends support to the notion that predation has been an important factor influencing patterns of patch use in these rodents. In any case, it is apparent that heteromyids respond strongly to the patch structure of both their microhabitats and their food supplies, even though the relative importance of these factors remains a topic of vigorous discussion.

IV. PATCHES AND POPULATIONS

Perhaps the most evident features of populations are their fluctuations in size through time. How are the temporal dynamics of populations related to spatial patchiness? Several theoretical models have addressed the dynamics of populations in heterogeneous environments (see Levin, 1978; McMurtie, 1978; Chesson, 1981; Nisbet and Gurney, 1982), generally predicting that spatial heterogeneity should enhance population persistence and stability. The linkage with stability, however, is perhaps more tenuous than that with persistence, especially on a large scale. Rosenzweig and Abramsky (1980), for example, have speculated that vole cycles might result from the erosive effects of immigration of individuals from other local populations on the adaptation by voles in a specific local population to chemical compounds in the plants encountered there. The breakdown of precise local adaptation to plant cues might then lead to reproduction at inopportune times, overloading the food supply and thus triggering a population decline. In this scenario (for which there is little direct evidence), environmental patchiness drives the population fluctuations. Wolff (1980) suggested that snowshoe hare (*Lepus americanus*) populations may be cyclic in the north because patches of suitable habitat provide refuges from obligate predators during the population crash, and thus protect the population from local extinction. The hares then move out of these patches into other nearby habitat types during the population increase phase of the cycle. In more southerly locations, the hares do not cycle. Wolff proposed that the wide spacing of suitable habitat patches prevents population buildup because dispersal is prevented by habitat discontinuities. In addition, the predators in southerly locations are facultative and have a greater diversity of lagomorph prey species available; they are thus less likely to cause a dramatic decline in hare numbers in any local area. Hansson (1979) has suggested a similar role for landscape patch interspersion in cyclic northern populations of *Microtus*.

Both of these cases are supportive of den Boer's (1968, 1981) concept of "spreading of risk." Den Boer suggested that a large population that is subdivided into many subpopulations distributed over different patches in a heterogeneous landscape will be more persistent and less likely to suffer extinction than a similar population occupying a homogeneous environment. While populations in local patches may suffer extinction, the separation and asynchrony of the dynamics of other subpopulations will buffer the population as a whole, and such populations will provide the sources for subsequent recolonization of the empty patches.

As Andrewartha and Birch (1954) observed 3 decades ago, distribution is the spatial aspect of abundance. Variations in local population dynamics in different places give rise to complex patterns of variation and patchiness in the spatial (and temporal) distribution of a species. Taylor and colleagues (Taylor and Taylor, 1977; Taylor et al., 1978, 1980; Taylor and Woiwod, 1980, 1982) have emphasized the dynamic nature of spatial distributions of densities within populations, suggesting that density-dependent controls on the movement behavior of individuals (attraction to suitable habitat patches balanced by a tendency to avoid high-density areas) may produce characteristic patterns of spatial variance for each species over its range. The "power law" that has emerged from these studies has been criticized for ignoring the influences of chance and environmental stochasticity, for basing its mechanism on features of optimal foraging, and for marshalling evidence from broad geographic areas to support the inferred operation of a process at a much more local scale (Anderson *et al.*, 1982; Hanski, 1980, 1982a). Hanski (1982b) has offered a somewhat different view, proposing that there is a bimodality to the frequency distribution of species among locations in a region, most of the species occurring either at very few of the sites or at most of them. This pattern is related to population abundance. The high-frequency species (''core'' species) are also those that are abundant where they occur, while the low-frequency species (''satellite'' species) are generally uncommon as well. It is unclear to what degree such patterns may be biologically real versus artifactual, however, since populations that are rare may inevitably also be spatially clumped (Cornell, 1982) or subject to greater sampling inaccuracy that produces the same result (Titmus, 1983).

Both Taylor's and Hanski's models bear some relation to the conception of population abundance/distribution proposed by Fretwell and Lucas (1969). In this model, individuals settling in an area select habitat patches in which their immediate fitness is maximized. The quality of a habitat in their model is a decreasing function of the density of the individuals already present. Therefore, a point will be reached at which the ''best'' habitat patches contain so many individuals that their overall suitability (fitness potential) is now equivalent to that of other lower-quality but unoccupied patches. As the total population size increases, then, individuals will begin to occupy a broader range of habitat patches, with the lower-quality patches supporting lower densities. Lundberg *et al.* (1981) found support for some features of this model, although their test was carried out in populations breeding in artificial nest boxes, possibly confounding density effects. For two species of sparrows breeding in a Wisconsin grassland, I found that habitat features in the territories of the individuals that arrived early differed significantly from those in territories of individuals arriving a few weeks later, whose options in selecting breeding habitat were more restricted (Wiens, 1973). In the absence of any measures of individual fitness or reproductive success, however, I was unable to ascribe quality differences to the early versus late territories.

In the Fretwell-Lucas model (and, to a degree, the Taylor *et al.* model as well), individual dispersal behavior and habitat selection are optimal. In nature, however, this may not always be so. In particular, dispersal from a patch may occur both when the population density in the patch approaches the carrying capacity of its resources and well before such saturation densities are attained. Dispersal in these two situations is likely to differ qualitatively (Lidicker, 1975). Either type of dispersal will be influenced by the landscape matrix of suitable patches, which may impose barriers to movement. Further, some habitat patches in a landscape represent ''sinks'' in which individuals may be able to survive (at least for a short time) but are unable to reproduce. Individuals may colonize these patches, as well as more suitable habitat patches, from the more favorable ''source'' patches. Indeed, the source population patches may provide the critical reservoirs to recolonize suitable habitat in which local subpopulations have suffered local extinction, as visualized by den Boer (see Fritz, 1980). Such a ''source–sink'' structuring of populations in a patchy landscape may have several important consequences: (a) the dynamics of the source populations may dominate those of the much larger meta-population; (b) the source populations, by providing most dispersers, may contribute differentially to the overall population gene pool and reduce the possibilities of genetic differentiation of local subpopulations; and (c) if density-dependent factors such as behavioral dominance, social structuring, or territoriality impose limits on the subpopulations occupying source patches, many of the individuals produced there may be forced to disperse elsewhere, some to sink patches. The sink populations may thus at times be more dense than the source populations, destroying the habitat quality–density relationship that is predicted by the Fretwell-Lucas model and that forms the foundation of some wildlife habitat management programs (States, 1976; Wiens and Rotenberry, 1981a; Daly, 1981; Van Horne, 1983). Time lags in responses to changes in patch quality will obscure such relationships further. Thus, while there is clearly a linkage between population density and patterns of patch occupancy and spatial distribution, that relationship is likely to be far more complex and multifactorial than existing theory would lead one to believe.

Population density and resource patchiness interact to affect the social structuring of populations as well. In house mice (*Mus musculus*), for example, the territorial system and social dominance structure are more rigid in demes organized about a single food patch than in populations with more dispersed food, in which the social system is more flexible and high densities are attained (Noyes *et al.*, 1982). Several lines of investigation have converged on the suggestion that when resources occur in a patchy distribution with high variance in patch quality and/or high within-patch resource renewal rates, mobile animals exploiting the resources will tend to occur in aggregations (Horn, 1968; Wiens, 1976; Waser and Wiley, 1980; Waser, 1981). Individuals with no other constraints on their movements, such as predator avoidance, should be expected to feed independently of one another from one food patch to the next when the amount of the resource available at a patch is not much greater than a single individual needs, but when patches contain far more food than is required by an individual, individuals will join together in social aggregations. Evidence for the reality of this resource–social grouping linkage in arid and semi-arid environments abounds—herds of ungulates exploiting seasonal flushes of grass production in the Serengeti (Sinclair and Norton-Griffiths, 1979), flocks of parrots or finches descending upon fields of ripening grass or grain (Crook, 1965; Newton, 1972; Wiens and Johnston, 1977; Rowley, 1974), and the like—although detailed information on the costs and benefits to individuals of joining groups versus remaining solitary and defensive is unavailable and (like virtually all cost–benefit information) will be difficult to obtain.

V. PATCHES AND COMMUNITY STRUCTURE

Just as population properties are a consequence of the dynamics of the individuals that make up the populations, so community dynamics result from the behavior of their constituent populations. In a sense, communities represent aggregations of species that are integrated over many patches in space, each patch undergoing its own phase of development, disturbance, extinctions, and dynamics. It is because each species is immersed in different patch dynamics and different interactions with its environment and with other species at different places in its range, and because the species that co-occur at any location themselves integrate patch dynamics and interactions over differing scales of space and time, that the investigation of community dynamics and their explanation in terms of simple generalities have proven so difficult (Wiens, 1983). Nonetheless, the effects of patch structure on community features have received considerable theoretical attention, primarily with reference to how patchiness enhances the persistence of predator and prey populations, by providing prey with spatial refugia from predation, or the coexistence of competitors, by permitting spatial differentiation of resource use (Hanski, 1981, 1983; Hastings, 1978; Murdoch, 1977; Weiner and Conte, 1981; Pacala and Roughgarden, 1982; Tilman, 1982).

Rotenberry and I have considered how the patterns of communities are related to heterogeneity in grassland and shrubsteppe bird communities (Rotenberry and Wiens, 1980; and unpublished data; Wiens and Rotenberry, 1981b; Wiens *et al.,* unpublished data). If we examine the correlations between the distribution and abundance of birds at a "continental" scale, including the range of conditions from tallgrass prairies through arid shrublands, several sets of species show clear patterns: species such as dickcissels (*Spiza americana*) and grasshopper sparrows (*Ammodramus savannarum*) are closely associated with habitats of substantial vertical structuring but low horizontal patchiness (tallgrass), while sage sparrows and Brewer's sparrows clearly affiliate with high vertical structuring and high horizontal heterogeneity (shrubsteppe). When we reduce the scale of investigation to a "regional," within-shrubsteppe level, however, the patterns for the shrubsteppe species are no longer statistically evident. No clear affiliations of species with variations in either vertical or horizontal heterogeneity emerge, nor do the different species seem to differ in their habitat occupancy patterns. If we ask how individuals of coexisting species differ in their use of patches on a local scale (e.g., within a single 10-ha plot), the answer is "not much." On two study plots in central Oregon, both sage and Brewer's sparrows used vegetation patches for foraging in a clearly nonrandom manner, but there were no discernable differences between the two species in their patch use (Table 1). We have interpreted these patterns, from the continental to the local scale, as indicative of independent, individualistic responses of species and individuals to features of their habitats. At the most general level, species do differ in the broad geographic patterns of their distribution and habitat selection, but once a more local view is adopted, the species appear to vary in abundance independently of the details of habitat structuring or of the occurrence of other species. The studies of Holmes and his colleagues of breeding insectivorous birds in hardwood forests (Holmes *et al.,* 1979; Robinson and Holmes, 1982) and the experimental manipulations of rodent habitats performed by Abramsky (1978) would seem to lend broader support to this individualistic view of species' habitat responses.

Many of the studies of microhabitat or patch use by heteromyid rodents in southwestern U.S. deserts have provided clearer evidence of differences between coexisting species. In habitats containing both *Dipodomys* and *Perognathus* species, for example, the former generally orient their activities more toward open areas between shrubs, while the latter use the shrub patches more intensively (Section III,D). In at least some communities, however, the differences in microhabitat use between species are not fixed, but may vary as a function of the densities of the species (M'Closkey, 1981). Most heteromyid community studies have interpreted the patterns of species differences as resource partitioning resulting from competitive interactions between species. A major role for predation in determining community composition, however, was apparent in Thompson's (1982b) experimental manipulations of habitat structure. Thompson placed small cardboard shelters in open spaces between desert shrubs, thereby modifying the distance to cover in the habitat (and thus the profile of predation risk) but not the distribution of food resources. In response to this manipulation, *Dipodomys* declined in abundance, an intermediate-sized *Perognathus* became established, and two *Peromyscus* species invaded the community. In another experimental manipulation of coexisting *Dipodomys* species, the removal of one species failed to produce the broadening of habitat use by the other anticipated from competition theory, but instead was followed by recolonization of the removal areas by immigrants of the removed species (Schroder and Rosenzweig, 1975). The results were interpreted as evidence of competition in the past, which had acted to constrain the microhabitat preferences of the species. An alternative explanation, that the species responded to habitat features independently of one another, unaffected by either present or past interactions, seems equally tenable.

In other studies, the linkage between patterns of patch use by coexisting species and competitive processes seems less ambiguous. Glass and Slade (1980a,b), for example, documented the shifting patterns of spatial overlap and exclusion between cotton rats (*Sigmodon hispidus*) and prairie voles (*Microtus ochrogaster*) in different patches of old-field habitat in Kansas. Here the cotton rats aggressively excluded voles from favored habitats and produced spatial segregation during the reproductive season of the rats. At that time, the voles were forced to occupy marginal habitat patches that were unsuitable for the rats. During the nonreproductive season, however, cotton rat densities were low, the frequency of interspecific encounters was reduced, and aggression diminished. The voles then expanded their habitat occupancy to co-occur with the rats in localized syntopy. Persistence of the species was facilitated by the patchiness of the habitat, which provided refuges for the voles during the rats' reproductive period. Here, as in the community studied by O'Farrell (1980), the aggressively dominant species was more specialized in habitat affinities than the subordinate species and evidenced a greater degree of spatial segregation.

In another study, Smith and Vrieze (1979) examined the patterns of habitat-patch use by three rodent species in the Florida Everglades. There the three species co-occurred on tropical hardwood hammocks during the wet season, when they all bred. Segregation into different habitats occurred during the dry, nonbreeding season, when one of the species remained on the hammocks, another moved off the hammocks to occupy the prairie interstices, and the third disappeared, presumably to seek more mesic habitats elsewhere. Again, the patch structure of the landscape apparently facilitated the persistence of this community.

VI. VERTEBRATES AS PATCH PRODUCERS

The relationship between patchiness and vertebrates in arid and semiarid environments involves not only the effect of patchiness on vertebrates but also the reverse. Vertebrates may have a variety of influences upon the production of patch structure in an environment. By burying caches of seeds, for example, granivores such as rodents may enhance germination probabilities (Reichman, 1979), especially if only a portion of the caches are later recovered by the animals (Johnson and Jorgensen, 1981). Such clumping of seeds may subsequently lead to aggregation in the distributions of seedlings as well. By dispersing seeds and other propagules between patches in their foraging movements, animals may also increase the survival probabilities of the seeds (Howe and Smallwood, 1982) or enhance the persistence and stability of the plant populations (through a ''spreading of risk'') (DeAngelis et al., 1979). More often, however, the influences of vertebrates on habitat patch structure are mediated through disturbances of one sort or another.

Some vertebrates alter the patch structure of a habitat by direct modification of the soil rather than by grazing vegetation. Fossorial mammals such as gophers or badgers may create patches of freshly disturbed soil through their digging and burrowing activities, and if the animals are colonial (e.g., prairie dogs, rabbits), these small-scale patch disturbances may in turn be clumped at a larger scale. In western Nebraska, areas disturbed by pocket gophers (*Geomys bursarius*) have more bare soil and litter and less basal coverage of vegetation than undisturbed areas (Foster and Stubbendieck, 1980). On these mounds, as on the soil mounds created by badger (*Taxidea taxus*) burrowing in tallgrass prairies, there is an orderly and distinctive pattern of revegetation following disturbance, with species of annual ''fugitive'' plants being followed in time by a series of perennials (Platt, 1975; Platt and Weis, 1977). Areas of gopher or badger activity may thus contain a mosaic of patches of differing ages and successional status, increasing tremendously the heterogeneity of the vegetation as a whole and maintaining a state of nonequilibrium patch structuring.

Perhaps the most conspicuous disturbance effects in these environments (especially in grasslands), however, result from grazing. In North America it is difficult to comprehend how grazing by large, mobile, native herbivores might have affected the structure and composition of the vegetation, since virtually all of these are now extinct and have been replaced by cattle. In African grasslands and savannas, however, a diverse complement of native herbivores persists. Elephants, for example, may have striking effects on the coverage and species composition of woody vegetation in areas in which they forage at high densities (Laws *et al.*, 1975; Anderson and Walker, 1974). Many of the grazing herbivores in these systems are selective in their feeding. This may lead to an orderly sequence of changes in habitats that are produced by the grazing of herbivores that, in turn, occupy an area in a seasonal succession of species dictated by their food preferences (Bell, 1970). Studies in the Serengeti have clearly documented that food is in short supply and unpredictable in location during the dry seasons, leading many of the ungulates to migrate seasonally through foraging areas; a host of other adaptations accompanies the adoption of this strategy (Pennycuick, 1979; Jarman and Jarman, 1979). The native grasses of the Serengeti express a variety of adaptations in response to such grazing pressures (McNaughton, 1979).

In contrast to such native herbivores, domestic cattle are restricted in their movements both by their more sedentary nature and by human enclosures. Much of their effect in generating patchiness thus occurs at a landscape scale, as a consequence of the interspersion of pastures or ranges that are subjected to different grazing intensities or seasonalities. Within pastures a finer scale of heterogeneity may be produced if the animals' grazing activities are spatially concentrated, as by topography, the distribution of palatable forage plants, or a localized water source. Where grazing is intense (on whatever scale), it has well-documented effects upon the stature of the vegetation, the relative coverages of plant species, water infiltration rates, and a variety of other system properties (Risser *et al.*, 1981). Not all consequences of grazing are necessarily detrimental, however. Domestic cattle, for example, produce seemingly endless quantities of dung, which is generally deposited in small heaps. These create local concentrations of nutrients returned to the system, and these enriched patches often exhibit clearly greater productivity or altered floristic composition from the surrounding sward.

Smaller herbivores may also influence the patch structure of the environment, although of course on a smaller scale. Grazing in low- to moderate-density *Microtus* populations, for example, may be concentrated in areas close to runway systems, creating zones in which the vegetation differs in structure from that located in lacunae in the runway network. Even such small herbivores, however, may also have landscape-level effects. Thus, in areas in which patches of suitable wet grassland habitat are interspersed in a mosaic with coniferous forests undergoing regeneration, voles may move seasonally from the grassland patches, causing considerable damage to seedlings in nearby regeneration areas (Hansson, 1977).

Other foraging activities may also influence the patch structure of the food resources. Any form of nonrandom search, in fact, is likely to affect prey distribution patterns by creating uneven patterns of resource depletion. These effects may be most obvious in central-place foraging systems, especially when the focus is a colony location (e.g., breeding colony or roost; Orians and Pearson, 1979). Depletion of resources in the immediate vicinity of the central place may be severe, although this effect will be reduced if the size of the overall feeding area is large (Andersson, 1978). Organisms foraging without relation to a central place may also affect resource dispersion. Desert granivores such as ants and (especially) rodents appear to select large seed clumps preferentially, which are then depleted. This has the consequence of significantly reducing the mean seed-clump size and produces a skewing toward small patches in relation to control areas from which the granivores are excluded (Reichman, 1979). The density of seeds in the soil bank is thus reduced and the distribution of those seeds that remain is altered. This, of course, affects the nature of the resource available for subsequent foraging, but it may also have effects on the plant community by altering the species composition of the seed bank and by reducing potential seedling competition in the clumps (Brown et al., 1979; Reichman, 1979).

The exploitation of seed clumps by granivores or of grasses by selective herbivores is often accomplished by area-restricted search, and a general consequence of this behavior is to increase the overall heterogeneity of resource distribution (Gill and Wolf, 1977; Pleasants and Zimmerman, 1979; Walker, 1981). In a sense, the existence of patchiness in resource distribution favors the development of area-restricted search behavior, but the consequence of this behavior is to concentrate feeding in localized patches, thus depleting them in relation to the surroundings and creating patchiness in turn. What sets the limits to this positive feedback system? One possible mechanism is suggested by the marginal value theorem. If, indeed, individuals cease foraging in a patch when its value is decreased to the average value for the habitat as a whole (Charnov, 1976), such a strategy would over time tend to reduce the difference in quality between patches until all were virtually identical, at which point foraging would presumably cease to be patch oriented. The assumptions of the marginal value theorem, however, are suspect (Section III,C). Part of the answer, instead, may relate to resource renewal rates. The potential for area-restricted searching to create resource patchiness would seem to be limited to situations in which renewal rates are intermediate. If renewal rates are rapid relative to consumption or patch revisitation rates, the effects of a feeding episode will be rapidly erased, and whatever form of patchiness characterized the system before the foraging will remain. If, on the other hand, renewal rates are relatively slow, patches suffering depletion will remain so for long periods, leading individuals to forage in previously unforaged areas. All patches will thus eventually be likely to suffer a similar degree of depletion, fostering increased spatial homogeneity in resource levels.

VII. HABITAT FRAGMENTATION

Patch dynamics assume a very real and immediate practical significance when they relate to the management of populations, habitats, or entire landscapes. Discussions of conservation issues increasingly contain references to fragmentation of natural habitats and its effects, how fragments should be arrayed with respect to one another, and what their optimal area might be. The issues have generated considerable controversy (see Whitcomb et al., 1976; Cole, 1981; Higgs, 1981; Frankel and Soulé, 1981; Simberloff and Abele, 1982; Margules et al., 1982). No clear concensus has emerged, and it is becoming increasingly apparent that it is inappropriate to apply general theory to situations in which details of the patch dynamics of the environment or idiosyncracies of the patch responses of organisms are of critical importance (McCoy, 1983; Haila *et al.*, 1982; Pickett and Thompson, 1978; Simberloff and Abele, 1982). Surely part of the reason that different groups of birds, mammals, and lizards differ in the way they are affected by habitat fragmentation in the Australian wheatbelt region (Kitchener *et al.*, 1982; Humphreys and Kitchener, 1982) is related to the numerous differences in the natural histories of these animals.

These concerns are so critical, because human activities are fragmenting natural ecosystems into fewer and smaller pieces at an accelerating pace. Such fragmentation reduces the average size of patches of a given natural habitat, increases the distance between patches, decreases the ratio of interior to edge area within patches, and increases (at least initially) the landscape diversity of an area through the creation of new patches of disturbances that may undergo succession. Species differ in their potential sensitivity to these effects (Fig. 4). Species that are habitat specialists will generally occupy only a small portion of the patch types in an area, and their range of patch occupancy will be further restricted to patches above a certain size if they specialize on patch interiors rather than edges. Because larger organisms require bigger home ranges and have lower individual fecundity, on average, their abundance per unit area in suitable habitat patches will be relatively low; the same patterns characterize organisms occupying higher trophic positions as well. Low abundance may also contribute to a reduced frequency of patch occupancy (Hanski, 1982b). If the species are sedentary or dispersed only over short distances, or if populations have low recruitment rates, the rate of colonization of patches created by local extinctions or habitat changes will be low. Habitat specialization and low population densities further reduce patch colonization rates. All of these consequences—low abundance in occupied patches, low frequency of patch occupancy, and low rates of patch colonization—increase the probability that a population residing in a patch will suffer local extinction. With increasing fragmentation of a landscape, stochastic effects become more important and may enhance the likelihood of local extinctions further; with a reduction in the number of suitable patches in a region, regional extinction thus also becomes more probable. Fragmentation will also lead to local extinctions as a simple consequence of species-area relations, for as the area of habitat is reduced some of the species that were previously rare will eventually become extinct, through faunal "relaxation" (Diamond, 1972; Western and Ssemakula, 1981). Further, some extinctions may have "ripple effects" through the system; the disappearance of a large predator species, for example, may permit prey populations to increase to levels at which they overgraze resources or become more intensely competitive with one another (Frankel and Soulé, 1981). Because of the greater role of stochastic effects and the inevitable time lags in recruitment and patch colonization, populations (much less communities) in a fragmented landscape are not likely to be in overall equilibrium, and theory or management practices founded upon equilibrium or deterministic assumptions are not likely to perform very well.

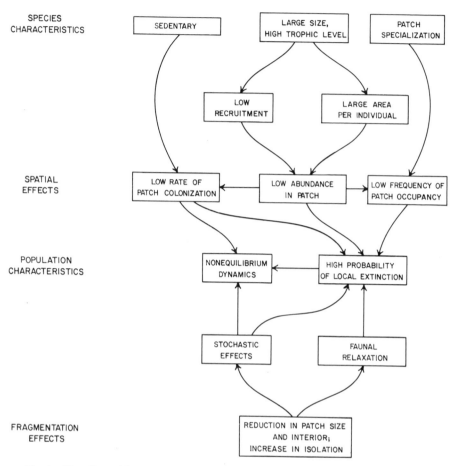

Fig. 4. The effects of fragmentation of natural habitats into smaller, more isolated patches (bottom of figure) on the probability of extinction of local populations of a species (middle of figure), as a function of various characteristics of the species (top). See text.

Some insight into the complications that may beset attempts to monitor population responses to habitat fragmentation emerges from studies of sites in which patches of the habitat have been drastically altered. At one of our shrubsteppe study sites, for example, a range fire swept over one of our plots, reducing sagebrush coverage from 8 to less than 2%. Our studies at a more general level had revealed a close correlation between densities of sage sparrows and sagebrush coverage, so we expected a severe reduction in their population level the following year. Densities were lower, but not to the extent anticipated; even in the following year, numbers had not reached the level that would be predicted on the basis of sagebrush coverage alone (Rotenberry and Wiens, 1978). At another of our sites, the habitat was subjected to ''range improvement'' (spraying, disking, drilling of crested wheatgrass) midway through our 7-year study period. There sagebrush coverage was reduced from 27 to 4%. Again, sage sparrow densities did not respond as expected, although densities were lower 2 and 3 years following the treatment than before. We believe that this "inertia" in population response to habitat changes or fragmentation is at least partially a consequence of site tenacity of the birds. Individuals that have previously bred successfully at a site may return and establish territories the following year, even though the habitat has undergone dramatic changes (see Hildén, 1965). Some individuals may breed successfully even under the altered conditions and return in subsequent years. The consequence is a lag in the density component of faunal relaxation, indicating once again that projections of population dynamics based on simple bird–habitat correlations (or species–area relationships) are likely to overlook important details of the patch responses of organisms.

VIII. CONCLUDING CAUTION

Clearly, patches and their effects must receive greater recognition in both the models we develop and the field investigations we conduct. There can no longer be any excuse for naively presuming nature (or even small parts of it) to be homogeneous. This is all well and good, of course, but the operational difficulty of detecting, defining, and describing patches in a way that accords with the patch perceptions of the organisms under study remains a major hurdle. We unavoidably approach the study of patch dynamics from an anthropocentric perspective, seeing the boundaries and discontinuities in nature that seem important to us. There is more to this than simply coping with the fact that we view structure on a scale different from that of an aphid or ant, for we also tend to emphasize those factors that accord with our own dominant senses, primarily vision. Other organisms, however, may perceive environmental mosaics in quite different ways. Unless we adopt an organism-centered view of the environment, we are unlikely to discern or measure the elements of patch structure or dynamics that are *really* important, and instead may well document apparent "patterns" that are little more than artifacts, products of our misperception of reality. The solutions to this problem must be founded on a detailed knowledge of the natural history of the organisms and must look to discontinuities in the behavior of the organisms in space to reveal the possible dimensions of environmental patch structuring as the organisms perceive it.

ACKNOWLEDGMENTS

The National Science Foundation has supported the research in grassland and shrubsteppe ecosystems that heightened my awareness of patches and their effects, most recently under Grant No. DEB-8017445. John Rotenberry participated in much of this research. He and Shawn Crowley, Luke George, Bob Pietruszka, Bea Van Horne, and Steve Zack offered comments on a draft of this chapter. Yevonn Ramsey prepared the figures. The framework of Fig. 4 matured following discussions with Richard Forman and Paul Risser.

RECOMMENDED READINGS

Crawford, C. S., and Gosz, J. R. (1982). Desert ecosystems: their resources in space and time. *Environ. Conserv.* **9,** 181–195.

den Boer, P. J. (1981). On the survival of populations in a heterogeneous and variable environment. *Oecologia* **50,** 39–53.

Haggett, P., Cliff, A. D., and Frey, A. (1977). "Locational Analysis in Human Geography." Edward Arnold, London.

Hansson, L. (1979). On the importance of landscape heterogeneity in northern regions for the breeding population densities of homeotherms: a general hypothesis. *Oikos* **33,** 182–189.

McNaughton, S. J. (1979). Grassland–herbivore dynamics. *In* "Serengeti: Dynamics of an Ecosystem" (A. R. E. Sinclair and M. Norton-Griffiths, eds.), pp. 46–81. Univ. Chicago Press, Chicago.

Van Horne, B. (1983). Density as a misleading indicator of habitat quality. *J. Wildl. Mgmt.* **47,** 893–901.

Wiens, J. A. (1976). Population responses to patchy environments. *Ann. Rev. Ecol. Syst.* **7,** 81–120.

Part III

ADAPTATIONS OF PLANTS AND ANIMALS IN A PATCH DYNAMIC SETTING

Chapter 11

The Response of Woody Plants to Disturbance: Patterns of Establishment and Growth

CHARLES D. CANHAM[1] and P. L. MARKS

Section of Ecology and Systematics
Cornell University
Ithaca, New York

[1]Present address: New York Botanical Garden, Institute of Ecosystem Studies, Cary Arboretum, Box AB, Millbrook, New York 12545.

THE ECOLOGY
OF NATURAL DISTURBANCE
AND PATCH DYNAMICS

I. INTRODUCTION

Hypotheses on the coexistence of plant species (Connell, 1978), niche differentiation (Grubb, 1977), and resource partitioning (Denslow, 1980a) in plant communities have relied heavily on the requirement for some form of disturbance during the life cycles of many plant species. In general, disturbances reduce the dominance of a site by established individuals and create openings for colonization and growth by new individuals. Where regrowth of vegetation following a disturbance is rapid, competition for those openings should be severe.

In this chapter, we examine traits that affect the ability of woody plants to respond to openings created by natural disturbances. Our approach is to consider traits that determine the types of disturbances in which a species is most likely to be successful in reaching reproductive size. We focus primarily on trees of closed forests, particularly the deciduous and mixed conifer–hardwood forests of eastern North America with which we are most familiar. However, many of the patterns we describe below are common to trees, shrubs, and lianas in any habitat that can support a closed forest canopy.

II. NATURAL DISTURBANCES AND RESOURCE AVAILABILITY

One of the principal effects of natural disturbances is to alter the availability of resources for plant growth. There are at least two mechanisms by which disturbances can temporarily increase the availability of light, water, and soil nutrients. The first is simply the reduction in rates of uptake or use of resources due to the loss of biomass. This effect is most apparent in the enhancement of light levels in canopy openings (March and Skeen, 1976; Chazdon and Fetcher, 1984). A second mechanism is the decomposition and mineralization of nutrients held in organic matter. Increased insolation at the soil surface and reduced transpiration following large-scale windthrow, for example, may increase nutrient availability by increasing the rate of decomposition of soil organic matter (Bormann and Likens, 1979). Fires will often increase the availability of some nutrients despite losses via groundwater, smoke, and volatilization (Christensen and Muller, 1975a; Grier, 1975; Krebs, 1975; Sanchez, 1976; Stark, 1977; Chapin and Van Cleve, 1981). Severe disturbances may actually reduce the rate of supply of water and certain nutrients through physical degradation of the site (see, e.g., Nye and Greenland, 1964; Garwood et al., 1979) or through disruption of nutrient cycling (Christensen, 1977; DeBano and Conrad, 1978). However, in closed forests, most forms of canopy disturbance produce a temporary increase in some of the resources necessary for the establishment of new individuals or the growth of understory plants.

In general, there is a positive relationship between disturbance size or intensity and the availability of resources for plant growth. An increase in the size of an opening in a closed forest, for example, will increase the amounts of both diffuse and direct radiation in the understory (see, e.g., Minckler et al., 1973). The intensity of a disturbance (as measured by the reduction in biomass per unit area) should also be directly related to the degree of reduction in rates of transpiration and interception of water and the uptake of nutrients. However, for a given size of disturbance, the magnitude and even direction of change in the availability of water and nutrients can be affected by changes in many factors, such as the amount of water lost in surface runoff and evaporation. The relationship between disturbance size and resource availability may be strongest for small-scale disturbances that primarily cause local reductions in the interception of light and the uptake of water and nutrients, without substantial physical disruption of the site.

An important feature of any increase in resource availability produced by a disturbance is its transient nature. As biomass is reestablished at a site, the relative availability of resources for future colonists will, in general, decline. Tilman (1982) has presented an analysis of plant competition for levels of resources available when the rate of uptake by an intact community has equilibrated with the rate of supply of a resource. The transient pulses of resources produced by many natural disturbances represent a distinctly different pattern to which plant species can respond. Traits traditionally associated with the response of woody plants to disturbance (e.g., rapid growth rates, small sizes, and early reproduction) appear to facilitate the exploitation of such pulses of resources.

III. ESTABLISHMENT OF WOODY PLANTS FOLLOWING DISTURBANCE

In communities where there is rapid regrowth of vegetation following a disturbance, the availability of resources for colonization should reach a peak soon after a disturbance. Consequently, the first plants that become established after a disturbance should enjoy greater availability of resources than plants that become established later. Establishment of seedlings of many species of woody plants is often limited to a brief period following disturbance (c. 1–5 years), particularly when reestablishment of biomass is rapid (Oliver, 1981). The competitive advantages of early arrival time and large propagule size have been demonstrated for herbs (see, e.g., Black, 1957; Holt, 1972). Early arrival time should be particularly critical for species of woody plants that are intolerant of shade.

A. Seed Production and Dispersal

Patterns of seed production and dispersal vary widely among woody plants. One of the most conspicuous patterns of seed production and dispersal—early and copious production of light, wind-dispersed seeds—is generally correlated with the ability to respond to large disturbances (Baker, 1974). Many factors other than seed crop size and seed weight, of course, will influence the availability of seeds in a disturbance, and we can ask to what extent other aspects of seed production and dispersal lead to differentiation in the response of woody species to disturbances. The factors considered below are fluctuations in annual seed production and the mode, seasonal timing, and duration of seed dispersal.

Relative constancy in annual production of seeds should enhance the likelihood that seeds are available for colonization of recent disturbances. However, there is substantial annual fluctuation in the size of seed crops within and between species. The frequency of annual seed crop failures or near failures in temperate trees appears to be higher in large-seeded trees (e.g., species of *Quercus*) than in small-seeded trees (e.g., species of *Populus*) (data of Godman and Mattson, 1976). In some cases, seed crop failures are the result of problems such as the adverse effects of weather on some aspect of reproduction (suggested by the data of Godman and Mattson, 1976), high energetic costs of repeated production of large crops of large seeds, or indirect effects of extreme weather or animal feeding that reduce or eliminate sexual reproduction by reducing overall plant productivity. In contrast, the synchronous production of large numbers of seeds at irregular intervals may satiate seed predators (Janzen, 1971).

The mode of seed dispersal should also influence the amounts, kinds, and distribution of seeds within disturbances. Following cutting and burning of mature tierra firme forests of the Amazon basin, bird- and bat-dispersed seeds of *Cecropia* and other woody pioneers were clumped, presumably reflecting roosting site preferences (Uhl *et al.*, 1981; McDonnell and Stiles, 1983). In an early study, Watt (1925) described the tendency of seedlings of *Fraxinus excelsior* (wind dispersed) to be concentrated near the center of forest gaps, while seedlings of the heavier-seeded *Fagus sylvatica* (bird and mammal dispersed) were more common near the periphery of gaps. On the other hand, where dispersal distances are greater, plants with animal-dispersed seeds may achieve higher densities than plants with wind-dispersed seeds. For example, it is common to observe a greater density of bird-dispersed rather than wind-dispersed woody plants in the central parts of large abandoned agricultural fields in the northeastern United States (P. L. Marks, personal observation).

It seems reasonable to expect frugivorous birds that feed in open habitats to carry seeds of fleshy-fruited pioneer trees and shrubs from one large opening to another (see, e.g., Marks, 1974). However, in forests where regrowth following large disturbances is rapid, it is questionable whether the partnership between bird and plant works in the expected way; by the time a large blowdown (or its equivalent) contains plants producing fleshy fruits that would attract open-habitat, frugivorous birds, conditions at the soil surface may no longer be favorable for the establishment of pioneer plants. Moreover, the expected partnership assumes prompt germination of seeds following excretion or disgorgement by birds. Many of the fleshy-fruited invaders of temperate forest openings and fallow agricultural fields, however, have seeds that will not readily germinate when freshly collected (e.g., *Prunus pensylvanica, Rhus typhina, Cornus racemosa*). Undoubtedly, passage through a bird facilitates germination of such seeds (Krefting and Roe, 1949). However, for bird-dispersed seeds that have poor germination when fresh, it remains to be shown whether seeds commonly arrive soon enough, and in suitable condition, to germinate and produce seedlings with a reasonable likelihood of growing to maturity in disturbances where subsequent regrowth is rapid. It may turn out that bird dispersal of seeds functions more to maintain input to buried seed pools in closed forests (Uhl *et al.*, 1981).

A final aspect of dispersal—the seasonal timing and duration of seed dispersal— should also influence the availability of seeds in an opening. Seeds of a number of temperate floodplain trees (*Populus deltoides, Salix* spp., *Ulmus americana*) are dispersed in late spring. Seeds of another major floodplain tree, *Platanus occidentalis,* ripen in the fall but do not normally disperse until the following spring (Fowells, 1965). The combination of spring dispersal and rapid germination in these floodplain species should contribute to the establishment of seedlings on fresh alluvium deposited by spring floods.

The duration of seed dispersal from woody plants varies enormously, from species whose seeds are released relatively synchronously (within a week or so) to species some of whose seeds remain on the parent plant for the better part of a year. Examples of relatively synchronous release of seeds are those trees with extremely short-lived seeds (*Populus, Salix*) and at least some temperate zone fleshy-fruited species (Thompson and Willson, 1979) in which flowering and fruit ripening are relatively synchronous. Examples of temperate tree species with a prolonged duration of seed dispersal are *Liriodendron tulipifera,* species of *Fraxinus,* and species of *Betula,* all of which often release seeds from fall until spring, despite synchronous flowering. Release of seeds throughout much of the year is also accomplished by species, common in the tropics, that flower (on the same plant) through all or much of the year (Frankie *et al.,* 1974). *Magnolia grandiflora* is a warm temperate example. For wind-dispersed species, it seems likely that extended release of seeds from a given plant will result in greater variability in the distance and direction of seed dispersal because winds of different directions and velocities will be involved. A wider distribution of seeds around a parent plant should increase the probability that some of the seeds will find their way into an opening. In moist tropical forests in which canopy openings and seed germination can occur throughout much of the year, extended release of seeds from a given plant would also seem to have an advantage in placing fresh seed in a recent opening. On the other hand, plants that rely on animals as pollinators or dispersers may have greater reproductive success if flowering, fruiting, and seed dispersal are synchronous (Bazzaz and Pickett, 1980).

B. Seed Storage and Germination

It is possible to recognize three general patterns of seed storage and germination.

1. No or minimal delay between dispersal and germination. The salient feature of this pattern is the lack of a persistent buried seed pool. As a result, colonization of a disturbance must come from either a pool of suppressed seedlings (if one exists), the sprouting of surviving roots or boles, or more or less continuous production and dispersal of seeds from beyond the disturbance area.

2. Extended delay between dispersal and germination, with germination being triggered by some aspect of disturbance. The salient feature of this pattern is the extended storage of seeds in the soil. A key question, for which there is currently inadequate information, concerns the extent to which repeated input of seeds is required to maintain the seed pool from one disturbance to the next.

3. Intermediate delay between dispersal and germination. Here innate seed dormancy is long enough to compensate for poor seed production in one to several years, but short enough (c. 5–7 years) so that frequent input of seeds is required to maintain a soil seed pool from one disturbance to another. The remainder of this section expands on the above patterns.

Species with little or no delay between dispersal and germination include shade-tolerant trees of the eastern United States, more than half of 180 rain forest species reported by Ng (1978) for Malaya, the floodplain trees of the northeastern United States, and some pioneer trees (e.g., *Populus*). In cases in which there is a delay between dispersal and germination, the delay is normally keyed to seasonal aspects of climate. The proximate cause of the delay is either an unfavorable environment at the time of dispersal or innate properties of the seed. For example, in a seasonal tropical forest, seeds dispersed either toward the end of the rainy season or in the dry season did not germinate until the next rainy season (Garwood, 1983). Most seeds of woody plants subsequently germinated within the first 2 months of the 8-month rainy season (Garwood, 1983).

In the northeastern United States, seeds of *Acer rubrum* are shed in spring not long after flowering. After dispersal, they are sensitive to the ratio of red/far red light, presumably due to the presence of phytochrome (Smith, 1972; Marquis, 1973; Bazzaz and Pickett, 1980; Cook, 1980). Plant canopies absorb much of the incident light in the red wavelengths (660 nm), which reduces the red/far red ratio at the soil surface and prevents germination beneath a full canopy. Thus, only the seeds that fall within canopy openings germinate during the season of dispersal. *Acer rubrum* seeds that do not germinate during their first growing season germinate the following spring before trees are in leaf and before the next year's crop of seeds has been shed (Marquis, 1973). Thus, unlike other species whose seeds do not remain alive for more than 1 year, *A. rubrum* has an extra growing season during which its seeds germinate only in response to disturbance.

The second pattern of seed storage and germination involves woody species that have a substantial delay between dispersal and germination; the seeds of such species remain alive in the soil for periods ranging from years to decades. Many pioneer trees fit this pattern. When combined with a germination response that is keyed to an effect of a disturbance, extended storage of seeds in soil can virtually ensure that seedlings will be present soon after the formation of a canopy opening (Bazzaz, 1979). In species in which germination of buried seeds is triggered by an opening, live seeds should tend to accumulate in the soil until a disturbance occurs, subject to losses from consumption, rot, or a low rate of germination in the absence of disturbance. In one study of buried seeds of the pioneer tree *Prunus pensylvanica* in a 60-year-old forest in New Hampshire (Marks, 1974), contributions to the soil seed pool came in unknown proportions from plants that had previously grown on the site and from bird dispersal of seeds from other populations. Loss of seeds was high, particularly due to consumption by small mammals that consumed an average of c. 100 seeds per square meter over an estimated 3–5 decades since the last disturbance. Other losses of seeds from the buried seed pool included as many as 45 seeds per square meter that succumbed to either embryo abortion prior to dispersal or rot in the soil and 12–15 seeds per square meter that germinated in the shade over a period of several decades. Despite all of the losses, there were still about 40 viable *P. pensylvanica* seeds per square meter on average. Although some germination of *P. pensylvanica* occurs in the absence of disturbance, most of the seeds in the soil do not germinate until triggered by some aspect(s) of the environment of an opening. The specific germination trigger is still under investigation. Several studies (Cheke *et al.*, 1979; Hall and Swaine, 1980; Holthuijzen and Boerboom, 1982) indicate that many tropical pioneer trees have seed storage and germination characteristics similar to those just described. Phytochrome control of germination has been described in the tropical pioneer trees *Cecropia obtusifolia* and *Piper auritum*, whose seeds otherwise behave more or less like those of *P. pensylvanica* (Vázquez-Yanes and Smith, 1982).

The third pattern of seed storage and germination involves short-term storage in the soil (c. 3–7 years) followed by germination that is not necessarily cued to disturbance. *Prunus serotina* provides an example. Unlike those of *P. pensylvanica,* seeds of *P. serotina* remain alive in the forest floor for only 3–5 years (Wendel, 1972, 1977; Marquis, 1975a). Observations of germination of *P. serotina* seeds of known age under natural conditions (Wendel, 1972, 1977; Marquis, 1975a) showed that the vast majority of an initial cohort of seeds germinated after 3 years in the absence of a disturbance. Like *P. serotina,* at least some temperate zone gap trees (*Fraxinus americana, Liriodendron tulipifera*) have seeds capable of being stored for several years in the soil (Leak, 1963; Clark and Boyce, 1964; Marquis, 1975a). It would be interesting to know whether the seeds of these trees accumulate in the soil until a disturbance occurs, or whether some (variable) fraction of an initial cohort germinates each year without regard to disturbance, as suggested for *P. serotina* by the results of Marquis (1975a) and Wendel (1977). One major effect of this pattern of short-term storage in the soil and relatively continuous germination of seeds is a pool of buried seeds that provides a buffer against annual fluctuations in seed production and dispersal.

C. Seedling Establishment

Since the work of Salisbury (1942), it has been known that there is a correlation between seed size and the ability of different species to establish seedlings under a closed forest canopy. Large-seeded plants are more likely to become established in closed forest than are small-seeded plants for at least two reasons. Because of access to seed reserves, seedlings of large-seeded plants are more independent of the physical environment than are seedlings of small-seeded plants. It has been shown, for example, that the ability of small seedlings to grow in shade is closely related to seed size (Grime, 1966), and Ng (1978) has commented on the ability of first-year seedlings of large-seeded trees in Malaya to persist in closed forest for up to 6 months. The second influence concerns leaf litter and humus. In those forests in which there are matted leaves resting on a surface organic layer, small-seeded plants are unlikely to become established in large numbers, even under the favorable conditions following a disturbance, if the litter layer is intact. This is because matted leaves act as a physical barrier to initial establishment, and frequent desiccation of the surface organic layer, particularly in the warm, sunny environment of a large opening, causes high mortality of small seedlings (Marquis, 1965; Putz, 1983). However, virtually all disturbances involving fire and most large wind-caused disturbances will result in at least some exposed mineral soil conducive to the establishment of small-seeded plants. It is also worth noting that although small-seeded plants have a greatly reduced likelihood of becoming established as seedlings in the presence of thick litter and humus layers, the subsequent growth of seedlings is much greater in soil with the humus layer intact (Marquis *et al.,* 1964). Thus, for small-seeded plants in certain kinds of forests, conditions that promote establishment (i.e., exposed mineral soil) differ from those that promote subsequent rapid growth of seedlings (i.e., an intact humus layer).

In conclusion, we might ask: How precisely are the traits that govern seedling establishment coordinated with respect to particular types or intensities of disturbance? There are notable examples of mechanisms that tie the production, dispersal, storage, and especially the germination of seeds to disturbances that create conditions favorable for the growth and reproduction of a species. The cueing of germination of buried seeds to conditions associated with particular types or intensities of disturbance involves substantial specialization in the mode of establishment following a disturbance. At the same time, however, most woody plants seem to have at least some capacity to place seedlings in the favorable environments created by most disturbances, and that capacity does not necessarily involve close coupling of particular life history stages with the size, the kind, or even the occurrence of a disturbance. Consider the mode of seed dispersal. It is difficult to argue that a particular mode of dispersal is ideally suited to one kind or size of disturbance. Temperate woody pioneers, for example, include species with small, short-lived, wind-dispersed seeds (*Populus*) and others with large, long-lived, bird-dispersed seeds (*Prunus*). The release of seeds from serotinous cones provides an example of the cueing of seed dispersal by a disturbance, but the effect of serotiny is largely to allow the accumulation of a protected seed pool (in this case, stored in cones rather than in the soil) in advance of a fire. Thus, the colonizing ability of woody plants can result from many combinations of different traits. Many woody species depend on buried seed pools or advance regeneration to provide a pool of individuals present at a site in advance of a disturbance. Other species rely on annual production of widely dispersed but short-lived seeds, while still others compensate for fluctuations in annual seed crop size by maintaining a buried seed pool for a few years.

Interestingly, species of *Quercus* not only tend to have large year-to-year variation in seed production but also lack the ability to compensate by accumulating individuals as buried seeds. Thus, factors such as seasonality, seed predation, and the relative unpredictability of both seed dispersal and suitable disturbances appear to place constraints on the degree of specialization in the mode of establishment of woody plants following disturbance.

IV. GROWTH AND RESOURCE USE BY WOODY PLANTS FOLLOWING DISTURBANCE

In communities where there is rapid regrowth of vegetation following a disturbance, we would expect strong competition among plants that either sprout from roots (or boles) or become established from seeds. The number of plants present at a site soon after a disturbance will generally far exceed the number that can eventually reach reproductive size, and we would thus expect differential growth rates to be an important aspect of competition during the early stages of the recovery of biomass. A seedling with a higher rate of growth in a given microsite should prevail over a more slowly growing, neighboring seedling established at approximately the same time. This outcome is to a significant degree determined by the ability of the faster-growing seedling to increase its level of resource use, often at the expense of its neighbors (Harper, 1977).

A high relative growth rate in a particular disturbance is likely to be the product of a large number of physiological and morphological traits. Aspects of the physiology of plant response to natural disturbances have been reviewed by Bazzaz (1979) and Bazzaz and Pickett (1980). Much of the physiological research on plant carbon gain has focused on factors affecting the photosynthetic response of a unit area of leaf tissue to different environmental conditions. However, a similar pattern of photosynthetic response in two different species does not necessarily imply similar levels of whole plant carbon gain, since, for example, differences in the magnitude of respiration by shoots and roots will influence whole plant growth rates (Evans, 1975). In woody plants, in which the ratio of nonphotosynthetic to photosynthetic tissue can be relatively high, respiratory losses can be substantial (Yoda *et al.*, 1965; Whittaker and Woodwell, 1967; Kira, 1975). Moreover, there are often substantial differences between species in the patterns of allocation of net photosynthate to different structures within a woody plant (see, e.g., Logan and Krotkov, 1968; Cannell and Willett, 1976). Models of water use efficiency, for example, suggest that different sizes and shapes of leaves require differences in the amount of photosynthate that must be allocated to roots (Orians and Solbrig, 1977; Givnish, 1979). Thus, even when unit photosynthetic and respiration rates of different tissues are known, their net effect on whole plant growth rates will be mediated by differences among species in plant growth form. The extensive literature on plant growth analysis (Evans, 1972; Hunt, 1978), although primarily dealing with herbaceous and crop species, underscores the effect of plant growth form on net growth rates. Even parameters such as unit leaf rate (the amount of biomass produced per unit of leaf area and time) are determined by a combination of physiological and morphological traits (Causton and Venus, 1981). In general, whole plant growth rates should be strongly influenced by the effects of plant growth form on such factors as the uptake of water and nutrients, the interception of light, and the allocation of carbon to the maintenance of roots and shoots.

A. Rate versus Efficiency of Growth and Resource Use of Woody Plants

The response of woody plant growth rates to increasing size or intensity of disturbance is shown for four hypothetical species in Fig. 1A. The growth curves are defined by three features: first, the minimum size of disturbance required for net growth of seedlings and saplings; second, the growth rate of a species in large disturbances in which resources may be overabundant; and third, the shape of the curve describing growth rates for intermediate sizes of disturbances. Studies of the relationships between shade tolerance and the growth rates of trees in different light regimes suggest that there is generally a direct relationship between the growth rate of a tree species in large disturbances and the minimum size of disturbance required for net growth of seedlings (Grime, 1966; Grime and Hunt, 1975; Marks, 1975). Shade-intolerant tree species tend to have higher growth rates in the open than shade-tolerant species (Grime, 1966; Grime and Hunt, 1975; Marks, 1975). Studies of the response of tree seedlings to different light levels indicate that growth rates increase rapidly as light levels increase above the minimum required for net growth of seedlings (Logan, 1965, 1970; Strothman, 1967; Logan and Krotkov, 1968). However, particularly in the case of shade-tolerant species, growth rates may level off or even decline at light intensities well below that of full sunlight (Logan, 1965; Canham, 1984).

In general, competition between seedlings should act to restrict the success of a species to sites at which its seedlings have higher rates of growth than other seedlings. Because the growth rates of most species show a general increase over a wide range of disturbance sizes, the growth rate of a species in a disturbance should not necessarily be directly correlated with its competitive ability in a given disturbance. For example, although a shade-tolerant tree species may have a higher growth rate in a large blowdown than in a small canopy gap, it may have a much greater competitive ability in a small canopy gap. The hypothesis that a seedling with a higher growth rate should outcompete a more slowly growing seedling established at approximately the same time makes no predictions about the portions of a gradient in disturbance size or resource availability in which a species should be most successful in competition with a variety of other species.

In order to make such predictions about those disturbance sizes in which a species should compete most successfully with other species, we will use the concept of growth efficiency. The gradient of disturbance size in Fig. 1A corresponds to a gradient in resource availability, particularly in the case of light levels for different sizes of openings in closed forests (see, e.g., Chazdon and Fetcher, 1984). The efficiency of growth for a species in a given disturbance can be defined as the quantity of biomass produced per unit of biomass and time relative to (divided by) the quantity of the most limiting resource available in the disturbance per unit area and time. If the gradient of disturbance sizes is scaled to correspond to an approximately linear gradient in resource availability, then the growth efficiencies for the four hypothetical species in Fig. 1A are shown in Fig. 1B. The curves for species A in both Fig. 1A and Fig. 1B are drawn to represent species that may be capable of responding to a disturbance in the absence of faster-growing species or if the establishment of faster-growing species is delayed.

The portions of the disturbance size gradient in Fig. 1A in which a species has higher growth rates than the other species are the portions where the species has its highest growth efficiency (Fig. 1B), rather than its maximum growth rate. This suggests the hypothesis that a species should be most successful in responding to disturbances for which it has its highest efficiency of growth, rather than its highest potential growth rate, simply because at its sites of highest efficiency it is more likely to have higher rates of growth than species whose maximum growth efficien-

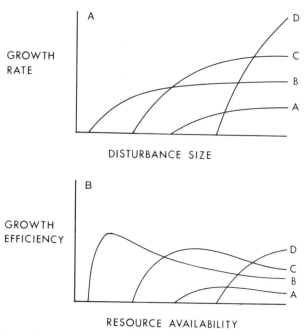

Fig. 1. (A) Generalized patterns of the relative growth rates of four hypothetical tree species with increasing size of disturbance. (B) The growth efficiencies of the four hypothetical species shown in (A) over a corresponding range of resource availability. Growth efficiency is defined as the growth rate divided by the level of the most limiting resource available at a site. If the gradient of disturbance size is scaled to represent a linear gradient of resource availability, then the growth efficiency at any point on one of the curves in (A) is equal to the slope of a line drawn through the origin and the point on the curve.

cies occur at higher or lower levels of resource availability. In more general terms, we are suggesting that the types or intensities of disturbances for which a species is most likely to have a higher rate of growth than other potential competitors are those in which the species is most efficient at exploiting the available resources.

The growth rates and efficiencies depicted in Fig. 1 are for the response of a species to the transient pool of resources available following different disturbances in a single environment. In general, the effect of a particular disturbance should vary at different sites. For example, the effect of a small canopy gap on understory light levels will be influenced by the slope, aspect, and latitude of the site. The degree of change in the availability of nutrients and soil water in and around a gap should also vary for different sites. There have been few studies that have explicitly considered the effects of differences in the quality of a site (as reflected in the levels of resources available to undisturbed portions of a community) on the response of woody plants to disturbance (see, e.g., Smith, 1983). It is likely that the growth rates and growth efficiencies of a species following a particular size or type of disturbance will vary in different environments.

The curves in Fig. 1 are clearly hypothetical and are intended to illustrate the relationships between disturbance size, growth rate, and growth efficiency in a relatively simple case. They are most applicable to sets of species and ranges of disturbance intensity in which a single resource such as light is limiting for all the species considered. Even when species are limited by different resources or combinations of resources in different disturbances, we would still expect a species to be most successful in responding to those disturbances for which it has its highest growth rate relative to resource availability (e.g., in the disturbances in which it has its highest growth efficiency). Note that by our definition of growth efficiency, a species with a higher efficiency of growth than other species at a particular site also has a higher growth rate for that site than do other species. Thus, the hypothesis that competition should act to restrict the success of species to sites at which they have their highest efficiency of growth is not in conflict with the postulated importance of growth rate per se. Rather, the hypothesis makes a prediction about the portion of a gradient of disturbance size or resource availability in which a species should be most successful in reaching reproductive size.

By focusing on the efficiency of growth and resource use by woody plants, it is possible to evaluate the effects of different growth forms on the potential growth rates of species in different environments. To this end, the overall efficiency of growth can be partitioned into two components: a resource uptake efficiency and an internal use efficiency, in which uptake efficiency is defined as the rate of uptake of a resource relative to the rate of supply of that resource at a site. Internal use efficiency is then the rate of biomass production relative to the rate of uptake of the resource. So,

Growth efficiency = resource uptake efficiency × internal use efficiency

where

$$\text{Resource uptake efficiency} = \frac{\text{resource uptake}}{\text{resource availability}}$$

and

$$\text{Internal use efficiency} = \frac{\text{biomass production}}{\text{resource uptake}}$$

Warren Wilson (1981) has proposed a modification of the traditional growth analysis equations (Evans, 1972) to incorporate equivalent terms for the efficiency of utilization of light by crops. Overall growth efficiency will reflect compromises between the efficiency of uptake of resources and the efficiency of internal use of resources by different growth forms. For example, a growth form with a pattern of leaf display that allows a high rate of absorption of light at high light levels (and thus high uptake and growth efficiencies) can be expected to have a low growth efficiency at low light levels in part because of high metabolic costs of supporting excess leaves and branches (and thus a low efficiency of internal use of resources). Aspects of the morphology and physiology of plants that affect these efficiencies are likely to be important in understanding the differentiation of species along gradients of resource levels created by natural disturbances.

B. Aboveground Growth Patterns and the Uptake and Use of Resources by Woody Plants after Disturbance

Allocation of carbon to nonphotosynthetic tissues in woody plants implies that the rate of increase in growth with increasing plant size is limited by the degree to which woody plants can increase their effective leaf display and light absorption. While the fraction of available light absorbed by a canopy is largely a function of leaf display, the quantity of biomass produced for each unit of light absorbed will be determined, in part, by the respiratory costs of maintaining roots and shoots. In the rest of this section, we review studies that suggest that differences in the efficiency of uptake of light and the internal use of photosynthate due to differences in branching and leaf display may have a significant influence on the competitive ability of woody plants following disturbances. We will concentrate on aboveground patterns because of the scarcity of information about roots.

Horn's (1971) monograph on the relationship between leaf display and the successional status of trees is perhaps the most widely cited work on the ecological significance of leaf display in woody plants. He argued that trees with sparsely packed leaves on many layers of branches should have higher rates of total canopy photosynthesis at high light levels than trees with uniformly packed leaves in one or a few layers (Horn, 1971). The converse should be true at low light levels. In general, late successional temperate trees show lower numbers of effective layers of leaves, whereas species characteristically found in high light regimes following large disturbances have higher numbers of layers of leaves.

The predictions of Horn's model are based largely on the efficiency of uptake and the use of light by different patterns of leaf display. Although a single layer of densely packed leaves could have an uptake efficiency equal to or greater than that of several layers of more sparsely packed leaves, the efficiency of the internal use of high light levels will be higher for the multilayer if individual leaves are light saturated at considerably less than full sun and the overall interception of light is partitioned among a greater total leaf area. However, Horn's model does not consider the consequences of allocation of photosynthate to the roots and shoots required for the support of different patterns of leaf display. Subsequent studies have begun to consider the influence of different branching patterns on the efficiency of interception and use of light (see, e.g., Whitney, 1976; Honda and Fisher, 1978; Steingraeber *et al.*, 1979).

While Horn's work dealt largely with the display of leaves in whole crowns, there is an extensive literature on environmental influences on individual leaves (see, e.g., Parkhurst and Loucks, 1972; Taylor, 1975; Givnish and Vermeij, 1976; Givnish, 1979). Temperate tree species that grow rapidly in large disturbances and correspondingly high light levels tend to have more vertically oriented leaves than later successional species (McMillen and McClendon, 1979). Higher leaf angles in high light regimes should allow greater penetration of light to leaves deeper in the crown and should thus allow a greater effective area of leaves displayed per unit area of ground. Thus, high leaf angles in a crown with a high leaf area index should increase the overall efficiency of the use of high light levels for the same reasons that multilayered leaf displays have a high use efficiency at high light levels. High leaf angles can also reduce the heat load on leaves, and thereby potentially decrease water stress and the reduction of photosynthesis due to stomatal restriction of gas exchange (McMillen and McClendon, 1979). The correlation between leaf angle and light levels does not appear to be as strong in moist tropical forests, where a number of species that respond to large disturbances have large, nearly horizontal leaves (Whitmore, 1975).

Studies have also investigated the effects of branching patterns on the efficiency of interception of light by woody plants. Honda and Fisher (1978) and Fisher and Honda (1979) have modeled the effects of branch angle and length on the efficiency of leaf display in *Terminalia catappa*. Measured branch angles in trees of *T. catappa* were close to angles predicted by a branch simulation model for the maximization of effective leaf area (Honda and Fisher, 1978). A series of studies of branching patterns in forest shrubs (Pickett and Kempf, 1980; Kempf and Pickett, 1981; Veres and Pickett, 1982) indicate that different species can minimize leaf overlap through a variety of patterns of branching.

There have been a number of attempts to use stream-ordering techniques to describe the architecture of branching of woody plants (Leopold, 1971; Oohata and Shidei, 1971; Whitney, 1976; Steingraeber *et al.*, 1979; Pickett and Kempf, 1980; Steingraeber, 1980; Veres and Pickett, 1982). According to the stream-ordering system developed by Horton (1945) and modified by Strahler (1957), terminal leaf-bearing shoots are considered first-order branches. Whenever two branches of equal order meet, the resulting branch is assigned the next highest order. The resulting ordering of both physical and biological branching patterns reveals several fairly consistent patterns. The most often cited pattern is for the numbers of branches of successively higher orders to decrease geometrically. In addition, the mean diameter of branches of successively higher orders increases approximately geometrically. The average ratio between the number of branches of successive orders is termed a "bifurcation ratio."

The relationship between bifurcation ratios and the efficiency of leaf display by woody plants was first explored by Leopold (1971), who suggested that leaves could be displayed in a horizontal plane using the shortest total length of branches if the branch system had a low bifurcation ratio. The total length of branches required for the display of leaves in a vertical, cylindrical crown would be minimized by a branch system with a high bifurcation ratio (i.e., many short first-order branches off a few nearly vertical second-order branches). Whitney (1976) subsequently reported that for 16 woody species in the northeastern United States species characteristically found in open, high light regimes had higher bifurcation ratios than shade-tolerant species. He proposed that high bifurcation ratios were characteristic of the branching patterns of species with multilayered leaf display. Whitney (1976) and Oohata and Shidei (1971) also suggested that bifurcation ratios were species-specific constants. However, Steingraeber *et al.* (1979) and subsequent workers (Pickett and Kempf, 1980; Veres and Pickett, 1982) have shown that for a number of species individuals from different light regimes had systematic differences in bifurcation ratios.

The use of bifurcation ratios to describe branching patterns in trees has been criticized on the grounds that the ordering system does not reflect the actual process of development of the shoot systems in woody plants (Borchert and Slade, 1981; Honda *et al.*, 1981). However, the very features that led to these criticisms by developmental biologists provide the motivation for the recent interest of ecologists in bifurcation ratios. Both within and between species, bifurcation ratios of whole shoot systems and of separate branches within the crowns of various species have been shown to be correlated with both light intensity and growth vigor (Steingraeber *et al.*, 1979; Pickett and Kempf, 1980; Borchert and Slade, 1981; Veres and Pickett, 1982). Despite these results, the relationship between bifurcation ratios and the efficiency of the uptake or use of light is unclear. Dimensions of a branch system that are likely to affect both the pattern of leaf display and light interception (i.e., branch angles) and the metabolic costs of producing and maintaining branches (i.e., branch density, length, and surface area) cannot be currently predicted on the basis of bifurcation ratios.

The relationships between the uptake of light by the canopy of a woody plant and the costs of that uptake in terms of allocation of photosynthate to the production and maintenance of roots and branches are only poorly understood. However, there are a number of ways in which branching patterns can differ in the magnitude of allocation of photosynthate to the support of leaves. The costs of producing branches are reflected in both the synthesis of new shoot material and the maintenance respiration of existing branches (Penning de Vries, 1972, 1975). Studies of the respiration of the branches of woody plants indicate that maintenance respiration rates are correlated with the surface area or diameter of branches (Yoda *et al.*, 1965; Whittaker and Woodwell, 1967; Kinerson, 1975; Yoda, 1983). Denser wood should have higher costs of synthesis of a unit volume of tissue (Penning de Vries, 1972), but the higher wood density may allow smaller diameter twigs to provide sufficient mechanical and hydraulic support for leaves with lower annual maintenance respiration due to the smaller average diameter of branches. King (1979) suggested that the stout but low-density twigs of some early successional trees may have low net costs over the relatively short lifetime of a branch (and thus a high internal use efficiency) because initial synthesis costs per unit volume of wood are low. Because the branches are rapidly self-pruned as the trees increase in height, the potentially high long-term maintenance costs of these branches are avoided (King, 1979). Dense, slender branches in many late successional species may have lower net costs than stouter twigs of the same weight if the slender branches are functional for longer periods of time before being self-pruned. Early successional tropical trees with horizontal layers of large, often compound leaves may represent an extreme example of this trend. Large leaf sizes and compound leaves allow some tropical pioneers to display large total leaf areas with a minimum of allocation to secondary branch growth (Givnish, 1978). Rapid height growth in these species would seem to favor minimizing allocation to lateral branches during juvenile stages.

The studies discussed in this section suggest that a high rate of growth following a given disturbance (and thus a high efficiency of growth for that disturbance) will often involve a pattern of branching and leaf display that is less efficient in exploit-ing other disturbances. Our hypothesis that interspecific competition should act to restrict species to disturbances in which they have their highest efficiency of growth is largely based on the apparent specialization of plant architecture and physiology to particular ranges of disturbance intensity. A woody plant with a maximum growth efficiency for a particular combination of resources may well have a higher rate of growth in a larger disturbance because of an increase in the availability of one or more of those resources. But if the overall growth efficiency is lower, there is a greater chance that there is another species with physiological and morphological traits that are more efficient in exploiting the increased concentration of resources in the larger disturbance.

Genotypic variability, phenotypic plasticity, and physiological and morphologi-cal acclimation should all act to expand the range of disturbances in which a species has both a high rate and a high efficiency of growth. A number of studies indicate that species that respond to large disturbances have a considerable ability to accli-mate to changes in environmental conditions and resource availability (Bazzaz, 1979; Bazzaz and Carlson, 1982; Fetcher et al., 1983). However, there are com-parable examples of plasticity and acclimation in the morphology and physiology of species that respond to small disturbances (see, e.g., Logan and Krotkov, 1968; Steingraeber et al., 1979).

Quantifying the effect of plant architecture on the growth of woody plants follow-ing disturbance will require considerable research on whole plant physiology and morphology under different levels of resource availability. There has been surpris-ingly little empirical work on the efficiency of the interception of light by different patterns of leaf display in woody plants. More research is also needed on variation in the respiration rates and metabolic costs of the branching systems of different species. Although most empirical studies of the ecological significance of plant architecture deal solely with aboveground patterns, Orians and Solbrig (1977), Caldwell (1979), and Givnish (1979) have used cost/benefit models to explore the allocation of photosynthate to roots required for different patterns of leaf display. More research is needed on the effects of differences in the allocation of photosynthate to roots versus aboveground growth on the ability of woody plants to respond to disturbance.

There have been relatively few studies that have explicitly considered the effect of site differences on the response of woody plants to disturbance (see, e.g., Smith, 1983). Work on the effect of soil fertility on shade tolerance in an understory herb (Peace and Grubb, 1982) suggests that the effect of soil fertility on the response of tree species to canopy gaps should be investigated. The existence of interactions between site quality and gap response would provide an additional dimension for resource partitioning in gap phase species.

Of more general significance is a greater understanding of the effect of actual growth rates per se on the ability of woody plants to respond to disturbances. We have focused mainly on growth in terms of net aboveground carbon gain. However, differences in the allocation of carbon to specific aspects of growth (e.g., branch diameter growth, height growth, or lateral spread of the crown) should affect the competitive status of a plant (Grime, 1979). The crown form in densely grown trees of *Populus tremuloides* was close to that predicted from a biophysical model for maximizing height growth while maintaining a minimum of structural support (King, 1981). However, rapid height growth in these individuals appears to have been achieved at the expense of a higher susceptibility to windthrow than many later successional species (King, 1981). In contrast, saplings of both *Fagus grandifolia* and *Acer saccharum* show little increase in height growth rates as gap size increases in old-growth forests of northern New York (Canham, 1984). However, lateral growth of the crowns of both species increases continuously across a broad range of gap sizes. Rapid height growth in woody pioneers would seem to be of particular significance in habitats where herbaceous species are capable of establishing a dense layer that can persist long enough to eliminate subordinate individuals of shade-intolerant species. The significance of height growth may be more problematic in shade-tolerant species, in which canopy recruitment may depend on the response of saplings to repeated periods of both suppression and subsequent release in small gaps.

V. DISCUSSION

In many forests, natural disturbances are a principal cause of mortality in woody plants (White, 1979). The variety of traits that reduce the susceptibility of woody plants to mortality by frequent and often severe disturbances suggests that differential mortality has been an important selective process in environments subject to recurrent disturbances. Even if the patterns of mortality following disturbances were entirely nonselective, natural disturbances would still play a major role in the ecology and evolution of woody plants by providing a diverse set of conditions for seedling establishment and plant growth. It is well known that pioneer species require disturbances in order to complete their life cycle. But recent work suggests that even some climax species may depend on the pulses of resources produced by small-scale disturbances to complete their life cycles. In old-growth forests of the Adirondack Mountains of New York, the growth rates of saplings of *A. saccharum* and *F. grandifolia* were too slow for saplings to reach the canopy within the expected life span of either species (Canham, 1981, 1984). However, both species showed significant increases in growth rates in small canopy gaps (Canham, 1984). The significance of shade tolerance in these two species appears to be twofold. Understory saplings have a clear advantage over smaller seedlings in exploiting small canopy gaps before they are closed. However, particularly in the case of *F. grandifolia,* the relatively slow growth rates of saplings even in gaps suggest that many beech trees reach the canopy through multiple episodes of gap formation. The ability of understory saplings to persist for long periods under a closed canopy and subsequently to respond to new gaps is an important element of this mode of canopy recruitment.

Seeds of woody plants are dispersed, germinate, and become established in a much wider range of conditions than the range in which they can successfully reach reproductive size. There are well-known examples of mechanisms that tie the dispersal, germination, and establishment of seedlings to disturbances that create conditions favorable for growth and reproduction of a particular species. Pine trees with serotinous cones release many years of seed crops only after the heat from a fire allows the cone scales to open. Similarly, at least some chaparral shrubs (*Rhus laurina* and *R. ovata*) have dormant seeds that are stored in the soil until stimulated to germinate by the heat of a fire (Stone and Juhren, 1951). And in still other woody plants, germination of stored seeds is coordinated with the occurrence of a canopy opening by means of the light-sensitive pigment, phytochrome. Such examples of specialization, however, should not obscure the fact that many woody plants routinely become established following disturbances as a result of less specialized but perhaps more common traits that allow relatively continuous production, dissemination, and germination of seeds. For example, new plants may come from recently dispersed seeds (e.g., species of *Populus*), from pools of short-lived buried seeds (e.g., *Prunus serotina*), or from a layer of shade-tolerant seedlings (e.g., *A. saccharum*). In addition, most woody angiosperms and a few gymnosperms have the ability to produce sprouts and other kinds of vegetative responses to disturbances. Thus, following disturbance of many kinds of woody communities, a variety of mechanisms, only some of which show specialization with respect to the kind, size, or occurrence of a disturbance, allow the establishment of new individuals and the persistence of residual plants.

It is worthwhile to ask why examples of specialization in patterns of woody plant establishment following disturbance are not more frequent. The absolute frequency of cases of precise cueing of the germination of woody plants to disturbance is unknown. However, even in known cases of phytochrome-induced germination, there appear to be limitations on the ability of seeds to limit germination to disturbances that are large enough to allow growth to reproductive size. While seeds of *Cecropia obtusifolia* germinate in response to canopy openings, many seeds apparently germinate in openings that are smaller than the minimum size required for growth to maturity (Vázquez-Yanes and Smith, 1982). The inability of seeds to sense an environment suitable for growth to reproductive size may be more common in woody plants, in which there is a greater disparity between the size of the immediate environment of a seedling and the eventual amount of space required by an adult than it is for smaller herbaceous species. Apart from the cases involving inducement of germination by the environment of a disturbance, the other major mode of specialization during establishment seems to be limited to habitats where disturbances occur at highly predictable times. Spring dispersal of seeds by temperate floodplain species should be particularly effective in the deposition of seeds on any fresh alluvium left by spring floods. However, many forms of disturbance are predictable only in their occurrence somewhere within a habitat in a given interval of time. To the degree that rapid colonization of a disturbance confers a subsequent competitive advantage, a relatively generalized pattern of dispersal and germination in which a habitat is flooded with seeds that germinate throughout climatically favorable seasons would seem to have advantages in habitats where there is considerable unpredictability in the precise timing and location of disturbances. Similarly, for shade-tolerant tree species, widespread establishment of seedlings in a forest understory prior to a disturbance should confer an advantage in initial height over seedlings that germinate after a small gap opens in the canopy.

It has been frequently noted that pioneer and early successional species have higher potential growth rates than species capable of responding to small disturbances (Grime, 1965; Grime and Hunt, 1975; Marks, 1975). Rapid growth rates are often cited as a basic correlate of r-selection in plants (Grime, 1979). We suggest that rapid growth rates should confer a selective advantage even in species that respond to minor disturbances, but that their growth rates are rapid only in proportion to the relative availability of resources in small disturbances. To the degree that a particular set of morphological and physiological traits will result in high growth rates for only a limited range of disturbance sizes or resource availability, the ability of a woody plant to respond to a disturbance should be more closely correlated with growth efficiency (i.e., growth rate relative to resource availability) than with growth rate per se.

In contrast to many of the combinations of traits involved in seedling establishment, the traits that determine woody plant growth appear to be more specialized in their response to disturbance. The studies discussed in Section IV suggest that a particular growth form will have a maximum growth efficiency in a fairly narrow range of disturbances or levels of resource availability. Factors such as patchiness in the patterns of seedling establishment, differences in the timing of establishment, and regrowth of surviving plants can act to reduce the importance of differential growth rates and efficiencies in the regrowth of vegetation following disturbance.

However, for disturbances that result in a pulse of seedling establishment, differences in the efficiencies of uptake and use of resources by different growth forms should be an important factor in the ability of a woody plant to reach reproductive size.

ACKNOWLEDGMENTS

We thank R. E. Cook and D. S. Gill for helpful comments on the chapter.

RECOMMENDED READINGS

Chapin, F. S., III, and Van Cleve, K. (1981). Plant nutrient absorption and retention under differing fire regimes. *In* "Fire Regimes and Ecosystem Properties," *Gen. Tech. Rep. WO—U.S., For. Serv. [Wash. Off.]* **GTR-WO-26,** pp. 301–321.

Garwood, N. C. (1983). Seed germination in a seasonal tropical forest in Panama: a community study. *Ecol. Monogr.* **53,** 159–181.

Leopold, L. B. (1971). Trees and streams: the efficiency of branching patterns. *J. Theor. Biol.* **31,** 339–354.

Marquis, D. A. (1975). Seed storage and germination under northern hardwood forests. *Can. J. For. Res.* **5,** 478–484.

Vazquez-Yanes, C., and Smith, H. (1982). Phytochrome control of seed germination in the tropical rain forest pioneer trees *Cecropia obtusifolia* and *Piper auritum* and its ecological significance. *New Phytol.* **92,** 477–485.

Chapter 12

Responses of Forest Herbs to Canopy Gaps

B. S. COLLINS, K. P. DUNNE, and S. T. A. PICKETT

Department of Biological Sciences and Bureau of Biological Research
Rutgers University
New Brunswick, New Jersey

THE ECOLOGY
OF NATURAL DISTURBANCE
AND PATCH DYNAMICS

I. INTRODUCTION

Available information indicates that creation of canopy gaps in forests is a widespread and important phenomenon (White, 1979). Single or multiple tree canopy gaps are created in old-growth, deciduous forests each 100 years or so (Runkle, 1981), and about 1–2% of the forest area will be exposed by new gaps each year (Runkle, 1981, and Chapter 2, this volume). The fall of a canopy tree as a result of senescence, environmental catastrophe, or, more likely, a combination of the two frees space, alters resource availability within the forest, and changes the immediate environmental influences on forest organisms. Such changes have been shown to have a profound effect on woody species in forests (Canham and Marks, Chapter 11, this volume). Dispersal of propagules (Thompson and Willson, 1978), germination of seeds (Marks, 1974), survival of seedlings (Hibbs, 1982), growth of trees (Trimble and Tryon, 1966), plant architecture (Veres and Pickett, 1982) and reproduction (Hibbs and Fischer, 1979) are important phenomena that are sensitive to gap formation.

Because gaps can affect so many processes, the temporal and spatial patterns of gap creation interact with species strategies to determine forest richness and structure (Denslow, 1980a; Oliver, 1981; Pickett, 1980; Whitmore, 1978, 1982). Much of tree biology, and the structure of the woody component of forest vegetation, can be understood in terms of gaps and disturbance regimes (Grubb, 1977; Whittaker and Levin, 1977). Both the tree layer and shrubs and woody vines show a pronounced response to gap formation (Crawford, 1976; Veblen and Ashton, 1978).

In spite of the clear importance of disturbance and gaps in forest ecology, there is almost a complete lack of information about the role of gaps in the ecology of forest herbs. Indeed, aside from general information on distribution, phenology, and, more recently, pollination biology and demography, forest herbs are poorly known ecologically. A rich and productive herb layer is a conspicuous component of temperate deciduous forests, especially those on mesic sites. The herb layer is important in energetic and trophic interactions (Bormann and Likens, 1979), nutrient flow (Blank et al., 1980; Muller, 1978), and tree regeneration (Maguire and Forman, 1983). Here, we will assemble the scattered information on herbs related to their response to canopy gaps and their participation in forest gap dynamics. We will focus our discussion on herbs of the mesic deciduous forests of eastern North America and Europe, although we will sometimes draw on information about herbs in other forest types. We will discuss herbs of all phenological guilds and life history types, including spring ephemerals, summer green, and evergreen species.

It is appropriate to separate consideration of mesic forest herbs from that of woody plants (Canham and Marks, Chapter 11, this volume) or herbs of grasslands (Loucks et al., Chapter 5, this volume) for several reasons. First, the lack of persistent vertical structure, which distinguishes herbs from woody vegetation, delimits their potential responses to disturbance. Unlike trees, many herbs have narrow phenological windows. Additionally, horizontal architecture, clonal spread, and reliance on a newly restructured shoot system each year are aspects of herb biology that differ conspicuously from that of woody plants. Second, mesic forest herbs differ from herbs of either permanently open communities, such as grasslands, or ephemeral open communities, such as old fields (Zangerl and Bazzaz, 1983). Unlike herbs of prairies, forest herbs are usually unequal competitors with the structural dominants of the community and are relatively intolerant of physical stresses. Their capacities for resource uptake and deployment undoubtedly differ from those of prairie plants (cf. Blank *et al.*, 1980; Parrish and Bazzaz, 1982).

Our first task will be to review the environmental conditions that can exist in canopy gaps. Second, we will examine the degree and mechanisms of responses of herbs to varying environmental conditions. Both ecophysiological and life history responses will be considered. Third, we will show or predict the responses of herbs to forest gaps. The opportunities of the various herb guilds to respond to gaps and the constraints to immediate or evolutionary responses will be important issues. Finally, we will discuss the implications of forest patch dynamics for the ecology of forest-dwelling herbs. Community organization and the balance of the various life history guilds, species coexistence, and population structure are important issues in this section. This chapter will indicate the importance of herbs in forest patch dynamics. We hope to stimulate research on forest herbs that will further understanding of their relationships with natural disturbance.

II. GAP ENVIRONMENT

Other authors have considered the relationship of gap environment and the response of woody plants (Runkle, Chapter 2; Brokaw, Chapter 4; Canham and Marks, Chapter 11, this volume). This section will present information on gap environment that is relevant to herbs. We first enumerate the characteristics of gaps and then present the available information on conditions in gaps that might affect herbs.

The physical characteristics of gaps, such as size and shape, affect all environmental parameters of gaps and are thus of major importance. Furthermore, the various characteristics may be combined in different ways to produce heterogeneity within and among gaps.

A. Size, Shape, and Depth

One principal characteristic of gaps is size. The importance of size is demonstrated by the failure of light-demanding trees to establish or succeed in gaps smaller than some minimum area (Kramer, cited in Whitmore, 1978). Some investigators have divided forest trees into categories of small versus large gap preference (see, e.g., Denslow, 1980a). Generally, an opening in the canopy less than about 5 m across is not considered a gap (Brokaw, 1982b).

Size alone is a rough indicator of the gap environment, but other factors also have strong influences on conditions within gaps. The height and patchiness of adjacent canopy determine whether a gap of a particular area will have an environment much different from that of the closed forest. Small gaps in either tall or open canopies have little effect. Gap shape also determines the impact of an opening of a given area. Long, narrow gaps will have much less influence on the understory than will more isodiametric ones.

The depth the gap reaches in the forest profile will also have an impact on gap environment. In studying tree regeneration, a gap is considered effective if it reaches to within 2 m of the ground (Brokaw, 1982b). Clearly, this is a convenience based on the human scale; the biology of forest herbs may require a different definition. Whether saplings, shrubs, or tree seedlings remain after formation of canopy gaps can have a significant impact on forest herb response.

Gaps may alter the substrate as well as some or all of the forest profile. If trees are uprooted, bare soil and often parent material will be exposed in the gap. In small gaps, this may amount to a large fraction of the substrate. Even if trees are snapped off, which is a common occurrence (Runkle, 1981), the substrate conditions may be altered. Leaf and branch litter, as well as the conspicuous downed trunks, may be clumped in part of the gap. The success of herbs in gaps may depend on the thickness and distribution of such debris.

B. Modification of Gap Characteristics

The above are primary characteristics of gaps; their dimensions and spatial patterns can have much influence on herb response. But these factors may be further modified. Topography may enhance or diminish the impact of gaps. Steepness and aspect of the slope, along with canopy height, will determine the path and duration of insolation in and around the gap, and thus will affect moisture availability and plant and substrate temperature in the gap. Orientation of asymmetrical gaps alone or in conjunction with topography will determine the environment.

Further complexity of the gap environment may depend upon the presence of tree and shrub seedlings or sprouts. How long the gap lasts before closure by growth of either the surrounding canopy or embedded woody plants can also influence the degree of herb response.

This survey of the major characteristics of gaps, and of the circumstances that modify the impact of these primary characteristics on the gap environment, illustrates the complexity that may be found in canopy gaps. Not only is the collection of gaps in a forest heterogeneous, but each gap is itself a mosaic that changes from the center to beneath the closed canopy. We are not aware of any published studies that address this whole range of variation, but there is a literature that shows, in broad outline, the nature and degree of environmental alteration associated with gaps. We review this in the following sections.

C. Light

The most conspicuous environmental change in forest gaps is the increase in light. For example, in mixed hardwood and oak–hickory forest in southern Illinois,

Minckler and Woerheide (1965) discovered that as gap diameter, measured as a ratio of diameter to intact canopy height (D : H), increased, so did light intensity:

Gap size (D:H)	Ambient light (%)
0:1	10
0.5:1	20–45
1:1	45–70
2:1	65–90

Other studies also demonstrate increased light intensity in gaps (Jackson, 1959; P. C. Miller, 1969a,b; Skeen, 1976). Even without leaves, deciduous canopies reduce the passage of light. Leafless European beech and oak stands reduce light by about 30% (Geiger, 1965); gaps would thus be brighter than the rest of the forest even before leaf flush.

No measurements of light quality have been made over a range of gap sizes, but measurements in stands thinned to different degrees (Marquis, 1973) suggest that, in small gaps, light reaching the herbaceous layer would be enriched in far red (730–740 nm), green (500–600 nm), and blue (400–500 nm) light. Larger gaps, in which part of the forest floor was exposed to direct insolation, would present a complex pattern of filtered and unfiltered sunlight. Patchiness of insolation remains unmeasured.

D. Temperature

Soil temperature also changes in gaps. Large clearcuts have higher maximum and lower minimum soil temperature than closed forest (McGee, 1976; Pontailler, 1979). Ash and Barkham (1976) found the April maximum daily soil temperature to be as much as 5°C higher and the minimum 3°C lower in cleared areas. The June daily maxima were as much as 13°C higher and the minimum 2°C lower in cleared areas. Surface soil temperature in gaps of one to a few trees depends on whether the particular spot is exposed to insolation (Pontailler, 1979), but maxima in summer may be 4–5°C higher in gaps than in adjacent forest (Marquis, 1973; Pontailler, 1979; Skeen and March, 1977). Mean daily minima are sometimes lower in both summer and winter in gaps than beneath a closed canopy; however, winter maxima depend more on the degree of soil cover by mosses, herbs (Pontailler, 1979), and litter or snow (Federer, 1973). Tip-up mounds are more likely to freeze than are other sites on the forest floor (Federer, 1973).

Patterns of air temperature in gaps are complex and appear to depend on gap size, insolation, and wind patterns. Guntenspergen and Stearns (1984) monitored the air temperature regimes in forest clearings in a northern hardwood forest in Wisconsin. During summer, mean monthly maximum temperature increased with increasing clearing size until a D : H ratio between 1.5 and 2.0 was reached. When clearing size increased beyond a D : H of 2.0, the maximum temperature decreased due to increasing wind. Pons (1977) found the April maximum daily air temperature to be 1–4°C higher and the minimum to be 0–3°C lower in 1-year-old clearings. The June maximum daily temperature was 5–10°C higher and the minimum was 0–3°C lower in these openings. Air temperature was higher and fluctuated more in two natural gaps in old-growth oak–beech forest than beneath the intact canopy (Pontailler, 1979).

E. Soil Moisture

Soil moisture differs between gap and canopy sites. Throughfall, evaporation, and transpiration, which interact to determine the amount of soil moisture, very likely differ in and around gaps. Since intact canopies can intercept a large fraction of rainfall and, in small rainfall events, can return a great deal to the atmosphere by evaporation from leaf surfaces, gaps may permit more precipitation to reach the forest floor. Sites opened by oak wilt mortality had higher soil moisture than unaffected areas (Auclair, 1975). Minckler and Woerheide (1965) found that available soil moisture in the top 18 in. increased from 5% beneath intact canopy to 10% at gap edge to 20% at the center of a large opening. Other studies have also shown less soil moisture in intact forest than in thinned sites (Marquis, 1973) and some gaps (Pontailler, 1979). In spite of the high quantity of rainfall in large gaps, the surface 2.5 cm of the forest floor may be quite dry (Marquis, 1973). The northern sides of gaps may be drier than elsewhere due to insolation and subsequently to greater growth and transpiration of woody vegetation (Pontailler, 1979). Tip-up mounds tend to be drier than the remainder of the forest floor (McLintock, 1959). Vapor pressure deficit is higher in gaps than beneath closed canopy (Pontailler, 1979; Schulz, 1960).

F. Nutrients

It is difficult to say how gaps might affect the availability of nutrients to herbs. In spite of the large amount of work that has been done with forest nutrient dynamics (Bormann and Likens, 1979; Vitousek, Chapter 18, this volume), none has focused on gaps as opposed to areas having intact canopies. Nutrient availability may increase in gaps due to increased decomposition of organic matter, and increased nutrient release may in turn increase mass flow of nutrients to herbs. However, because nutrient availability is controlled by many interacting and often antagonistic factors (Gorham et al., 1979), it is impossible to predict with certainty whether nutrients available to herbs would increase in all gaps. Because herbs that are active during the summer apparently compete with trees for nutrients (Peterson and Rolfe, 1982), we expect that gaps may enhance nutrient availability to some herbaceous plants.

G. Animals as Resources

A final type of resource that might differ between gaps and closed canopies is the presence or activity of mutualists. Little information is available on the levels of such resources, but a few examples indicate feasible effects. In sunflecks, the activity of itinerant pollinators is greater than in the full shade of the canopy (Beattie, 1971). Since much dispersal of herbs is also accomplished by insects (Beattie and Lyons, 1975; Handel, 1976), the higher temperatures in gaps may enhance dispersal as well as pollination. Thompson and Willson (1978) discovered that bird-disseminated fruits presented experimentally in gaps were more likely to be removed than controls beneath closed canopy.

III. RESPONSES TO THE ENVIRONMENT

In both field studies and under more controlled laboratory conditions, herbs have been found capable of shifts in metabolism and life history in response to differing physical environments. In this section, we review known responses of herbs to several environmental parameters that may differ in gaps compared to intact canopy.

A. Temperature

The yearly timing of herb phenology, including the date of plant emergence and the length of time to maturation and senescence, may be influenced by soil and air temperature regimes. *Viola sororia* emerges from its overwintering state when soil temperature exceeds 10°C and air temperatures are >0°C (Solbrig *et al.*, 1980). Other spring herbs, *Claytonia caroliniana* and *Erythronium americanum,* appear when the soil warms above 0°C (Vezina and Grandter, 1965). Acceleration of shoot growth and development of *E. americanum* immediately follow snowmelt, and development of photosynthetic tissue and flowering correlate with air temperature (Muller, 1978).

Temperature regimes and the diurnal fluctuations of temperature may also affect assimilation and growth in some herbs. When *Fragaria vesca* from a clearing edge were exposed to various controlled day/night temperatures (Chabot, 1978), net photosynthetic rates were highest in plants kept in moderate (20/10°C and 30/20°C) temperatures, and maximum assimilation was seen between 15° and 30°C. Diurnal temperature amplitude also affected net photosynthesis; the highest assimilation rates were noted when day/night temperatures differed by 10°C.

The daily temperature amplitude of 10°C also correlated with the greatest total biomass gain in *Fragaria.* Biomass was allocated to asexual reproduction when diurnal temperatures fluctuated by 10°C and when plants were kept at moderate (30° day/20°C night) temperatures. Flowering, however, occurred only under 20°/10°C temperatures (Chabot, 1978). It appears, therefore, that altered temperature regimes in gaps may affect both herb growth and reproduction.

B. Light

Herbs may be grouped into three general categories relative to their photosynthetic response to varying light environments: inflexible *sun* plants are genetically adapted to high-intensity light environments; inflexible *shade* species are able to assimilate only in low-intensity light; and *light-flexible* herbs are physiologically plastic over a variety of light intensities. Most woodland herbs fit into one of these general categories. A few species, however, have photosynthetic characteristics of more than one group. For example, sun clones of *Solidago virgaurea* (Bjorkman, 1968), *Geum rivale* (Bjorkman and Holmgren, 1966), and *Solanum dulcamara* (Gauhl, 1976) adjust to either low or high light by varying the concentration of ribulose-1,5-bisphosphate (RUBP). In contrast, shade clones of these species cannot increase the concentration of carboxylating enzymes and so cannot adjust to high-intensity light.

Herbs such as *Maianthemum canadense* and *Oxalis montana*, which develop beneath a closed forest canopy, are examples of obligate shade species (Sparling, 1967). Light-flexible species are exemplified by *S. dulcamara;* clones from both sun and shade habitats can adjust to varying light intensities (Clough *et al.,* 1979).

Light quantity has been shown to influence both assimilation and growth in *F. vesca* (Chabot and Chabot, 1977). When plants from a clearing edge were grown in low (25 E/m²/sec), medium (150 E/m²/sec), and high (650 E/m²/sec) light (Chabot, 1978), light compensation points increased with the treatment light intensity, and saturation light intensities were higher for plants under medium and high light. Expressed on a dry weight basis, net photosynthesis was greatest in the low-light treatment; however, on a leaf area basis, photosynthetic rates increased with light intensity.

The differing light environments produced differences in the biomass yield and leaf morphology of *Fragaria* (Chabot, 1978). Aboveground biomass increased directly with the light intensity of the growing treatment. Additionally, both sexual and asexual reproduction were confined to the medium- and high-light regimes, and the proportion of biomass allocated to flowering was greatest in high light intensity. The assimilation pattern was accompanied by progressively thicker and more densely packed leaves under high light intensity.

Variations in the light environment have been found to elicit changes in the growth patterns and reproductive potential of other herbs. In *Uvularia perfoliata* populations transplanted into North Carolina hardwood forests, pine woods, or meadows (Whigham, 1974), 22% of the emerged plants in the hardwood forest flowered as opposed to 13.9% in the pine woods and meadow, and significantly more plants reproduced vegetatively under the hardwood canopy. These results suggest that increased light or other environmental conditions of the pine woods and meadow may be adverse to *Uvularia.*

Light quantity has also been shown to be a major factor controlling the growth and distribution of the European bluebell, *Scilla non-scripta,* a woodland plant absent from open sites (Blackman and Rutter, 1950). Over a 4-year period, shading to 0.5–0.6 and 0.2 daylight lead to heavier bulbs (1.64 and 1.80 final weight/initial weight versus 0.87 for control unshaded plants).

Neither leaf, stem, nor petiole dry weight of the woodland perennial *Veronica montana* differed significantly when plants were grown in unshaded (25–45

W/m^2), shaded (10–20 W/m^2), or enhanced far red conditions. The root dry weight of *V. montana* was greater in unshaded plants (Fitter and Ashmore, 1974). However, in the open habitat congenor, *V. persica,* reduced light quality was associated with leaf, petiole, and root dry weight decrease, while light reduction accompanied by increased far red light caused a further decrease in leaf dry weight (Fitter and Ashmore, 1974). These results suggest that light quantity or quality comparable to that beneath the canopy may be detrimental to the growth of open-habitat species.

Following coppicing of European forests, shoot numbers of the bulbous perennial *Narcissus pseudonarcissus* may increase sharply for 4 years (Barkham, 1980b). By censusing individually marked plants, Barkham (1980a) found that the probability of an adult *Narcissus* flowering was greater in open sites than under hardwood canopy. Further, more seeds were produced in the open, and there was a greater chance that vegetative offspring would reach adult size in their first year. The proportion of vegetative offspring was three times greater in open sites (Barkham, 1980a).

Plants of another European bulbous perennial, *Allium ursinum,* under a relatively open (receiving 65% of full radiation) *Stellario–Carpinetum* canopy had bulbs five and a half times heavier than those of plants on a south-facing *Asperulo–Fagetum* forest (receiving 40% of full radiation), and these individuals had bulbs twice as heavy as those of *Allium* on a north-facing slope that received only 6% of full daylight (Ernst, 1979). These results indicate that increased light in the more open forest may correlate with increased growth of *Allium.*

Laboratory studies of *S. dulcamara,* which is found in both high and low light environments in the field, indicate that light quantity may influence plastic growth responses (Clough *et al.,* 1979). Plants in 100% light had specific leaf weight (SLW) 170% greater than that of plants in 4% light. Individuals in the low light treatment put a greater proportion of their mass into leaves and roots (68% and 13% as compared to 53% and 21% in high light plants).

In general, it appears that variations in the light environment influence photosynthesis and assimilation in many woodland herbs. Less directly, changes in the light regime may correlate with changes in growth and allocation to sexual and asexual reproduction.

C. Water Relations

The responses of woodland herbs to variations in soil moisture or relative humidity have been little studied in the United States. *S. dulcamara* exhibited differences in both physiological and growth responses when watered either once per day or once per 5 days (Clough *et al.,* 1979). Plants from the high water treatments had $\Psi = -3.8$ bar at both the beginning and the end of the treatment, while water-stressed plants had $\Psi = -13.1$ bar on the fourth day without water. The total leaf conductance was 29% greater in the well-watered treatment; there was a small increase in both chlorophyll content and photosynthetic unit (PSU = P700/total chlorophyll) density, and a small decrease in SLW and PSU size. In the plants watered once per day, 56% of the total dry weight was allocated to leaves and 17% to roots, while in the water-stressed plants 52% was allocated to leaves and 28% to roots (Clough *et al.,* 1979).

In Czechoslovakia, Elias (1978, 1981) has studied the water relations of spring and summer herbs in temperate deciduous forests. In two oak–hornbean woods, light during the spring ephemeral growing season was c. 30–50% of full, and soil moisture was close to field capacity (Elias, 1981). The specific leaf area (SLA) of the spring herbs varied between 30 and 70mm^2/mg dry matter. Their water saturation deficit ranged from 0.3 to 10.0% and their leaf water and osmotic potentials ranged from -0.18 to -0.63 MPa over 3 days. The water-holding capacity of the plants was low. In the laboratory they lost water quickly under desiccation, and, on a dry mass basis, maximal transpiration was high, exceeding 23.35 mg/g/min (Elias, 1981).

In a similar study, the water relations of 7 spring ephemerals and 16 summer herbs in a mesic forest with summer drought were investigated (Elias, 1978). Over the growing season, soil moisture in the upper 0.1–0.2 m, where herbs were rooted, fell from 20–25% in April to c. 12% in the midsummer drought, and then varied in response to rainfall, eventually rising again to 24% by the end of October. Water deficits of the herbs were correlated with soil moisture in the top 0.1 m; deficits declined from 35–60% at 10% soil moisture to <10% at 17–23% moisture. Leaf water deficits of the spring herbs were low, ranging from 0.3 to 9.96% over the seven species measured. However, deficits of the summer herbs ranged from 10 to 60% during the summer drought, and these plants had wilted leaves and aerial shoots. Throughout a daily cycle, maximum water deficits of the summer herbs correlated with or lagged behind the maximum air temperature, and the daily deficit could vary by 10%.

The results of these few studies indicate that herbs vary in assimilation, growth, and allocation patterns in response to fluctuations in soil moisture.

D. Nutrients

Researchers of woodland herb nutrient relations have focused almost exclusively upon seasonal responses of plants rather than upon individual responses to varying nutrient availabilities. VanAndel and Jager (1981) investigated the nutrient relations of six herbs of differing life-history strategies that colonize woodland clearings in Europe. In the annual rosette plant *Senecio sylvaticus,* N, P, and K resources were initially put into leaves and later into flowers. In perennial species, late season reserves were put into the overwintering parts: in *Digitalis purpurea* (a biennial rosette), most N, P, and K resources were in the rosette leaves, while in *Rumex acetocella* and *Chamaemerion angustifolium,* neither the dry weight nor the N, P, and K content of leaves increased after the 13th week of growth; rather, the mass of the perennating roots increased.

A similar pattern of nutrient accumulation and allocation was seen in the European bluebell *Scilla non-scripta* (Blackman and Rutter, 1949) and in *E. ameri-*

canum in hardwood forests of the northeastern United States (Muller, 1978). In *Scilla,* the N, P, and K that rapidly accumulated during leaf expansion were allocated to leaves and inflorescences. With senescence and seed ripening, nutrients were withdrawn into the overwintering bulb (Blackman and Rutter, 1949). In bulbs of *E. americanum,* a spring ephemeral, concentrations of K, Ca, and Mg increased slightly from January 27 through April 14, while N increased 3.3–4.1%. During rapid spring growth, the concentrations of these nutrients in the bulb decreased and then remained stable until leaf senescence, when concentrations of all elements in the bulb and runners increased (Muller, 1978).

There are no complete studies of how alterations in the nutrient environment may affect ion accumulation or allocation in woodland herbs of the eastern United States. With increased levels of N, P, and K, *S. non-scripta* has greater nutrient uptake. However, if light is reduced to 20–22%, there are significant decreases in N and P absorption. Further, shading enhances the percentage content of N, P, and K in bulbs and shoots, although the total amount of these nutrients in the shoot is lower (Blackman and Rutter, 1949). These findings suggest that greater light intensity and soil nutrient concentrations could result in larger nutrient concentrations in some herbs.

IV. HERB RESPONSES TO GAPS

In this section, we will use information presented in the previous two sections to discuss possible responses of herbs to canopy openings. We will consider both responses of already established herbs and the role of gaps in establishment and maintenance in the community of typically nonforest herb species. For this discussion, we will assume an "ideal" gap and will predict maximal herb responses. Constraints on these conditions will be discussed in a subsequent section.

Possible responses of an herb species to the suite of resources within a gap will be governed by the genetic capabilities of the plant. In the foregoing discussion of observed herb responses to environmental factors, three broad categories of herbs—sun, light flexible, and shade—were delimited. Since these categories reflect genetic capabilities of the plants, possible responses of each group to a canopy gap will be discussed separately. Although the categorization is stated in terms of herb response to light, we assume that the capacity to use other resources varies in a similar way. For example, maximum rates of assimilation and degree of plasticity of growth are often correlated with demand and flexibility in using nutrients and water (Bazzaz and Carlson, 1982; Parrish and Bazzaz, 1982).

A. Sun Herbs

Herbs in this category have a metabolism adapted to high-intensity light environments. In the temperate deciduous forest, they are the spring ephemerals that mature and begin to senesce before canopy closure (Sparling, 1967). Other, more extreme sun plants are species typically associated with open environments, such as roadside and old-field herbs.

In the leafless spring forest community, a single- or multiple-tree gap may alter the temperature, moisture, and, to a lesser extent, light environment of forest herbs. However, the quality and magnitude of the plant response to these altered conditions may depend both upon the degree of resource alteration (which may be related to gap size) and upon critical levels of the resource required by the plant. For example, in normal rainfall years, spring ephemerals may respond little to the more widely fluctuating moisture regime within a gap (Marquis, 1973; McLintock, 1959; Pontailler, 1979), since spring forest soils may generally be close to field capacity (Elias, 1981). Nor may these plants respond to the light environment within gaps, since there may be little difference, particularly on overcast days, in light intensity between the gap and adjacent leafless canopy area, and herbs may already be light saturated.

In contrast, wider temperature fluctuations within a gap may result in altered phenological timing, assimilation, and resource allocation patterns in spring herbs that initiate growth as the soil temperature rises above freezing. Plant emergence and growth to maturity may occur earlier, or be more rapid, within gaps where the soil may reach minimum growth temperatures earlier in the year. This effect is suggested by observations of two spring plants discussed previously, *V. sororia* (Solbrig *et al.*, 1980) and *E. americanum* (Muller, 1978), in which emergence and growth to maturity are correlated with air and soil temperature.

However, earlier maturation of plants within gaps has various potential consequences. Earlier-maturing plants within gaps could flower either earlier than or concurrently with neighboring plants beneath the canopy. The response seen may depend upon the degree of temperature difference between gap and canopy. Muller (1978) observed that *Erythronium* flowering correlated strongly with air temperature, and Jackson (1966) noted that flowering was advanced on warmer slopes.

Plants that flower earlier in gaps may or may not set more seed than nongap individuals. Although pollinators, which are known to be more active in sunflecks (Beattie, 1971), may have greater activity in gaps, the offset of flowering time may result in poorer development of search images for the earlier plants and, hence, lower reproductive success.

The probability of flowering (Barkham, 1980a; Chabot, 1978; Muller, 1978; Whigham, 1974), number of seeds set (Barkham, 1980a), asexual reproductive potential (Barkham, 1980a; Chabot, 1978), and rate of or total vegetative growth (Barkham, 1980a; Blackman and Rutter, 1950; Ernst, 1979; Fitter and Ashmore, 1974) of herbs may be influenced by the light environment. In gaps, light intensity differences from adjacent canopy area per se may have little effect on photosynthesis and biomass accumulation in spring ephemerals. However, gaps present open, high-light environments that may be photosynthetically exploited for a longer time than closed forests. Herbs that emerge and mature earlier in gaps may senesce either before or after their counterparts beneath the canopy. Muller (1978) observed that 5.0% of an *E. americanum* population in a wooded strip were mature and 58.0% were senesced on a given date, while in an adjacent cut strip, 36.4% of the plants were mature and only 24.3% were dead. This suggests that open-growth plants may remain photosynthetically active longer than those beneath a canopy. However, Muller did not indicate the comparative emergence times of the two populations. Thus, it is not possible to conclude with certainty that *Erythronium* beneath an intact canopy have a shorter photosynthetic phase. In other species, a different response has been observed. Leaves of both *Dentaria diphylla* (Sparling, 1967) and *Anemone nemorosa* (Goryshina *et al.*, 1981) that expand in high light often yellow quickly, while shade-developed leaves persist into summer.

The length of the mature (positive C balance) photosynthesizing leaf phase may affect total seasonal assimilation. For those species that persist longer with a positive C balance in a gap, total seasonal biomass accumulation may be increased. In spring herbs, in which the life history stage is size dependent (Bierzychudek, 1982), this increase in assimilation may be reflected in greater vegetative growth or clonal spread, or by an increased probability of flowering. However, for those species that senesce earlier than or concurrently with plants beneath a leafless canopy, seasonal biomass accumulation may be no greater in gaps. Consequently, the rate of vegetative or clonal growth and the percentage of flowering plants may not differ between gap and canopy environments.

For the spring ephemerals, which may respond primarily to the temperature and secondarily to the light environment in gaps, we predict that canopy openings may trigger changes in seasonal phenology, growth, and reproductive potential. For other nonforest sun herbs, gaps may elicit a different suite of responses. In these plants, for which very high light intensity is required for assimilation and biomass accumulation, canopy openings may provide patches for establishment and growth. In addition, gaps large enough to provide the required light environment may be necessary for continued representation of these species in the forest community. In beech–maple forest one- to two-tree gaps were required for the establishment of disturbance-associated plants such as pokeberry and elder. In smaller openings, the common woodland herbs *Podophyllum peltatum* and *Thalictrum thalictroides* were seen (Gysel, 1951). In a northern hardwoods forest, importance values of the forest herbs (*E. americanum, C. caroliniana,* and *Viola* spp.) were greater 50 years after cutting in stands in which groups of trees had been harvested than in single-tree harvest areas. Forest "invaders" such as *Taraxacum erythrospermum* and *Ranunculus abortivus* were seen only in group harvest stands (Metzger and Schultz, 1981).

B. Light-Flexible Herbs

Light-flexible herbs are photosynthetically plastic over a range of light intensities (Sparling, 1967). In the forest, they include both species that are spatially dispersed over sun and shade patches and species that are temporally widely dispersed and experience both leafless and fully developed canopy. Examples of this latter group are the spring–summer herbs, including *Trillium* spp. (Sparling, 1967; Taylor and Pearcy, 1976; Vezina and Grandter, 1965), *Podophyllum peltatum* (Sparling, 1967; Taylor and Pearcy, 1976), and *Viola* spp. (Sparling, 1967), as well as some forest evergreens such as *Hepatica* spp. (Kawano *et al.*, 1980). Architecturally, the light-flexible herbs are diverse; they range from forms with small, dissected leaves and much vertical stratification, such as *Dicentra* spp., to monolayered forms with large, horizontally aligned leaves, such as *Podophyllum*.

For the spatially flexible herbs, a canopy gap may trigger a corresponding physiological and perhaps architectural shift, resulting in photosynthetic metabolism less designed to maximize light interception and more layered, less horizontal architectural forms. The growth rate and reproductive potential of flexible herbs in gaps may exceed those of plants in canopied areas due to the potential for greater assimilation in a higher light environment. For example, the seed output of *Aster acuminatus* depends in part, on light intensity (Pitelka *et al.*, 1980).

The temporally flexible, or spring–summer, herbs usually emerge early in spring, mature with canopy closure, and senesce in early summer (Sparling, 1967; Taylor and Pearcy, 1976; Vezina and Grandter, 1965). These plants show a characteristic shift from sun to shade photosynthetic physiology with canopy development (Sparling, 1967; Taylor and Pearcy, 1976). For such plants, a gap may elicit responses similar to those predicted for the high light–adapted spring ephemerals. Emergence, maturation, and flowering may be temporally advanced due to earlier soil and air warming, and adequate soil moisture and nutrient flux in gaps.

Over the season, the lack of canopy development may permit elongation of the high light–adjusted phase of photosynthesis in temporally flexible herbs. Such elongation, combined with the potential for earlier emergence and maturation, may permit a greater seasonal C gain in openings. Thus, we may predict greater seasonal biomass accumulation, reflected in increased vegetative growth, clonal spread, and/or sexual reproduction (Pitelka *et al.*, 1980) in temporally light-flexible herbs in gaps. Additionally, an architectural shift to multilayered or less prostrate plant forms associated more with competition than with optimal light interception (Givnish, 1982) may be seen in gaps.

Other temporally flexible herbs include those evergreen species that invest in a partial or full complement of leaves each year and that have photosynthetic characteristics of both sun and shade plants over a year. In deciduous forests of Japan, *Hepatica nobilis* has characteristics of a photosynthetically flexible evergreen (Kawano *et al.*, 1980). By extrapolation of this plant's life history characteristics, we may make some general predictions about evergreen herbs in gaps. First, leaf replacement or new leaf growth, which may occur in spring, could commence earlier in gaps. Second, flowering, if temperature dependent, could be temporally advanced. Finally, as discussed for the spring–summer herbs, a gap may prolong the spring high light–adjusted phase of assimilation. Consequently, leaf number, size, or asexual or sexual reproduction may be greater in openings than beneath canopied areas.

C. Shade Plants

Among the woodland herbs, shade-adapted plants are those such as *Oxalis* spp., which mature and senesce beneath a closed canopy (Sparling, 1967). These plants are usually small and grow close to the ground, with a horizontal leaf arrangement and little vertical stratification. They are physiologically adjusted to low light environments. Many, when placed in bright light, show signs of photolability (Boardman, 1977).

For these plants, a gap may provide more amenable moisture, temperature, and nutrient flux than a canopied environment. However, the increased summer light intensity, especially in multiple tree gaps, may be detrimental to these plants, causing death or lowering total seasonal assimilation. We predict that these species will be absent from all but smaller gaps unless they are sheltered by taller herbs or woody plants.

V. CONSTRAINTS ON HERB RESPONSES

In this section, we identify and discuss constraints on predicted maximal responses of herbs to canopy gaps. Physical and temporal characteristics of gaps, which determine the timing and amount of resource release, and the density and composition of the local flora, are the most important factors. These constraints may act singly or in combination, and their effects may vary depending on the herbs present in the gap.

A. Gap Origin and Longevity

Gaps may occur by either the snapping or overturning of a canopy tree. However, only in the latter case is a tip-up mound formed. The tip-up mound and its accompanying pit provide varying microsites for plant colonization, nutrient flux, and water availability, and may result in an herb layer composition quite different from that in gaps lacking a mound.

A second factor that may influence herb composition or growth is the length of time a gap is open to colonization or herb response. This period is delimited by the gap closure rate. The closure rate of a gap is determined by the opening size and shape, and by lateral and vertical growth rates of affected tree species (Canham and Marks, Chapter 11, this volume; Crawford, 1976; Hibbs, 1982). For long-lived perennial forest herbs, in which growth and reproduction during the current season are dependent upon stored reserves, there may be a lag time between gap formation and plant response. Effective gap closure, by lateral growth of surrounding trees or by vertical growth of trees, shrubs, or other herbs, may limit the available time for plant response.

Herbs in larger, more isodiametric gaps may have a longer available response time than those in smaller or more irregularly shaped gaps. An example of a narrow temporal window of response to gaps appears in *Viola fimbriatula*. Only the cohort established immediately after blowdown of two trees grew large enough to flower. Later cohorts were unsuccessful (Cook and Lyons, 1983).

B. Canopy Density

The magnitude of resource release during gap formation may largely determine the degree and type of herb response. In gaps that form in either open or multi-layered forests, the degree of difference on the forest floor between gap and canopied environments may be small, and there may be little herb response. In contrast, in gaps that have formed in monolayered or densely canopied communities, greater herb response might be seen.

C. Plant Density and Composition

Because most of the individuals that will respond to a particular gap were present prior to gap formation, the initial density and diversity of the herbaceous layer are primary considerations affecting herb response. In communities already saturated with one or more species, the ability of an herb to respond to an opened canopy by increased height or clonal growth may be limited by competition from neighboring plants. The effects of such competition may be regulated in part by the density and composition of neighbors. When *Fragaria vesca* was grown in competition with varying densities of barley, total biomass and allocation to reproduction were depressed below those of *Fragaria* plants grown alone (Chabot, 1978). A dense herb layer may additionally limit herb response by preventing establishment of invaders through depletion of resources or physical occupation of space.

D. Herbivores

Additional constraints on the ability of herbs to respond to a gap may be the removal or destruction of plant biomass caused by selective herbivory or trampling. In the southern Appalachians, introduced wild boar can reduce herb cover to <5% of the expected value, with a shift in plant population structure toward smaller, nonflowering individuals (Bratton, 1974, 1975). In many eastern forests, white-tailed deer cause extensive damage through browsing and trampling; they are known to affect tree regeneration (Aldous, 1941, 1952; Bjorkbom and Larson, 1977; Graham, 1958; Hough, 1949; Marquis, 1974; Stoeckler *et al.*, 1957; Tierson *et al.*, 1966; Webb *et al.*, 1956), and may determine stand structure and composition (Little and Somes, 1965; Ross *et al.*, 1970).

VI. CONCLUSIONS

We have reviewed the literature concerning gap environment and the physiology of forest herbs in order to determine the nature of herb responses in gaps and to make predictions about the impact of gaps on herb populations in mesic deciduous forest. We have concentrated on forest herbs because their biology differs from that of woody forest plants and from that of herbs of either permanent or ephemeral herbaceous communities. The potentially long life spans of forest herbs, especially of species that spread clonally, coupled with the frequent formation of gaps in mesic forests, may have provided strong selection pressure for herb response to gaps.

Specific predicted responses of each herb guild are summarized in Table 1. We suggest the following general conclusions as guides for establishing future hypotheses concerning the interactions of gaps and forest herbs.

1. Changes in environmental signals and resources that may cue or elicit herb response vary with the physical characteristics of the gap and are potentially quite large at the herb level in some gaps.

2. The major physiological response capabilities of herbs may allow them to be divided into the following groups: (a) sun herbs, which demand high levels of light; (b) light-flexible herbs, which adjust physiologically, anatomically, and architecturally to changing light levels accompanying seasonal canopy closure (this group may include both species that emerge in spring and persist into the summer and evergreen species); and (c) shade herbs, which are adapted to low light levels only. These categories emphasize differential adaptation to the light environment, but it is reasonable to expect differences among the groups in adaptive response to other resources, as has been shown for herbs of other communities (see, e.g., Parrish and Bazzaz, 1982.

TABLE 1

Predicted Responses of Herbs to an Ideal Gap[a]

	Sun herbs	Light-flexible herbs	Shade herbs
Water uptake[b]	0	+	+
Nutrient uptake[b]	+/0[c]	+/0	0
Seedling establishment	+	+	0/−
Leaf duration	+/0	+/0	0/−
Assimilation	+/0	+	−
Pollination	+/0	+	+
Flowering	+/0	+	0/+
Seed set	+/0	+	0
Clonal growth	+/0	+	0/−
Architectural shift	0	+	0
Survivorship	0	+	−

[a]Herbs are divided into general categories explained in the text. Responses have been categorized as positive (+), lacking (0), or negative (−).

[b]Predictions concerning water and nutrients are especially preliminary due to a limited data base.

[c]Two symbols in a column suggest that the response will depend on details of herb biology that are not group specific.

3. Not all herbs respond positively to gaps due to (a) genetic limitations of some groups, (b) minimum thresholds of gap size for sufficient resource release, (c) environmental stress or resource depletion in some gaps, (d) competition with woody seedlings, sprouts, or shrubs, and (e) altered numbers or activity of mutualists or consumers.

The wide variety of intrinsic and extrinsic influences on herb response to gaps make prediction of herb–gap interactions difficult in the absence of long-term, multivariate studies. Nevertheless, it appears that the relationship to gaps in mesic deciduous forests is an important aspect of herb ecology that can no longer be neglected.

ACKNOWLEDGMENTS

This chapter reflects work carried out with support of NSF DEB 80-03504.

RECOMMENDED READINGS

Bjorkman, O., and Holmgren, P. (1966). Photosynthetic adaptation to light intensity in plants native to shade and exposed habitats. *Physiol. Plant.* **19,** 854–889.

Cook, R. E., and Lyons, E. E. (1983). The biology of *Viola fimbriatula* in a natural disturbance. *Ecology* **64,** 654–660.

Elias, P. (1978). Water deficit of plants in an oak–hornbeam forest. *Preslia (Praha)* **50,** 173–188.

Sparling, J. H. (1978). Assimilation rates of some woodland herbs in Ontario. *Bot. Gaz.* **128,** 160–168.

Chapter **13**

Adaptations of Insects to Disturbance

TIMOTHY D. SCHOWALTER

Department of Entomology
Oregon State University
Corvallis, Oregon

I. INTRODUCTION

The importance of many insect species as agents of vegetation destruction is generally recognized. However, the dependence of insect survival and reproduction on environmental conditions, especially disturbance, is receiving increased attention. Disturbances influence insect population densities, often through changes in host condition, in ways that subsequently influence interactions and processes at the guild, community, and ecosystem levels of organization. Furthermore, insect responses to environmental changes could represent regulatory mechanisms contributing to ecosystem stability. Consideration of the consequences of adaptive behavior for populations, communities, or ecosystems does not require that these levels of organization represent units of selection. Rather, selection of individual attributes

THE ECOLOGY
OF NATURAL DISTURBANCE
AND PATCH DYNAMICS

235

influences higher-level interactions and processes in ways that further influence individual fitnesses. Hence, an understanding of insect responses to disturbance can contribute to our views of community and ecosystem organization.

This chapter emphasizes the adaptive nature of insect responses, particularly of herbivores, to disturbance. I will discuss (a) insect responses to changing environmental conditions, especially disturbance, and (b) consequences of these responses for communities and ecosystems. Throughout this chapter, the terms "community" and "assemblage" will be used to denote the total biota and a taxonomic group, respectively. I will demonstrate that selection among individual insects by disturbance can lead to interactions at the community level that could mitigate disturbance and contribute further to individual fitnesses.

II. FACTORS INFLUENCING INSECT RESPONSES TO DISTURBANCE

Insect species respond differently to disturbance dynamics, depending on their relative ability to locate and exploit disturbed patches. Some insects are particularly adapted to exploit disturbed patches and become much less abundant during community development in infrequently disturbed patches; other insects are virtually eliminated by disturbance. These differences are reflected in the dynamics of insect assemblages within the spatial and temporal framework of the disturbance regime.

This section emphasizes the mechanisms and cues responsible for insect population responses to disturbance and to postdisturbance changes in community organization. The principal factors governing these responses are insect dispersal and host selection behavior, and resource quality and quantity.

A. Dispersal and Host Selection Behavior

Insect dispersal has been examined extensively but only recently has been viewed as adaptive behavior leading to maximum colonization of available habitats (Clark, 1979; Matthews and Matthews, 1978; W. G. Wellington, 1980). Successful selection of suitable hosts is essential to population establishment at a site. These behavioral attributes represent evolutionary adaptations that enhance species persistence within a constantly changing environment and are, therefore, the primary determinants of species responses to disturbance.

1. Dispersal

Disturbances often significantly reduce or eliminate insect populations locally. Furthermore, disturbance characteristics determine the pattern of occurrence, in time and space, of suitable patches. Therefore, insect populations are faced with both the threat of local extinction and the need to locate suitable resources that vary in abundance over space and time (Matthews and Matthews, 1978; W. G. Wellington, 1980). Population survival within such a dynamic environment requires dispersal of the population through space and often through time. The consequences of poor dispersal ability are illustrated by examination of leafhopper (*Errhomus* spp.) populations following the eruption of Mt. St. Helens in 1980 (P. W. Oman, personal communication). Females of *Errhomus* are brachypterous and incapable of flight; other life stages, except for adult males, also lack means of long-distance dispersal. The volcanic eruption greatly reduced insect and spider populations in many noncrop areas within the zone of heavy ash fallout between the volcano and Ritzville, Washington. *Errhomus* populations, especially those dependent on broad-leaved plant species that suffered more severe damage, did not reach predisturbance levels within this zone by 1981, as did some more mobile insects and spiders. Hence, the small *Errhomus* populations may be vulnerable to further reduction through competition, predation, or other environmental factors. Because most studies of dispersal have concentrated on species with the greatest dispersal capabilities, studies such as Oman's should increase our understanding of the adaptive significance of dispersal.

Few insects are as vulnerable to disturbance as *Errhomus* spp. Most insects have evolved behavioral mechanisms that ensure population dispersal and colonization of suitable patches. The probability that a suitable site will be colonized is a function of the number of colonizing individuals, which is the product of the individual probability of arrival and the number of dispersing individuals (Price, 1975; Schowalter *et al.*, 1981b).

The individual probability of arrival is determined by the dispersal mechanism, i.e., the agent and direction of transportation. Many insects noted for long-distance dispersal capability rely to a large extent on wind transport. For example, the dispersal behavior of migratory locusts (*Schistocerca gregaria* and *Locusta migratoria*) has been found to depend on wind patterns. Locust swarms remain compact not because of directed flight but because randomly oriented locusts reaching the swarm edge reorient toward the body of the swarm. The swarm is displaced downwind into equatorial areas where converging air masses rise. Rising air leads to precipitation and consequent vegetation growth favorable to the locusts (Matthews and Matthews, 1978). Many other insects disperse by riding on larger animals (phoresy). For insects relying on wind or animal agents, the probability of arrival in a suitable patch is determined largely by the degree of behavioral adaptation to predictable weather patterns (as in the case of the migratory locust) or to the dispersal patterns of phoretic hosts (Krantz and Mellott, 1972). However, many insects are attracted to sources of host odors. Certain wood-boring species can be attracted to suitable patches from considerable distances by smoke or host scent plumes emanating from burned, cut, or injured trees (Mitchell and Martin, 1980; Raffa and Berryman, 1980). Such directed dispersal increases the probability of locating suitable patches.

The number of dispersing insects is determined by several factors, particularly dispersal strategy, population density, and nutrition (Price, 1975; W. G. Wellington, 1980). Typical r-strategists, characterizing temporary habitats, produce large numbers of offspring and show a high rate of disperal. Although the individual probability of arrival in a suitable patch is often low, the large number of dispersing individuals maximizes colonization of available habitats. The number of dispersing individuals is also related to population density. Because survival and fecundity are often density dependent (Coulson, 1979; Price, 1975), dispersing individuals may have higher fitness than nondispersing individuals at high population densities (Price, 1975). Finally, certain studies indicate that nutritional factors influence the number of dispersing insects. Wagner *et al.* (1981) and W. G. Wellington (1980) reported that populations of some insects show considerable variation in vigor as a result of variation in food quality and quantity and of maternal partitioning of nutrient resources to progeny. Obligatory flight distance in these insects appears to be determined by the amount of nutrient reserves: dispersing individuals respond to external stimuli only after depleting these reserves. Hence, less vigorous individuals tend to remain within or near the patch, whereas more vigorous individuals tend to disperse to more distant patches. As a consequence of interaction between these factors, dispersal tends to peak while favorable resources still promote population growth and vigor.

Although I have emphasized dispersal in space, dispersal in time is also an important mechanism for some insects, particularly those exploiting plant reproductive structures. Fruit and seed production by many plant species fluctuates erratically from year to year at a given site. Several holometabolous fruit-and-seed predator insects can undergo an extended diapause in the pupal stage. A proportion of each cohort becomes active during each of several successive years (Hedlin *et al.,* 1982). Thus, portions of these insect populations are dispersed through time to years of favorable fruit and seed production.

2. Host Selection

Population establishment within a colonized patch depends on successful location of suitable resources. As with dispersal, disturbances have provided selective pressures on insects to develop a variety of mechanisms for locating suitable resources.

Some dispersing insects transported passively by wind or animal agents apparently rely on chance to deposit them on suitable food resources (Matthews and Matthews, 1978), but most insects are apparently capable of actively searching for suitable feeding sites, at least within the immediate vicinity of the landing site. Many insects fly randomly to both host and nonhost materials (Raffa and Berryman, 1980). By sampling a large number of potential host materials, populations of such insects maximize the probability of eventually colonizing suitable resources.

Selection of suitable food material involves visual and chemical cues. Visual cues include wavelengths of light and host silhouettes. For example, aphids are attracted to young foliage by longer-wavelength yellows and greens (Matthews and Matthews, 1978); the mountain pine beetle (*Dendroctonus ponderosae*) is attracted to silhouettes of tree boles (Raffa and Berryman, 1980). Chemical cues include host allelochemics, nutritional factors, and insect pheromones. The importance of plant biochemistry to host selection and population growth by insect herbivores has been reviewed by Blum (1980), Matthews and Matthews (1978), and Mattson (1980).

Plant allelochemics and nutritional status provide taste and scent cues that indicate plant suitability for herbivore development. These cues differ among plant species and vary with plant age and condition. Plant chemicals are often used by insect herbivores as precursors in the production of pheromones and may also synergize the attractiveness of pheromones (Blum, 1980; Matthews and Matthews, 1978; Raffa and Berryman, 1980). Mating pheromones provide such powerful attraction to suitable hosts that eventually these chemicals may be used to interfere with host selection behavior and thereby control populations of insect herbivores (Matthews and Matthews, 1978; Sower, 1980).

B. Resource Quality and Quantity

Whereas insect colonizing ability is determined by dispersal and host selection behaviors, subsequent population growth and persistence are determined by resource quality and quantity within patches. Resource quality and quantity are functions of disturbance impact. The factors that are particularly important are vegeta-

TABLE 1

Arthropod Biomass and Percentage of Representation by Functional Groups on Foliage during 1978 on an Undisturbed Watershed and on a 2-Year-Old Clearcut Watershed at Coweeta Hydrologic Laboratory, North Carolina

Functional group	Undisturbed		Clearcut	
	kg/ha	%	kg/ha	%
Aphids and aleyrodids	0.002	1.6	0.702	25.6
Other sucking phytophages	0.218	16.3	0.482	17.6
Caterpillars and sawfly larvae	0.211	15.8	0.276	10.1
Orthoptera	0.252	18.9	0.024	0.9
Leaf-feeding beetles	0.080	6.0	0.062	2.3
Leaf miners	0.008	0.6	0.005	0.2
Flower feeders	0.018	1.3	0.201	7.3
Bark and wood borers	0.209	15.7	0.126	4.6
Saprovores	0.017	1.2	0.053	2.0
Ants	0.031	2.3	0.153	5.6
Predaceous beetles	0.009	0.7	0.078	2.8
Lacewings	0.012	0.9	0.002	0.1
Predaceous flies and wasps	0.013	1.0	0.021	0.8
Spiders	0.137	10.3	0.319	11.6
Phalangida	0.015	1.1	0.013	0.5
Predaceous Heteroptera	0.063	4.7	0.175	6.4
Parasitoid flies and wasps	0.008	0.6	0.024	0.9
Adult aquatic insects	0.010	0.7	0.004	0.1
Miscellaneous	0.002	0.1	0.025	0.9
Total	1.335	99.8	2.745	100.3

tion and litter structure (i.e., abundance and species composition) and vegetation and litter condition (i.e., attractiveness and suitability for insect reproduction and development). These factors change during the course of community organization.

1. Vegetation and Litter Structure

Changes in vegetation and litter structure are determined largely by disturbance intensity and duration. Acute or low-intensity disturbances, such as those caused by weather fluctuations, may not appreciably alter vegetation or litter structure. On the other hand, chronic and/or high-intensity disturbances, such as long-term drought or wildfire, may alter vegetation and litter structure drastically, such as from forest to annual vegetation and from leaf litter to scorched logs (Schowalter et al., 1981a).

Changes in vegetation and litter structure result in changes in the structure of insect assemblages. For example, both aboveground and forest floor arthropod assemblages changed dramatically following clearcutting of a mixed-hardwood watershed at Coweeta Hydrologic Laboratory in North Carolina (Table 1). Prior to clearcutting, the watershed supported an oak–hickory forest with a dogwood–rhododendron understory and moist, shaded leaf litter; after clearcutting, the watershed supported a black locust–dogwood–grape community with abundant flowering forbs and unshaded logs and other woody debris (Abbott and Crossley, 1982; Boring et al., 1981). Consequently, insect and other arthropod species exploiting early successional dominants or woody debris became more abundant after clearcutting, whereas insect and other arthropod species exploiting late-successional dominants or cool, moist leaf litter became less abundant (Schowalter et al., 1981c; Seastedt and Crossley, 1981).

2. Vegetation and Litter Condition

Plants possess an array of chemical and mechanical defenses against herbivory (Blum, 1980; Mattson, 1980), but for a given species these defenses vary with plant age and condition (Mattson, 1980; Sturgeon, 1979). Plant suitability for insects appears to depend on the net nutritional value remaining after deducting the expense(s) of detoxifying allelochemics and/or extracting nutrients (Fox and Macauley, 1977). Insects can discriminate and exploit gradients or differences in the nutritional value of host tissues. Disturbance often increases host suitability. Plants surviving disturbances are often stressed by injury, exposure, or resource imbalances (Schowalter et al., 1981a; Smith, 1981). Such plants tend to reallocate resources among different metabolic pathways, often at the expense of biotic reservoirs such as tubers, wood, or defensive mechanisms as plant needs change (Mattson, 1980; Waring, 1982; Waring et al., 1980).

Aphids are particularly adapted to exploit small changes in nutrient translocation rates in host plants. Aphids have extremely high rates of parthenogenic reproduction. These insects can respond rapidly to changes in host condition, as demonstrated by the characteristic pulses in aphid populations in undisturbed forests during spring and fall (Fig. 1), when host plants are actively translocating nutrients to or from sites of photosynthetic activity (Schowalter et al., 1981c; Van Hook et al.,

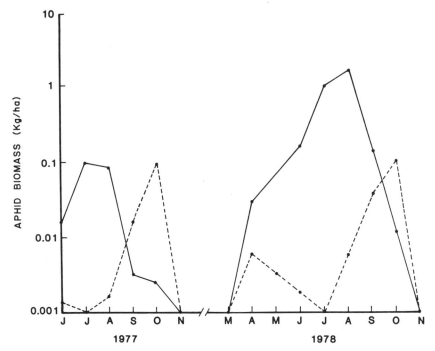

Fig. 1.   Seasonal trends in aphid biomass during 1977 and 1978 on an undisturbed mixed-hardwood watershed (----) and on an adjacent watershed clearcut during 1976–1977 (—) at Coweeta Hydrologic Laboratory, North Carolina.

1980). Sustained nutrient fluxes in rapidly growing or stressed plants contribute to continued aphid population growth, as observed on the clearcut at Coweeta (Fig. 1; see also Schowalter *et al.*, 1981c).

Many species of defoliating insects respond to changes in host condition indicated by changes in foliage nutrient and allelochemic content. However, whereas the short response time of aphid populations permits seasonal pulses in aphid populations in response to host physiology, the lower reproductive rates of defoliating insects result in a longer response time. Population growth in these species usually follows several seasons of altered host conditions. Examples of defoliator responses to foliage biochemistry or to host stress include studies of desert grasshoppers (Schowalter and Whitford, 1979, and references therein) and of forest Lepidoptera and Hymenoptera (Pitman *et al.*, 1982; Stoszek *et al.*, 1981; Tilman, 1978; W. G. Wellington, 1980; Zlotin and Khodashova, 1980).

Bark beetles and many other wood-boring insects are noted for their ability to locate and exploit severely stressed or injured trees (Coulson, 1979; Pitman *et al.*, 1982). Most species are unable to exploit unstressed hosts because the small nutrient reserves in the bark and wood tissues of unstressed trees are well defended by mechanisms such as the oleoresin system of conifers (Mattson, 1980; Sturgeon,

1979). Hence, populations of these insects increase only when the abundance of stressed, injured, or dead hosts increases as a result of competition, flooding or drought (Schowalter *et al.*, 1981a,b), ice storms (T. D. Schowalter, personal observation), windthrow and fire (Cole and Amman, 1980; Schowalter *et al.*, 1981a), or previous insect or pathogen activity (Berryman and Wright, 1978; Goheen and Cobb, 1980).

Litter condition similarly determines the structure of saprophage assemblages. The primary determinants of litter condition are nutritional value and allelochemic content, temperature, and moisture. Increased exposure of litter to extremes of temperature and moisture, as a result of canopy-opening disturbances, reduces litter arthropod populations at the soil–litter surface (Abbott and Crossley, 1982; Seastedt and Crossley, 1981). However, disturbances may also create favorable microhabitats within which population growth of litter arthropods is enhanced (Santos and Whitford, 1981; Seastedt and Crossley, 1981).

3. Community Development

The rate and direction of community development are primarily functions of disturbance frequency and severity. Less severe disturbances change species densities and biomass relations as a result of differences in individual tolerances (Schowalter, 1981); more severe disturbances prevent further community development and often restart succession (Schowalter *et al.*, 1981a; Sprugel and Bormann, 1981; White, 1979; Woods and Whittaker, 1981). The sequence of establishment, proliferation, and decline of plant species determines the sequence of establishment, proliferation, and decline of host-specific insect species. Changes in host age, abundance, and condition through time and space produce predictable shifts in species dominance in insect assemblages as well as in vegetation.

Long-term trends in insect assemblages (Table 1) may reflect functional responses to changing vegetation structure (MacMahon, 1981) and/or changing defensive strategies of vegetation (Mattson, 1980). Early seral vegetation is dominated by ruderal plants with small biomass and high productivity (Boring *et al.*, 1981). Aphids and other sucking arthropods exploit the abundance of young, succulent plant tissues with high nutrient levels. Ants can forage efficiently on small plants for foliage tissue, extrafloral nectaries, aphid honeydew, and prey species encountered while foraging for these or the above resources (Messina, 1981; Schowalter, 1981; Tilman, 1978). Mutualism between ants and homopterans promotes homopteran population growth via protection from predators, sooty-mold contamination, and host decimation by defoliators, and promotes ant population growth via enhanced nutrition from feeding on honeydew (Messina, 1981; Schowalter *et al.*, 1981c). Defoliator populations are subject during early succession to efficient foraging by predators (Miller, 1980; Tilman, 1978) and/or to host allelochemics (Mattson, 1980); degrader-decomposer populations are dependent on the abundance and exposure of residual litter (Abbott and Crossley, 1982; Seastedt and Crossley, 1981). As canopy height and ground coverage increase, wood production and reduced ant foraging efficiency reduce aphid abundance (Schowalter,

1981; Schowalter *et al.*, 1981c). Reduced predator foraging efficiency, a changing biochemical regime, and increasing plant competitive stresses favor the population growth of defoliator species (Tilman, 1978). Reduced exposure of the soil–litter interface and increased plant mortality resulting from competitive interaction promote population growth of degrader-decomposer species (Abbott and Crossley, 1982; Seastedt and Crossley, 1981).

Succession in insect assemblages also occurs on individual host species relative to the season and with increased host size and age. Most insect species occur at particular times during the growing season (Fig. 1; see also Mattson, 1980; Miller, 1980; Schowalter *et al.*, 1981c) or at a particular age and/or size of their hosts (Cole and Amman, 1980; Connell and Slatyer, 1977; Miller, 1980; Schowalter *et al.*, 1981b,c). These short-term trends often accelerate host decline and replacement by other plant species and may represent important mechanisms of community organization (Connell and Slatyer, 1977; Schowalter, 1981). Therefore, the remainder of this chapter will be devoted to a consideration of mechanisms by which insects can influence community organization.

III. INSECT PROPAGATION OF DISTURBANCE EFFECTS

Disturbances can trigger exponential population growth in insect species that exploit disturbance conditions. Insect population growth can be dramatic and can contribute significantly to the impact of disturbance, often in ways that promote continued insect population growth.

In this section, I will focus on the ways in which insects can influence community development. In particular, I will discuss increased plant mortality resulting from insect population growth and the resulting predisposition of such patches of dead trees to subsequent disturbance.

A. Increased Plant Mortality

Insect herbivore population growth is regulated during nominal environmental conditions by host quality and quantity (Mattson, 1980) interacting with predator and parasite foraging efficiency (Miller, 1980; Schowalter and Whitford, 1979; Tilman, 1978) at levels that cause little mortality to host populations. Disturbance, however, can disrupt regulatory mechanisms and initiate insect herbivore population growth. The time lag before regulatory factors can reestablish control of herbivore population growth depends on a threshold population size. Before the population reaches this threshold, density-dependent and density-independent factors can dampen the perturbation with a minimum time lag; when the population exceeds this threshold, regulatory mechanisms become effective only after a long time lag (Berryman, 1981; Southwood and Comins, 1976). For example, control of southern pine beetle (*Dendroctonus frontalis*) population growth in east Texas apparently is determined by the population size reached during favorable weather conditions in the spring. Schowalter *et al.* (1981b) found that *D. frontalis* populations with fewer than 100,000 adults by early June disappeared during the hot, dry summer, but populations exceeding this threshold by early June maintained sufficient abundance during the summer to permit continued colonization of pine trees.

Exponential population growth permits insect populations to become agents of vegetation suppression or mortality. Insect-induced patterns of vegetation suppression are based on host selection behavior and reflect the initial structure of both the vegetation and the associated insect assemblage. Mortality can result directly from insect feeding activity or indirectly from insect transmission of plant diseases.

Aphids, adelgids, and other rapidly reproducing insects that feed on phloem and xylem fluids must ingest large amounts of water and photosynthates to obtain adequate amounts of amino acids and other limiting nutrients (Mattson, 1980; Schowalter *et al.*, 1981c). Consequently, large populations of these insects have a major impact on host water balance, productivity, and resource allocation, especially when feeding rates are stimulated by ants (Schowalter *et al.*, 1981c, and references therein). Schowalter *et al.* (1981c) calculated that ant-tended aphids on the clearcut at Coweeta consumed plant material amounting to 23% of the foliage standing crop biomass, 25% of the foliage standing crop of potassium, and 300% of the foliage standing crop of sodium. This direct impact of sap-sucking insects is amplified by host injury and deformation, resulting from repeated piercing during feeding site selection and/or from injection of salivary toxins, and by transmission of plant diseases (Furniss and Carolin, 1977). These effects can kill host plants but more often contribute to host mortality by weakening hosts and facilitating colonization and population growth by other phytophagous insects (Furniss and Carolin, 1977).

Defoliator populations increasing on chronically stressed hosts can remove up to 100% of host foliage, thereby disrupting photosynthesis (Knight and Heikkenen, 1980). The impact of this disruption is greatest for evergreen conifers that are not adapted for rapid foliage replacement: a single complete defoliation often kills these trees (Knight and Heikkenen, 1980). On the other hand, deciduous conifers, hardwoods, and grasses are adapted for regular foliage loss and replacement, and often survive several years of repeated defoliation (Knight and Heikkenen, 1980; MacMahon, 1981); defoliation may even stimulate primary productivity of these plants (Schowalter, 1981, and references therein). Defoliating insects can cause additional injury to the host as a result of host abscission of defoliated twigs, insect transmission of plant diseases, and facilitation of host colonization and population growth by other insects, especially bark beetles.

Populations of bark beetles and other wood-boring insects grow in trees severely stressed by prior injury, disease, or herbivory (Berryman and Wright, 1978; Coulson, 1979). Bark beetles of the family Scolytidae are particularly important mortality agents in forest ecosystems. These insects require dead phloem for reproduction and development, and have evolved means of facilitating host decline and death (Coulson, 1979). Large populations of these insects can kill host trees directly through gallery excavation and tree girdling, but host death is often accomplished more through the transmission of mutualistic fungi (*Ceratocystis* spp.) that disrupt water transport within the tree. These fungi synergize the impact of bark beetle populations, permitting adequate populations of several bark beetle species to colonize and kill living trees.

Insect populations released from regulation and growing exponentially in disturbed patches often spread into surrounding patches. As a population grows, density-dependent dispersal ensures that distant patches will be colonized, thereby enhancing population contributions to future gene pools. However, most individuals tend to disperse shorter distances into surrounding patches (Fig. 2; see also Clark,

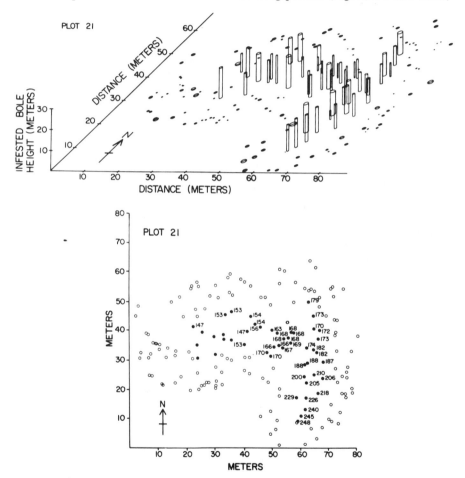

Fig. 2. Spatial and temporal framework of a *Dendroctonus frontalis* infestation (plot 21) in east Texas during 1977. (Upper) Cylinders represent attacked trees and are proportional to the size of trees in the infestation; two-dimensional ellipses represent unattacked trees within 10 m of attacked trees. (Lower) Day numbers (Julian calendar) of attack initiation for trees attacked (closed circles) after sampling began. Open circles represent unattacked trees within a 10-m strip around the infested trees. (Reproduced from Schowalter *et al.*, 1981b, by permission of the Society of American Foresters.)

1979; Schowalter *et al.*, 1981b). The resulting wave of vegetation suppression and mortality can cover areas of up to several million hectares (Clark, 1979). Intense herbivory by any insect species can weaken the host and facilitate successive host colonization by other insect species exploiting more highly stressed plants. The resulting heterotrophic succession accelerates host decline, death, and decomposition.

B. Patch Predisposition to Disturbance

Insect-induced vegetation mortality opens the canopy and creates fuel accumulations that predispose the patch or surrounding patches to subsequent disturbances. The nature of the ensuing disturbance determines the future growth and migration of insect populations and the future rate and duration of community development within the patch. Therefore, the pattern of insect population interactions with the disturbance regime is a major determinant of patch dynamics, as illustrated by the following two examples.

Historically, the high frequency of fires during summer droughts was responsible for the persistence of fire-tolerant pine forests in the southeastern United States (Schowalter *et al.*, 1981a). The southern pine beetle is a periodically important predator of pine trees over 30 years old in these forests (Coulson, 1979; Schowalter *et al.*, 1981b). Low-density populations of these beetles are normally restricted to scattered injured or stressed pines, because even with the aid of blue-staining fungi, small numbers of colonizing beetles are unable to overcome the oleoresin defense system of unstressed pines (Coulson, 1979). Disturbance, however, creates patches of injured or stressed trees; such patches often neighbor beetle refuge trees. Reduced distance between refuge trees and vulnerable trees, particularly in stands with a high density of pines, results in reduced exposure of beetles to mortality agents, increased efficacy of pheromone communication between colonizing and dispersing beetles, and sufficient aggregation of beetles, and of blue-staining fungi, to overcome host defenses (Schowalter *et al.*, 1981b). The beetle population advances between neighboring pine trees, initiating a wave of pine mortality (Fig. 2).

Forest regeneration occurs in the wake of pine mortality (Fig. 3). In the absence of subsequent disturbance, beetle-induced mortality of the dominating pines accelerates the transition to fire-intolerant hardwood forest, with the youngest hardwoods in the immediate wake of the mortality wave and progressively older hardwoods extending away from the wave (Schowalter *et al.*, 1981a). Southeastern forests, however, are subject to frequent disturbance. Severe ice storms and wind storms injure or stress exposed vegetation behind the wave and along its edge, increasing vegetation vulnerability to insects and other mortality agents and contributing to continued wave advance. Within the time frame of log degradation, the probability is high that thunderstorms will eventually ignite a standing pine snag during hot, dry summers, particularly in large patches of dead pines resulting from extensive wave advance (Schowalter *et al.*, 1981a). Fire, fueled by abundant woody litter, probably terminates the advance of the insect-induced mortality wave but provides new beetle

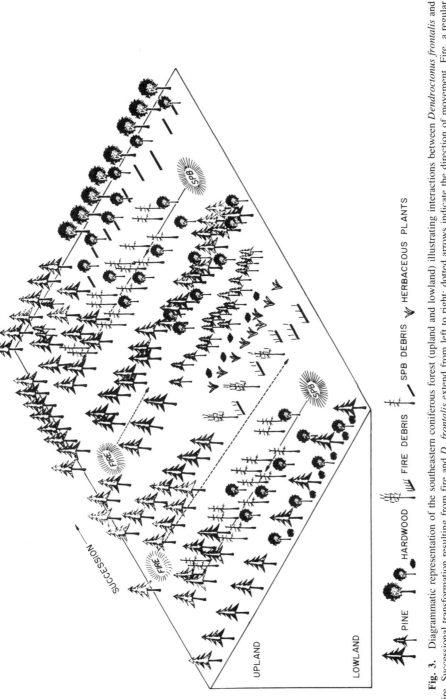

Fig. 3. Diagrammatic representation of the southeastern coniferous forest (upland and lowland) illustrating interactions between *Dendroctonus frontalis* and fire. Successional transformation resulting from fire and *D. frontalis* extend from left to right; dotted arrows indicate the direction of movement. Fire, a regular feature of the generally dry, well-drained uplands, invades generally moist, poorly drained lowlands, where drought or *D. frontalis* creates favorable conditions. *Dendroctonus frontalis* depends upon fire for regeneration of pine stands. The hardwood climax forest (far right lowland) results from freedom from fire and can be reduced by fire. (Reproduced from Schowalter *et al.*, 1981a, by permission of the Entomological Society of America.)

refuges in fire-scarred pines. Fire also destroys fire-intolerant hardwoods and re-establishes the pine forest, which again becomes increasingly vulnerable to insect-induced wave mortality after 30 years (Fig. 3). Similar wave regeneration patterns reflecting insect interactions with disturbance apparently result from mountain pine beetle (*Dendroctonus ponderosae*) outbreaks (Cole and Amman, 1980) and spruce budworm (*Choristoneura fumiferana*) outbreaks (Flieger, 1970).

Insect interactions with plant diseases may cause host mortality to become continuous over long time periods. Fungi attacking living plants often predispose their hosts to colonization by insects (Goheen and Cobb, 1980; Lewis, 1979) and often can remain viable for many years in soil, in relic diseased root systems, or in resistant plants. Black-stain root disease fungus (*Ceratocystis wageneri*) is transmitted between adjacent host trees in western coniferous forests, creating expanding rings of dead hosts. Although this fungus can be transmitted by root contact or grafting, transmission is accelerated by several species of root-feeding beetles attracted to and injuring stressed trees and surrounding ones (Goheen and Cobb, 1978; Witcosky, 1981). Subsequent tree-to-tree spread of the fungus by root grafts and beetle dispersal establishes a wave of host mortality. Insect activity can be promoted by disturbance-induced injury (e.g., precommercial thinning) and stress to plants in neighboring patches (J. J.Witcosky and T. D. Schowalter, unpublished data). Although regeneration occurs in the wake of the advancing wave, the fungus can remain established at a site for many years and can cause continuous mortality to successive generations of host species as a consequence of insect dispersal (Witcosky, 1981; J. J. Witcosky and T. D. Schowalter, unpublished data). Host replacement by nonhosts would cause an eventual decline in insect and fungus abundance during the period preceding reestablishment of host species. A similar pattern of wave mortality and regeneration can be inferred from the dynamics of oak wilt (*Ceratocystis fagacearum*) transmission by nitidulid bark beetles (Menges and Loucks, 1984).

IV. INSECTS AS REGULATORS OF ECOSYSTEM PRODUCTIVITY

As shown in the preceding discussion, insect responses to the immediate impact of disturbance on resource quality and quantity can cause extensive plant suppression and mortality. Consequently, insect herbivores have been viewed as disruptive agents and have been the targets of extensive control measures. However, although insect responses to disturbance obviously contribute to individual fitness, these responses also concentrate herbivory on weakened or injured plants, thereby favoring disturbance-tolerant plants and promoting development of communities that mitigate disturbance severity. Hence, insect adaptation to disturbance can influence processes at higher levels of organization, perhaps contributing to the inclusive fitnesses of individuals.

Insect adaptations to disturbance could contribute to ecosystem stability in two ways: (a) insect assemblages regulate ecosystem productivity by accelerating changes in plant species biomass relations in response to changes in plant–soil nutrient–light relations, and (b) insect assemblages regulate ecosystem productivity by regulating the reliability of denudation. Because our attitude toward insect herbivory in ecosystems and our perception of mechanisms governing community organization and stability determine our ecosystem resource and pest management strategies, this last section draws attention to the potential regulatory importance of insect adaptations to disturbance.

A. Community Organization

Although the search for mechanisms governing community organization has motivated ecologists for at least the last century, host-specific herbivory by insects has been recognized as an important mechanism only within the past decade (Cole and Amman, 1980; Connell and Slatyer, 1977; Coulson, 1979; Flieger, 1970; Mattson and Addy, 1975; Schowalter, 1981; Schowalter et al., 1981a; Springett, 1978). Changes in insect assemblages, induced by changes in vegetation quality and quantity, appear to regulate rates of ecosystem processes: ecological succession, energy flow, and nutrient cycling emerge as the coupled manifestations of herbivore control of resource requirements by vegetation relative to resource availability (Schowalter, 1981; Schowalter et al., 1981a). Water, mineral nutrients, and light can become limiting to plants as a result of increasing biomass and/or changes in weather patterns (Fig. 4). Concentration of herbivory on those plants or plant species most stressed by resource limitation reduces plant growth and hence nutrient requirement; continued weakening of these plants through heterotrophic succession accelerates plant decline, decomposition, and replacement. Although thinning may eventually occur as a result of plant competition, insect-induced plant mortality accelerates changes in plant competitive relationships and reduces the time lag to equilibration of resource availability and resource demand by vegetation.

In some ecosystems, environmental conditions prevent establishment of competing vegetation, and insect regulation of resource demand occurs through changes in the age distribution of vegetation. For example, pines are tolerant of fire but intolerant of resource limitation relative to hardwoods and some other conifers. Frequent fire maintains pine forests by reducing interspecific competition for resources, but intraspecific competition for light and nutrients often leads to senescence of overstory pines and prevents pine seedling establishment. Because large pines are rarely killed by fire, insect-induced mortality of stressed overstory pines may be a major mechanism in promoting pine regeneration and productivity in such ecosystems (Schowalter et al., 1981a).

In most ecosystems, insect-induced vegetation mortality appears to adjust the competitive relationships of plant species (and, hence, total resource demand) to the ambient conditions of resource availability for uptake by vegetation. When resource availability exceeds demand, e.g., following plant dormancy or on newly exposed sites, insect responses to the increased nutritional quality of host plants apparently

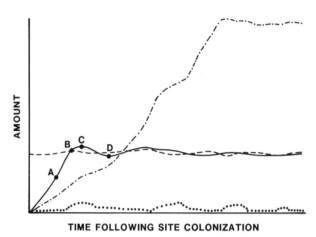

TIME FOLLOWING SITE COLONIZATION

Fig. 4.   Trends in resource supply (----), resource demand (—), and vegetation (----) and consumer (••••) biomass in ecosystems through time. Rapid gross primary productivity following site colonization (A) finally reaches the carrying capacity established by the resource supply (B). Plants respond to the increasing resource deficit (C) by reallocating resources and thereby initiating consumer responses. Consumer-induced plant mortality reduces resource demand (D) and facilitates establishment of younger plants or competing vegetation. Consumer regulation of the resource supply : demand ratio accelerates succession and ultimately predisposes the community to denudation by disturbance.

tend to stimulate host growth and/or favor the establishment of more productive plants; as increasing biomass or disturbance results in reduced resource availability relative to biotic demand, insect responses to host stress tend to reduce host growth and/or favor the establishment of more competitive or stress-tolerant plant species (Connell and Slatyer, 1977; Mattson and Addy, 1975; Schowalter, 1981; Schowalter *et al.*, 1981a; Zlotin and Khodashova, 1980). Acceleration of changes in plant species biomass relations by insects reduces the time required to balance resource demand and resource availability, and tends to stabilize the community against intense intraspecific and interspecific plant competition that can lead to community senescence and disruption (Schowalter, 1981; Schowalter *et al.*, 1981a).

B. Disturbance Reliability

The duration of community development can be regulated to a great extent by insect-mediated acceleration of plant decline. Insect responses to changing conditions often appear to predispose patches to disturbance at regular intervals relative to community development and thereby promote regeneration from surviving individuals or propagules of ruderal plant species. Increased regularity of denudation thus minimizes the time lag to community recovery. Vegetation mortality resulting from insect activity appears to be most predictable in ecosystems that would otherwise be denuded by disturbances at irregular intervals.

Desert, tundra, and grassland ecosystems are subject annually to high wind, temperature, and moisture extremes and/or fire. The frequency of these disturbances has led to vegetation adaptations to minimize disturbance impact: community recovery following denudation occurs primarily as resprouting from extant root systems (MacMahon, 1981). In these ecosystem types, insects tend to promote plant establishment and nutrient conservation, e.g., through seed dispersal and plant protection (Messina, 1981; Tilman, 1978), but usually do not cause extensive vegetation mortality.

Denuding disturbances tend to occur at irregular intervals in forest ecosystems. Recovery in these ecosystems is facilitated by ruderal plant species (disturbance-adapted species characterizing more frequently denuded patches) that stabilize soil/litter processes but that are competitively replaced during long disturbance-free intervals. Hence, plants and seeds of such species could have become rare by the time a denuding disturbance occurs, thereby increasing community recovery time. Insect population responses to changes in vegetation structure and condition tend to restart succession at intervals sufficiently short to maintain populations of ruderal plant species. The patch pattern of disturbance and insect activity also contributes to ecosystem stability by minimizing the area disturbed at any time, thereby facilitating complete patch colonization by populations from surrounding patches. Hence, the disturbance regime characterizing any ecosystem shapes the gene pools of resident populations and may, over evolutionary time, select species assemblages that interact to mitigate the effects of disturbance.

Increased understanding of insect responses to disturbance can contribute to our view of mechanisms controlling ecosystem processes and to our approach to managing destructive insects and ecosystem resources. Current methods of managing ecosystem resources often promote insect-induced vegetation mortality. For example, landscape-level monocultures of crop species provide insect herbivores with an extensive source of attractive host odors and hosts sufficiently stressed by cultivation techniques or by poor adaptation to environmental conditions to promote insect population growth. Increased diversity of crop species, genotypes, or age classes at the patch level would tend to minimize insect effects on vegetation. Fire exclusion frequently results in competitively suppressed vegetation and consequent insect population irruptions (Cole and Amman, 1980; Flieger, 1970; Schowalter et al., 1981a). Controlled burning would help restore the disturbance regime to which the community has adapted and thereby mitigate conditions that trigger insect activity. Thus, an understanding of insect adaptations to patch dynamic environments will enable us to predict better the consequences of our ecosystem management strategies.

V. CONCLUSIONS

Insect dispersal and host selection behavior represent adaptation to a heterogeneous environment. Insects respond to changes in host quality (i.e., chemistry) and quantity (i.e., abundance), reflecting the impact of disturbance on the community. Disturbances can generate perturbations of sufficient magnitude in insect assemblages to result in extensive vegetation suppression and patch predisposition to subsequent disturbance. Insect responses to changing conditions within patches contribute to individual fitness and also appear to stabilize ecosystem productivity by (a) regulating plant–soil nutrient–light relations through changes in plant age structure and/or species biomass relations, and (b) increasing the reliability of disturbance and thereby mitigating its impact. The regulatory importance of insects appears to be a function of disturbance frequency and severity, and supports the view of ecosystems as cybernetic systems with mechanisms for stabilizing ecosystem productivity.

ACKNOWLEDGMENTS

This chapter was supported by the Oregon Agricultural Experiment Station, ORE 00258, and is Oregon Agricultural Experiment Station paper number 6664. R. N. Coulson (Texas A&M University), D. A. Crossley, Jr. (University of Georgia), G. W. Krantz (Oregon State University), P. W. Oman (Oregon State University), E. M. Hansen (Oregon State University), E. J. Rykiel, Jr. (Texas A&M University), and C. S. Schowalter, as well as the editors, critically reviewed the manuscript. I especially thank P. W. Oman for permission to cite unpublished data.

RECOMMENDED READINGS

Flieger, B. W. (1970). Forest fire and insects: The relation of fire to insect outbreak. *Proc. 10th Annu. Tall Timbers Fire Ecol. Conf.*, pp. 107–120. Tall Timbers Res. St., Florida.

Mattson, W. J. (1980). Herbivory in relation to plant nitrogen content. *Annu. Rev. Ecol. Syst.* **11**, 119–161.

Schowalter, T. D. (1981). Insect herbivore relationship to the state of the host plant: biotic regulation of ecosystem nutrient cycling through ecological succession. *Oikos* **37**, 126–130.

Schowalter, T. D., Coulson, R. N., and Crossley, D. A., Jr. (1981). Role of southern pine beetle and fire in maintenance of structure and function of the southeastern coniferous forest. *Environ. Entomol.* **10**, 821–825.

Schowalter, T. D., Pope, D. N., Coulson, R. N., and Fargo, W. S. (1981). Patterns of southern pine beetle (*Dendroctonus frontalis* Zimm.) infestation enlargement. *For. Sci.* **27**, 837–849.

Wellington, W. G. (1980). Dispersal and population change. *In* "Dispersal of Forest Insects: Evaluation, Theory and Management Implications" (A. A. Berryman and L. Safranyik, eds.), Proc. Int. Union For. Res. Organ. Conf., pp. 11–24. Washington State Univ. Coop. Ext. Serv., Pullman.

Chapter **14**

Within-Patch Dynamics of Life Histories, Populations, and Interactions: Selection Over Time in Small Spaces

JOHN N. THOMPSON

Departments of Botany and Zoology
Washington State University
Pullman, Washington

I. INTRODUCTION

Populations are often distributed as a patchwork within habitats, corresponding to patterns of past disturbances, colonization, heterogeneity in the physical environ-

THE ECOLOGY
OF NATURAL DISTURBANCE
AND PATCH DYNAMICS

ment, and interspecific interactions. In terrestrial communities, plants form the basic patchwork structure (Watt, 1947; Pickett, 1976, 1980; Whittaker and Levin, 1977; Grubb, 1977; Huston, 1979; Denslow, 1980a; Gilbert, 1980) within which biological interactions occur, whereas in marine communities, algae, angiosperms, or a variety of sessile animals form the patchwork (Levin and Paine, 1974; Connell, 1978; Paine and Levin, 1981). Population structure at all trophic levels changes over time within each patch, and the within-patch dynamics of populations and interactions can be both a cause and a consequence of natural selection within a patch.

In this chapter, I analyze (a) how demography of plants can vary with respect to position within a patch, generating a pattern of within-patch dynamics that is a consequence of different but co-occurring patterns of survivorship and reproduction; (b) the extent to which within-patch dynamics can be a consequence of the sorting of genotypes within a patch over time; and (c) the degree to which interspecific interactions and selection can vary within a patch over time as the patch changes in age and size and in the reproductive and genetic structure of individuals. I will focus primarily on the relationship of within-patch dynamics and selection, using examples mostly from terrestrial plants and insects.

II. WITHIN-PATCH DYNAMICS, LIFE HISTORIES, AND SELECTION

A. Single-Species Patch Dynamics

The development of a new patch depends upon a disturbance in a habitat, which allows colonization. Disturbances vary widely between habitats in origin, areal extent, frequency, and intensity (Connell, 1978; Pickett and Thompson, 1978; White, 1979; Pickett, 1980; Bazzaz, 1983), and the initial structure of a new plant patch depends upon these variables plus the availability of colonizers, residual propagules, or regenerating plants. As a patch of plants develops, peaks in numbers, and then dissipates, population structure within the patch changes in the distribution of ages, sizes, and reproductive states. These changes over time within each patch have been documented for a taxonomically diverse group of plants (Watt, 1947; Ford, 1975; Thomas and Dale, 1974; Sprugel, 1976; Stergios, 1976; Harper, 1977; Mohler *et al.*, 1978; Thompson, 1978; Yeaton, 1978; Law, 1981; Winn and Pitelka, 1981), although the pattern of change in age structure among long-lived plants remains poorly known (Kerster, 1968; Schaal and Levin, 1976; Harper, 1977).

The pattern of change within patches is generally analyzed by comparing histograms of size or age over time. We know little, however, about the pattern of change in structure at different positions within patches. The overall histogram of sizes or ages for a patch may mask two or more demographic patterns that vary based upon position within the patch. Furthermore, interactions between plants and insects can vary with position within a patch, so that plants at different positions are subject to different selection pressures. Without an understanding of position effects in within-patch dynamics of populations and interactions, it is difficult to obtain an accurate indication of the variance in the life histories of neighboring conspecifics. And it is therefore difficult to sort out how variable selection might act on the life histories of individuals at all trophic levels within a patch.

For example, patch structure for the monocarpic perennial herb *Pastinaca sativa* (Umbelliferae) changes drastically over time (Thompson, 1978). The wind-dispersed seeds of *Pastinaca* colonize small disturbance sites and, under the rapidly changing conditions in old fields in Illinois, the resulting patches expand and then dissipate over a period of several generations. A seed colonizing a new site usually grows into a large, robust plant that, in turn, sets many seeds. Most of these seeds fall near the parent plant. As a result, plants at the center, growing under conditions of high conspecific density, take several years to develop and often just barely reach the threshold of rosette mass necessary for flowering, whereas plants at the periphery tend to be large (Fig. 1). Eventually, the patch dissipates and other plant species invade the site.

Demography at the periphery of the patches differs from that at the center. Large rosettes, which are most common at the periphery of patches, have a higher likelihood of surviving until the next spring than small rosettes (Thompson, 1978). The large rosettes produce predictably more flowers, the largest plants producing more than 50 times the number of seeds produced by the smallest plants (Thompson, 1978). Moreover, larger plants produce larger seeds (S. D. Hendrix, personal communication), which may differ from smaller seeds in competitive ability, although this has not yet been shown for *Pastinaca*. So the pattern of seed production is highly biased with respect to position within a patch. Therefore, the life histories of individuals vary greatly within a patch at any one point in time as well as over a number of generations. Plants at the center of a patch are probably under a different

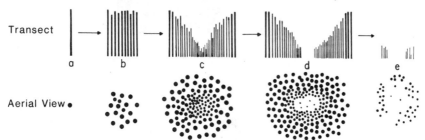

Fig. 1.   Idealized model of within-patch dynamics of the monocarpic perennial *Pastinaca sativa* over several generations in an abandoned agricultural field in east-central Illinois. Only flowering individuals are indicated. Colonization of a new area by a seed results in (a) a large, isolated plant; (b) progeny develop into a small patch of large individuals; (c) over time, the center of the patch becomes increasingly dense, leading to a decrease in the size of the plants in the interior of the patch; (d) the center of the patch becomes a mass of rosettes with few flowering individuals, while larger plants grow on the periphery; (e) continual site deterioration, invasion by other species, and longer development times lead to dissolution of the patch. (From Thompson, 1978.)

selective regime than are plants at the periphery; plants maturing in the early years of patch development are probably under a different selective regime than are their offspring.

An analysis of within-patch structure and dynamics based only on histograms without respect to position would neglect different patterns of life histories within the same patch. Histograms would show only variance, without any indication of how that variance is partitioned within the patch. Figure 2 compares structure across one *Pastinaca* patch with a histogram representation. If the different-sized plants were distributed throughout the patch without any pattern of variation from center to periphery, then the histogram would be a useful representation. Here, however, it masks a level of variation in how the patch is structured.

Not all plant patches have an internal structure as pronounced as that of *Pastinaca* growing on rich soil in Illinois. The within-patch structure of *Pastinaca,* however, indicates that life histories, and potentially selection on them, can be highly variable depending upon the position within a single patch.

The interspecific interactions that develop within patches can increase the variation in selection within a patch. *Pastinaca* individuals in the largest size class have a much higher probability of being attacked by the oecophorid moth *Depressaria*

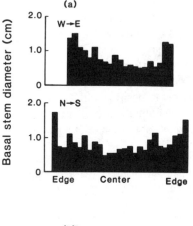

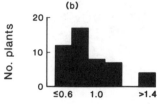

Basal stem diameter (cm)

Fig. 2.   (a) Variation in the basal stem diameter of flowering individuals in two directions across one patch of *Pastinaca sativa*. Plants were sampled every 0.75 m (from Thompson, 1978). (b) A histogram of the same plants, ignoring the position of plants within the patch.

pastinacella than do plants in smaller classes. This pattern holds when the usually large, isolated plants are compared with plants in dense patches (Thompson and Price, 1977), as well as within patches and among isolated plants (Thompson, 1978). The larger the plant, the more umbels it produces. Since umbels are produced asynchronously, and since *D. pastinacella* successfully oviposits mostly on plants with unopened umbels, larger plants are available for oviposition for a longer time during the spring oviposition season than smaller plants. Hence, the vast majority of the large plants at the periphery of a patch are attacked by the moths, whereas only a small fraction of the plants in the center of well-developed patches are attacked. This pattern of insect–plant interaction emphasizes the potential differences in the selective regime within a patch.

Another umbellifer, *Lomatium farinosum*, also shows distinct within-patch variation in demography. In this species, the outcome of interactions with herbivores varies with position within the patch. *Lomatium farinosum* is a small perennial herb that grows in steppe communities in eastern Washington, and the plants are generally restricted to patches of lithosolic soils (Schlessman, 1980). Within a patch, however, soil depth varies. Plants growing in relatively deep soil (10 cm or more) suffer a significantly higher mortality than plants on shallow, rocky soil. A large percentage of the deaths on deep soil results from mammalian activity, especially digging, creation of runways, and feeding on roots. Plants growing on the shallow, rocky soil, in contrast, often have their globose roots tucked between basaltic rocks. Although mammals eat the leaves and flowers of these plants, the plants survive because the roots are left undamaged. Of 100 flowering individuals tagged in 1980, 88% of those growing on shallow, rocky soil were still alive 2 years later, whereas only 9% of those on deep soil survived. The pattern was the same in both years: 5–7% of plants on shallow, rocky soil died each year, whereas 58–79% of plants on deep soil died (Thompson, 1983a).

There are therefore at least two, not one, demographic patterns within a single patch of *L. farinosum*. Overall, 42% of the plants survived for the 2 years after tagging. But this mean percentage is not very useful for understanding the actual dynamics within the patch. Risk of mortality differs greatly among individuals depending upon soil depth, as does the potential effect of mammals as selective agents on the life history of these plants. Moreover, there is an indication that plants surviving on deep soil have a greater chance of flowering in 2 consecutive years than plants growing in shallow soils (Thompson, 1983a). How selection is likely to act on life histories under these conditions will depend upon the proportion of individuals in deep as compared with shallow, rocky soil in the patch and patterns of gene flow within the patch (see, e.g., Levin and Wilson, 1978).

These studies of the within-patch dynamics of umbellifers and their interactions with herbivores indicate that selection may be variable in both space (*L. farinosum* and *P. sativa*) and time (*P. sativa*) within a single patch. The outcomes of interspecific interactions cannot be viewed as constants even at the very local level of the patch. Although it is conceptually easier to concentrate on the mean pattern when studying demography, the evolution of life histories, and the evolution of

interactions, attention to the local variance in selection on life histories and outcomes of interactions is likely to generate a more thorough understanding of the
within-patch dynamics of populations and interactions.

B. Multispecies Patch Dynamics

Few patches in terrestrial communities are actually monospecific stands of one
plant taxon and its associated herbivores and other animals. Watt's (1947) classic
studies stand out as a model for understanding how species diversity can be maintained within habitats partly through the process of patch dynamics. When several
plant species colonize a new disturbance site, they often occupy particular positions
within the patch. Nonetheless, analyses of within-patch species diversity and the
dynamics of diversity have generally not sorted out position effects. Yet relationships between patch size and species composition (Hartshorn, 1978, 1980;
Denslow, 1980a,b; Whitmore, 1982) suggest that an understanding of within-patch
dynamics in species diversity demands analyses of positions of individuals and
species within patches. Furthermore, the internal heterogeneity of substrates within
patches virtually guarantees that some position effects will be masked if patch
structure is analyzed only with respect to the number of species and their relative
abundance of individuals. For instance, herbaceous species differ in their ability to

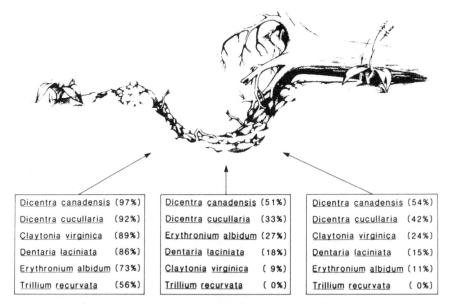

Dicentra canadensis (97%)	Dicentra canadensis (51%)	Dicentra canadensis (54%)
Dicentra cucullaria (92%)	Dicentra cucullaria (33%)	Dicentra cucullaria (42%)
Claytonia virginica (89%)	Erythronium albidum (27%)	Claytonia virginica (24%)
Dentaria laciniata (86%)	Dentaria laciniata (18%)	Dentaria laciniata (15%)
Erythronium albidum (73%)	Claytonia virginica (9%)	Erythronium albidum (11%)
Trillium recurvata (56%)	Trillium recurvata (0%)	Trillium recurvata (0%)

Fig. 3. Probabilities of colonization of treefall sites by spring herbs in east-central Illinois. Given is
the percentage of sites at which a species occurring on the soil surface within 1 m of the edge of a pit also
colonized a particular position within the pit. Only the six most common species are shown. (Data from
Thompson, 1980.)

colonize different parts of pits created by the uprooting of trees in the eastern deciduous forest of North America. *Dicentra canadensis* is the most successful colonizer of the centers of pits among the spring flora of Illinois, whereas *Trillium recurvatum* and *Claytonia virginica* are among the least successful colonizers (Fig. 3).

III. ECOLOGICAL GENETICS OF WITHIN-PATCH STRUCTURE AND DYNAMICS

If selection varies in space or time within a patch, then the genetic structure of the group of plants within the patch may vary accordingly, unless prevented by gene flow. Moreover, any change over time in genetic structure may in turn be partly responsible for within-patch dynamics of populations and interactions.

Individuals of the annual herb *Veronica peregrina*, which grows in vernal pools in California, differ genetically between the center and the periphery of the pools (Linhart, 1974). When grown in a greenhouse, plants from the periphery of one pool weighed significantly more, produced more branches, devoted a higher percentage of their biomass to vegetative parts rather than seeds, and produced more but individually lighter seeds, which had a lower percentage of germination, than plants from the center of the pool. Plants from another pool showed similar but nonsignificant differences. The same pattern apparently occurs in two other vernal pool species, *Boisduvalia glabella* and *Downingia concolor* (Linhart, 1976).

Variances in most phenotypic characteristics are higher for periphery-derived plants of *V. peregrina* than for center-derived plants. Increased variances could indicate more variable selection (or reduced selection) at the periphery of patches. Alternatively, increased variances could indicate differences in heterozygosity and developmental homeostasis. Increased phenotypic variability has been associated with degree of homozygosity in some domestic plants and animals (Lerner, 1954), and in natural populations of lizards, aquatic bivalves, and the monarch butterfly (*Danaus plexippus*) (Soulé, 1979; Kat, 1982; Eanes, 1978). Within some populations, homozygous individuals are more phenotypically variable than more heterozygous individuals (Eanes, 1978; Mitton, 1978). A decrease in heterozygosity from the center to the edge of a patch could conceivably occur in some plants if almost all pollination is restricted to the patch, simply because peripheral plants are surrounded by fewer conspecifics than central plants. (Of course, the opposite pattern of heterozygosity could hold if most pollen transfer involving peripheral plants is between patches, whereas pollen transfer involving central plants is within patches.) Within-patch distribution of heterozygotes, however, is known for few species (Schaal, 1975).

The genetic variation in the within-patch structure of *V. peregrina* is an evolutionary response to strong selective differences between the center and the edge of the pools. Even without such pronounced differences imposed by the physical environment, within-patch dynamics in some populations may reflect change over time in the genotypes of individuals. Currently, however, the extent to which within-patch dynamics in size, age, or reproductive structure is a sorting out over time of genetically different individuals is poorly known. In an elegant suite of studies, Bazzaz et al. (1982) demonstrated genotype-specific survivorship curves in experimental populations of *Phlox* grown in a glasshouse. The pattern of thinning in crowded flats was determined by the genetic makeup of the plants. Survivorship curves varied with both the cultivar and nutrient status of the soil, and the pattern of thinning in crowded flats was determined partly by the genetic makeup of the plants. Genetic changes found in other plant populations during self-thinning (Clegg and Allard, 1973; Antonovics, 1976) and comparisons of genotypes among plants at different stages (seeds, seedings, and adults) (Bradshaw, 1972; Allard et al., 1977; Clegg et al., 1978; Hamrick, 1982) indicate that within-patch dynamics can be as much a genetic as an ecological process.

In *Liatris, tuna, Daphnia, Sceloporus* lizards, and *Colias* butterflies, heterozygosity increases with the age of individuals at one to many loci (Schaal and Levin, 1976; Fujino and Kang, 1968; Hebert et al., 1972; Tinkle and Selander, 1973; Watt, 1977) and, in several species of mussel, with size (Koehn et al., 1973, 1976; Tracey et al., 1975). Average heterozygosity increases with age in the perennial herb *Liatris cylindracea* within a single patch, suggesting that mortality depends upon genotype (Schaal and Levin, 1976). Furthermore, rate of development and fecundity are positively related to individual heterozygosity. Plants flowering at 2 years of age have significantly higher average heterozygosity than plants that are still vegetative at 2 years. Significant relationships between growth rate and number of heterozygous loci in individuals have also been shown among clones of quaking aspen (*Populus tremuloides*) (Mitton et al., 1981), and within cohorts of the American oyster (*Crassostrea virginica*) (Singh and Zouros, 1978; Zouros et al., 1980). Each oyster cohort was the group of oyster larvae settling on a scallop shell (and thereby delimiting the patch) within a 3-week period. After 1 year of growth, the mean weight of oysters with seven heterozygous loci (among the loci tested) was 71% greater than that of oysters with no heterozygous loci; mean weight increased monotonically with the number of heterozygous loci for intermediate numbers of heterozygous loci (Zouros et al., 1980).

This relationship between heterozygosity and life history in a wide variety of taxa suggests that a significant cause of the pattern of within-patch dynamics in some populations is a sorting out of genotypes over time as well as differential growth and reproduction among the remaining genotypes. The coupling of studies on the within-patch distribution of plants of different ages, sizes, and reproductive states with studies of patterns of genetic change over time could do much to bridge the gap between population genetics and population ecology. Not all species vary in genetic structure with increasing size of individuals—e.g., ponderosa pine (*Pinus ponderosa*) and lodgepole pines (*P. contorta*) in Colorado (Linhart et al., 1981; Mitton et al., 1981)—but enough do vary among those that have been studied to suggest that the situation is likely to be widespread.

IV. ECOLOGICAL GENETICS AND THE WITHIN-PATCH DYNAMICS OF INTERACTIONS

As within-patch structure varies over time through changes in species composition, age, size, or genetic structure, the interactions within and between species can also change (Gilbert, 1980; Price, 1980; Thompson, 1980). The interactions can change in (1) probability of occurrence, (2) mechanics, and (3) outcome. "Mechanics" of an interaction refers to the way the interaction occurs, such as the plant part fed upon by an insect or the length of time an herbivore feeds on one plant. These changes in interactions can result from changes over time in the age, size, or reproductive state of individuals and changes in the genetic structure of the population. The interactions, in turn, can be a major cause of these populational changes within a patch. Very little is known quantitatively, however, about variation and within-patch dynamics of interactions, but several examples follow that suggest the wide variation in interactions that can occur within patches over time.

A. Probability of Encounter

Predictable patterns of variation in the probability of encounter are known for a wide variety of interactions between animals and plants both within and between patches (Root, 1973; Janzen, 1975; see also review in J. N. Thompson, 1982). For example, fleshy fruits are removed at a significantly faster rate from light gaps than from surrounding areas under closed canopy in Illinois. The result is reproducible using different species of fruit at different times within a year (Thompson and Willson, 1978) as well as between years (Willson and Melampy, 1983). In summer and early fall, fruits not eaten by birds soon after ripening have a high probability of being eaten by invertebrates (Thompson and Willson, 1978), so that the rate at which fruits are found by birds is an important variable in understanding these interactions. Since each light gap is eventually closed by the encroaching canopy, the probability of interaction between birds and fruits undoubtedly changes within a patch over time.

Even within a single patch, the probability of encounter can differ between the periphery and the center. This difference in encounter rate can result from an "edge effect," that is, the tendency of individuals at the periphery of a patch to have a greater likelihood of interspecific interaction than individuals at the center. Additionally (or alternatively), this difference can result from differences in the demography of individuals at the periphery as compared with the center. In the umbellifer *Lomatium grayi*, a perennial herb restricted to patches of shallow, rocky soil in the steppe of eastern Washington and surrounding areas, a higher proportion of established individuals at the periphery of some patches reaches larger sizes than plants at the center. Larger plants have a higher probability of flowering than smaller plants, and large flowering plants have a higher probability of being attacked by the oecophorid moth *Depressaria multitidae* than small plants. Therefore, the lifetime probability of a plant being attacked at least once by *D. multifidae* can be higher at the periphery than at the center (Thompson and Moody, 1985). These examples suggest the value of analyzing the probability of encounter in interactions in a dynamic framework that includes analyses of within-patch structure over time.

B. Variation in Mechanics of Interactions

Lomatium grayi is attacked by *D. multifidae* in a variety of ways. This moth species is the most variable of the *Depressaria* species studied to date in the way it feeds on its host plant (Thompson, 1983b,c,d). Larvae feed in leaf axils and on leaves when on small vegetative plants; on axils, leaves, and flowers when on moderately sized plants in flower; and on all of these parts plus the internal tissues of floral stalks when on plants large enough for a larva to bore into the stalk (Thompson, 1983d). Therefore, the way in which the larvae interact with the plants depends on the age, size, and reproductive states of plants within the patch, and the within-patch population structure changes over time. It is not yet known if some of the variation in interactions within patches of *L. grayi* is also related to genetic variation in either the moths or the plants and to changes in their genetic structure over time.

C. Variation in Outcomes of Interactions

The effects of interactions on the fitness of individuals can vary with age, size, and reproductive state. As patches change in demographic structure over time, the outcomes of interactions are also likely to change. Concomitant with these changes can be changes in the genetic structure of populations, which also affects the outcomes of interactions. Genetically based changes in the outcomes of interactions over time are most likely to occur within patches among relatively sessile species such as plants and parasites. Reciprocal transplants of clones of *Trifolium repens* taken from different sites within a pasture show that *Trifolium* is less affected by interspecific competition within patches of the type from which it is taken than from unlike patches into which it is transplanted (Turkington and Harper, 1979). Local adaptation of *Trifolium* within patches has affected the outcome of interspecific competition.

Similarly, insects parasitic on plants can adapt to their hosts within patches. In fact, Price (1980) argued that the evolution of genetic systems in parasites is best understood when analyzed in the context of the within-patch dynamics of populations. Phytophagous insect populations can differentiate genetically within patches of long-lived host plants, as Edmunds and Alstad (1978) demonstrated for the black pineleaf scale (*Nuculaspis california*) on *P. ponderosa*. As a result, the outcome of these interactions changes within a patch over time. Intertree and intratree transfers of the scales showed that scales transferred between trees had much lower survival than scales transferred within trees. Hence, the scale population within a patch is actually a series of demes, each tree harboring a separate deme. The density of scales within trees increases with tree age (actually size) in this case, presumably because the scales become increasingly adapted to the tree. Furthermore, some trees are more resistant to scales than others, so that scale density varies greatly among trees. Similar variation in quality occurs even within trees for the aphid *Pemphigus betae* feeding on *Populus angustifolia* (Whitham, 1978, 1981; Zucker, 1982). Differentiation at some loci occurs between demes of the membracid bug *Enchenopa binotata* on adjacent conspecific host trees, although there is no evidence of major intraspecific differentiation in response to individual trees (Guttman *et al.*, 1981). As Guttman *et al.* (1981) note, however, some of the differences they found in allelic frequencies may be a sampling artifact, since only five nymphs from each of four branches were sampled on each tree.

Because the genetic structure of both plant and insect populations can change within patches over time, a study of insect–plant interactions in a 50-year-old patch may provide very different results from a study in the same patch when only 10 years old. Even without genetic changes, interactions can change in outcome as species composition and plant architecture change within a patch (Lawton, 1983). The size and duration of clones of *Aphis* spp. on individual plants can depend upon the availability of ants within a patch and the presence of other aphids that may compete for the ants (Addicott, 1978a,b). Analyses of both genetically based and plastic changes in the probability, mechanics, and outcomes of interactions within single patches over time can show the limits to the available heritable variation in interactions, thereby providing the necessary raw material for study of the evolution of the interactions (J. N. Thompson, 1982).

D. Interactions Amid Changes in Within-Patch Diversity

Among insects that can feed on several plant species, the genetic structure of an insect population can depend upon the mix of host plant species available at any point in time. Multilocus genotype frequencies of the largely parthenogenetic fall cankerworm (*Alsophila pometaria*) vary widely among stands with different mixtures of host trees, including red maple (*Acer rubrum*), red oak (*Quercus rubra*), flowering dogwood (*Cornus florida*), and black birch (*Betula lenta*) (Mitter *et al.*, 1979). The proportion of sexual individuals also varies among localities, although the extent to which these differences are genetically determined is unknown. Since even local patch structure is not constant over time, the genetic structure of the moth populations should be expected to vary over time within, as well as between, patches.

Within-patch changes in insect genotypes have been correlated with within-patch changes in vegetational composition for the spittlebug (*Philaenus spumarius*). This species has seven color morphs in Finland, corresponding to seven different alleles at one locus (Halkka *et al.*, 1975a). The frequency of color morphs can differ between small meadows (80–300 m²) separated by very short distances; some of the differences in morph frequency correspond to differences in the host plant composition of the meadows (Halkka, 1978). On the island of Stora Vastra Langgrudet, significant differences in morph frequency between populations separated by only 5 m remained nearly constant for 9 consecutive years, as did morph frequencies within each population. The vegetation structure in these small meadows remained fairly stable in composition over the 9 years of study. When a meadow is disturbed, however, the morph frequencies can change drastically. On the treeless island of Mellanspiken, *Philaenus* populations occur within small patches of grasses and other herbs located in depressions and crevices in the granite. Vole populations reached very high densities in 1969, and a flood overran much of the island in 1975, changing the vegetational composition of the island. Within each of two grass patches separated by only 12 m, the frequency of morphs has varied widely over time since the vole eruption. The two patches differed in *Philaenus* morph frequency every year over a 9-year period, and frequencies changed over time within each meadow (Halkka *et al.*, 1975b; Halkka, 1978).

V. CONCLUSIONS

A large amount of variation in life history, genetic structure, and interactions can occur within single patches of plants and their associated animals. The process of within-patch dynamics is a result of (1) plastic responses of individuals living under different conditions at different positions within a patch, (2) differential survival of individuals depending upon the distribution of genotypes and heterozygosity within and among individuals (at least in some species), and (3) varying selection over time as interactions change within and among species. The size, frequency, and intensity of physical disturbances, together with the life histories of the organisms, will determine how these factors will combine to produce the pattern of the within-patch dynamics of a population. Studies of within-patch structure in plant populations indicate that the shapes of survivorship curves and the within-patch dynamics of populations may often depend upon the mix of genotypes and their positions within a patch.

The interactions between species within patches are as dynamic as the populations themselves, and studies of the patch dynamics of interactions are likely to provide a more solid basis for understanding the evolution of interactions (J. N. Thompson, 1982). Analyses of variation in interactions have three components: (1) variation in the probability of encounter between species, (2) variation in the mechanics of an interaction (e.g., plant parts fed upon by a larva), and (3) variation in the outcome of the interaction. All three types of variation are found in interactions among species within single patches as population structure and selection change over time.

ACKNOWLEDGMENTS

I am grateful to D. W. Biedenweg, R. N. Mack, S. T. A. Pickett, P. W. Price, and P. S. White for very helpful comments on an earlier draft of the manuscript. This work was supported by NSF grants DEB79-18492 and DEB82-19884.

Chapter **15**

Animal Population Genetics and Disturbance: The Effects of Local Extinctions and Recolonizations on Heterozygosity and Fitness

ROBERT C. VRIJENHOEK

*Department of Biological Sciences
and Bureau of Biological Research
Rutgers University
New Brunswick, New Jersey*

THE ECOLOGY
OF NATURAL DISTURBANCE
AND PATCH DYNAMICS

I. INTRODUCTION

Most animal species are not distributed uniformly in time or space. Instead, they are often fractured into a mosaic of scattered colonies that fluctuate in size with varying environmental conditions. Natural disturbances contribute directly to this patchiness by reducing local population sizes, or indirectly by altering resource abundance and quality. For many species, regular seasonal disturbances lead to predictable fluctuations in population size and distribution. Superimposed on these regular cycles are unpredictable regional disturbances that may result in further reductions in population size and even in local extinctions. The gaps in a species distribution created by natural disturbances facilitate the accumulation of both adaptive and nonadaptive genetic differences among local populations. Ultimately, the maintenance of genetic continuity across such gaps depends on a number of characteristics that may be unique to each species. Local population sizes, local rates of increase, and the patterns and modes of dispersal together play a major role in determining the distribution and longevity of discrete patches in the genotypic landscape (Wright, 1977, 1978; Roughgarden, 1979).

Unfortunately, few studies of animal populations have been undertaken for sufficiently long periods of time to observe directly the effects of population crashes, extinction–colonization cycles, and migration on the genetic structure of natural populations. Instead, we are often forced to infer the existence of such historical events from the present-day distribution of gene frequencies. Such inferences are weak if based only on a few genotypic characters, morphological phenotypes, or life history traits, particularly if those characters are subject to local selection pressures or environmental effects. However, molecular genetics has made it possible to examine multilocus genotypes of natural populations in a wide variety of situations (Lewontin, 1974). Since nonselective factors such as population crashes, local extinction–colonization cycles and migration should have similar effects on the distribution of allelic variance at all neutral gene loci, the inclusion of a large sample of gene loci in molecular population studies permits one to examine the internal consistency of the data with regard to a particular hypothesis of population structure (Lewontin and Krakauer, 1973; Selander and Whittam, 1983).

I have been studying natural disturbances and patterns of genetic variation in small freshwater fish of the genus *Poeciliopsis* (Poeciliidae) for over 10 years. These fish are of particular interest in the context of patch dynamics because they inhabit a highly fluctuating environment, the partially isolated springs and temporary arroyos of the Mexican Sonoran desert. In this chapter, I will introduce concepts concerning the effects of natural disturbances on genetic variability within local populations and use my *Poeciliopsis* data to illustrate key points. It is not my aim to review the theoretical and empirical literature on spatial and temporal processes that can act to preserve genetic variation in heterogeneous environments (see Levin, 1968; Hedrick *et al.*, 1976; Gillespie, 1977; Maynard Smith, 1978; Roughgarden, 1979). Readers interested in the spatial aspects of genetic population structure should refer to the review by Selander and Whittam (1983). Instead, my approach stresses temporal variation in the size of patchily distributed, local populations and its consequences for heterozygosity, immediate fitness, and long-term adaptive potential.

II. NATURAL DISTURBANCES AND GENETIC DRIFT

It is well known that in the absence of evolutionary pressures, such as differential selection, migration, and mutation, the genetic variance in a large, randomly mating population will remain stable from one generation to the next. However, in small, isolated populations, stochastic changes in gene frequencies (genetic drift) inevitably result in an overall reduction in genetic variability (Section IV). Populations living in patch dynamic environments are often subject to bottlenecks in population size due to natural disturbances such as fire, drought, or flood. Alternatively, bottlenecks may occur when a small number of founders establish a colony as a normal aspect of patch turnover. The stability of gene frequencies within isolated patches is inversely related to the bottleneck population size (Wright, 1969; Crow and Kimura, 1970). Depending on its intensity, natural selection can constrain random variation in gene frequencies, but one should not assume that natural selection and genetic drift are strict antagonists. In small populations, even relatively deleterious alleles can drift to appreciable frequencies, leading to a general depression of fitness.

Wright (1969) and Crow and Kimura (1970) enumerate several interrelated aspects of the breeding system and demography that act to reduce the genetically effective size of a population (N_e) below the census number of adults. Inbreeding and phenotypic assortative mating increase the genetic correlation between individuals in the local population. As an extreme example, for a population of self-fertilizing hermaphrodites, N_e would be one-half the census number of adults. Differential fertility among the adult age classes must also be taken into account (Felsenstein, 1971). Furthermore, the adult sex ratio must be considered if imbalances occur in reproductive contributions due to differential survivorship of the sexes or asymmetrical mating systems such as polygyny or polyandry. For example, in a harem-forming species in which 1 male mates with 10 females on the average, N_e is reduced to one-third the total number of breeding adults. Natural disturbances can further reduce N_e by increasing the overall variance in reproductive output among individuals of both sexes. In populations at numerical equilibrium, the mean number of gametes an individual contributes to the next generation and its variance will both equal 2. Under such a stable regime, N_e equals the number of breeding adults, but patch dynamic populations are often reduced significantly below their equilibrium numbers. If the ratio of the mean replacement number and its variance should be larger than 1, N_e is reduced (Crow and Morton, 1955). However, the magnitude of this reduction in growing populations is a function of the intrinsic rate of increase. Thus, the effects of natural disturbances on N_e will depend not just on the census number of adults that survive a local population crash or the number of founders replacing an extirpated colony, but also on the patterns and alterations of their mating system, age class structure, fertility schedule, and sex ratio.

As a local population fluctuates in size, its average effective size is best repre- sented by the harmonic mean of N_e in each generation (Crow and Kimura, 1970). For example, if a population increased from an average of 1000 individuals to 1 million once every 10 generations, its harmonic mean N_e would be 1111 over that period. Conversely, if a population maintained 1000 individuals for 9 generations and suffered a catastrophic drop to 10 survivors for 1 generation, the harmonic mean N_e would be 91.7 over the same 10-year period. Thus, population crashes have a far greater impact on the effective population size than population flushes.

In addition to the potential for inbreeding depression in small, isolated popula- tions, natural disturbances can also reduce some colonies below a critical number necessary for lek formation or other social interactions that stimulate mating (Allee *et al.*, 1949). This requirement for social facilitation of breeding may have hastened the extinction of the passenger pigeon (Terres, 1980). Such species should have high local extinction rates in disturbed environments. To persist in a patch dynamic environment, they would require larger propagule sizes and high migration rates to initiate successful colonies and would also require a greater number of survivors following periodic population crashes. One would expect many of the highly social, colonial breeding birds to be underrepresented on small islands and in disturbed habitats, unless of course they also migrate as large groups.

Frequent local extinctions and colonizations act to reduce N_e for a large popula- tion that is composed of many partially isolated colonies (Maruyama and Kimura, 1980). If no colony extinctions occur, N_e is simply a product of the mean effective size of the colonies (N) and the number of colonies (n), or nN. However if the extinction–colonization rate is high, N_e for the total population is reduced substan- tially. Thus, a consequence of frequent extinction–colonization cycles is the loss of genetic diversity among local populations. Even without local extinctions, the prin- cipal effect of the gene flow that accompanies migration is to increase the genetic diversity within local populations and to decrease the diversity among them. A variety of gene flow and migratory patterns have been modeled for both discrete and continuously distributed populations; see Wright (1931), Malecot (1955), Endler (1977), and Maruyama and Kimura (1980) for discussions of their effects on the distribution of genetic diversity within and between populations.

It is commonly assumed that relatively small amounts of gene flow can coun- teract the loss of genetic variation within colonies and random divergence among colonies. Although this is true for neutral alleles at single loci, tight linkage and limited migration can enhance each other in maintaining associations between non- allelic genes in a subdivided population (Ohta, 1982). Disturbance-induced population crashes, distributional gaps, and founder events can create associations between alleles at different gene loci, or between favorable alleles and new mutants at nearby loci (Hill and Robertson, 1966). The decay of these associations, or linkage disequilibria, is a function of the recombination rate between the affected loci. Randomly produced linkage disequilibria between closely linked genes can persist for a long time even in the face of considerable gene flow (Ohta, 1982). These random associations and their possible fitness effects in a subdivided population provide a cornerstone for Wright's (1931, 1932) shifting balance model of evolution (Section VII).

III. GENETIC DRIFT AND HETEROZYGOSITY

The average rate of loss of neutral alleles due to genetic drift is $1/(2N_e)$ per generation. A similar loss of variation occurs over all loci, thereby reducing total heterozygosity in the population. Thus, the level of heterozygosity in a particular generation (H_t) can be stated as a function of the initial heterozygosity (H_0), effective population size (N_e) and time in generations (t) (Crow and Kimura, 1970):

$$H_t = H_0 (1 - 1/2N_e)^t \qquad (1)$$

Figure 1 displays the rates of loss of heterozygosity for different effective population sizes. Thus, populations that have maintained a large size since the last genetic bottleneck should contain more variation than small populations (Soulé, 1976).

Bottlenecks and founder events in single generations have similar effects on

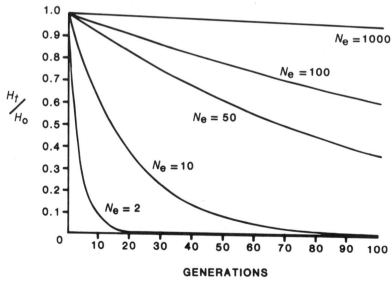

Fig. 1. The decay of heterozygosity, H_t/H_0, for different effective population sizes (N_e).

heterozygosity depending on how long it takes for the population to recover its equilibrium size. In organisms with a high intrinsic rate of increase, the decline in heterozygosity may be minimal, due principally to the loss of rare alleles during the short period when the population is small (Nei *et al.*, 1975). The potentially deleterious genetic effects of natural disturbances should be exacerbated in low-fecundity organisms because of the long time it takes for them to recover an equilibrium population size, coupled with the prolonged effects of inbreeding in very small colonies. Furthermore, in an age-structured population, natural disturbances may affect some age classes more than others, thus leading to considerable time lags and numerical fluctuations in the recovery process. The genetics of such age-structured populations are complicated, but fortunately they can be approximated using the discrete generation models of classical population genetics (Charlesworth, 1980).

In addition to the simple loss of mean heterozygosity in patch dynamic populations, population subdivision can have manifold effects on the multilocus genotypic structure of the total population. Randomly induced linkage disequilibria increase the variance in heterozygosity over that of a panmictic population because of a significant between-subdivision component (Brown *et al.*, 1980). Even for a genus under strong balancing selection, a drift event that precipitates random changes at one locus can alter the adaptive relationships among genes at many other loci, thus necessitating many generations before a new multilocus equilibrium is achieved. In a classical experiment with laboratory populations of *Drosophila*, Dobzhansky and Pavlovsky (1953) found that the equilibrium frequencies for a pair of chromosomal inversions were highly variable in bottleneck populations but significantly less variable in large populations. In the bottleneck populations, the randomly ''drifted'' genetic background established different sets of selected equilibria for the chromosomal polymorphism. Following bottlenecks and founder events, the adaptive trajectories taken by different demes are unique and dependent upon the base of variation that chance has left as the starting material (Wright, 1931).

IV. HETEROZYGOSITY AND FITNESS

The decline of allelic diversity associated with population bottlenecks mimics inbreeding in its consequences. Animal and plant breeders have long been aware of the depression of growth, viability, and fertility associated with systematic inbreeding of organisms that ordinarily outbreed. Comprehensive treatments of this subject are provided in the volumes by Lerner (1954) and Wright (1977). Falconer (1960) noted that ''the fitness lost on inbreeding tends to be restored on (out)crossing . . . ; heterosis is simply inbreeding depression in reverse.'' Unfortunately, the exact relationships between genic heterozygosity, heterosis, and inbreeding depression are not completely understood. According to the overdominance hypothesis, heterosis results from the multiplicative effects of generally superior heterozygous genotypes at many loci. The loss of these superior heterozygous genotypes results in inbreeding depression. However, there is little experimental support for extensive genic overdominance in quantitative genetic studies (Gowen, 1952; Comstock, 1977; Wright, 1977; Falconer, 1977). According to the "dominance theory," heterosis results from dominant wild-type alleles masking the expression of numerous mildly deleterious, recessive mutations that accumulate naturally throughout the chromosomes. In large natural populations, these deleterious alleles are ordinarily maintained at low frequencies due to genetic recombination and purifying selection (Muller, 1964; Felsenstein, 1974; Maynard Smith, 1978). Inbreeding unmasks this mutational gene load by increasing the proportion of homozygous genotypes in the population. Furthermore, mutations may occasionally "hitchhike" to appreciable frequencies along with closely linked beneficial traits in populations that are subjected to severe bottlenecks or intense selection, which is itself a bottleneck (Hill and Robertson, 1966).

If the overdominance hypothesis were generally applicable, inbreeding depression would inevitably be tied to increased homozygosity. Yet homozygosity per se is not necessarily deleterious since plants frequently employ breeding systems (e.g., selfing) that rapidly lead to complete homozygosity. In fact, inbreeding depression is not found in plants that ordinarily self (Young and Murray, 1966). Apparently, natural selection has purged their mutational loads (Wright, 1977; Charlesworth and Charlesworth, 1979). Although less frequent, selfing is known to occur in animals such as the snail *Rumina decollata,* which is completely homozygous for electrophoretic markers (Selander and Kaufman, 1973). Strains of the selfing fish *Rivulus marmoratus* are isogenic for histocompatibility genes (Kallman and Harrington, 1964). I surveyed 12 strains of these fish (obtained from E. Harrington) for electrophoretic variation at 31 gene loci. As expected, I found no heterozygosity within the strains, although there were allelic differences among some of the geographic strains (Vrijenhoek, 1985). Thus, it is possible, over the long term, for selection to produce inbred lineages with high fitness in nature. However, unless the genetic stock has a history of inbreeding or bottlenecking in nature, this goal is more difficult to attain in the time scales we observe in the laboratory. It takes many generations of recombination and selection to purge the hitchhiking gene load due to genetic linkage. As observed in Wright's (1977) famous guinea pig experiments, in the process of inbreeding most lineages simply die out, but sometimes a lucky few pass through this severe bottleneck to establish reasonably vigorous, homozygous lineages. However, even these surviving lines have lower fecundity and resistance to stresses than their outcrossed relatives.

With simple dominance, the benefits of heterozygosity should reach a threshold in large outbreeding populations that are at equilibrium for the mutational gain and selective purging of deleterious mutations (Vrijenhoek and Lerman, 1982). Along the horizontal axis of Fig. 2, point P represents the threshold level of heterozygosity expected to maximize fitness in a theoretically large panmictic population without a history of bottlenecks. A rapid reduction of heterozygosity through inbreeding or genetic drift would result in a loss of fitness (the area to the left of point P). Thus, random genetic drift in patch dynamic environments may keep most partially iso-

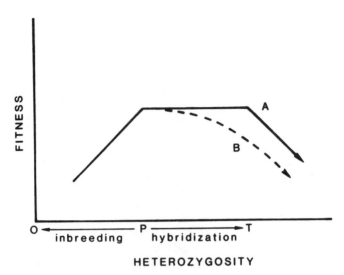

Fig. 2. The relationship between heterozygosity and fitness. See the text for an explanation of the symbols. [After Vrijenhoek and Lerman (1982) and Shields (1982).]

lated demes below this threshold level of heterozygosity and its associated fitness. Hybrids produced by outcrossing between local populations of a patchily distributed species might recover the fitness characteristic of a theoretically large panmictic population at point P, but no greater. Also, hybrids between geographically distinct races (subspecies) and species also would not be expected to show an increase in fitness beyond that found in a large, continuous population at point P. In fact, if the hybridizing populations are sufficiently distinct genetically, and locally well adapted, the F_1 hybrids might even suffer a decrease in fitness due to incompatibilities between highly integrated developmental programs (to the right of point T in Fig. 2). Incompatibilities between coadapted genomes might be amplified in the recombinant progeny of backcross or F_2 generations (Dobzhansky, 1950; Wallace and Vetukhiv, 1955). This accelerated decline in the fitness of recombinant progeny (the dotted line in Fig. 2) has been called "outbreeding depression" by Shields (1982).

I wish to draw a distinction between the present model and the one advocated by Shields (1982) who argues that low-fecundity animals optimize their fitness through philopatric behaviors that promote inbreeding. Such a breeding structure ensures the "copy fidelity" of a locally well-adapted genotype and prevents outbreeding depression due to crosses between such philopatric groups. I agree that outbreeding depression occurs in crosses between well-differentiated geographic races, semispecies, and species, but I find little evidence for Shields' contention that the local demes we identify electrophoretically represent genetically differentiated, philopatric units practicing this form of local fitness optimization. It is safer to start with the null hypothesis that the patterns of genetic structure we see in most electrophoretic studies are the neutral products of genetic drift and geographic subdivision (Selander and Whittam, 1983) rather than local selection. Failure to reject the neutral hypothesis, as is most often the case, leaves little justification for more elaborate adaptationist scenarios.

To argue in favor of philopatry and optimal inbreeding, Shields downplays the relationship between heterozygosity and immediate fitness (Shields, 1982, Fig. 16). Yet this relationship is well supported by agricultural and experimental genetic studies with a variety of animals and plants (Lerner, 1954; Falconer, 1960; Wright, 1977). Unfortunately, attempts to relate heterozygosity and important fitness characters (e.g., growth rates, survival and fertility, developmental stability) in natural populations are fraught with greater difficulties. Such studies suffer not only from the problems of separating the effects of heterozygosity at specific gene loci from chromosomal effects, but also from local environmental effects. Furthermore, most often little or nothing is known about the history of disturbances that created the present-day distribution of gene frequencies and levels of heterozygosity in patchily distributed species. However, I must repeat the caveat of Frankel and Soulé (1981), since one tends to focus on the positive results of a few studies. Negative results often go unreported.

A. Growth Rates and Survival

Electrophoretic studies of the American oyster, *Crassostrea virginica,* provide some of the strongest evidence for a relationship between heterozygosity and a fitness-related character within a natural population; at the same time, they provide a good example of the difficulty in determining the actual foci of natural selection. Singh and Zouros (1978) and Zouros et al. (1980) reported that body weight was positively correlated with heterozygosity. The more heterozygous individuals grew faster and exhibited significantly less variation in growth rates. These researchers argued that single-locus overdominance acts additively over these loci in contributing to fast growth. Koehn and Shumway (1982) reported that heterozygosity at these enzymatic gene loci in oysters is correlated with respiratory efficiency. They postulated that higher metabolic efficiency in the more heterozygous oysters permits increased allocation of energy to growth. If oysters occur as continuously distributed and randomly mating, large populations, no linkage disequilibrium should exist between the enzyme-determining gene loci and other loci. Thus, it might be safe to infer that heterozygosity at individual enzyme loci contributes additively to metabolic efficiency. But unfortunately, neither group of researchers could exclude the effects of heterozygosity over larger blocks of genes since detailed studies of the breeding system, population structure, and patterns of migration had not been performed for the oyster populations. In fact, natural disturbances such as epidemic diseases may decimate local oyster populations, leaving a patchy distribution of susceptible and resistant survivor populations (Ford and Haskin, 1982). Zouros et al. (1980) reported slight heterozygous deficiencies at most gene loci surveyed, a result that seems at odds with their conclusions of heterozygote superiority for individual enzyme loci, unless there is an undetected structure to these oyster populations or a reversal in the direction of selection for heterozygotes at different life history stages (e.g., negative selection or fast growth rates in planktonic larvae and positive selection in postsettling juveniles).

Relationships between individual heterozygosity, growth rate, and survivorship within populations have been observed in several other animals, but similarly little information on the history of natural disturbances or population structure is available. Laboratory and field studies of the tiger salamander, *Ambystoma tigrinum*, revealed that higher growth rates were often associated with heterozygosity (Pierce and Mitton, 1982). The positive relationships between heterozygosity and increased size in three other marine pelecypods (*Mytilus edulis* and *Modiolus demissus:* Koehn *et al.*, 1973, 1976; *Mytilus californicus:* Tracey *et al.*, 1975) may be manifestations of the same phenomenon, but in the absence of demographic information, one cannot exclude the possibility that heterozygosity contributes to higher survivorship. Demographic studies of the butterfly, *Colias philodice eriphyle* (Watt, 1977), and the sagebrush lizard, *Sceloperus graciosus* (Tinkle and Selander, 1973), found an increase in heterozygosity associated with increased age. Samollow and Soulé (1983) reported higher survival rates of heterozygous juvenile toads, *Bufo boreas,* over the winter, a period of severe stress. However, no correlation between heterozygosity, viability, or fecundity was found in experimental populations of *Drosophila* (Mukai *et al.*, 1974), and a negative correlation with survival and growth rate was found in rodent populations (Gaines *et al.*, 1978).

Similar relationships have been observed in several plants. Individual heterozygosity was positively correlated with high mean growth rates in quaking aspen (*Populus tremuloides*) but not in lodgepole pine (*Pinus contorta*) and ponderosa pine (*Pinus ponderosa:* Mitton *et al.*, 1981). Ledig *et al.* (1983) report that in pitch pine (*Pinus rigida*), the heterozygosity–growth rate relationship was strongest in older stands and in stands that have been subjected to the least predictable environments. A careful demographic study of an herbaceous perennial, *Liatris cylindracea,* found that heterozygosity was related to increased size, age, and seed set (Schaal and Levin, 1976).

B. Developmental Stability

Lerner (1954) attributed developmental homeostasis "to the self-regulatory mechanisms of the organism which permit it to stabilize itself in fluctuating inner and outer environments." He observed that inbreeding depression in agricultural and experimental populations is often correlated with an increased frequency of phenodeviants (developmental errors), and thus he proposed that generalized developmental stability also depends upon heterozygosity. Several recent studies of heterozygosity have revealed relationships with developmental stability in natural populations of animals. Fluctuating asymmetry of bilateral characters is thought to be an appropriate measure of developmental stability, since it reflects the degree to which specific genotypes are affected by environmental stresses during growth and development (Van Valen, 1962). Soulé (1979) found a negative relationship between average heterozygosity and fluctuating asymmetry in island populations of side-blotched lizards, *Uta stansburiana*. He interpreted this negative relationship as evidence for greater developmental stability associated with higher levels of genomic heterozygosity. Unfortunately, Soulé had no knowledge of the history of bottlenecks, founder events, or other natural disturbances affecting these populations. He could only assume that the electrophoretic estimates of heterozygosity were accurate reflections of overall heterozygosity for the populations. Such an assumption is reasonable when comparing partially isolated colonies in a structured population, since nonselective processes such as genetic drift and migration would affect the variance structure at all neutral loci to a similar degree (Lewontin and Krakauer, 1973). Estimates of heterozygosity based on a few loci are less predictive of total genomic heterozygosity for individuals (Mitton and Pierce, 1980), unless there is a hidden substructure to the population (e.g., assortative mating or admixture). Nevertheless, individual levels of heterozygosity were negatively correlated with deviant morphological phenotypes in natural populations of the killifish, *Fundulus heteroclitus* (Mitton, 1978), and monarch butterflies, *Danaus plexippus* (Eanes, 1978). A study of rainbow trout (*Salmo gairdneri*) revealed a relationship between heterozygosity and fluctuating asymmetry within a genetically variable hatchery stock (Leary *et al.*, 1983). Thus, whatever the electrophoretic gene loci are marking for us—their own heterozygosity, chromosomal heterozygosity, or genomic heterozygosity—they are providing some index to the genetic diversity in the populations.

V. NATURAL DISTURBANCES AND *POECILIOPSIS* POPULATIONS

I have been following gene frequency changes in *Poeciliopsis monacha* for over 10 years (Vrijenhoek, 1972, 1979; Vrijenhoek and Lerman, 1982). Natural disturbances clearly affect the genetic structure of this sexually reproducing species. During the long dry season, these fish are crowded into disconnected residual pools and small streams that are fed by natural springs. Mark-release-recapture studies showed that most of these pools contain only a few hundred fish (Eisenbrey and Moore, 1981). In most years, the streams desiccate completely and local extinctions occur. During the rainy season, populations grow and disperse from permanent springs to occupy the formerly dry arroyos. The major factors affecting the genetic structure in this species are the number of refugia during the dry season and the rates of extinction and migration among these springs.

Additional demographic factors affect N_e within local populations of these fish. The age-class structure of survivors of the population crashes can have a dramatic effect because fecundity is directly related to size and presumably age. A few large surviving females can contribute most of the recruitment in a flush-phase population, exacerbating the effects of reproductive variance on N_e. Also, sex ratios can become extremely unbalanced. Monarchistic dominance hierarchies develop among males, preventing most males from mating (McKay, 1971). Also, male mortality is high under stressful conditions such as crowding and low food quality (Schultz, 1977). I have frequently observed pools containing only 1 male for every 99 females, potentially reducing N_e to 3.96% of the actual number of adults. Occasionally, pools contain no surviving males at all. However, reproduction is sustained by long-term sperm storage. Thus, each female is likely to represent two genetically distinct individuals, herself and the sperm of at least one male. Since most females are multiply inseminated, each female probably carries even higher potential genetic diversity (Leslie and Vrijenhoek, 1977). Similar mechanisms for preserving genetic diversity have been found in snails that inhabit patch dynamic environments (Murray, 1964; Mulvey and Vrijenhoek, 1981). Does the potential for genetic drift in such environments favor genetic storage mechanisms that act to buffer fluctuations in N_e?

A. Heterozygosity and Genetic Drift

A study of genetic variation based on 25 gene loci revealed considerable polymorphism in Rio del Fuerte populations of *P. monacha;* the average heterozygosity per individual (*H*) was about 4.5% of the gene loci (Vrijenhoek, 1979). This level of variation is undoubtedly due to a high density of relatively stable headwater springs that sustain *P. monacha* populations and to occasional migration between them. In contrast, the neighboring rivers each have only one known population of *P. monacha,* and heterozygosity is significantly lower; *H* = 1.7% in the Rio Mayo and 0% in the Rio Sinaloa (Vrijenhoek, 1979).

The genetic consequences of the extinction–colonization process were observed in one headwater tributary of the Rio Fuerte in 1975. A total of 5 out of the 25 genetic loci examined were polymorphic, but significant differences in allelic frequencies occurred between the fish populations inhabiting pools only a few hundred meters apart (Fig. 3). Those fish inhabiting pools and arroyos continuous with the mainstream shared identical allelic frequencies and a relatively high level of heterozygosity (*H* = 6.9%). Recent field observations on marked fish reveal that daily migrations of as much as 50 m are not uncommon (R. Schenck, personal communication). Gene flow between these pools maintains the homogeneity of gene frequencies in the mainstream pools (Fig. 3: AC and AT). However, a steep waterfall at the base of the Arroyo de los Platanos (PL) prevents significant upstream migration. The population above the waterfall exhibits distinct frequencies at all five polymorphic loci and has a significantly reduced heterozygosity level (*H* = 4.3%). The fish in the nearby Arroyo de Nauchapulon (NA) have allelic frequencies intermediate between those of PL and the mainstream populations (AC and AT). The NA site dries completely in years of severe drought and is then repopulated by migrants from the mainstream (AC and TA) and PL. Using the allelic frequencies and Bernstein's (1931) formula for genetic admixture, I estimated that about 19% of the 1975 NA population were immigrants from PL and the rest were from the

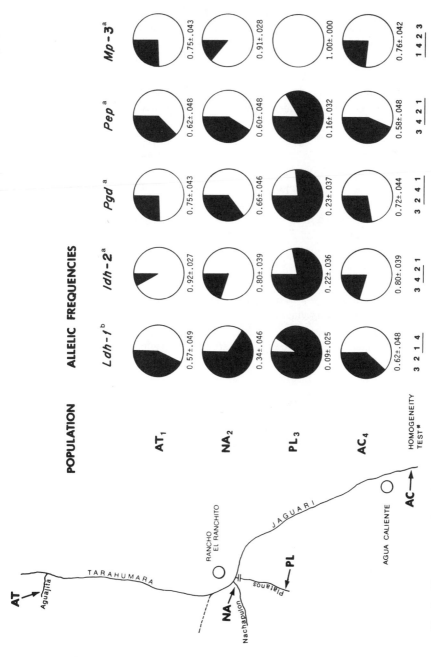

Fig. 3. The Arroyo de Jaguari and its associated tributaries. Allelic frequencies for the five polymorphic loci in *Poeciliopsis monacha* are represented as pie diagrams. Homogeneity of allelic frequencies over the four sites was tested using the chi-square contingency method. The site numbers that are underlined are not significantly different based on the 0.01 rejection criterion. [From Vrijenhoek (1979).]

277

mainstream populations (Vrijenhoek, 1979). The respective proportions have varied in subsequent years, but the NA population is generally a mixture. The dynamics of local extinctions, colonizations, and admixture in these desert pools has created considerable linkage disequilibrium in the Rio Fuerte population of *P. monacha* (Vrijenhoek, 1979). The *Ldh-1* and *Idh-2* loci are 37.5 map units apart in linkage group I (Leslie and Vrijenhoek, 1980), and their alleles are often associated in the admixture populations. The overall genetic structure of this species is characterized by strong local differences in allelic frequencies and multilocus associations among partially isolated spring populations.

A severe drought during the spring of 1976 extirpated the population of *P. monacha* in the uppermost portion of the Arroyo de los Platanos. Subsequently, this site was colonized by fish from the more persistent population in the lower portion of this stream. Given a minimum generation time of 3–4 months in *Poeciliopsis,* no more than six generations could have intervened between the 1976 sample and our next collection in the spring of 1978. The levels of heterozygosity in these populations before and after the drought are listed in Table 1. A drop in heterozygosity from about 4.0% to 0.1% requires a harmonic mean N_e of about 1.1 individuals per generation [Eq. (1)]. Even if there were several thousand *P. monacha* in the upper Platanos in 1978 following six generations of reproduction, the results are consistent with colonization by a single pregnant female and a gradual population recovery. *Poeciliopsis monacha* is a relatively low-fecundity organism; furthermore, the females of this species are highly cannibalistic on newborn fish. Thus, it lacks the high intrinsic rate of increase necessary to escape genetic drift and inbreeding following a founder event.

TABLE 1

Mean Proportion of Heterozygous Gene Loci Per Individual in Populations of *Poeciliopsis monacha* Collected in the Upper and Lower Reaches of the Arroyo de los Platanos and the Nearby Mainstream into Which It Flows[a]

Year	Locality		
	Upper	Lower	Main
1975	0.043	[b]	0.069
1976	No fish	0.044	0.081
1978	0.001	0.045	0.070
1980	—0—	0.065	0.086

[a] Estimates of heterozygosity are based on a sample of 25 gene loci as in Vrijenhoek (1979).

[b] No sample collected because a mark-release-recapture study was in progress by W. S. Moore.

B. Genetic Drift and Developmental Stability

The extinction–colonization event that severely reduced heterozygosity at our electrophoretic marker loci in the upstream Arroyo de los Platanos population of *P. monacha* would also affect the variance at many other gene loci (Cavalli-Sforza, 1966; Lewontin and Krakauer, 1973). Thus, our electrophoretic estimates of heterozygosity should accurately represent the whole genome. To determine whether this loss of genomic heterozygosity affected developmental stability, Vrijenhoek and Lerman (1982) compared the founder population with the more heterozygous populations downstream. We examined fluctuating asymmetry in eight discrete morphological characters in three samples from the heterozygosity cline: upstream Platanos ($H = 0.1\%$), downstream Platanos ($H = 4.5\%$), and mainstream Jaguari ($H = 7.0\%$; Table 1). There was a concordant and statistically significant increase in fluctuating asymmetry for all eight characters in the highly homozygous founder population when compared with the more permanent populations downstream. Thus, the eight bilateral characters all exhibited less phenotypic stability in the founder population. However, it is possible that the upstream environment was in some way more stressful to developing *P. monacha* than the downstream environments.

C. Cloning and Developmental Stability

To determine whether the increased fluctuating asymmetry in the founder population of *P. monacha* was due to a gradient in environmental stress rather than a loss of developmental homeostasis, we examined a sympatric clone of the asexually reproducing fish, *Poeciliopsis 2 monacha-lucida*. This clone also experienced the extinction–colonization cycle in the upper Platanos, but due to its asexual mode of inheritance, it suffered no loss in heterozygosity. Thus, *P. 2 monacha-lucida* provided a genotypically uniform control ($H = 52.0\%$) for gradients in environmental stress in this system. It showed no consistent upstream trends in asymmetry over the eight characters. Some characters increased, some decreased, and some did not change. Also, for any particular character the highly heterozygous clone revealed no increased homeostatic ability beyond that of the more heterozygous sexual *P. monacha* populations occurring downstream (Vrijenhoek and Lerman, 1982). However, its clonal mode of reproduction protects it from the loss in heterozygosity associated with founder events. Does this protection provide a competitive advantage for clones in a patch dynamic environment?

D. Growth Rates, Developmental Stability, and Fitness: A Relationship?

It would be desirable to establish the relationship between developmental stability and other fitness characters such as growth, fertility, survival, and competitive abilities in natural populations. For the most part, we can only make inferences about fitness from the correlation between growth and reproductive characters in agricultural and experimental studies (Lerner, 1954). Our studies with *Poeciliopsis* in the Arroyo de los Platanos are suggestive of such a relationship, however. The fish community of this stream consists mostly of the sexual species, *P. monacha,* and a single clone (I) of the triploid hybrid biotype, *P. 2 monacha-lucida.* A second clone (II) is present at low frequency (Vrijenhoek, 1978). Prior to the disturbance-induced extinction–colonization cycle that reduced heterozygosity in *P. monacha,* this species numerically dominated the fish community of the entire stream (Thibault, 1974; Vrijenhoek, 1978; see also Fig. 2). However, for the 6 years since the recolonization of the upper Platanos, *P. monacha* has rarely exceeded 10% of the fish. Clone I has nearly displaced it. No similar shifts in numerical abundance were observed in the downstream portions of this stream, where heterozygosity levels in the *P. monacha* population have remained stable (see Vrijenhoek, 1984a,b, for more details on the dynamics of this system). Is it possible that the failure of developmental stability in the homozygous founder population of *P. monacha* is functionally related to other fitness traits such as growth, survivorship, and fecundity? Could this loss affect the outcome of competition between *P. monacha* and clone I? Laboratory studies with closely related species of *Drosophila* (Latter and Robertson, 1962) revealed that inbreeding depression has a strong effect on competitive abilities. Also, Garten (1976) found that the aggressive behavior of the field mouse, *Peromyscus polionotus,* was correlated with heterozygosity. These findings are tantalizing and suggest a foundation for attempts to relate patch dynamics to genetic variability and community structure (Section VII).

    Perhaps a general relationship exists between genomic heterozygosity, growth rates, and developmental stability in sexual, outcrossing species such as *P. monacha.* Heterozygosity may lead to greater metabolic efficiency, permitting increased energy allocation to somatic growth, as is apparently the case in oysters (Koehn and Shumway, 1982). It is possible that individual enzymes have small, additive overdominant effects up to some threshold of maximum efficiency (Berger, 1976; Zouros *et al.,* 1980), but this seems less likely than the intriguing model proposed by Kacser and Burns (1981). Based on the simple dominance of wild-type isoalleles over enzymatic mutants that are mostly recessive, heterozygous individuals should achieve approximately the same ''flux'' or output rate in a multienzyme pathway as an individual that lacks such mutations. Perhaps the same process affects developmental stability. For example, meristic characters such as the fin rays, scales, and teeth we measured in the *Poeciliopsis* study have specific developmental stages, or ''phenocritical periods,'' in the terminology of Harrington (1968), during which they are susceptible to environmental disturbances. If development is slowed directly (or even indirectly because of metabolic inefficiency), those phenocritical periods are extended over longer time intervals, increasing the window of vulnerability for an environmentally induced developmental accident. Thus, a loss of heterozygosity associated with population crashes, founder events, or extinction–colonization cycles could have manifold effects on fitness. By rapidly increasing the homozygosity of deleterious recessive mutations without the opportunity for recombination and purifying selection, metabolic efficiency could be reduced and development slowed, consequently making the phenotype more susceptible to environmental stresses. This condition is temporary until migration reestablishes heterozygosity, or selection reestablishes a balanced genotype, or both.

The *Poeciliopsis* example only suggests that genetic drift can have a severe impact on the immediate fitness of sexually outcrossing, natural populations. However, without experimental tests, we cannot exclude other unforeseen factors. Fortunately, tests of this hypothesis are possible in many plants and animals, and they are currently underway with *Poeciliopsis*. First, I have begun examining the heterozygosity–growth rate relationship under controlled environmental conditions in the laboratory. Second, during March 1983, I transplanted highly heterozygous samples of about 30 *P. monacha* from the mainstream into each of several pools in the upper Platanos. I hope to track changes in heterozygosity, developmental stability, and community dynamics as a test of this hypothesis.

VI. BREEDING SYSTEMS AND PATCH DYNAMICS

Local extinctions and colonization are a major feature of patch dynamic environments. Organisms that are particularly well adapted to exploitation of such temporary environments are often referred to as "weedy" species (Baker, 1965, 1974). Modifications of the breeding system are commonly associated with colonization ability in plants (e.g., selfing and cloning; see Rice and Jain, Chapter 16, this volume). Such mechanisms provide "reproductive assurance" under low population density and thereby diminish the propagule size necessary for successful colonizations (Tomlinson, 1966; Ghiselin, 1974). However, selfing and cloning are not equivalent in their effects on heterozygosity and genotypic diversity. Inbreeding by a genetically variable colonist inevitably results in a loss of heterozygosity. Genetic diversity in inbreeders is largely contained in the differences between lineages and local populations, rather than within individuals, but the species-wide diversity is not necessarily lower than that in sexual outcrossers (Jain, 1976; Brown and Marshall, 1981). The rapid proliferation of genotypically different inbred lineages has been suggested as a potentially adaptive feature for colonizing species invading coarse-grained, multiple-niche environments (Stebbins, 1957; Rice and Jain, Chapter 16, this volume). Selfers that derive from populations with a history of inbreeding need not suffer the consequences of inbreeding depression, since their load of deleterious recessive mutations may already be very low (Young and Murray, 1966). Similarly, sexual outcrossers that have a history of genetic drift and population substructure might also be purged of their mutational loads. Thus, they might have a temporary advantage over a highly heterozygous, panmictic species that is suddenly thrust into a patch dynamic environment.

Angus and Schultz (1983) argued that a highly inbred line (39 generations of sib mating) of the sexual outcrosser *Poeciliopsis lucida* does not suffer from inbreeding depression, at least in its tolerance to temperature stresses in a laboratory environment. This line derives from an isolated natural population that exhibits no electrophoretic heterozygosity at the 25 marker loci (R. C. Vrijenhoek, unpublished data); however, other inbred lines of *P. lucida* from more polymorphic populations are not so robust (Angus and Schultz, 1983). We have never been able to produce a robust inbred line of *P. monacha*. Why has *P. monacha* not developed stable homozygous lineages as a result of the extinction–colonization cycles it constantly faces? The upper tributaries of the Rio Fuerte probably do not provide the opportunity for sufficient isolation before local populations are swamped by migrants from nearby springs. The isolated single populations of *P. monacha* in the Rio Mayo and the Rio Sinaloa may achieve such stable genotypes. Studies of inbreeding depression based on sib matings versus outcrosses are warranted as a test of this hypothesis.

True cloning preserves intact the genotypes of the founders. Patch dynamics will not affect the heterozygosity within clonal lineages, since the genotype is fixed; however, they will reduce clonal diversity as a result of random drift among lineages. Perhaps an advantage of cloning is the protection it provides against inbreeding depression within a lineage. Thus, hybrid clones might have a temporary advantage over their sexual counterparts during the short-term fluctuations in patch dynamic environments. This advantage will persist as long as the sexuals are isolated from gene flow and thus forced below the heterozygosity threshold (Fig. 2) by genetic drift. The transplantation experiments with *Poeciliopsis* (Section V,D) should address this hypothesis.

Interspecific hybridization and polyploidy also occur commonly in asexual animals and plants (Schultz, 1969; Stebbins, 1971). These processes may provide additional molecular diversity that confers broad tolerance of environmental variation, or "general-purpose genotypes" (Baker, 1965; Bulger and Schultz, 1979, 1982; Brown and Marshall, 1981). Since asexual reproduction is a derived condition in animals and higher plants, any one clone can express only a fraction of the variability in its genetically variable sexual ancestors. Yet, a recurrent production of new clones from genetically variable sexual ancestors might provide a multiplicity of clonal genotypes that could partition resources in a multiple-niche environment more efficiently than the sexually outcrossing ancestors (Roughgarden, 1972; Vrijenhoek, 1979, 1984b). Both selfing and cloning provide "copy fidelity" of the genotype, permitting the rapid fixation of unique, locally adapted lineages that are sheltered from recombinational degradation (Shields, 1982). However, in a patch dynamic environment, the longevity of clonal or inbred lineages that depend on specific genotype–environmental interactions is limited by the turnover rate of patches and the organism's ability to escape and colonize identical patches when and where they occur.

VII. LOCAL FITNESS AND COMPETITION

Ecologists have not paid sufficient attention to the role that genetic variability might play in competitive interactions within and between species. Admittedly, community dynamics are sufficiently complex that we should be reluctant to add another layer of variability to our models. Nevertheless, local population fitness might affect community dynamics, as suggested above for *Poeciliopsis*. I have focused primarily on the potentially negative effects of genetic drift associated with natural disturbances. Should the population survive and reach numerical saturation, it may be faced with a new set of challenges, including intense intraspecific competition. Wallace (1981) proposed a "self-thinning" model in which genetic variation lessens the impact of intraspecific competition on individual fitness. Under exploitive competition, every individual in a genetic monoculture may end up stunted, resulting in a precarious survivorship potential for the population as a whole. However, in a phenotypically diverse assemblage, some individuals might get a genetic head start in the contest for resources. Thus, the population as a whole might realize the individual sizes, fecundities, and survivorship capabilities characteristic of less crowded conditions. Although this argument appears to be group selectionist, it is not really different from a form of "interdemic selection" envisaged by Wright (1940; see also Section VIII). Obviously, stunted demes with low total productivity will contribute less to dispersal and colonization in patch dynamic environments. Also, this phenomenon can be based on individual selection if individual levels of heterozygosity contribute to gaining a head start in the scramble for limited resources. The same individual heterozygosity will regenerate genotypic diversity in the next generation as a result of the recombinant processes of sexual reproduction.

The number of individuals that can be maintained under crowded conditions will also be affected by the efficiency with which the resource spectrum is exploited. If the niche breadth of a species has a significant genetic component, then an assemblage of diverse phenotypic specialists might exploit a resource base more efficiently than a single general-purpose phenotype. Roughgarden (1972) partitioned the niche breadth of a population into two components: the between-phenotype component, due to phenotypic (and presumably genotypic) differences in resource utilization among individuals; and the within-phenotype component, due to the plasticity in resource use by individual phenotypes. Natural disturbances that reduce genetic variability underlying the between-phenotype component of niche breadth could also reduce the carrying capacity of the affected species and thereby alter its competitive relationships with other species. Furthermore, the reductions in fertility and survivorship associated with inbreeding depression could also alter competitive relationships between species without altering niche breadth.

Ultimately, the relationships between competing species can be viewed in terms of the "Red Queen" model of Van Valen (1973), wherein a genetic improvement in the adaptedness of one species is perceived as a deterioration in the environment of its competitors. If disturbance-induced losses in genetic variability alter the evolutionary responsiveness of participants in such a race, one can turn the Red Queen upside down: a genetic deterioration in the adaptedness of one species may be perceived as an improvement in the environment of its competitors. One can continue to speculate in this regard, but unfortunately, few data exist regarding the effects of genetic variability on multivariate interactions of a community. Some evolutionary theory and a better understanding of this complexity are beginning to emerge, but one should proceed cautiously with the faithful assumption that genetic variability is all that is needed to reach optimal coevolutionary solutions to these complex interactional problems (Lewontin, 1979; Futuyma and Slatkin, 1983; Thompson, Chapter 14, this volume; Rice and Jain, Chapter 16, this volume). Certainly, one cannot overlook the role of phenotypic plasticity in rapid adjustments of resource utilization and life histories for many plants faced with novel challenges (Baker, 1965). The same processes should not be overlooked in animals. A potentially rewarding avenue of zoological research would be to consider the range of phenotypic responses that can be produced by individual genotypes from genetically variable versus inbred populations reared under a broad range of environmental conditions and biotic challenges. The use of genotypic clones and inbred strains would be helpful in this regard (Annest and Templeton, 1978; Vrijenhoek, 1984b).

VIII. LONG-TERM PROCESSES: THE SHIFTING BALANCE

Acting alone, in small isolated demes, genetic drift leads to a deterioration in the average fitness of the total population (Wright, 1929). However, it is erroneous to consider the long-term evolutionary effects of genetic drift outside of the context of migration and natural selection. Wright (1932) proposed his "shifting balance" theory to integrate these three processes, and he emphasized their concerted role in adaptive evolution.

In large panmictic populations, natural selection acts on the average effects of genes against all existing genetic backgrounds (Fisher, 1930). Although mass selection in large populations increases the average fitness of the population, the particular genetic trajectory a population follows is conditioned by its initial set of gene frequencies. Numerous peaks exist in the adaptive landscape created by the interactions among the many genes and their pleiotropic effects on fitness. Separated by valleys, some of these peaks are higher than others. Wright (1932) identified a limit to adaptation imposed by mass selection in large panmictic populations. If the initial genetic composition of a population places it in the foothills of a lesser adaptive peak, natural selection will nevertheless force the population up that hill. It cannot descend into the adaptive valleys and gain access to higher peaks. However, Wright emphasized that most populations are subject to the kinds of natural disturbances that lead to distributional gaps. Genetic drift in partially isolated colonies will push some local populations into the adaptive valleys, where they may fall by chance into the attractive domain of higher peaks. Here again, selection will improve their average fitness. Thus, the shifting balance between random drift that continuously forces populations into the valleys and selection that again forces populations up the hills permits a greater sampling of the genotypic landscape than could occur by mass selection alone in large panmictic populations. Overreproduction by local populations that achieve the highest peaks will result in the diffusion of new multilocus genotypic complexes to surrounding colonies. The differential fitness of demes in terms of emigration and extinction rates (consider these analogous to birth and death rates) leads to ''interdemic selection'' (Wright, 1932), a rapid mechanism for the spread of new adaptive gene combinations through the total population.

The shifting balance process is accelerated in populations composed of numerous small colonies that are subjected to repeated local extinctions and colonizations (Wright, 1940). However, its effectiveness would depend on the turnover rate of the patches in this mosaic. If patches turn over more rapidly than selection can produce newly stabilized, local genotypic combinations, most colonies might generally reside in the adaptive valleys. Furthermore, a high turnover rate of patches would increase the genetic homogeneity of the population as a whole (Ohta, 1982) and thus retard unique responses to local selection pressures. Such conditions should favor the ''general-purpose genotypes'' (Baker, 1965) that can effectively tolerate the variable conditions of a coarse-grained environment without the need for genetic adjustments (see Section VI and Rice and Jain, Chapter 16, this volume).

In summary, the evolutionary consequences of population crashes, local extinctions, and colonizations in patch dynamic environments may be twofold. Over the short term, local fitnesses may decline as a result of genetic drift, but over the long term, the increased variance among populations of a patchily distributed species might accelerate the rate of adaptive evolution through interdemic selection. Ultimately, the outcome of these shifts between degradative and constructive processes depends on the time scale of natural disturbances leading to population subdivision and the balances among genetic drift, migration, and natural selection.

ACKNOWLEDGMENTS

I thank my colleagues T. Whittam and M. Douglas, and students R. Schenck and J. Graham, for their helpful discussions and comments on an earlier draft of this chapter. C. Leck, R. E. Loveland, and S. T. A. Pickett suggested some helpful references. I am indebted to Karen Kotora for her careful proofreading of the manuscript and preparation of the figures. The *Poeciliopsis* research reported here was supported by grants from the National Science Foundation (DEB 79-16620 and DEB 82-12150). I am indebted to the Secretaria de Pesca for issuing the fish-collecting permits (13 and 4962) that allowed me to undertake the field studies in Mexico.

RECOMMENDED READINGS

Crow, J. F., and Kimura, M. (1970). ''An Introduction to Population Genetics Theory.'' Harper & Row, New York.

Frankel, O. H., and Soulé, M. E. (1981). ''Conservation and Evolution.'' Cambridge Univ. Press, London and New York.

Roughgarden, J. (1979). ''Theory of Population Genetics and Evolutionary Ecology: An Introduction.'' Macmillan, New York.

Wright, S. (1977). ''Evolution and the Genetics of Populations. Vol. III: Experimental Results and Evolutionary Deductions.'' Univ. of Chicago Press, Chicago, Illinois.

Wright, S. (1978). ''Evolution and the Genetics of Populations. Vol. IV: Variability within and among Natural Populations.'' Univ. of Chicago Press, Chicago, Illinois.

Chapter **16**

Plant Population Genetics and Evolution in Disturbed Environments

KEVIN RICE[1] AND SUBODH JAIN

Department of Agronomy and Range Science
University of California
Davis, California

I. INTRODUCTION

A long-standing view in plant ecology proposes that the structure of a plant community can be described as a spatial and temporal mosaic of patches (Watt, 1947). Relatively recent descriptions of the dynamics of plant community structure have focused on the role of disturbance in structuring this patch mosaic (King, 1977; Grubb, 1977; Pickett, 1980; Mooney and Godron, 1983; also several authors in this volume). Plant population ecologists have begun to study within- and be-

[1]Present address: Department of Botany, Washington State University, Pullman, Washington 99164-4230.

THE ECOLOGY
OF NATURAL DISTURBANCE
AND PATCH DYNAMICS

tween-patch variation in demographic parameters such as survivorship and fecundity (Werner and Caswell, 1977; Gross and Werner, 1982; Thompson, 1983a; Chapter 14, this volume) and in species coevolution (Turkington and Harper, 1979). Patchiness of a species' distribution, characterized in terms of patch sizes, interpatch distances, and patch turnover (local extinction and recolonization), must be governed by the patterns of disturbance. Such a complex demography of substructured plant populations in patchy environments must be accompanied by a similarly dynamic and spatially complex genetic structure. Thus, we are concerned with the evolutionary changes in heterogeneous environments (Levins, 1968) in which the relative roles of the microevolutionary parameters of selection, migration, and genetic drift are dependent upon the degree of population subdivision in space and time.

What has not been widely realized is that many patch dynamic statistics of interest to plant ecologists are equivalent to many parameters of population structure used by geneticists. In order to analyze the genetic structure of a particular plant population, one needs to consider the spatial and temporal distributions of individuals, summarized by deme size and mating system statistics, and the role of disturbance and patch formation in determining gene frequency patchiness. This, in turn, influences the outcome of selection and random drift. A population ecologist interested in evolutionary aspects of plant populations must therefore consider the influence of the genetic structure of populations in patchy environments on the potential for localized adaptation as well as interdeme selection, and therefore on the population's persistence or growth.

Gap size is a parameter frequently measured in studies of the patch dynamics of plant communities. Gap size is important to community ecologists because of its effect on species diversity within a patch (Denslow, Chapter 17, this volume). Similarly, the size of a patch may often determine the effective population size, which, in turn, can have a strong influence on factors determining genetic diversity (heterozygosity or allelic diversity per locus) within a patch. The spatial distribution of patches or gaps is another parameter that should be considered by both ecologists and geneticists. When coupled with estimates of dispersal rates, such measurements allow ecologists to make inferences about plant life history strategies (Noble and Slatyer, 1980; Denslow, 1980b) or to estimate the degree to which localized recruitment into recently formed openings determines the patterns of vegetation (Hubbell, 1979; Stearns and Crandall, 1981; Rice, 1984). The same kinds of measurements provide population geneticists with information on the potential amounts of gene flow that may be occurring between demes located in contrasting patches (Dickinson and Antonovics, 1973; Endler, 1977; Antonovics and Levin, 1981; Jain, 1975), and whether high levels of polymorphisms, heterotic gene complexes, or highly plastic (all-purpose) genotypes evolve under patch dynamic environments (Zangerl et al., 1977; Hedrick et al., 1976).

Patch turnover rates and frequency of disturbance have received much attention in both theoretical and empirical studies of the maintenance of species diversity in animal and plant communities (Connell, 1978; Paine and Levin, 1981; Denslow, 1980b, and Chapter 17, this volume). For a plant species that is dependent on gaps in the vegetation for survival and reproduction, the disturbance frequency may also determine the number of times populations of that species experience genetic bottlenecks because of a lack of open sites for colonization.

Ideally, we need data on genetic changes in substructured populations for which we have demographic records and observations on migration rates. However, although nearly 25 years have passed since the "birth" of population biology, most genetic studies provide only casual information on demography, and vice versa (see reviews in Harper, 1977; Solbrig, 1980; Mooney and Godron, 1983). Patch dynamics is largely unaddressed, even in the studies of colonizing species, which are definitely relevant here; therefore, we shall briefly review some very general population genetic features of colonizing species. This will provide us with a series of potentially testable hypotheses and will identify population parameters influencing the effective size and genetic variation of populations that must be quantified in order to test these evolutionary hypotheses. Seldom do we find the necessary ecological and genetic work on the same species and sites, so that most ecological genetic ideas remain conjectural, especially in relation to the regulation of plant population numbers. The review of population genetic models in the context of patch dynamics illustrates the highly interactive nature of evolutionary forces such as migration, drift, and interdemic selection. We also argue for less reliance on broad generalizations and for a greater emphasis on long-term empirical investigations.

II. LINKING POPULATION ECOLOGY TO POPULATION GENETIC THEORY

The success of population biology in generating and testing hypotheses concerning the evolution of plant populations in patch dynamic environments will depend a great deal on (a) how well the demography of plant populations is integrated with the "demography" of disturbance processes (Paine and Levin, 1981; Mooney and Godron, 1983), and (b) how often investigations of plant populations in patchy environments consider both the dynamics of individuals and the dynamics of genes. To illustrate the relevance of ecological studies in patch dynamic environments to population genetics, we now examine two specific examples in detail: estimates of patch sizes in badger mounds, and patchiness of gene frequency distribution in rose clover, a successful colonizer in California grasslands.

Platt (1975) and Platt and Weis (1977) focused on the colonization dynamics of fugitive prairie plants in an environment disturbed by periodic badger mound formation. Much of the information gathered in this investigation provides valuable insight into the forces shaping the genetic structure of fugitive plant populations.

An important parameter in both empirical and theoretical population genetic studies is the effective population size. The effective population size or neighborhood size (N_e) has a significant influence on the relative importance of genetic drift versus selection in the evolutionary process (Kimura and Ohta, 1971; Vrijenhoek, Chapter 15, this volume). Using results from Platt (1975) and Platt and

TABLE 1

Population and Patch Statistics for Species within a Fugitive Prairie Plant Guild[a]

Population parameter	Species					
	Mirabilis hirsuta	Verbena stricta	Solidago rigida	Asclepias syriaca	Apocynum sibiricum	Oenothera biennis
Mean population size per patch (N)	88	105	135	100	49	?
Mean number of seeds per plant	130	320	450	600	600	?
Variance in number of seeds per plant	20,000	20,000	25,000	4000	6250	?
Variance to mean ratio	154	62	56	7	10	?
Mean "optimal" soil moisture (g H_2O/g soil)	0.091	0.109	0.125	0.185	0.236	0.112
Optimal soil moisture "range" (mean ± 2 SD)[b]	0.027 ± 0.155	0.039 ± 0.179	0.039 ± 0.211	0.097 ± 0.273	0.146 ± 0.326	0.072 ± 0.152
Percentage of open patches within a soil moisture range[c]	60	78	85	66	43	58

Mean minimum distance (m) between patches with mean optimal moisture	3	5	7	16	30	5.5
Mean maximum dispersal distance	0.4	1.0	4.9	13.8	25.7	1.8
Neighborhood area (N_a in m²)[d]	2.0	10.2	265.9	3019.1	8494.9	26.4
Number of patches within N_a[e]	1	6	71	>250	≥150	19
Number of patches within optimal soil moisture range within N_a	0.6	4.6	60.5	—	—	(11)
Neighborhood size ($N_a × N$)	88	630	9585	>25,000	≥7350	760

[a] Calculated from data in Platt (1975) and Platt and Weis (1977).
[b] Optimal moisture ranges are inferred from frequency distributions of plant species along a moisture gradient [Table 4 in Platt and Weis (1977)].
[c] Number of patches within optimal soil moisture ranges derived from Fig. 5 of Platt and Weis (1977).
[d] Neighborhood area was calculated as the area of a circle whose radius was a distance that included 95% of dispersed seed. Dispersal data are from dispersal frequency distributions in Fig. 6 of Platt and Weis (1977).
[e] Number of patches within neighborhood area calculated from Fig. 7 of Platt and Weis (1977). Patch frequency distribution for M. hirsuta was also used for Oe. biennis.

Weis (1977), estimates were made of N_e for plant populations in a guild of six fugitive species (Table 1). The impact of effective population size for a species dependent on newly opened gaps depends on the rates of patch renewal. A disturbance rate of 4.5 new patches per year was found for a section of the Cayler Prairie (Platt and Weis, 1977). This rate seems low enough to cause significant genetic "bottlenecking" in species such as *Mirabilis hirsuta* and *Oenothera biennis,* which have limited spatial dispersal. The low availability of open patches is further accentuated by the fact that newly created patches are open to recruitment for only a short time, since after 2–3 years the plants already resident on the patch prevent further colonization (Platt, 1975). However, it should be noted that seeds of *Oe. biennis* can remain viable in the soil for 80–100 years (Kivilaan and Bandurski, 1973) and that *M. hirsuta* has a long reproductive life span (Platt, 1976). Both of these life history attributes can act to reduce bottleneck effects caused by low rates of patch renewal.

Another factor that can potentially influence the effective size of a population is the distribution of reproduction among adult members of the population (Crow and Morton, 1955). Most models in theoretical population genetics assume a Poisson distribution of fecundity among the reproductive members of a population. This requires the variance-to-mean ratio in progeny per reproductive individual to be equal to 1. If this ratio is much greater than 1, the effective population size is reduced and inbreeding levels within a population are affected. The reproductive data from the badger mound study indicate that large inequalities in reproductive output among individuals within a population result in variance-to-mean ratios of progeny output consistently greater than 1. There is also significant interspecific variation in the value of this ratio, suggesting differences among species of potential evolutionary responses to selective pressures.

Several populations of rose clover (*Trifolium hirtum*) have been studied during its recent colonization of disturbed ruderal areas adjacent to many grassland sites. Patches are established by both short- and long-range seed dispersal. Patch sizes were assessed by a census of seed banks and established plants over successive years, whereas polymorphisms at four morphological marker loci measured genetic variation (Jain and Martins, 1979, and unpublished data). The relationship between patchiness and patch dynamics, and polymorphism, was studied in two sites in northern California. These were an undisturbed roadside population (Auburn) of rose clover patches growing in size, and a small right-of-way area disturbed heavily by people (Placerville). Generally, neighboring patches exchanged pollen through frequent pollinator movement, and especially at the Auburn site, the larger rapidly expanding patches showed significant increases in genetic polymorphisms (PI) (Table 2). The Placerville patches generally decreased in size and showed a decrease in PI. Successful patches had apparently higher outcrossing rates and higher levels of heterozygosity. New patches were continually established through animal-aided seed dispersal, but the patch size increase and seed output were not correlated. Patch extinctions are difficult to evaluate, since they must be verified by long-term observations on seed banks. Overall, the successful colonization by rose clover in disturbed patchy environments seemed to involve rapid genetic changes.

TABLE 2

Population Size (N), Growth Rate (R), and Polymorphism Index (PI) in Some Rose Clover Patches Scored at Two Sites Over a 3-Year Period[a,b]

Patch no.	Year 1		Year 2			Year 3			
	N_t	No. of seed/pl.	N_{t+1}	$R_1 = N_{t+1}/N_t$	PI	N_{t+2}	$R_2 = N_{t+2}/N_{t+1}$	PI	ΔPI
Placerville site									
1	38	40	185	4.9	0.075	32	0.2	0.084	+
2	50	52	98	2.0	0.044	30	0.3	0.030	−
3	47	68	14	0.3	0.035	—	—	—	Extinct?
4	19	92	43	2.3	0.029	6	0.1	.005	−
5	26	102	72	2.8	0.070	18	0.2	0.015	−
6	6	24	16	2.7	0.047	—	—	—	Extinct?
7	38	115	90	2.4	0.108	18	0.2	0.020	−
8	42	146	63	1.5	0.048	3	0.1	0.009	−
9	185	31	292	1.6	0.149	14	0.0	0.010	−
Auburn site									
1	4	90	21	5.2	0.016	111	5.2	0.067	+
2	10	170	79	7.9	0.180	108	1.3	0.157	−
3	20	120	99	5.0	0.103	87	0.9	0.148	+
4	2	207	16	8.0	0.042	44	2.8	0.067	+
5	2	20	11	5.5	0.026	32	3.0	0.018	−
6	17	194	67	4.0	0.112	46	0.7	0.116	+
7	16	250	132	8.2	0.133	135	1.0	0.183	+
8	40	140	55	1.4	0.068	187	3.4	0.129	+
9	3	280	1	0.3	0.000	7	7.0	0	Monom
10	10	210	2	0.2	0.000	10	5.0	0	Monom
11	3	125	0	0	0	—	—	—	Extinct?
12	4	810	26	1.5	0.015	17	0.7	0.009	−

[a] Unpublished data of S. K. Jain and P. S. Martins.

[b] N_t, N_{t+1}, N_{t+2} are numbers of plants in a patch, scored over 3 years. R_1 and R_2 are estimated rates of population growth. ΔPI is the change in the polymorphism index.

III. GENETICS OF COLONIZING PLANT SPECIES

As noted above for rose clover, population studies on weedy successional or colonizing species are quite relevant to a discussion of the evolutionary responses of plant species to natural disturbance. The first attempt to analyze comprehensively both the genetic and ecological attributes of colonizing species was made during a symposium held in 1965 (Baker and Stebbins, 1965). The basic approach has been to analyze successful colonizing species as a group in an attempt to establish common genetic and ecological features. Currently, there are numerous theories and generalizations on the origin of various colonizing strategies. These are related mainly to mating systems, adaptive genetic polymorphism, ecotypic differentiation, phenotypic plasticity, and reproductive strategies (Brown and Marshall, 1981; Jain, 1983).

Several authors have concluded that uniparental reproduction (selfing or apomixis) is a common feature of successful colonizers. It has been proposed that the major advantage of this mating system is that it assures reproduction even in environments with a scarcity of potential pollinators (Baker, 1974) and the rapid fixation of successful genotypes. However, certain examples suggest that higher rates of outcrossing may evolve in a colonizing species (Jain and Martins, 1979). In fact, Jain (1983) and Brown and Marshall (1981) concluded from literature reviews that colonizing species are highly variable in their ecological and genetic features, including the mode of reproduction and the mating system. A species may increase self-fertilization to maintain a high reproductive rate and to avoid segregational load, or alternatively may evolve higher outcrossing rates to produce genetic variability that may promote exploration of a patchy environment.

Founder effects, due to the reduced number of individuals normally involved in the establishment of new colonies, have frequently been postulated to reduce the genetic variation of these colonies and to influence the potential for successful colonization (Moran and Marshall, 1978; Jain et al., 1981). However, certain examples suggest that the original levels of genetic diversity can be rapidly regenerated in new colonies after the initial bottleneck. Comparative studies in *Avena barbata* (Jain, 1978; Clegg and Allard, 1972) and *Bromus mollis* (Brown and Marshall, 1981) showed similar average diversity between the colonial and source populations. In a comparison of the genetic structure of populations of colonizing *Lupinus succulentus* Dougl. with noncolonizing *L. nanus* Dougl., Harding and Barnes (1977) reported the colonizing populations of *L. succulentus* to have an excess of heterozygosity over the expected levels. Brown and Marshall (1981) suggested that large amounts of linkage disequilibrium between loci may even promote a postbottleneck increase in the frequency of alleles favoring outcrossing or recombination.

Ecotypic differentiation is a common feature of many plant species. Populations of colonizing plant species also show differentiation, both locally (Jain, 1975) and over large geographic areas (Hancock and Bringhurst, 1978). Ecotypic differentiation might evolve in colonizing species if genetic divergence in different patches is also accompanied by adequate variation within patches to allow selective changes. Snaydon (1980) reviewed several examples of rapid ecotypic differentiation on a highly localized scale. However, McWilliam *et al.* (1971) found that even after 90 years of cultivation over a large region of Australia, *Phalaris aquatica* L. (syn. *P. tuberosa* L.) showed no ecotypic differentiation for a wide range of morphological, developmental, and agronomic characters. Apparent genetic similarity throughout an ecologically heterogeneous region was also observed for *A. barbata* populations in central California (Jain, 1975). A comprehensive study of population differentiation in colonizing plant species must include the rate and range of local gene flow through pollen and seed, estimations of neighborhood size, the spatial scales of genetic and nongenetic components of variation in both morphology and life history, and measurements of environmental heterogeneity.

Baker (1974) postulated that phenotypic plasticity, mainly in traits involving survival and reproduction, is a common feature of colonizing plant species. Several comparative studies on genetic versus plastic adaptive strategies have been conducted on pairs of related species. Wu and Jain (1979) found that although *Bromus rubens* L. is genetically much less variable than *B. mollis* L., *B. rubens* exhibits more individual plasticity in response to different environments. Similarly, *A. barbata* demonstrates more phenotypic plasticity than the more polymorphic *A. fatua* (Jain, 1975). Brown and Marshall (1981) suggested that opportunities may arise for successful colonization by some species in a patchy and naturally disturbed environment, leading to concomitant increases in both genetic polymorphism and phenotypic plasticity.

The study of reproductive strategies has been an important topic in plant population biology, especially in relation to colonizing plant species (Harper, 1977). Studies on reproductive effort in plants have stressed the importance of r-selection in colonizers during their initial establishment and of K-selection in species living in undisturbed habitats or closed communities (Gadgil and Solbrig, 1972; Solbrig and Simpson, 1974). However, the measurement of reproductive effort in plants and the use of such information in the analysis of colonizing strategies remain problematic (Hickman, 1975). Hickman (1975) found that substantial variation in reproductive allocation in *Polygonum cascadense* over various habitats was due entirely to developmental plasticity. In fact, the current plethora of generalities on the optimal life history of colonizers needs a careful reexamination.

Nearly 10 years of comparative studies in six annual grassland species have focused on the nature of 25 genetic structure and life history traits (Jain, 1975; Foin and Jain, 1977), ranging from visual scores to responses in large, controlled experiments. These species vary widely in several ecogenetic traits, in spite of the fact that they all presumably share a similar habitat, represent a "similar" history of introduction into California, and possess the same predominantly selfing system.

This brief review of the genetic and ecological characteristics of colonizing species makes it clear that we are far from being able to characterize neatly the common attributes of all or most colonizing plant species. In fact, as noted by Brown and Marshall (1981), the continuing fascination with the ecology and genetics of colonizing species may be due to the diversity of life histories and evolutionary strategies that are responsible for the successful establishment of a population in local patches vis-à-vis a new geographic region.

IV. MODELS OF GENETIC VARIATION
IN PATCHILY DISTRIBUTED SPECIES

Patchy environments are most likely to generate a pattern consisting of small populations (local demes) partially isolated from each other because of limited gene flow between them. Either disturbances or certain drift-related phenomena could generate local extinction and recolonization cycles. Population genetic theory of drift and founder effects is fairly well developed for predicting the fate of new mutants, chromosomal variants, and decay of variation, and more recently, for studying interdeme selection for group traits. Vrijenhoek (Chapter 15, this volume) has reviewed the concepts of genetic drift, effective population size, and metapopulation, all of which were incorporated in Wright's (1970) theory of three-phase, shifting balance in evolution. Wright postulated evolutionary changes to be rapid in subdivided, small populations exposed to phases of (a) random differentiation among demes, (b) mass selection favoring locally adapted genotypes, and (c) interdeme selection allowing migration, selection, and drift to interact. The two important genetic features of colonizing species' patch dynamic landscapes include recurrent cycles of founder events in newly created patches and migration events leading to new colonies or recolonization. In what follows, we assume the role of disturbance to be important in creating patchiness, i.e., subdivided, small populations.

There are two important questions here: Can we predict the patterns of genetic variation in patchy, i.e., spatially subdivided and temporally unstable, environments? What kinds of adaptive changes result from evolution in such environments? To answer the first question, one must look at the population genetic models in an order of increasing complexity. For instance, the loss of genetic variation within small, isolated populations is predicted by a straightforward stochastic process. The effect of random drift, particularly in relation to population size bottlenecks, has been theoretically examined (Nei et $al.$, 1975; Motro and Thomson, 1982). Heterozygosity is reduced, and thus average genetic variation in small populations (small, measured by harmonic $\bar{N}$) is reduced. Motro and Thomson (1982) further emphasize these losses of variation by showing the persistent effects of bottlenecks. This, combined with the studies of restricted dispersal discussed below, provides the basis for large, persistent mosaics of genetic variation in patchy environments.

By introducing an island model of migration in which all subpopulations receive migrants representing the average gene frequency of all subpopulations pooled, i.e., the metapopulation, one finds that a small amount of migration can counteract drift effects. Thus, one readily accepts the general notion that migration promotes the maintenance of genetic variation. However, as noted below, several models show that too much or too little migration may not help retain variation. Spatial heterogeneity and temporal variation in environment, and therefore selection, patch persistence, rates of new recolonization, and rates of population growth, all need to be considered in some detail before one can trust such intuitive generalities. As Felsenstein (1976) put it, one makes death-defying leaps of faith in making numerous theoretical assumptions about equivalence of various population structures and of different migration characteristics.

Specific models of migration make an enormous difference in the distribution pattern of genetic variation. In an elegant computer simulation study, Turner *et al.* (1982) showed that under restricted pollen and seed dispersal, patchy genetic variation developed quickly and persisted for many generations even in large populations. Heterozygosity was distributed at the edges of such patches. This study also showed gene frequency patchiness to be greater than predicted from dispersal models. Rai and Jain (1982) empirically discovered a similar result in *A. barbata* populations. Selective forces would have dramatic local effects, as shown by Levin (1981) in his simulation of the rates of spread of favorable genes.

Several other extensions of this theory provide new insights. In a general model of polymorphism in patchy environments, Gillespie (1975, 1978) showed stochastic variation in fitnesses to be important for maintaining genetic variation under selection–migration balance. Multilocus selection was treated in the theory of evolution of sex in heterogeneous environments by Maynard Smith (1978). Spatial variations in the rate of outcrossing had large effects on genetic heterozygosity within populations (Brown, 1979). Hedrick *et al.* (1976) emphasized the critical role of autocorrelation between successive environments; the conditions for stable polymorphisms within demes are not determined by the temporal patterns, which, however, do yield different allelic distributions. Rates of migration might often have to be intermediate, since too much or too little migration among demes could fail to provide for polymorphism (Gillespie, 1975, 1978). For example, Slatkin (1981a,b) showed that if the local extinction rate is greater than the migration rates among neighboring demes, the fate of new mutants would be more dependent on the extinction time scale (i.e., patch turnover rate) than on the rates of migration.

In dealing with variation in spatially heterogeneous treatments, Roughgarden (1979) reviewed the theory of patchy environments and the role of selection–migration balance in maintaining clines or pockets of local adaptations. Of the many complex results, we mention just two. A comparison of two models of multiple niche selection differing in the role of population density and the stage of selection provided different outcomes quantitatively—an example of subtlety and the need for a critical analysis in theory. Even with the same selection types in different demes, one may see a mosaic spatial pattern of gene frequencies (presumably due to low migration, founder effects, and/or history). Density dependency in natural selection and the relative roles of intra- and interspecific competition were also briefly treated by Roughgarden (1979). Since numerous variables enter into these models, we cite the work of Hedrick (1981) on the probability of establishment of new genotypes in the form of chromosomal variants in finite populations as an elegant example of factor analysis. Here, too, we are dealing with a species living in patchy environments; its small populations undergo inbreeding and patch turnover. Hedrick (1981) summarized the effects of four factors, singly and pairwise, (meiotic drive, advantage of new homokaryotype, inbreeding, and drift) and concluded that inbreeding in concert with either a selective advantage of the new karyotype or with genetic drift is potentially important in chromosomal evolution and thereby in the origin of new taxa, whereas other combinations appear to be far less effective. Such an approach allows the interactions among various evolutionary features to be analyzed quantitatively.

We assume that somehow patches represent sufficiently different microhabitats so that locally different genotypes are selected. Then we superimpose environmental variation such that selective values vary among successive generations. Disturbance-created patchiness may well fulfill these conditions. Hedrick (1983) may be consulted for an excellent introductory treatment of evolution in heterogeneous environments. He uses simple models to illustrate the following: (a) Spatial heterogeneity is more likely to maintain polymorphism than temporal heterogeneity. (b) One may distinguish two kinds of selection among multiple niches: hard selection, which may involve density-dependent fitnesses such that interdeme selection [*sensu* Wright (1970)] may operate, and soft selection, in which the same number of adults enter the mating pool in each generation. These models differ in the role of density dependency and a relationship between the weighted mean population fitness ($\bar{W}$) and population numbers (i.e., ecological genetic aspects of selection). (c) The temporal variation pattern is important; correlations of environmental changes in different subpopulations are significant, as are dominance, multiple allelism, heterozygote advantage, and the migration structure of the model. Single versus multilocus treatment of selection and drift makes an important difference in relation to the theory of recombination systems (Maynard Smith, 1978). For example, the evolutionary advantage of sex may arise in patchy environments in which sibcompetition within local populations relies on sexual generation of numerous new gene combinations. In contrast, Vrijenhoek (Chapter 15, this volume) provides evidence for the maintenance of highly homeostatic, adaptively favored heterozygotes under an asexual system. The diversity of such findings derives from the variety of selective models.

In small, partially isolated subpopulations, interdeme selection was proposed by Wright (1970) to be conducive to rapid evolutionary changes. Wade and McCauley (1980) defined "populational heritability" as the fraction of the observed betweendeme phenotypic variance that is heritable in the sense that "propagules chosen at random from a "parental" deme resemble one another and that parent in their populational growth characteristics." Their research on *Tribolium castaneum* had involved varying population migration rates among demes and artificial selection for productivity. According to a model by Slatkin (1981b), a response to group selection occurred even under such theoretically unlikely conditions as high rates of random extinctions and moderately high migration rates (up to 12%). Jain (1984), in particular, referred to breeding and recombination systems as group-selected traits (Wade and McCauley, 1980); for example, rose clover data suggest the possible evolution of higher outcrossing rates in new colonies.

In summarizing this theoretical material, we can offer an observation and a caveat. Theory is rapidly developing to deal with evolution under heterogeneous environments, but the results so far are rather sketchy. One must be cautious when trying to compress conclusions from a rather varied lot of models into a very few general, biologically testable statements. We still have no simple predictions on the pattern of variation, much less on the nature of adaptive outcomes.

V. GENETIC VARIATION: OBSERVATIONS IN PLANT POPULATIONS

With these theoretical ideas in mind, we might expect to find genetic variation within and among populations living in a patchy environment to provide some clues to (a) the role of genetic variation in adaptability, (b) the relative magnitudes of selective versus other factors, and (c) optimality of demographic and competitive processes in life history evolution. Prior to the development of electrophoretic assays, the literature on variation in plant populations emphasized the genecology of a few selected traits such as flower color, pubescence of leaves, leaf shape, flowering time, or seed morphs. Often such assays were aimed at ecological patterns described over large geographic scales, but were somewhat lacking in genetic analyses and microgeographic detail. Currently, we have allozyme variation data on numerous species, including colonizers, populations with different histories, and so forth. Hamrick *et al.* (1979) surveyed these studies under various schemes of species groupings and found that "species with large ranges, high fecundities, an outcrossing mode of reproduction, wind-pollination, a long generation time, and from habitats representing later stages of succession have more genetic variation." Brown (1979) analyzed many examples of inbreeders and outbreeders and concluded that their patterns of interpopulation differentiation and heterozygosity within populations vary due to many complex factors affecting mating systems. Partial inbreeding due to restricted neighborhood size is important in many outbreeding species.

Brown and Marshall (1981) reviewed evolutionary features of colonizing species and suggested that colonizing species are somewhat depauperate in allozyme variation. As Zangerl *et al.* (1977) noted, however, there are conflicting theories about the trends in variation along successional gradients or central to peripheral areas along geographic gradients. In the studies of microgeographic differentiation in plants reviewed by Brown (1979), the spatial units of sampling are so heterogeneous that patch dynamic relationships between variation and distances (or gradients) are presently difficult to compare among species.

The slender oat (*A. barbata* Brot.) has been a subject of detailed population studies. Surveys of genetic variation have shown a regional pattern with climatic determinants, a cline with evidence of coadapted gene complexes explained by a moisture gradient, and a potential example of hitchhiking as an alternative to the coadaptation argument (Hedrick and Holden, 1979). Experiments using founded colonies demonstrated the role of natural selection. At least three spatial scales of sampling provided parallel evidence for selection. Finally, variation in patchy environments was analyzed by taking samples every 15 cm along linear transects. Localized measurements of gene flow and plant densities gave estimates of neighborhood size in the range of 40–400 plants. The relative sizes of neighborhoods in several populations were correlated with the patchy distribution of different genotypes within short distances, but patch sizes had a wide range among different sites. Highly localized gene flow seemed to account for the observed pattern of highly patchy variation even when as in many cases, the dispersal curves for both pollen and seed were platykurtic. In another study, 20 new roadside colonies were scored along with 48 well-established grassland sites. Data on genetic polymorphism, scored at two morphological loci, showed peripheral, isolated roadside colonies of slender oat to be significantly less polymorphic than the large central populations in continuous stands (Table 3). The role of random drift (founder effect) was evident in the genetic structure of such roadside colonies, which were, however, not monomorphic. Multilocus associations also suggested a large Hill-Robertson effect in generating gametic disequilibria. Such isolates, with varying amounts of elapsed time since the founder events, offer useful material for a study of population dynamics of patches, as discussed earlier in the rose clover example. Here too, we still lack the demographic information and the multiseason analyses of genetic changes.

The work of Schaal and Levin (1976) on the demographic genetics of *Liatris, Cylindracea,* now a classic, reported a positive correlation between stand age, reproductive potential, and mean heterozygosity; this suggested strong heterosis. The authors proposed that if the habitat deteriorated, heterosis would be magnified and would maintain genetic variation in the face of population bottlenecks. Thus, a powerful buffer against untoward demographic events exists. Ledig *et al.* (1983) found an age-dependent positive relation between tree growth and heterozygosity in

TABLE 3

Genetic Variation at 11 Allozyme Loci in *Avena barbata* Populations

| Habitat type | No. of sites | PLP | | H^a | D'^b | | D_{ST}/H^c |
		$\bar{X}$	Range		Range	$\bar{X}$	
Grassland	22	29.8	11–48	1.85	0.82–2.14	0.20	16
Roadside isolates	14	12.5	0–26	0.73	0–1.40	0.44	37

[a] H is the mean within-subpopulation diversity index defined as $H_j = 1 - \Sigma p_{ij}^2$ for the jth subpopulation.

[b] Monomorphic loci were ignored in computing D' values.

[c] D_{ST}/H is a measure of between-subpopulation differentiation (Brown, 1979).

pitch pine. Furthermore, they stated that disturbance (e.g., fire) helps the offspring of a few chance survivors to establish new colonies, often with inbreeding and relaxed selection, followed by crossing among them to regenerate high heterozygosity and expansion of stands. During this recovery, sib-competition should favor even greater heterozygosity.

Hiebert and Hamrick (1983) found bristlecone pine to have higher heterozygosity and more variation than most conifers, which they postulated to be due to stable and large population sizes, higher outcrossing rates, and localized microhabitat adaptation in a spatially and temporally heterogeneous environment. However, heterogeneity was not sufficiently described and did not seem to affect locally stable population sizes.

Ledig and Conkle (1983) studied variation in a narrow endemic, Torrey pine, in which variation was very low (as expected). However, these authors further noted that red pine and western red cedar, with wider ranges than Torrey pine, are only slightly more variable. Guries and Ledig (1982) reported pitch pine to be highly variable, with high within-population and rather little between-population variation for allozyme loci. The authors claim that it is a fugitive species, experiencing bottlenecks in population size as a result of fire, but presumably migration rates are high. Thus, a review of variation statistics (Guries and Ledig, 1982) shows a wide range of estimates. We find that population biologists know rather little about the ecology and history of different species, and we must be wary of allozyme variation statistics in selection arguments. Lacking such information, we tend to look desperately for examples that fit some simple theoretical expectations, and rather overzealously, we too often accept simplistic explanations.

VI. VARIATION, SELECTION, AND ADAPTATION

The bulk of population genetic theory deals with the maintenance of polymorphisms in a population or species; the role of natural selection is often specified by the relative fitnesses assigned to various genotypes. When these fitness sets are varied in relation to specific environmental conditions, one could argue that the genetic changes and the fate of variation are somehow linked to certain measures of adaptedness. The weighted mean population fitness ($\bar{W}$) has important properties relative to the local stability of equilibrium gene frequencies. However, in the ecologists' context of adaptedness, the measures of well-being of a population are interpreted in terms of reproductive success, relative abundance or persistence in time, or all of these together. Accordingly, many attempts have been made to find relations between $\bar{W}$ and values of r (intrinsic growth rate in the logistic equation) or population regulation mechanisms. In fact, this was defined as the first goal of population biology in aiming to correlate kinds and numbers within populations.

Populations living in patchy environments may allow local selection to favor different genotypes in different patches (given a rapid evolutionary rate or sufficient patch duration); this would lead to genetic differentiation (high F_{ST}, etc.). Polymorphism is assured in the metapopulation (patches pooled together), whereas within patches polymorphism would be maintained by migration, overdominance, and other selective forces. Even then, we may not be able to specify how this variation has an adaptive role in the ecological-demographic well-being of the species. Intuitively, we often argue that variation is useful in the continuing response to new environmental challenges. However, note that Reddingius and Den Boer's (1970) model assures persistence of a species in a patchy environment by spreading the risk, i.e., a subdivision–dispersal–recolonization process, without any genetic shifts.

The traits likely to evolve in patchy environments include certain migration characteristics and features of the recombination–sexuality system. These apply to both nongenetic, persistence models (Reddingius and Den Boer, 1970) and models of polymorphism (Felsenstein, 1976; Slatkin, 1981a). Vrijenhoek (Chapter 15, this volume) and others (see, e.g., Zangerl *et al.*, 1977) find high heterozygosity (genetic homeostasis) under high temporal heterogeneity, asexuality, or an *Oenothera*-type recombination system (selfing, translocation heterozygote) on the one hand. On the other hand, heterozygosity can be associated with sexuality or open recombination systems, according to Maynard Smith's model (1978) of sib-competition under spatial heterogeneity and changing environmental conditions (including temporal heterogeneity). Likewise, a given life history feature might also be advantageous in contrasting situations: rapid population increase in newly founded patches, high r-selection, opportunism (Stearns and Crandall, 1981), and wide dispersal as opposed to dormancy. Interdeme selection would act on most of these features as we have discussed, i.e., through optimization to make the metapopulation more fit.

Finally, it should be noted briefly that many changes are currently taking place in evolutionary thinking. The origin of variation is subject to new molecular probes; polygenic mutation rates might be very high; regulatory genetic variation needs to be understood; a strict population genetic approach to evolutionary rates may need modification; and the origin of species (macroevolution) certainly is not just a continued gene substitution process. In this changing scenario, patch dynamics, as it introduces a large role for random drift and heterogeneity of local units of evolutionary change, has an important role to play. We need to incorporate founder–flush theory; demographic genetic features of selection, migration, and drift; the role of inbreeding in subdivisions; and interdeme selection. Thus, it is not business as usual—a new level of complexity is compellingly introduced. A contrasting view, however, denies the role of random drift but assumes that transitional and isolated patches show loss of phenotypic stability, leading rapidly to new evolutionary lines (see Maynard Smith, 1983, for a review of new developments in evolutionary thinking).

VII. SUMMARY

In this chapter, we have briefly reviewed theoretical aspects and selected empirical studies on the population ecology and population genetics of plants existing in patchy environments in an effort to define a common conceptual framework upon which to build a synthesis of the two disciplines. The initial growth and long-term success of such a synthesis would appear to be largely dependent on the compatibility of the experimental evidence and modeling efforts generated within both ecology and genetics. Unfortunately, at the present time, studies of various aspects of the evolutionary process seem to be rather rigidly compartmentalized into largely independent conceptual paradigms, each with its own well-defined theoretical structure and base of empirical observations. The inability of much of the theory to predict specific adaptive responses to spatial and temporal heterogeneity may be due, in part, to the often vague characterizations in many models of the nature of the environmental variability.

However, numerous developments in evolutionary biology draw attention to the usefulness of work on species living in disturbed environments. Their population genetic structure may invoke rapid, observable genetic changes; their colonizing features may allow life history analyses jointly with genetic studies; and certainly, they link various processes at the population and community levels of biotic organization. Thus, patchy environments offer the exciting prospect that both spatial and temporal aspects of environmental heterogeneity might be effectively captured within a definable set of patch statistics that can be measured and perhaps even manipulated experimentally in the field. Variation in the magnitude and direction of selective processes may be much easier to measure in a patchy environment because of the often abrupt changes in both biotic and abiotic conditions across patch boundaries and among patch types. Quantifying environmental heterogeneity within a series of patch parameters may force models examining evolutionary processes within variable environments to be more unambiguous in their predictions and therefore more testable in the field.

Studies examining ecotypic or phenotypically plastic variation within a species, across a gradient (natural or artificial) of a particular patch parameter, would undoubtedly provide new insights into the causes and consequences of life history variation in response to specific types of environmental variation. Populations of numerous colonizing plant species offer excellent research materials.

RECOMMENDED READINGS

Brown, A. H. D. (1979). Enzyme polymorphisms in plant populations. *Theor. Popul. Biol.* **15**, 1–42.
Brown, A. H. D., and Marshall, D. R. (1980). The evolutionary genetics of colonizing plants. *Proc. 2nd. Int. Congr. Syst. Evol. Biol.*, 351–363. Hunt Inst., Carnegie-Mellon, Univ., Pittsburgh, Pennsylvania.
Hamrick, J., Linhart, Y. B., and Mitton, J. B. (1979). Relationships between life history characteristics and electrophoretically detectable genetic variation in plants. *Annu. Rev. Ecol. Syst.* **10**, 175–200.
Hedrick, P. W. (1983). "Genetics of Populations." Van Nostrand-Reinhold, New York.
Hedrick, P. W., Genevan, M. E., and Ewing, E. P. (1976). Genetic polymorphism in heterogeneous environments. *Annu. Rev. Ecol. Syst.* **7**, 1–32.
Wright, S. (1977). "Evolution and the Genetics of Populations," Vol. III. "Experimental Results and Evolutionary Deductions." Univ. of Chicago Press, Chicago, Illinois.

Part IV

IMPLICATIONS OF PATCH DYNAMICS FOR THE ORGANIZATION OF COMMUNITIES AND THE FUNCTIONING OF ECOSYSTEMS

Chapter 17

Disturbance-Mediated Coexistence of Species

JULIE SLOAN DENSLOW

Departments of Zoology and Botany
University of Wisconsin
Madison, Wisconsin

I. INTRODUCTION

Disturbances affect the spatial and temporal heterogeneity of ecosystems and the relative abundances of the species present. Discussions of both species interactions in patchy environments and predator–prey interactions are thus applicable to the effects of disturbances on the structure and composition of communities. In these guises, disturbances have been the focus of a substantial literature (see reviews in Drury and Nisbet, 1973; May, 1976; Wiens, 1976; Grubb, 1977; Whittaker and Levin, 1977; White, 1979; Pickett, 1980; and other chapters in this volume). Discussions of disturbances are likewise interwoven with considerations of the causes and consequences of species diversity and ecosystem stability (see, e.g., May, 1973; Thiery, 1982). Several studies have concluded that disturbances and, similarly, predation may enhance species diversity (a) by lowering the dominance of one or a few species (thereby freeing resources for other less competitive species;

Paine, 1966; Janzen, 1970; Loucks, 1970; Connell, 1971, 1978, 1980; Lubchenco, 1978) and/or (b) by increasing environmental heterogeneity (thereby providing a basis for specialization and resource partitioning; Grubb, 1977; Platt and Weis, 1977; Denslow, 1980a; Tilman, 1982).

Although accumulating examples show that natural disturbances are often a critical factor in maintaining species diversity and that simple models of species interactions provide insights into the processes involved (Levins, 1968; Horn and MacArthur, 1972; Levin, 1974; Levin and Paine, 1974; Slatkin, 1974; Horn, 1976; Hubbell, 1980; Greene and Schoener, 1982; Tilman, 1982), effects on particular systems are not always predictable. In some communities and under some circumstances, disturbances are critical in the maintenance of coexisting species (Loucks, 1970; Platt and Weis, 1977; Lubchenco, 1978). In others, disturbance contributes to the elimination of species and to long-lasting changes in ecosystem structure, as in the conversion of forest to savanna at the forest–savanna boundary (Hopkins, 1965). A species richness curve that is "hump-backed" with respect to disturbance intensity has some support from both theoretical (Connell, 1978; Grime, 1979; Huston, 1979; Tilman, 1982) and empirical (Lubchenco, 1978; del Moral, 1983) studies of processes occurring in ecological time. However, the evolutionary mechanisms of these phenomena have been little explored. Moreover, the effects of disturbance intensity on community diversity are often related to the differences between natural (indigenous) and foreign (exotic) disturbance regimes. It is therefore useful to examine the mechanisms underlying the effects of indigenous and exotic disturbances on community structure.

II. PATCHES AND LANDSCAPES

Discussions and models of patch-related coexistence or predator–prey interactions often focus on the behavior of simple systems in the vicinity of equilibria (see, e.g., MacArthur and Wilson, 1967; Caughley, 1976; Hassell, 1976). At this scale, the study of natural disturbances to ecosystems is an inquiry into the behavior of ecosystems deflected from local equilibria and the processes that contribute to their return. It is the study of secondary succession in an individual patch (or in a closed system; Caswell, 1978). At this scale, disturbance may increase or decrease the number of species locally present.

How species diversity changes in a patch during the succession following a large-scale disturbance is likely to depend on the historic size (or intensity)-frequency distribution of disturbances to the ecosystem (Denslow, 1980b; Fig. 1). Communities commonly subject to small-scale disturbances (e.g., treefalls in forests) and rarely subject to large-scale disturbances (e.g., fire) are likely to contain relatively few species adapted to patches with the size and environmental conditions created by fire. Relatively few species are likely to establish in new fire scars, but diversity increases as succession proceeds and as germination and establishment conditions approach those in the old-growth forest (Denslow, 1980b). In contrast, commu-

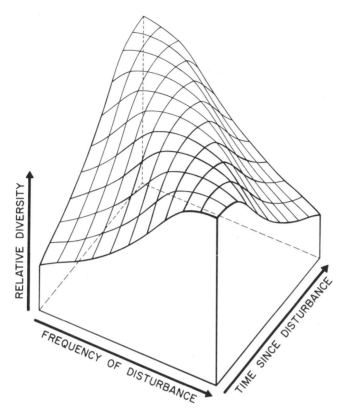

Fig. 1. Changes in species diversity during succession since a large-scale disturbance. Communities frequently subject to disturbances creating large patches are most diverse early in succession and become less diverse with time in the absence of disturbance. In communities rarely subject to large-scale disturbance, species diversity is greatest late in successional time. Diversity is relative to the maximum in each community.

nities that are frequently disturbed by fire, such as chaparral (Parsons, 1976; Keeley *et al.*, 1981) and Australian heath (Specht *et al.*, 1958), are more diverse soon after fire and less so thereafter. The seed germination and seedling establishment conditions of a larger proportion of these communities are found in the postfire environment. Diversity rises quickly following fire but falls as the community persists and fewer species are able to regenerate. In some communities (e.g., some north temperate forests; Loucks, 1970), species richness declines following an early peak composed of both early and late successional species.

At another level, disturbances affect global species diversity through changes in patch structure within the landscape (or open system; Caswell, 1978). Interactions between rates of disturbance, resource supply, propagule dispersal, establishment, and local extinction affect the processes occurring in individual patches and coexistence of species in the landscape. Although an open system might approach a stable age (stage) distribution of populations and patches, such would rarely be the case in a closed system. The open system is thus more amenable to modeling factors and processes affecting long-term species coexistence. It is in the context of an open system that the following discussion will consider the interaction of diversity and disturbance.

III. CHARACTERISTICS OF DISTURBANCES

I divide the effects of disturbances on ecosystems into three components: changes in environmental heterogeneity, changes in temporal heterogeneity, and changes in the relative abundance of species.

A. Disturbances as a Source of Spatial Heterogeneity

Disturbances to sessile communities typically open space, allowing the establishment of other individuals. If species differ in their colonizing abilities, the mere increase in heterogeneity caused by variation in availability of open space may permit the coexistence of some species (Tilman, 1982). The theory behind this coexistence, i.e., the trade-offs among reproduction-mortality schedules, dispersal effectiveness, and the ability to establish and occupy space, has not, however, been well developed. Life history theory addresses the responses of populations to rates of juvenile and adult mortality and hence to disturbance regimes (Stearns, 1976). However, role of the environmental conditions under which propagules must establish is rarely taken into account explicitly, although they are critical to seedling survival. Mortality associated with interactions within a population (density-dependent mortality) is not distinguished from mortality associated with habitat conditions or the presence of competing species (Stearns, 1977; Hickman, 1979). Although species characterized by different rates of invasion and establishment may coexist because fugitive species are able to occupy space in advance of longer-lived, more competitive species (Horn and MacArthur, 1972), important differences in establishment capabilities imply trade-offs associated with growth and reproduction under the environmental conditions of newly opened versus occupied space. Survivability is as important as getting there, as MacArthur and Wilson (1967) found for species colonizing oceanic islands. Assuming *no* differences in resource levels associated with openings of different sizes, the species most effectively dispersed and most rapidly established should occupy both large and small spaces. The important contrasts between r- and K-selected species are associated both with sources of mortality and with the availability and environmental quality of the space occupied, especially at the establishment phase of the life history. Differential responses of species within a disturbance regime arise because adaptive trade-offs are inherent in dispersal and establishment properties of propagules and in survival and reproduction of adults.

Although a primary effect of disturbance is the opening of space, that space is usually associated with a change in the availability of other resources—light and soil nutrients in terrestrial systems (Schulz, 1960; Denslow, 1980a), access to suspended particulate matter in aquatic systems (Sousa, Chapter 7, this volume)—or with a change in the physical environment—temperature, wind and water flow patterns (Chazdon and Fetcher, 1984; Sousa, Chapter 7, this volume). Moreover, habitat heterogeneity is increased because disturbances themselves vary in their effect on a site. Indeed, disturbances are the source of multiple levels of environmental heterogeneity, and thus potentially provide a complex basis for resource partitioning among coexisting species.

Functionally, habitat heterogeneity associated with disturbance patches depends on both environmental variability and the ability of species to exploit that variability. Depending on the scale of the disturbance and on the scales of important underlying environmental patterns, a disturbance may either increase or decrease environmental heterogeneity. To the degree that available species are differentially able to exploit disturbance patches, such disturbances will increase the potential number of coexisting species in the community. However, large-scale or frequent exotic disturbances may impose a functional homogeneity on a landscape. Fire, landslide, and windthrow result in such effective homogeneity in most tropical rain forests because relatively few tree species have evolved to exploit this (rare) habitat. In contrast, diversity increases following fire in chaparral or boreal forest, suggesting that in those areas heterogeneity may be increased. There, frequent fires provide a predictable resource for pioneer species; habitat specialization within the postfire landscape is more likely to occur than in communities in which the postfire landscape itself is rare (Denslow, 1980b).

In forests, clearing size has important consequences for environmental conditions within the gap. Large gaps involving the fall of several trees are brighter, hotter, and drier than gaps created by the fall of single trees or branches (Schulz, 1960; Whitmore, 1975; Denslow, 1980a; Chazdon and Fetcher, 1984); whether a tree breaks off, uproots, or dies standing affects the nature of microhabitats within the gap. Experimental studies have demonstrated fine-scale species differences in seed germination and seedling establishment requirements for several species groups (Werner, 1979; Gross and Werner, 1982; Fowler, 1982); it is thus likely that such microhabitat differences within and among clearings have differential consequences for the germination and growth of tree seedlings, although this has not yet been well documented (but see Orians, 1982). For instance, fast-growing pioneer species such as *Cecropia* (Moraceae), *Trema* (Ulmaceae), and *Macaranga* (Euphorbiaceae) will germinate only in gaps greater than 150–1000 m^2 (Brokaw, Chapter 4, this volume). In shorter vegetation, the correlation between patch size and physical environment is not as great, although other size-related differences may exist. Adult mussels can quickly fill small patches by "leaning" or by immigration (Sousa, Chapter 7, this volume). Propagules of new colonists are therefore likely to establish only in patches above some minimum size (Paine and Levin, 1981) or below some degree of isolation (Connell and Keough, Chapter 8, this volume; Sousa, Chapter 7, this volume). Similarly, small gaps (branch falls) in forests are filled primarily by lateral growth of adjacent standing trees rather than by growth of seedlings from below (Whitmore, 1975; Runkle, 1982).

The magnitude of environmental differences between disturbed patches and the surrounding matrix may influence the number of species able to coexist in a patchy environment. Presumably, more species can effectively subdivide a long gradient (large environmental differences) than a short one. For example, Ricklefs (1977) hypothesized that more species are able to partition the long environmental gradient represented by the extremes of large gap centers and the understory below closed canopy in a tall tropical rain forest than the shorter gradient of a temperate forest. Environmentally, open spaces in deserts or tundra do not differ greatly from the intact community. Many of the same species colonize the patches as constitute the mature community (Muller, 1940, 1952; Billings and Peterson, 1980). In temperate deciduous forests, environmental differences are smaller, and early and late successional communities are more similar in xeric than in mesic habitats (Peet and Loucks, 1977). Although large-gap specialists are a conspicuous part of tropical rain forest and temperate mesic forest floras, we lack data from experimental studies to determine whether this gradient is further subdivided. Surveys of environmental heterogeneity represented by disturbance patches would also help determine the steepness of such gradients in different ecosystems, i.e., the abundance of habitats represented along the length of the gradient. Vegetation transitions are likely to be more abrupt along steep gradients (few intermediate habitats) (Beals, 1969), but we expect that more species may be packed along gentle gradients (many intermediate habitats).

B. Disturbances as a Source of Temporal Heterogeneity

In addition to the nature and distribution of environmental patterns in space, habitat heterogeneity has an important temporal component—how rapidly environments change following disturbance and how frequently and predictably disturbances occur. Disturbances that open space for establishment of new propagules initiate successional sequences in which the composition or relative abundance of species changes (Connell and Slatyer, 1977), contributing to the temporal heterogeneity of the landscape. Depending on the scale and intensity of the disturbance and on the breadth of environmental differences between the new opening and the existing matrix, these successions may involve several changes in dominants and varying degrees of environmental change (see Section IIIA and reviews of succession in Gleason, 1927; Drury and Nisbet, 1973; Bormann and Likens, 1979; Denslow, 1980b; West et al., 1981).

Superimposed on the rate and sequence of successional change is a compositional variability associated with seasonal changes in the propagule pool. Seasonal sampling variation may account for much of the variability in composition among early successional communities (see, e.g., Sarukhán, 1964; Sousa, 1979b; Greene and Schoener, 1982). If disturbances creating large patches occur primarily during certain seasons (e.g., during winter storms on the northwestern Pacific coast; Paine and Levin, 1981), while those creating small patches occur year round, among-patch variation will be less for large than for small patches (e.g., on intertidal boulders; Sousa, 1979b).

In another dimension, the history of disturbance at a site has important implications for rates and patterns of succession there. Ultimately, disturbance frequency may have as strong a selective effect on species as do the physical characteristics of the habitat. Successional patterns during the first few months following burning in a tropical rain forest are illustrative. Fields in the slash-and-burn agricultural system of northern Colombia are products of different frequencies and intensities of disturbance (Denslow, 1978). Fields planted to corn and rice are usually cleared, burned, and planted, at intervals of 15–20 years. In the fallow interval, forest cover is reestablished. At another extreme, fields used for pasture are cleared of woody growth annually and grazed in the interim. Other fields are cleared at 3- to 5-year intervals and planted to minor crops. Although all fields are colonized by pioneer, "weedy" species, a disturbance frequency gradient best accounts for patterns of plant species composition in these fields. Important species in the frequently cut fields are characterized by relatively poor dispersal characteristics and good sprouting abilities, whereas those in the infrequently cleared fields produce bird-dispersed seed but rarely reproduce vegetatively. Successional processes also vary across the disturbance gradient, reflecting differences in growth and reproductive characteristics of the dominant species. Within the first year following clearing, vegetation in the annually cleared fields becomes less diverse and more stable in composition as vegetatively reproducing individuals crowd out those establishing from seed (Fig. 2). Vegetation in fields cleared from forest increases in diversity over time and shows no tendency toward stabilization in the first year following clearing. The majority of plants there establish from seed. Community composition and the patterns of early succession are thus a product both of the hot, bright, dry conditions in the newly cleared fields and of their disturbance histories (Denslow, 1978).

The history of a stand is thus of particular importance to the prediction of disturbance-related effects. Although present disturbance regimes may be similar to historic ones in some habitats, substantial changes have occurred in many communities due to recent human activities. This is especially evident in fire-dependent communities in which fire frequencies and intensities have changed due to modern fire control policies (Mutch, 1970; Taylor, 1973; Wright, 1974; Forman and Boerner, 1981; Minnich, 1983; Loucks et al., Chapter 5, this volume). In managed forests, the size-frequency distribution of treefall gaps may shift if large trees are selectively logged or if particular species are removed by cutting or disease (Knight, 1975; Foster, 1980; Shugart et al., 1981b; Foster and Brokaw, 1982). The species composition of communities subject to new disturbance patterns is likely to reflect both new and old regimes. Likewise, species specialized on the occupation of new bare spots may be locally eliminated from a community if disturbance is not sufficiently frequent to provide new establishment sites within the lifetime of existing adults (see, e.g., Paine, 1979; Gross, 1980). Increased disturbance frequency and intensity, for example, through agricultural activity, often brings with it both the

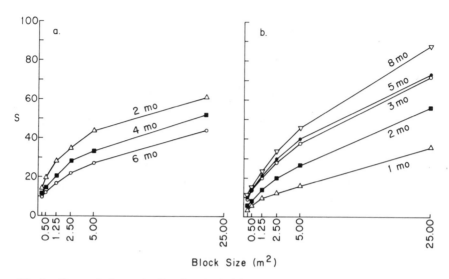

Fig. 2. Changes in the species diversity of colonizing weeds in two Colombian old fields. Curves are species-log area plots for successive samples of permanent quadrats. (a) Field E2 was cleared from pasture land that had been cleared annually for the past several years. (b) Field A3 was cleared from forest that had been cleared about 20 years previously. [From Denslow (1978).]

invasion of exotic weedy species (see, e.g., Bard, 1952; Kellman and Adams, 1970), which are better adapted to the new situation than are local species, and the local extinction of species unable to compensate for high mortality rates.

C. Predation as a Disturbance

The effects of increased mortality on species diversity depend on the relative abundance of the species concerned and on the frequency and intensity of the disturbance. Many patch-generating disturbances in communities of sessile organisms locally reduce the dominance of one or a few species through effects that are largely (but not completely) independent of the sizes and densities of the organism involved [e.g., wave battering on rocky intertidal communities (Paine and Levin, 1981) or boulder rolling on encrusting algae (Osman, 1977; Sousa, 1979b)]. Diversity is enhanced because species replaced by the dominants in long-undisturbed sites may be able to establish and reproduce in the newly opened patches.

As a source of mortality, predation operates as a special case of a patch-forming disturbance in that the predator is usually selective for some members (species, size classes) of the community. If predation causes a differentially higher mortality in dominant species, resources or space are released for exploitation by other species in the community, and diversity is likely to be enhanced, as it is with abiotic patch-forming disturbances. However, if the predator selects rare members of the community then diversity is decreased (Fig. 3; Harper, 1969; Lubchenco, 1978).

Reduction in the dominance of locally abundant species is thought to contribute substantially to the maintenance of high species richness in diverse communities. In tropical rain forest, Janzen (1970) and Connell (1971) have attributed some of the high tree species diversity to seed and seedling thinning by species-specific insect predation, although Hubbell (1980) questions whether the effect is sufficiently large to account for more than a few coexisting species. The "paradox of the plankton" (how can so many species coexist in a presumably homogeneous environment? Hutchinson, 1961) may rest on very small-scale patchiness due to planktonic grazers (Brooks and Dodson, 1965; Richerson *et al.*, 1970), in addition to environmental variability (Tilman, 1982). In Wisconsin lakes, unusually good recruitment years for cisco (Rudstam, 1983) may, in effect, be a disturbance to their prey species. As the fingerlings mature and exploit prey of increasingly larger sizes, a wave of heavy predation pressure moves through the plankton community. Similarly, differential predation may contribute to the coexistence of species in otherwise very stable deep-sea benthic environments (Rex, 1981). Even in such a "homogeneous" environment, temporal and spatial variation in species-specific mortality rates may contribute to species coexistence.

If the population size of the predator is large or the frequency of disturbance high, mortality rates may be such that some species are unable to establish propagules or do not survive to reproductive maturity. Under these conditions also, diversity may

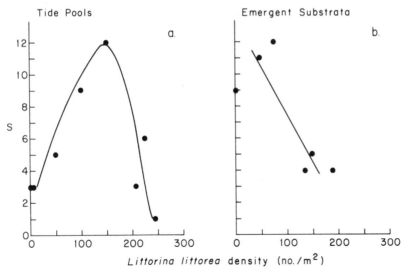

Fig. 3.   The effect of the snail *Littorina littorea* on the diversity of algae on the rocky shores of New England (a) in tide pools and (b) on emergent substrata. In tide pools the preferred algal species is *Enteromorpha*, a competitive cominant. Species diversity of algae is greatest at intermediate densities of *Littorina* and less where *Enteromorpha* is abundant or where grazing is so heavy that only highly resistant forms can survive. On emergent substrata, the preferred species—ephemeral algae—are subordinate. Any level of grazing by *Littorina* decreases the abundance of these species in comparison to the dominant perennial brown and red algae. [From Lubchenco (1978).]

be reduced. For example, the Atlantic coast sea urchin *Strongylocentrotus* grazes on the red alga *Chondrus,* which, in the absence of the urchin, outcompetes other algal species in the intertidal zone. Under moderate urchin densities, other algal species are able to settle in the patches cleared by the urchin and diversity is high. Under very heavy urchin populations, algal species diversity is again low because only a few unpalatable or ephemeral species are able to persist (Fig. 3; Lubchenco, 1978; see also Osman, 1977; Sousa, Chapter 7, this volume).

The recognized importance of intermediate levels of disturbance in the maintenance of species diversity (Connell, 1978; Huston, 1979; Lubchenco, 1978) originates in observations that either very low or very high levels of disturbance (or predation) result in local losses of species. However, ideal intermediate disturbance levels vary among ecosystems (Pickett and Thompson, 1978; White, 1979) and cannot be determined *a priori.* Determination of a disturbance regime to maintain the maximum number of nonexotic species in a given ecosystem must take into account the historical disturbance (or predation) regime of the community and thus the physiological, behavioral, and life historical characteristics of its component species. The processes maintaining species diversity can be understood only in terms of the interactions between disturbance regimes and the species' biologies (Denslow, 1980b; Drury and Nisbet, 1973; Pickett, 1976; Slobodkin and Sanders, 1969).

IV. ECOSYSTEM RESILIENCE

Resilience is the degree to which an ecosystem's long-standing composition and structure can be disturbed and yet return to that domain in which those processes and interactions function as before (Holling, 1973). A system disturbed beyond its limits of resilience will return to a new domain in the vicinity of a different structure characterized by altered composition and interactions of species and (perhaps) by a different disturbance pattern. In contrast, "stability" describes the frequency with which a system's long-standing composition is disturbed and its propensity to return if displaced within a domain of attraction (Holling, 1973). The critical consideration here is the change in community structure and species interactions due to some disturbance and whether or not those interactions resume on the removal of the disturbance. For our purposes, it is not necessary that an equilibrium be attained or that we define such an equilibrium.

Although multiple domains of attraction are characteristic of some natural systems, e.g., planktonic algal communities (Allen *et al.,* 1977), the shift to a new domain is frequently a product of an exotic disturbance. For example, when overexploited fish populations crash, new (lower) population levels may persist even after removal of fishing pressure (Smith, 1968; LeCren *et al.,* 1972). Logged and burned rain forest in the Far East is often invaded by the dominant grass *Imperata cylindrica* (Eussen and Wirjahardja, 1973; Whitmore, 1975), which persists as a stable, noninvasible association. Although such catastrophic disturbances are often human generated, this is not always the case. Mixing of North and South American mammals following the Pleistocene formation of the Panama land bridge established new associations and brought numerous extinctions of both North and South American forms (Marshall, 1981).

Unstable communities are often the most resilient (Holling, 1973) because they are likely to contain species adapted to variable environmental conditions and high rates of mortality. In general, species regularly subject to a variable physical environment are more likely to tolerate a novel stress than are species from very constant environments (Connell and Orias, 1964; Sanders, 1969; Slobodkin and Sanders, 1969; Thiery, 1982), with the result that unstable communities are more likely than stable communities to return to their previous composition and structure following some forms of exotic disturbance. Clearly, the extent and nature of the disturbance and the degree to which it mimics an indigenous one strongly influence the rate of return to the original community structure. For example, logging activities in boreal and tropical forests have strikingly different consequences, due in part to the different disturbance histories of the biomes. In Canadian boreal forest, the native spruce budworm–balsam fir interaction results in wide fluctuations in community structure and composition. Population explosions of the budworm are triggered by seemingly unpredictable bouts of dry weather and decimate extensive stands of mature balsam fir. A mixture of spruce and birch with an understory of shade-tolerant fir seedlings establish following the death of the adult fir, creating a forest less susceptible to budworm outbreaks than are fir-dominated old-growth stands. As the fir gradually replaces spruce and birch, the forest returns to its pre-outbreak structure (Morris, 1963; Holling, 1973). Succession after fire, also an indigenous disturbance to boreal forest (Heinselman, 1973, 1981a; Swain, 1978), and logging, an exotic disturbance, initiate successional patterns similar to those following budworm outbreaks, although the local regeneration patterns are a complex mosaic of species dependent on fire intensity, seed availability, and microenvironmental conditions (Heinselman, 1981a). Boreal forest thus appears relatively resilient to some forms of human exploitation.

In contrast, many tropical rain forests are rarely subject to such large-scale disturbance. In relation to the sizes of the canopy trees, disturbance patches are small—usually involving the fall of only one or two trees (Hartshorn, 1978; Brokaw, 1982a, and Chapter 4, this volume). Although we know little about natural fluctuations in the structure and composition of rain forest, natural, large-scale successions such as those typical of boreal forest are uncommon in most rain forests. Tropical rain forests are highly susceptible to logging or farming, which removes seed sources and dramatically changes soil structure (Gomez-Pompa *et al.*, 1972). Several hundred years may be necessary for the establishment of forest structure and composition similar to the original (Knight, 1975; Whitmore, 1975). However, by virtue of their topography or geographic location, some rain forests are subject to large-scale disturbances such as landslides and hurricanes (see, e.g., Whitmore, 1975; Garwood *et al.*, 1979; Crow, 1980; Uhl, 1982b), and the ecotone between Australian rain forest and tall open *Eucalyptus* forest is a dynamic one depending on fire frequency and severity (Ashton, 1981). Such forests are likely to contain a high proportion of trees with pioneer life history strategies and may be more resilient under pressures of human exploitation (especially logging) than more stable rain forests.

Continued heavy exploitation may lower the resilience of a community. Harvesting pressures on a multiple-aged population of fish differentially exploit the largest individuals; consequently, an increasing proportion of the recruitment to the next generation will be produced by a diminishing proportion of the population even though the commercial yield may remain high (Fig. 4). Near maximum yield, a small increment in fishing effort, or an additional environmental or biological stress on the population may effectively eliminate reproduction (Ricker, 1963). In contrast, a large proportion of the remaining individuals from populations of rapidly maturing species contribute to reproduction even under continuous fishing pressure. Such populations are more likely to be resilient than those of slowly maturing species. Reduced genotypic diversity is also likely to result from exploitation, habitat reduction, or an abrupt increase in the frequency of disturbance. All are likely to result in heavy mortality followed by strong selective pressures for rapidly maturing individuals adapted to the new competitive or environmental conditions. The short-term evolution of the population under strong new selective pressures

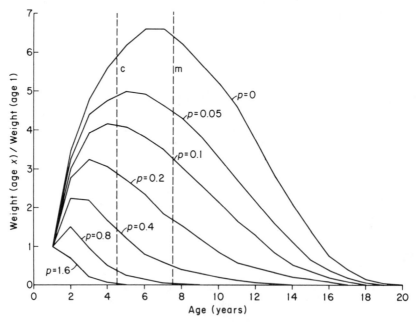

Fig. 4. Weight of the fish present at successive ages, for different rates of fishing (p), in terms of a unit weight of recruits at age 1. With constant recruitment, the equilibrium weight of the stock is proportional to the area under each shell. c, average minimum age limit of usable fish; m, average age at first maturity. Under increasing levels of exploitation, a smaller proportion of the population accounts for a large proportion of the recruitment. [From Ricker (1963).]

may mask a critical loss in genetic flexibility under other stresses such as disease or competition.

A common effect of catastrophic disturbance is homogenization of the environment to the extent that population refugia are eliminated. Two species can coexist on the same resource in a patchy environment if one is able to invade and reproduce on a patch before being eliminated (through competition or predation) by the other species (Huffaker, 1958; Horn and MacArthur, 1972; Levin, 1974; Levin and Paine, 1974; Levins and Culver, 1975). Habitat heterogeneity increases the probability that such refugia are indeed differentially exploitable by different species.

The importance of refugia in the composition of fish communities has been described in a study by Tonn and Magnuson (1982). Some shallow freshwater lakes of northern Wisconsin are regularly subject to winterkill when oxygen in the lakes reaches critically low levels. Large piscivores such as pike and bass are eliminated from these lakes unless they are able to emigrate to streams or connecting lakes during the winter months. Winterkill lakes (with no winter refugia and therefore free of piscivores) are dominated by mudminnows and several species of tolerant, soft-rayed minnows, whereas lakes containing piscivores during summer months are dominated by spiney-rayed species such as sunfish, bullheads, and yellow perch. The presence or absence of winter refugia determines the presence of top carnivores in these lake systems and therefore, indirectly, the composition of the prey species.

Overexploitation of fisheries stocks sometimes results when refugia, which had previously been the source of recruitment, are lost. Such was the case with the introduction of deep-water trap nets into the Great Lakes in the late 1920s (Smith, 1968). The traps successively eliminated local populations of whitefish, although the total catch remained high. By 1935, when trapping was prohibited below 80 m, stocking was insufficient to support a recovery. Introduction of the sea lamprey has maintained a low population of lake whitefish since that time. Local extinction of species in the face of invasion by an exotic predator (Zaret, 1979) and the persistence of a complex community (Cooper and Crowder, 1979) may be directly related to the availability of refugia from predation.

V. THE ROLE OF PROPAGULE DISPERSAL IN COMMUNITY STRUCTURE

The colonizer composition of a disturbance-created patch is directly related to the composition of the propagule rain reaching it. Strong seasonal and spatial variation among species colonizing disturbance patches is typical of many communities and is especially evident in species-rich communities (e.g., coral reefs: Sale, 1977; Talbot et al., 1978; Doherty, 1983; rain forests: Brokaw, 1982a; temperate prairies: Platt, 1975; Platt and Weis, 1977). The advantages accruing to the first arrivals appear to be substantial. Establishment and early growth take place in a virtually competition-free environment. Subsequent arrivals must not only compete for scarce resources, but, at least in communities of sessile organisms, must do so with a substantial size disadvantage in comparison to established individuals (see, e.g., Platt and Weis, 1977; Sousa, 1979b; del Moral, 1983). Some motile organisms (e.g., fish) must invade established territories of conspecifics (Sale, 1979). The production of numerous small, well-dispersed offspring is a common characteristic of species exploiting rare or ephemeral patches (MacArthur and Wilson, 1967; Gadgil and Bossert, 1970; Gadgil, 1971; Stearns, 1976). Although eventual convergence in community structure and composition occurs in many successions (Drury and Nisbet, 1973; Sousa, Chapter 7, this volume), early community structure and successional processes are often largely a function of the nature of those first colonizers (Keever, 1950; Bazzaz, 1975). Particularly good years for seed production or establishment, for example, can be traced through the size-class distribution of forest trees for hundreds of years and can affect not only the composition but the dynamics of the community as well (see, e.g., Shugart *et al.*, 1981b; McCune, 1982).

The swamping of a seed pool by propagules from an exotic species may contribute to the shift of an ecosystem to a new—and often less diverse—domain of attraction. Since the 1950s, the exotic shrub *Schinus terebenthefolius* (Anacardiaceae) has established large populations in newly abandoned farmlands of South Florida. Copious production of highly viable, bird-dispersed seeds and tolerance for a wide range of habitat conditions have facilitated the invasion of *Schinus* into recently disturbed, successional, and intact communities. Pineland is naturally swept periodically by ground fires and is the most susceptible to *Schinus* invasion (Ewel *et al.*, 1982). Where *Schinus* becomes an important component of the pineland ecosystem, it threatens to change both the patterns of regeneration following fire and the fire frequency, jeopardizing reproduction of many endemic species.

Although marine and terrestrial communities of sessile organisms are superficially similar in many ways (see, e.g., Connell, 1978), propagule dispersal is likely to be more limited in terrestrial than in marine systems. In the marine littoral zone, the probability of a species reaching a patch is a function of the composition of reproductive adults in the region and is less dependent on the composition of adults in the immediate vicinity of the patch (Paine and Levin, 1981; but see Crisp, 1979). To a colonizing species, the probability of encountering any given competitor is largely a function of its abundance and reproductive output in the region. If colonists of patches within a given season approximate a random sample from a well-mixed pool of propagules, the probabilities of competitor encounter are likely to be relatively constant from patch to patch.

Limitations on dispersal in a terrestrial environment impose an important constraint on patch dynamics (see, e.g., Platt and Weis, 1977; Gross, 1980). Although there are few data available on the ecology of tropical seeds and seedlings, it is likely that seed rain to a new tropical forest clearing comprises seed primarily from individuals fruiting in the vicinity of the patch (Whitmore, 1975). Seed dormancy capabilities and efficient dispersal mechanisms of some species spread seeds in time and space. Seed predation by host-specific insects (Janzen, 1970) and seedling mortality due to fungal infection (Augspurger, 1983) decrease clumping of seedlings in the immediate vicinity of the adult. Nevertheless, clumped distributions of adult trees are common among many species of rain forest trees (Poore, 1968; Hubbell, 1979). As in marine environments, seasonal and annual variation in seed rain (Frankie *et al.*, 1974; Foster, 1982) contributes to patch-to-patch variation in composition. Under such conditions, the probabilities of local extinctions, like local population aggregations, must be great. Competitive exclusion is less likely to proceed to completion where encounters are so variable and poorly predictable. Species present in the community are more likely to persist as scattered aggregations or individuals. Adaptive investments likely to increase dispersal effectiveness, and therefore rates of gap encounter, will be critical to reproductive success (Hamilton and May, 1977).

This population structure also contributes directly to speciation. Wright's (1931, 1982) "shifting balance" theory of evolution suggests that macroevolutionary steps favorable for incipient speciation are likely to occur when populations are structured as "numerous small colonies frequently subject to extinction and refounding by stray individuals from the more successful colonies" (Wright, 1982, p. 17). This structure of rain forest tree populations may enhance speciation rates and contribute to a high species diversity. Local invasions and extinctions related to oscillations in the disturbance regime probably also contribute significantly to species diversity in much the same way that has been hypothesized to account for the diversity of phytoplankton (Hutchinson, 1961).

The maintenance of many coexisting species is less likely in communities in which the composition of the seed rain and seed bank are more predictable. There, competitive interactions are more likely to proceed to completion because encounter rates are likely to be higher. This will occur when propagules are well mixed and distributed (as in marine intertidal communities), when climatic and competitive oscillations are damped by long dormancy times (as in xeric habitats); or when the occurrence of establishment sites is more predictable because disturbances are large in scale or frequent in time (as in fire-disturbed communities).

VI. CONCLUSIONS

Disturbances to biological communities affect the temporal and spatial heterogeneity of the habitat and are a source of mortality for component species. The community is here viewed as an open system [*sensu* Caswell, (1978)] in which interactions between the disturbance regime (as described by the intensities, frequencies, and variations in disturbances) and the biologies of the species (life history, physiology, behavior) determine the pattern of succession in any single patch. In ecological time, species diversity is likely to be maximized when the disturbance pattern resembles that historically characteristic of the community. Species with traits necessary to exploit patches forming at historic rates will persist under such circumstances. Greater or lesser rates of disturbance or disturbances of different intensity and impact will change the environment, duration, or formation rate of habitat patches, which may then no longer be adequate to maintain some species. At disturbance regimes only slightly different from the historic one, the consequence to community composition may only be a shift in relative abundance rather than loss of species. Catastrophic exotic disturbances, on the other hand, may shift the entire system to a new "domain of attraction" [*sensu* Holling, (1973)] in which population interactions and disturbance patterns are now relevant to the maintenance of a new (global) equilibrium. Such a shift is often associated with a loss of species diversity. Although factors contributing to community resilience are generally those that maintain heterogeneity of the system (e.g., maintenance of refugia, genetic heterogeneity, and a broad population base from which to recruit the new year class), heterogeneity per se does not guarantee resilience in the face of exotic disturbances. The most heterogeneous systems are often the least resilient (e.g., tropical rain forests and coral reefs).

The interaction of dispersal and establishment probabilities with the size, frequency, and persistence of disturbance patches is likely to contribute significantly to overall diversity patterns. Where propagule dispersal is widespread and effective, and/or disturbances are large in scale or frequent in occurrence, competitive interactions are more predictable and exclusion is more likely to proceed to completion. Fewer species are likely to coexist under such circumstances than in ecosystems in which dispersal is less effective and successful establishment is highly variable or in which disturbance patches (for propagule establishment) are small, rare, or isolated. In the latter case, competition for space may be intense, but the nature of the competitors is only poorly predictable and encounter rates for particular competitors are low. The rates of competitive exclusion are likely to be low. Species packing may be high and competition intense but diffuse. Both extinction and establishment of local aggregations of a species are expected to be common.

Where the period of change in disturbance regime is of the same order of magnitude as the life spans of the component species, the community is likely to be further enriched by relict species and by newly established species. Such communities are unlikely ever to be at equilibrium, because it is unlikely that disturbance regimes—especially those related to climatic patterns—remain constant long enough for slowly reproducing species to come to an equilibrium. Where individuals are short-lived in relation to the period of change in the pattern of disturbance, population structures will track the changes and are more likely to be close to some global equilibrium. Nevertheless, an equilibrium model is useful because it also emphasizes constraints on the number and characteristics of species a community can hold and on the availability of resources. Species persist within a biotic and abiotic milieu that includes the prevailing disturbance regime. Although, locally or in the short term, disturbances may reduce single-species dominance and free resources to be divided among more species, in the long term disturbance is an important selective factor affecting the evolution of those species (see, e.g., Pickett, 1980). Gap specialists are thus not generally less competitive than climax species; they fill a different niche and are competitively superior under different circumstances and at different resource levels. An intermediate level of disturbance (Connell, 1978; Huston, 1979; Hixon and Brostoff, 1983) does not so much permit the persistence of competitively inferior species as it preserves the patch structure and competitive conditions upon which some species are dependent for existence.

ACKNOWLEDGMENTS

I am grateful to S. C. Carpenter, T. Frost, M. J. Lechowicz, D. Mladenoff, T. C. Moermond, S. T. A. Pickett, D. Waller, and two anonymous reviewers for thoughtful criticism and/or for stimulating discussion during the writing of this chapter.

Chapter **18**

Community Turnover and Ecosystem Nutrient Dynamics

PETER M. VITOUSEK[1]

Department of Biology
University of North Carolina
Chapel Hill, North Carolina

I. INTRODUCTION

As McIntosh (1980, 1981) has pointed out, studies of ecosystem-level nutrient fluxes too often consider vegetation as one undifferentiated pool regardless of its composition and dynamics. Conversely, studies of vegetation often deal simplistically with resource availability, assuming (for example) that what is measured in a standard soil test actually represents nutrient availability.

In this chapter, I discuss the relationships among disturbance, patch development, and nutrient availability. I emphasize nitrogen and phosphorous because they are the elements most often limiting to plant growth in a wide variety of natural and managed plant communities. The total quantity or "availability" of some other

[1]Present address: Department of Biological Sciences, Stanford University, Stanford, California 94305.

THE ECOLOGY
OF NATURAL DISTURBANCE
AND PATCH DYNAMICS

element may correlate better with vegetation patterns or productivity (cf. calcium on the North Carolina Piedmont; Christensen and Peet, 1981), but plant growth is demonstrably nitrogen or phosphorus limited in most such situations. There is some evidence that nitrogen is more often limiting to plants in temperate and boreal forest zones, whereas phosphorus is more often limiting in the tropics (Vitousek, 1984).

I will discuss the mineralization (conversion, generally microbial, from organic to inorganic forms), transformations, uptake, and loss of nitrogen and phosphorus (Fig. 1), in addition to total amounts and instantaneously available pools in the soil. Most of the nitrogen and phosphorus in soils is not immediately available to organisms. In fact, total pools are often only weakly correlated with nutrient uptake or limitation (Shumway and Atkinson, 1978; Keeney, 1980; Powers, 1980). Measurements of instantaneously available pools (ammonium and nitrate for nitrogen, phosphate for phosphorus) can be similarly misleading. Competition for a limiting nutrient can cause extremely rapid turnover and very low pool sizes of a nutrient, but the amount of an element cycling through the available pool annually can still be

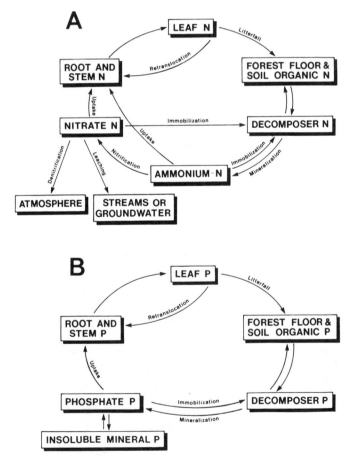

Fig. 1. The major transformations of nitrogen and phosphorus within intact forest ecosystems.

quite large. This situation is perhaps most extreme for phosphorus in lakes; tracer studies have shown turnover times for inorganic phosphate of less than a minute (Schindler et al., 1973). The same general pattern, though less extreme, is typical of ammonium and nitrate in temperate forest soils. Thus, without information on turnover, nutrient availability cannot be estimated from a knowledge of nutrient pool size alone.

The near absence of studies of nutrient cycling as affected by gap-phase turnover or larger-scale natural disturbance (with the exception of fire) precludes any direct review of this subject. Instead, I will examine the relatively well-studied changes in nutrient cycling that occur over the lifetime of a patch generated by clearcut logging. I choose clearcutting as an example because it generates well-defined patches and because a substantial amount of information is available. The effects of clearcutting will be compared to those of other disturbances, especially natural disturbances. Finally, the influence of patch dynamics on whole-ecosystem level nutrient budgets will be discussed. I will use the terminology of Bormann and Likens (1979) to characterize the phases of biomass dynamics and nutrient cycling that occur over the life of a patch. They identify four phases: reorganization, aggrading, transition, and steady state.

II. NUTRIENT CYCLING WITHIN A PATCH

A. Disturbance—The Reorganization Phase

Many studies of clearcutting have demonstrated that it usually leads to increased losses of nitrate and other nutrients to streamwater and groundwater (Vitousek and Melillo, 1979). The losses are often small, but rather large losses occur in a few sites (Bormann and Likens, 1979; Wiklander, 1981). A potential for nitrogen loss exists because the amount of nitrogen cycled annually (from plants through microorganisms to soil and back to plants) is 10–100 times greater than ammonium and nitrate pool sizes or annual losses (Rosswall, 1976). Cutting (or other disturbance) disrupts this cycle by interrupting plant uptake and generally by increasing decomposition and nitrogen mineralization as well (Glavac and Koenies, 1978; Matson and Vitousek, 1981). Removing the plants also leads to greater water flux through the soil, thus increasing nutrient losses by leaching and denitrification (Swift et al., 1975). Where large amounts of nitrate are lost in solution, calcium, magnesium, and potassium losses also increase as cations are displaced from soil exchange sites to maintain electrochemical neutrality (Nye and Greenland, 1960; Bormann and Likens, 1979).

Actual nitrogen losses after disturbance rarely approach the potential represented by nitrogen mineralization; immobilization of nitrogen by decomposers, uptake by regrowing or resprouting vegetation, and delays in the oxidation of ammonium to mobile nitrate combine to retain much of the mineralized nitrogen (Vitousek et al., 1982). Nitrogen losses are greatest and most rapid on relatively fertile (nitrogen-rich) sites (Wiklander, 1981); both immobilization and delays in nitrification are more substantial the more nitrogen-limited the site was prior to disturbance (Vitousek *et al.*, 1982).

This summary has emphasized losses of nitrogen rather than nitrogen availability per se because most studies report only losses. Even where losses are relatively low, however, increased soil temperature and moisture and decreased competition for nitrogen should cause a substantial increase in nitrogen availability following cutting (Matson and Vitousek, 1981).

Within-system cycling of phosphorus is also much greater than annual gains or losses in intact forests, but phosphorus has no important gas phase, and leaching losses of phosphorus rarely increase following disturbance. Like nitrogen, phosphorus is immobilized by decomposers. In addition, it is extremely immobile in most soils (Nye and Tinker, 1977), forming insoluble complexes with calcium at high soil pH and with aluminum, iron, and manganese at low pH. Only erosion causes substantially increased phosphorus losses from disturbed sites.

Phosphorus is also different from nitrogen in that most organic phosphorus in soil is ester bonded (carbon–oxygen–phosphorus), whereas most nitrogen is carbon bonded (Stewart *et al.*, 1983). Extracellular phosphatases secreted by decomposers can cleave the ester phosphate bond, but the organic molecule containing nitrogen must be broken down to mineralize nitrogen. Production of extracellular phosphatases is inducible; phosphorus mineralization can thus be tied directly to phosphorus demand, while nitrogen mineralization is coupled to decomposition (McGill and Cole, 1981). Except in strongly phosphorus-deficient sites such as the southeastern U.S. coastal plain (Pritchett, 1979) and many tropical and subtropical sites on old soils (Walker and Syers, 1976), phosphorus availability may not change substantially with disturbance and patch development. The appropriate information to test this suggestion (measurement of the active phosphorus pool by isotope dilution) is lacking, however.

Plant nutrient uptake (to aboveground components) generally returns to or above predisturbance levels in 2–5 years (Marks, 1974; Harcombe, 1977; Gholz, 1980). However, in sites where organic layers on the forest floor decrease after disturbance (Covington, 1981; Boone, 1982), net declines of organic matter and nutrients (negative net ecosystem production) can continue for several years after net primary production and nutrient uptake have returned to predisturbance levels (Bormann and Likens, 1979). As a consequence, nutrient availability can remain high. In addition, nutrient availability can remain high because the litter of pioneer species might decompose more rapidly (Melillo *et al.*, 1982), providing a higher substrate quality for nitrogen mineralization under a closed canopy (Matson and Vitousek, 1981). Nutrient release during decomposition may be particularly high when pioneer plants grow under high nutrient conditions (Vitousek, 1983).

B. Patch Development—The Aggrading Phase

Vegetation regrowth following disturbance eventually reduces soil temperature and runoff and reestablishes relatively closed nitrogen and phosphorus cycles

(Marks and Bormann, 1972), and the forest floor eventually becomes a net sink for organic matter and nutrients (Covington, 1981). Regrowth probably occurs more rapidly on relatively fertile sites (Vitousek *et al.*, 1982), although succession toward a species composition more typical of undisturbed forest can actually be arrested in more fertile sites (Harcombe, 1980).

The most rapid biomass and nutrient accumulation occurs relatively early in succession in both natural forests and plantations (Switzer and Nelson, 1972; Marks, 1974; Hansen and Baker, 1979; Marion, 1979; Sprugel, Chapter 19, this volume). Nutrient limitation can occur as a result of a decreased supply of nutrients or an increased demand for them; both processes combine to make nutrient availability relatively low in rapidly growing forests (Aber *et al.*, 1982). Low nutrient availability occurs in part because a large fraction of the nutrients in relatively labile pools are accumulated in biomass (living and dead) (Weetman, 1962b; H. G. Miller, 1981; R. F. Fisher *et al.*, 1982). It may also occur because plants growing under nutrient limitation become more conservative in their use of limiting nutrient(s). Such plants either operate with lower nutrient concentrations in active leaves (Van den Driessche, 1974; Birk, 1983) or retranslocate nutrients from senescing leaves more effectively (Miller *et al.*, 1976; Turner, 1977). This relatively efficient use of nutrients by trees causes litter produced by such trees to have high carbon:nutrient ratios (Hirose, 1975; Shaver, 1981; Vitousek, 1982). High ratios, in turn, cause increased nutrient limitation to decomposers, immobilization of nutrients, decreased nutrient availability to plants, and decreased losses from the system as a whole (Vitousek *et al.*, 1982). A decrease in nutrient availability could feed back to cause a further increase in nutrient use efficiency by the trees.

In general, then, the aggrading phase of patch development is marked by decreased nutrient availability. This conclusion is in accord with Grime's (1979) suggestion that nutrient stress increases with succession (at least up to a point), and with observations of increasing nutrient deficiency in many older plantation forests (H. G. Miller, 1981; R. F. Fisher *et al.*, 1982).

C. Stabilization? The Transition and Steady-State Phases

Following the phase of most rapid growth, a gradual decline in the rates of biomass and nutrient accumulation is generally observed in both natural forests and plantations (Switzer and Nelson, 1972; Gosz, 1980a,b; Peet, 1981; Sprugel, Chapter 19, this volume). In some cases, there may be an actual decline in total biomass and nutrient pools (negative accumulation rates), especially where a single shade-intolerant species strongly dominates the aggrading phase (Gosz, 1980a; Peet, 1981). This decline in storage, and the consequent (and hypothetical) increase in nutrient availability and loss, is termed the "transition phase" by Bormann and Likens (1979).

In other sites, net biomass accumulation may asymptotically approach some "equilibrium" level (net annual accumulation approaches zero). Peet (1981) showed that the same underlying processes could yield either a monotonic approach to equilibrium or an overshoot and decline (or continued or damped oscilliations) depending upon rates of tree establishment following disturbance and lags in forest regeneration upon canopy breakup. The rough constancy of biomass and element stores following the transition phase is termed the "steady-state phase" by Bormann and Likens (1979).

Whether or not a steady state can exist *within* a patch, and how nutrient availability and loss would be affected if it did, are not clear. Vitousek and Reiners (1975; Vitousek, 1977) demonstrated higher nutrient losses from old-growth (steady-state?) spruce–fir watersheds than from similar aggrading forest watersheds (Table 1); Leak and Martin's (1975) results and Bormann and Likens' (1979) simulations suggest the same result in northern hardwood forests. However, both Vitousek and Reiners (1975) and Bormann and Likens (1979) explicitly concluded that the reason for the increased losses (and presumably increased availability) of nutrients at steady state was that old-growth forests break up into smaller patches resulting from small-scale natural disturbance or treefalls. At any one time, most of these patches are assumed to be aggrading, while a few more recently disturbed ones are losing biomass and nutrients (the reorganization phase). The net result can be a rough constancy of biomass and nutrient stores averaged over a watershed. Christensen and Peet (1984) also reported evidence of increased resource availability in late successional pine forest (>80 years old); they too attributed the increase to the formation of patches of higher resource availability (perhaps single-tree gaps) during the transition phase, here from loblolly pine to hardwoods.

Would nutrient availability and loss in a patch increase without natural disturbance or before gap-phase turnover formed smaller patches within the forest? Net nutrient demand by living biomass is lower in older forests (H. G. Miller, 1981; Sprugel, Chapter 19, this volume), so it is possible that nutrient availability could increase without canopy turnover. Alternatively, nutrients could continue to be tied

TABLE 1

Concentrations of Major Ions (μeq/liter) in Streams Draining Old-Growth and Aggrading Spruce–Fir Forest Watersheds in New Hampshire[a]

Ion	Old-growth watersheds	Aggrading watersheds
Nitrate	53 (5)	8 (1)
Potassium	13 (1)	7 (0.5)
Magnesium	40 (5)	24 (2)
Calcium	56 (4)	36 (2)
Chloride	15 (0.3)	13 (0.3)
Sodium	29 (3)	28 (1)
Sulfate	119	123
Silica (μM)	75	86

[a] Mean growing season concentrations ($\pm$ standard errors) on nine unlogged and five aggrading watersheds are reported. Biologically important elements are accumulated by the aggrading forests, while nonessential or excess elements are not. [From Vitousek (1977).]

up in dead wood (Grier and Logan, 1977) or humus layers (Weetman, 1962b) until disturbance "activates" these pools. I know of no data showing increased nutrient availability or loss in old-growth forests without natural disturbance or canopy turnover, but I also know of no serious effort to obtain such data.

D. Other Disturbances

The pattern of nutrient cycling with patch development described above probably applies in general to any disturbance that creates a relatively large (≥ 1 ha?) patch. The type and severity of disturbance can significantly alter the reorganization phase, however. Even different timber-management methods can substantially change nutrient availability and loss as they affect either immobilization (through mechanical site preparation or prescribed fire) or vegetation regrowth (through herbicides) (Pritchett and Wells, 1978; Vitousek and Matson, 1984). Windstorms may have somewhat smaller effects on nutrient availability and loss than clearcutting; they add a large amount of immobilizing substrate to the soil, and vegetation sprouting or regrowth is not impeded. Conversely, fire volatilizes large amounts of carbon, nitrogen, and sulfur but mineralizes phosphorus and other nutrients. Both nitrogen and phosphorus availability are generally increased immediately following fire due to increased soil temperature and moisture and decreased immobilization (Raison, 1979). Nutrient losses to streamwater and groundwater are also increased following either wildfire or prescribed fire, although the increases are generally slight (Schindler et al., 1980; Richter et al., 1982).

Studies of nutrient cycling following natural disturbances other than fire are less common, perhaps in part because of the unpredictability of such disturbance. Swank et al. (1981) demonstrated increased nitrate losses in streamwater from partially defoliated southern Appalachian hardwood watersheds; they also reported increased pool sizes of available nitrogen in the soil. The defoliation apparently did not cause substantial tree mortality. Sprugel and Bormann (1981) reported changes in ammonium and nitrate pool sizes as a consequence of wave-form dieback of balsam fir; their results suggested that nitrification and probably nitrate leaching were increased in the dieback zone. Matson and Boone (1984) measured nitrogen availability in transects across waves of root rot–induced mountain hemlock dieback in the Oregon Cascades. They found very low nitrogen availability (as measured by both in situ and laboratory incubations) in old-growth forests ahead of the waves, increased availability in the dieback and young regrowth zones, and declining nitrogen availability with continued regrowth. In this last example, nitrogen availability may directly affect the probability of disturbance. Experimental studies demonstrated that mountain hemlock is more susceptible to mortality from root rot when the trees are grown under severe nitrogen limitation (Matson and Waring, 1984).

In all of the examples cited above, nitrogen availability appears to be increased by disturbance (natural or anthropogenic) and then to decline as the forest recovers. Further evidence for a long-term association between disturbance and increased nutrient availability can be drawn from the observation that a number of pioneer species germinate in response to increased nitrogen availability (Peterson and Bazzaz, 1978; Auchmoody, 1979).

III. PATCH SIZE AND NUTRIENT DYNAMICS

Heterogeneities far smaller than anything that could be defined as a patch in plant community ecology can control ecosystem-level nutrient flows. As an extreme example, denitrification in the interior of soil aggregates in generally aerobic soils can be the major pathway of nitrogen loss from intact forests (Myrold *et al.*, 1982). But how large must a patch be before it follows the dynamics outlined in the previous section?

Below some critical size (which itself may vary among regions), disturbance or tree mortality creates gaps that are filled in by root extension and canopy expansion of the remaining plants (Marks, 1974). After filling, the gap becomes (functionally) part of the patch from which it was derived. Nutrient availability and perhaps loss would still be increased temporarily within the area due to reduced competition and probably some increase in soil temperature and moisture.

Above the critical size, a disturbed area becomes a patch that follows the developmental dynamics outlined in the previous section. Single-tree gaps resulting from thinning in an aggrading forest probably fall below the critical size, but the fall of a single emergent canopy tree in some lowland rain forest sites clearly creates an opening large enough to maintain itself as a more or less discrete patch (Hartshorn, 1978). Williamson (1975) reported that individuals of tulip poplar (a shade-intolerant secondary successional dominant) are clustered at about the scale of the gap created by the fall of an old-growth canopy tree, suggesting that single-tree gaps can lead to patch formation in the temperate zone as well.

I know of no published nutrient cycling information on the scale of a single-tree gap. Cook and Lyons (1983) assumed that increased nutrient availability occurred within windfall-created gaps and affected the growth of an understory species they studied. Anderson and Swift (1983) speculated that nutrient availability is increased within the zone in which the tree crown (containing nutrient-rich leaves) falls but is temporarily decreased by microbial immobilization in the zone along the bole. Orians (1982) demonstrated that tropical tree seedling distributions can exhibit marked associations with either the crown or the bole zone of gaps. Direct studies of the effects of single-tree gaps on nutrient cycling and availability would be most useful.

IV. EFFECTS ON WATERSHED-LEVEL NUTRIENT DYNAMICS

An understanding of patch dynamics and patterns of vegetation turnover has proved critical to the interpretation of watershed-level nutrient budgets. The major watershed-ecosystem studies in the eastern United States (Hubbard Brook, New Hampshire, and Coweeta, North Carolina) both have intermediate-age (55–70 years postlogging) forests on their control watersheds [although both include smaller-scale heterogeneity—from the 1938 hurricane and subsequent salvage logging at Hubbard Brook (Cogbill, 1980) and from chestnut blight at Coweeta]. At first, their nutrient budgets were considered characteristic of mature forests; later, it was recognized that both are aggrading forests and that their nutrient budgets reflect that fact (Vitousek and Reiners, 1975; Bormann and Likens, 1979). The importance of net nutrient storage in biomass (living and dead) can be substantial; even rock weathering can be significantly underestimated if net cation storage in that pool is neglected (Gorham, 1961; Bormann and Likens, 1979).

As discussed above, an understanding of the patterns of vegetation turnover is equally important in determining whether a transition phase exists and in evaluating the nature of any steady state. A rough constancy in biomass and nutrient stocks may be observable only where vegetation turnover is frequent and turnover occurs in small patches relative to the size of the watershed (or other unit) under study. Any single patch may be either aggrading relatively slowly or degrading (reorganizing) rapidly (Woodmansee, 1978), and it may be only when the outputs of many aggrading patches are integrated with those of a few degrading patches that a steady state is recognizable. Such integration is possible in higher-elevation coniferous forest watersheds in the northeastern United States, where fir waves (Sprugel, 1976) and other small-scale turnover processes (Reiners and Lang, 1979) cause frequent canopy turnover on the appropriate scale. It may also be possible in some lowland tropical rain forests (Hartshorn, 1978) and moderately grazed grasslands (Woodmansee, 1978), where patches may be small and frequently formed. Steady-state conditions would not be observable in regions such as the upper Great Lakes or the Pacific Northwest, where (prior to extensive logging) large, relatively infrequent fires were the major agent of community turnover (Heinselman, 1981a; Franklin and Hemstrom, 1981). No reasonably sized study unit could integrate this large-scale variability into a rough steady state. Moreover, if large fires tended to occur in widely spaced drought years or series of years, temporal heterogeneity would also preclude the recognition of any steady state.

A final question is, to what extent can vegetation patches within a watershed interact to influence ecosystem-level nutrient dynamics? Nutrients leaching through the soil from disturbed patches toward streams or groundwater could be taken up by adjacent aggrading patches or by the riparian vegetation near streams if patch sizes are small and if percolating water is not too deep. In such a case, the watershed as a whole could retain nutrients more effectively than any individual patch, and natural vegetation made up of a mosaic of patches could be significantly more retentive than vegetation managed in large patches. Some experimental forest management schemes based on this possibility have been implemented (Hornbeck et al., 1975) or contemplated (Jordan, 1982). If these practices are useful and widely applicable, the management of patch dynamics would become an important way to manage the nutrient capital of terrestrial ecosystems.

Chapter **19**

Natural Disturbance and Ecosystem Energetics

DOUGLAS G. SPRUGEL[1]

Department of Forestry
Michigan State University
East Lansing, Michigan

I. INTRODUCTION

Changes in ecosystem energetics due to natural disturbance have been discussed much less than changes in nutrient cycling or species diversity. One reason for this may be the difficulty and expense of doing field research on energetic parameters. A comprehensive study of ecosystem energetics for even a single site is a massive undertaking, and extending the study to cover an entire successional sequence requires either a very felicitous experimental situation (Peet and Christensen, 1980; Cooper, 1981; Sprugel, 1984) or very large financial resources (Reichle, 1981). About a dozen studies have looked at temporal trends in the simpler energetic parameters such as biomass accumulation and net primary productivity (see below), but only one or two have considered less tractable parameters such as ecosystem

[1]Present address: College of Forest Resources, AR-10, University of Washington, Seattle, Washington 98195.

respiration or net ecosystem productivity. Moreover, nearly all of the studies have dealt with boreal coniferous forests, in which a high natural disturbance frequency and low human impact make it possible to find acceptable age sequences of stands. The following discussion will include some hypotheses concerning less-studied parameters and ecosystems, but the need for additional research is great.

A second reason for the lack of attention to ecosystem energetics is that, viewed in the short term, disturbance effects on ecosystem energetics may seem less drastic than effects on other ecosystem properties. It has become widely realized that disturbance is a major determinant of species diversity (Connell, 1978; Huston, 1979; Denslow, 1980b), if only because of the large number of "fugitive" species that ephemerally invade freshly disturbed areas in most regions. Postdisturbance changes in ecosystem nutrient dynamics may be equally impressive; nutrient loss may briefly increase many-fold (Bormann and Likens, 1979), and may counteract decades of nutrient accumulation (Vitousek and Reiners, 1975). By comparison, immediate disturbance effects on the more obvious ecosystem energetic parameters are much less dramatic. For example, net primary productivity usually declines briefly after disturbance, but only by about 50%—a far less impressive change than those seen in species diversity and nutrient cycling patterns. Parameters such as net ecosystem productivity may be less constant, but it is probably still true that postdisturbance changes in ecosystem energetics have been studied less than other postdisturbance changes at least partly because they seem less exciting.

This lack of study is unfortunate, because on a landscape scale, disturbance effects on ecosystem energetics may be just as important as effects on nutrient cycling and species diversity. Energy flow patterns change significantly during secondary succession, and these patterns, integrated over many patches of varying post-disturbance age, determine the energy flow characteristics of the landscape (Romme, 1982; Romme and Knight, 1982). Most successional studies have focused on the early or late stages of succession, when species composition and physiognomy are changing rapidly. However, understanding the energetics of a woody ecosystem may actually depend most on understanding energy flow changes during development and maturation of the initial woody stand, during which species composition may be fairly constant. These first-generation woody stands dominate many forest and shrubland landscapes under natural conditions because (a) the herbs and grasses that may establish immediately after disturbance are fairly ephemeral, and dominate the site only briefly before being replaced by woody species, and (b) natural disturbance frequencies in most forests and shrublands are high enough that the transition beyond the first generation to a second generation of more "tolerant" species may occur only rarely (White, 1979). [The eastern deciduous forest is an exception to this second generalization in that large-scale disturbances seem to be rare compared with small-scale, gap-forming disturbances (Runkle, 1982).] For most ecosystems, then, the energetic properties of the landscape mosaic as a whole (or of a single patch considered over a very long time) are controlled by changes in energy flow patterns during the first postdisturbance generation of woody plants.

The remainder of this chapter will examine some of these changes, and explore their implications for utilization and management of these ecosystems.

II. ECOSYSTEM CHANGES DURING STAND DEVELOPMENT

A. General Developmental Patterns

Oliver (1981) has described a sequence of developmental stages exhibited by most North American forests after disturbance. His terms and descriptions can also be applied to shrublands. Bormann and Likens (1979) developed more or less parallel terminology for hardwood forests under gap-phase and silvicultural disturbance regimes. Oliver's stages (with their Bormann and Likens equivalents in parentheses) are as follows:

1. Stand reinitiation ("reorganization phase"). After any major disturbance of a forest or shrubland, herbs and shrubs typically dominate the site for 1–10 years. Tree or shrub seedlings or sprouts also become established during this period and slowly increase in importance until they dominate the site. The stand initiation stage may be brief or protracted, depending on seedling and/or sprout density and on the growth rate of invading woody species.

2. Stem exclusion stage ("aggradation phase" in part). Eventually, the woody species monopolize the site resources so completely that further invasion by tree or shrub seedlings stops. This point is often termed "crown closure," although the limiting resource may actually be water or nutrients instead of light. During the stem exclusion stage, herbs and grasses are usually much reduced or absent, so physiognomic and species diversity is low.

3. Understory reinitiation stage ("aggradation phase" continued). After some time, the dominant woody species begin to "lose their grip" on the site (for reasons that are poorly understood) and understory species become reestablished. The understory normally includes some shade-tolerant herbs and may or may not include tolerant woody seedlings.

4. Old-growth stage ("transition phase"). If disturbance does not intervene, the first-generation trees eventually senesce and die. In some areas (particularly where natural disturbance is rare and/or usually small-scale), the first-generation trees are then replaced by more tolerant species, which either grow up into the degenerating canopy or replace fallen canopy trees by filling gaps (Runkle, 1982). In other areas, however, this replacement does not seem to occur; the ecosystem simply stagnates after the first-generation trees die off and regeneration does not occur (Rowe, 1961). Although it is ecologically interesting, such stagnation has been observed mainly in chaparral and conifer forest ecosystems where man has reduced the natural disturbance frequency; under natural conditions, fire or other catastrophic

disturbance would normally reset the successional clock before stagnation could occur. It may thus be of limited importance in understanding the dynamics of a natural landscape where disturbance is allowed to play its normal role.

B. Biomass Accumulation

The general patterns of biomass accumulation during stand development are well understood and seem to be fairly consistent among woody ecosystems. Total biomass typically follows a sigmoid curve, accumulating slowly at first, then more rapidly after the woody species gain dominance, and then more slowly again as stand biomass approaches a maximum during the understory reinitiation phase. Once the stand reaches the old-growth stage and the first-generation trees begin to senesce, living biomass may remain constant or decline slightly if dying trees are rapidly replaced by new trees from the understory, or may decline rapidly if regeneration is inadequate or absent. Bormann and Likens (1979) and Peet (1981) have discussed several possible patterns for biomass accumulation after large-scale disturbance in areas where small-scale disturbance is the norm.

Accumulation patterns for some biomass components are also fairly well understood. Foliage biomass increases rapidly at first and then levels off during the stem exclusion stage at some level apparently determined by light, water, or nutrient availability (Grier and Running, 1977; Gholz, 1982; Gray, 1982). It usually decreases slightly during the understory reinitiation stage (Kira and Shidei, 1967; Tadaki *et al.*, 1977), although this is not always the case (Sprugel, 1984). Foliage biomass may drop more significantly in stands approaching old age, particularly where senescence is pronounced (Kazimirov and Morozova, 1973; Rundel and Parsons, 1979). Where a transition to a second-generation stand of more tolerant species occurs, leaf biomass may increase moderately if hardwoods are replaced by conifers (see, e.g., Cooper, 1981), remain the same if there is no physiognomic change, or decrease moderately if conifers are replaced by hardwoods (see, e.g., Peet, 1981). If regeneration is inadequate or absent, foliage biomass presumably declines sharply as the older trees or shrubs die off without replacement (Rowe, 1961; Rundel and Parsons, 1979).

Bole wood and bark biomass exhibit a different pattern from foliage; they accumulate less rapidly at first, but continue to accumulate throughout the productive life of the stand. Because of its economic importance, bole wood production has been studied and modeled extensively and will not be reviewed further here (see references in Daniel *et al.*, 1979). Branch biomass seems to follow a pattern between those of foliage and boles; like foliage, it increases fairly rapidly in the early stages of stand development (although not as fast as foliage), but unlike foliage, it continues to increase throughout the development of the stand (but more slowly than bole wood). Root biomass trends have been very little studied, but limited data suggest that fine root biomass, like foliage, rises initially and then remains fairly constant, while coarse root biomass continues to increase as long as the stand is growing (Kazimirov and Morozova, 1973; Grier *et al.*, 1981).

C. Net Primary Productivity

Successional changes in net primary productivity (NPP) have been studied in a number of forest ecosystems, and several characteristic patterns have emerged (Fig. 1). NPP is almost always comparatively low in the stand reinitiation stage, and increases thereafter until it reaches a plateau at or about the time of crown closure (Bormann and Likens, 1979; Sprugel, 1984, and unpublished data). Most of the productivity in the stand reinitiation stage is in foliage and other deciduous tissues; comparatively little is devoted to building up permanent biomass. The cause of this postdisturbance NPP decrease is reasonably clear; NPP is ultimately controlled by site resources such as light, water, and nutrients, and for the first few years after severe disturbance the invading herbs, grasses, and woody seedlings are simply not large or numerous enough to utilize completely the site's resources. This is particularly true when the site is dominated by annuals, which must develop new root and shoot systems each year and cannot effectively utilize resources early in the growing season. (An exception might be noted in a desert, where water is limiting, but where the soil can retain water until rain-stimulated annuals are large enough to use it.)

Few studies have actually measured the productivity of both arboreal and herbaceous components of postdisturbance ecosystems. However, the data that are available suggest that productivity in the immediate postdisturbance period may be less than 50% of that in the mature forest (Bormann and Likens, 1979; Sprugel, 1984, and unpublished data). The magnitude of the decrease in any particular case will depend on the type and intensity of disturbance, the degree of damage to potential regeneration, and the seed pools of fugitive species available to colonize the site after disturbance. The length of the postdisturbance decline is also poorly documented, but it is reasonable to assume that for both forests and shrublands it lasts until the beginning of the stem exclusion period, when woody species regain full dominance. This may occur in as little as 5 years in areas where regeneration from buried seeds or sprouts is rapid and vigorous [e.g., in the pin cherry stands studied by Marks (1974)] or may take as long as 50 years in areas where weed competition is intense and regeneration is slow [e.g., in Douglas fir stands in the Pacific Northwest (Turner and Long, 1975)]. Any factors, such as farming, that destroy seed banks and sources of vegetative regeneration might prolong the period of lowered productivity (Bormann, 1982). If there are really such things as "arrested" or "deflected" successions (Egler, 1954; Abrams et al., 1985), in which chance events or human intervention allow development of quasistable grasslands where succession to forest would ordinarily occur, then these could represent an extreme or infinite prolongation of the postdisturbance productivity decline.

Once NPP has recovered from the postdisturbance drop, it probably remains relatively constant through the stem exclusion stage. In contrast to the stand initiation stage, the majority of the new tissue produced during this period is woody, although foliage still accounts for 20% (Kinerson et al., 1977; Grier et al., 1981) to 50% (Kazimirov and Morozova, 1973; Gray, 1982) of the total aboveground net productivity.

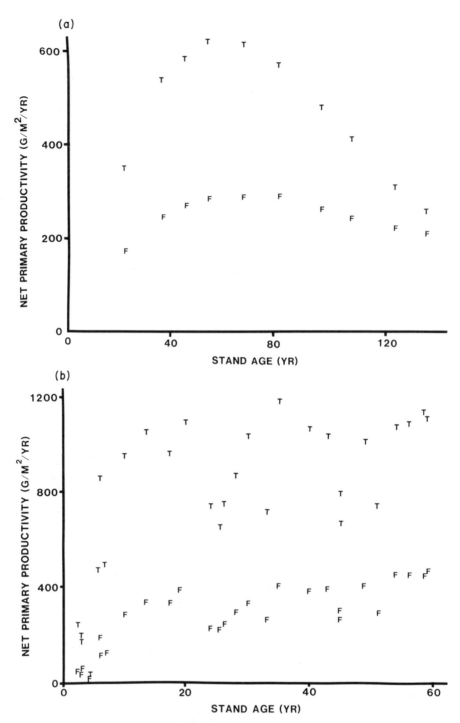

Fig. 1. Net primary productivity for two coniferous forest ecosystems through stand development. F = foliage productivity; T = total above-ground productivity. (a) Spruce (*Picea abies*) forests in Karelia, USSR (data from Kazimirov and Morozova 1973). (b) Balsam fir (*Abies balsamea*) forests in New York (data from Sprugel, 1984, and unpublished).

During the understory reinitiation stage, NPP probably drops slightly, reflecting the small decrease in foliage biomass. What happens next is rather interesting. In the majority of forest and shrubland ecosystems that have been studied through time, NPP apparently continues to decline throughout the later stages of stand development (Sirén, 1955; Drury, 1956; Kira and Shidei, 1967; Kazimirov and Morozova, 1973; see also Fig. 1a; Tadaki *et al.*, 1977; Van Cleve and Viereck, 1981; Cooper, 1981). These declines have sometimes been attributed to increased stand respiration (see Section II,D), but more often seem to be due to reduced nutrient availability, as evidenced in both conifer forests and chaparral by low nutrient concentrations in the foliage of older stands (Weetman, 1962a; Rundel and Parsons, 1980) and improved growth after fertilization (Hellmers *et al.*, 1955; Weetman, 1962a). The most drastic and best-documented declines occur in the cold boreal forests of northern North America and Eurasia (Rodin and Bazilevich, 1967; Kira and Shidei, 1967; Kazimirov and Morozova, 1973; Van Cleve and Viereck, 1981) and appear to be caused primarily by slow decomposition of conifer litter under cold, wet conditions. It has long been recognized that conifer litter decomposes rather slowly even under favorable conditions (Lutz and Chandler, 1946); where conditions for decomposition are poor, conifer litter may decompose so slowly that it builds up into a thick, insulating moss and litter layer on the forest floor (Sirén, 1955). This insulating layer keeps soil temperatures low all summer, which results in slow root growth and further declines in the rate of decomposition and mineralization. In many forests, a positive feedback loop develops in which lower leaf area leads to reduced transpiration and a higher water table, which in turn slows decomposition and mineralization still more (Heinselman, 1981a). If permafrost is present, the increased insulation may also result in a decrease in the annual thaw depth, decreasing the exploitable soil volume and further reducing tree productivity (Van Cleve and Viereck, 1981). If these feedback loops are not broken by fire or other disturbance, the affected areas may develop into unproductive bogs (Sirén, 1955; Drury, 1956).

Similar declines in productivity due to nutrient deficiencies in older stands have been reported in nonboreal conifer forests (Heilman and Gessel, 1963; Florence, 1965; S. N. Adams *et al.*, 1970), although the declines are generally less drastic where the added complication of a high water table does not intrude. H. G. Miller (1969, 1981) has provided an especially well-documented example in which low water retention of sandy soils in Scotland resulted in poor decomposition and mineralization of litter and growth declines in older Corsican pine (*Pinus nigra* var. *maritima*) stands.

Chaparral ecosystems are also supposed to exhibit productivity declines after age 20–30. Although the actual decrease in productivity seems never to have been quantified, leaf biomass and area are known to decrease in old *Adenostoma* chaparral stands (Rundel and Parsons, 1979), and it seems very likely that NPP does also. As in conifer forests, the main reason for this decline in chaparral growth seems to be a decrease in nutrient availability in older stands due to slow decomposition [although Schlesinger *et al.* (1982) have argued against this theory, at least for *Ceanothus* chaparral in the Santa Ynez Mountains]. Part of the cause is again sclerophyllous, decay-resistant foliage, but inhibition of decomposers by allelopathic toxins may also slow decomposition. The relative importance of the two factors is unclear; doubtless both contribute to some extent (Christensen and Muller, 1975a).

In all of these cases, once nutrient shortages have begun to develop, another feedback loop may occur as nutrient-starved trees produce nutrient-poor litter. This litter, which may also be rich in decomposition-inhibiting polyphenols, breaks down even more slowly and reduces nutrient availability still more (Davies *et al.*, 1964; Vitousek *et al.*, 1982). Although this pattern could theoretically develop under any vegetation, it seems particularly common under conifers, which have low carbon:nitrogen ratios and high polyphenol contents even under good conditions (Weetman, 1962a). Thus, under conifers, nutrient shortages and late-successional productivity declines seem to occur wherever conditions are poor for decomposition, whatever the cause.

Although old-age productivity declines have been found in a number of different studies in coniferous forests and "hard" chaparral, they are by no means universal. Sprugel (1984) found no evidence for a decline in productivity in aging balsam fir forests in the mountains of New York (Fig. 1b), perhaps because of continuous large inputs of nitrogen in cloud drip (Lovett *et al.*, 1982). There is also evidence that, unlike chaparral, California coastal sage scrub stands may not stagnate as they age, since there is continual regeneration of new stems in stands of all ages (Westman, 1981b). Finally, there do not seem to be any studies showing significant productivity declines in older angiosperm (hardwood) forests, although Moller *et al.* (1954) did observe a small decline in productivity of old European beech (*Fagus sylvatica*) plantations. There is less reason to expect declines in angiosperm forests because, as noted above, hardwood litter usually decomposes more readily than conifer litter and nutrients should not be sequestered except under extreme conditions.

Allocation of photosynthate among tissues also changes during the later stages of stand development. In general, foliage productivity is more constant than total aboveground productivity, so that when total productivity declines the proportion devoted to foliage increases. One of the more striking examples is the series of Russian spruce (*Picea abies*) stands studied by Kazimirov and Morozova (1973); foliage production accounts for about 45% of the aboveground production in vigorous young stands but over 80% in senescing ones (Fig. 1a). (Foliage production also declines in the older stands, but not as much as total production.) Where total production does not decline with age, it would be expected that foliage production would be fairly constant, although this has never been demonstrated. In at least one case, it seems not to be true; in the high-altitude fir forests of New York, foliage production actually seems to increase in older stands, probably as a result of increased loss of ice-laden foliage in winter storms (Sprugel, 1984; Fig. 1b).

There are few data bearing on the question of how productivity changes when early successional woody species are replaced by a new generation of more shade-tolerant trees or shrubs. Most ecologists seem to feel that late successional or "climax" forest communities are less productive than early successional ones, but the evidence for this generalization is surprisingly scarce. Horn (1974) has argued on geometric grounds that late successional, "monolayered" trees should be less productive than "multilayered" early successional trees, but has not provided productivity data to support his argument. Most of the data that do bear on this question come from forestry studies and are based on differences in the volume of bole wood produced (Smith, 1962; Crow, 1980). However, bole wood volume may be a poor index of total productivity, since early and late successional species may differ in wood density as well as in the proportion of photosynthate they devote to bole wood (Jordan and Farnsworth, 1980).

An additional problem in addressing this question is that it is often difficult to separate productivity declines due specifically to species transitions from those caused by changes in other ecosystem characteristics. Nutrient-related productivity declines that begin in the first generation will presumably continue in the second generation [for example, in the transition from white spruce to black spruce in some boreal forests, as described by Drury (1956) and Van Cleve and Viereck (1981)], but surely in this situation the species change is not the cause of a productivity decline that may have begun decades earlier. There do seem to be a few cases in which a productivity decline is directly attributable to a species change (Peet, 1981), but the number is so small that it remains to be seen if they are part of a general pattern or merely special cases.

D. Autotrophic Respiration and Gross Primary Productivity

In contrast to the rather large number of studies of NPP, autotrophic respiration (Ra) and gross primary productivity (GPP) have hardly been studied at all. It is possible to infer broad patterns of GPP from NPP trends, but patterns of respiration remain very difficult to generalize.

The early patterns of GPP and Ra seem reasonably predictable. GPP is probably low in the early stages of stand development (when NPP is also low), and increases throughout the stand initiation stage as the plants become larger and more able to exploit the site's resources. Ra probably also increases during this period as the amount of respiring biomass increases.

The course of GPP and Ra during the middle and later stages of stand development is less well understood. GPP probably reaches a maximum during the stem exclusion stage, when foliage biomass is maximal, and may decline slightly at the beginning of the understory reinitiation stage if foliage biomass also declines (Kira and Shidei, 1967). Where NPP continues to decline with stand age (i.e., where nutrients become increasingly restrictive in old stands), GPP doubtless does also. What happens to respiration during this period is less clear. It is widely believed that Ra increases with stand age (Odum, 1971; Horn, 1974; Whittaker, 1975), but this statement is usually referenced to Kira and Shidei (1967), who clearly stated it as a hypothesis rather than an empirical observation. The usual reason given for expecting Ra to increase with increasing stand age is that stand biomass increases with age, and more biomass means more respiration. This logic is superficially attractive but inherently flawed, because not all biomass is equal in respiratory demand. The most rapidly respiring aboveground tissues in woody plants are the leaves (Yoda et al., 1965), and as noted earlier, leaf biomass remains relatively constant or even declines slightly as stands age. In the woody tissues, respiration is greatest in the vascular cambium (immediately under the bark), so woody tissue respiration should be related to the surface area of the plants rather than their total weight. The increase in wood volume with stand age is irrelevant where respiration is concerned.

Unfortunately, there have been extremely few studies of changes in plant surface area with stand age. The few data that do bear on the subject suggest that, in some stands at least, branch surface area increases with stand age, while total bole surface area decreases (Whittaker and Woodwell, 1967; Sprugel, 1984). Since branch surface area is often several times greater than bole surface area, this suggests that total surface area increases (Whittaker and Woodwell, 1967). However, it does not address respiration trends, since boles of a given diameter respire significantly more rapidly than branches of the same size (Yoda et al., 1965). Where an attempt has been made (through much extrapolation) to estimate respiration rates of whole stands, the data suggest that respiration is more or less constant in the later stages of stand development (Moller et al., 1954; Sprugel, 1984), but more studies are needed.

E. Heterotrophic Respiration and Net Ecosystem Productivity

Net ecosystem productivity (NEP) is much talked about but very rarely measured. NEP is clearly an important ecosystem characteristic because it integrates the activities of producers, consumers, and decomposers; it summarizes the entire carbon/energy flux of the ecosystem in a single number. NEP has also been tied to ecosystem nutrient accumulation, which seems to depend at least partly on whether the ecosystem is accumulating or losing biomass (Vitousek and Reiners, 1975; Gorham et al., 1979). Finally, NEP is an important determinant of the world carbon cycle. Many statements about the connection between forest biomass and atmospheric CO_2 are actually statements about regional or worldwide NEP trends (see, e.g., Woodwell et al., 1978; Delcourt and Harris, 1980).

One of the problems in dealing with NEP is the confusion about definitions. Net ecosystem productivity is most commonly defined as the net change in total ecosystem biomass (living + dead) over time, or equivalently as the difference between gross primary productivity and total ecosystem respiration (autotrophic + heterotrophic) (Whittaker, 1975). (Since GPP − Ra = NPP, NEP is often measured as the difference between NPP and heterotrophic respiration.) This is the same as Odum's (1971) net community productivity, but differs from the usage in Whittaker et al. (1974), in which NEP was used to mean the change in *living* biomass over time. More recent studies (see, e.g., Covington, 1981) have made it clear that the two are by no means identical.

Averaged over a long period of time, NEP for a dynamically stable ecosystem in equilibrium with its natural disturbance regime should be zero (Sprugel and Bormann, 1981). That is, although there may be periods during the disturbance/regeneration cycle when NEP is positive or negative, over a long period of time they should balance out unless biomass is being imported or exported. (Positive or negative long-term NEP is possible in an ecosystem still undergoing primary succession and accumulating or losing biomass; see below.) However, even though they should balance out in the long run, the positive and negative deviations from zero may be interesting and instructive, and may give important clues to ecosystem function.

There do not seem to have been any studies in which changes in NEP have actually been measured over a successional sequence. NEP (calculated as the difference between NPP and heterotrophic respiration) has been estimated for several ecosystems (Whittaker and Woodwell, 1968; Harris et al., 1975; Grier and Logan, 1977; Gosz, 1980a), but never over the course of stand development. The best attempts at estimating successional trends in NEP have come from calculations of the change in living biomass + forest floor weight over time (see, e.g., Bormann and Likens, 1979; Pearson, 1982), although these are hampered by a lack of data on changes in soil organic matter. Soil organic matter may be relatively constant through succession, but the size of the soil organic pool is so large in many ecosystems that even small relative changes in it could be large absolute changes and might substantially affect overall NEP.

Despite the absence of measurements on successional NEP trends, it is possible to make educated guesses about probable patterns. Since it seems likely that the type of disturbance may significantly affect postdisturbance patterns, the following discussion will focus initially on recovery after a catastrophic windstorm, a fairly typical nonconsumptive disturbance. It will then explore possible differences in NEP patterns following fire, a biomass-destroying disturbance, and conclude with a brief consideration of other natural and human disturbances.

During a massive blowdown, no biomass is actually lost; the most significant change is that biomass is transferred from the living biomass pool to forest floor, dead wood, and soil organic matter pools (Fig. 2a–e). The first few years after this disturbance are generally assumed to be a period of significantly negative NEP (Vitousek and Reiners, 1975; Pearson, 1982). Living biomass accumulation is slow, particularly during the first part of this period, largely because of low NPP (Fig. 2a). Forest floor decomposition accelerates, because of an improved substrate (fresh litter deposited by the disturbance) and in some cases warmer temperatures and improved moisture conditions (Covington, 1981; Vitousek et al., 1982; Vogt et al., 1983; but cf. Edmonds, 1979). Soil organic matter decomposition probably also accelerates during this period for the same reasons. Since relatively little new litter is added to the forest floor during this period (after the initial postdisturbance pulse), forest floor volume declines sharply (Covington, 1981; Fig. 2b). Dead wood weights should also decrease, but more slowly (Lambert et al., 1980; Fig. 2c). Soil organic matter presumably decreases (Fig. 2d), although soil organic matter trends

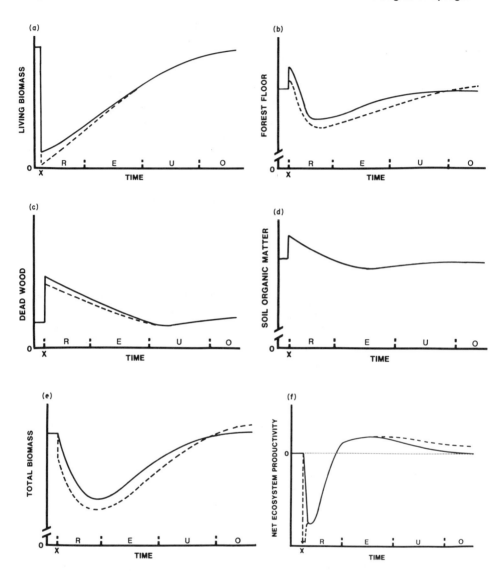

Fig. 2. Expected changes in ecosystem components through stand development. All graphs represent pools except (f), which indicates the rate of change in total biomass (net ecosystem productivity). Solid line indicates pattern expected in a deciduous angiosperm forest after a windstorm; dashed line shows pattern in an evergreen conifer forest after fire. Developmental stages: R = stand reinitiation; E = stem exclusion; U = understory reinitiation; O = old growth (Oliver, 1981).

are so little studied that nothing can be said about them with much confidence. In sum, though, the total biomass of the ecosystem probably declines during the first part of the stand reinitiation phase—rapidly at first, and then more slowly as woody plants take over and the recovery begins (Fig. 2e,f).

After the initial postdisturbance drop in total biomass, NEP probably rebounds and becomes positive for a number of years. Once the new forest stand is established, it usually accumulates biomass rapidly, particularly before leaf and twig biomass stabilize at the time of crown closure (H. G. Miller, 1981; Sprugel, 1984). Forest floor volume stabilizes and then begins to increase as decomposition slows down and the regenerating stand begins to produce significant amounts of litter. The dead wood pool probably continues to shrink, but the soil organic matter pool may begin to increase again after the postdisturbance decomposition pulse ends. With decomposition going on slowly and living biomass accumulating rapidly, NEP probably reaches a maximum during the later part of the stand reinitiation stage and the beginning of the stem exclusion stage.

Net ecosystem productivity probably decreases during the understory reinitiation and old-growth stages of stand development, although it is not clear whether or not it normally drops to zero. Living biomass accumulation typically slows down as the stand ages, and may reach zero if the stand achieves a self-reproducing steady state. Where severe stagnation occurs, living biomass may actually decline in old-growth stands (Grier and Logan, 1977; Gosz, 1980a), although this situation may occur only rarely under natural disturbance regimes. It is generally assumed that in hardwood forests the forest floor and dead wood pools reach a steady state in older stands, with decomposition balancing litterfall (Bormann and Likens, 1979). In conifer stands, however, the forest floor may continue to increase even after the stand becomes fully mature, accumulating nutrients all the time. Soil organic matter trends in old-growth forests are unknown, but may follow the same pattern as the forest floor. Thus, NEP probably declines to near zero in old, steady-state hardwood forests, but may continue to be positive in old conifer forests if increases in forest floor weight overbalance any decreases in living biomass (Viereck and Schandelmeier, 1980).

The situation after fire differs in several ways from that after a windstorm. First, fire actually removes large amounts of biomass from the system, rather than simply transferring it from one pool to another. In fact, if NEP is defined simply as the net change over time in total ecosystem biomass, then a fire could be considered as a brief period of extremely negative NEP (Fig. 2f). Since fire rarely consumes all of the aboveground biomass in a forest, there would also be an increase in the forest floor and dead wood pools from branches and boles killed by the fire. However, this would be partly counterbalanced if some of the forest floor were consumed by the fire. As in the windstorm situation, the amount of soil organic matter would increase due to the addition of roots of killed trees.

In the years following fire, there should normally be a period of negative NEP similar to that following a windstorm, and for the same reasons. Catastrophic fire is a severe disturbance that alters many ecosystem characteristics, and has long been known to be the most effective of all disturbances in stimulating a decomposition/mineralization pulse (Hesselman, 1917; Romell, 1935). This is particularly true where decomposition was formerly inhibited by volatile allelochemics or low pH (Christensen and Muller, 1975a). However, fire also destroys a significant amount of the fine and readily decomposable organic matter that would be added to the forest floor after a windstorm, thereby depriving the postdisturbance decomposition pulse of some of the fuel that might otherwise support it. Thus, the postdisturbance decomposition pulse after fire might be either smaller or larger than that after a windstorm. This question has not been adequately considered or modeled, and in the absence of data it is fruitless to try to guess which situation occurs. After the postdisturbance decomposition pulse ends, NEP should follow the same pattern after a fire as after a windstorm.

Fire and windstorm represent extremes of consumptive and nonconsumptive disturbances; most other types of disturbance are intermediate and should exhibit intermediate NEP patterns. Disease outbreaks should affect NEP much like wind, since in both cases no biomass is removed from the ecosystem. However, disease mortality is usually much slower than windthrow, which could cause differences in the NEP responses. Timber harvesting should be like fire in that substantial quantities of biomass are removed, but different in that the most nutrient-rich and readily decomposable material is usually left behind by harvesting, while in fire it is the most likely to be consumed. Also, ecologically sound timber harvesting (where erosion and soil compaction are minimized) is usually a less severe disturbance than fire, and would stimulate less decomposition (Romell, 1935). An insect outbreak might also be intermediate, since some biomass is lost in insect respiration, while some is deposited on the ground as frass and leaf fragments. Even more than disease outbreaks, insect irruptions take years to kill the target trees, and might therefore cause a substantially reduced decomposition pulse. Thus, each type of natural disturbance has some unique characteristics, but in broad outline the patterns seen after wind and fire should be typical.

The fact that in a "stable" ecosystem NEP must sum to zero if averaged over several disturbance cycles suggests some important interactions between the natural disturbance regime of a site and the course and endpoint of primary succession. One of the factors determining whether decomposition will balance production over a complete disturbance cycle is the amount of organic matter in the soil and the forest floor; all other things being equal, the more organic matter there is, the more will decompose over any given period of time. The amount of dead organic matter in the ecosystem thus "fine tunes" the long-term NEP balance, and primary succession can be said to continue until organic matter has reached a level at which gains and losses through the disturbance cycle balance out. If the ecosystem is still gaining or losing organic matter (in the long run), it has not stabilized and primary succession is still going on.

If the disturbance regime is one of the factors determining the equilibrium organic matter level of an ecosystem, it follows that different stable disturbance regimes should lead to different equilibrium organic matter levels. For example, in a stabilized ecosystem with a wind-dominated disturbance regime, the total decomposition over one disturbance cycle will, on the average, just balance the NPP during the same period. In a stabilized fire-dominated ecosystem, however, biological decomposition throughout the cycle will be less than NPP by the amount consumed by the fire. To the extent that decomposition during any period is determined by the amount of organic matter available to be decomposed, one would expect that a fire-dominated ecosystem should have less organic matter than a wind-dominated eco-system under comparable conditions. ("Comparable" is obviously an important qualifier here, since the disturbance regime on any given site is determined by a host of factors including climate, exposure, soil, and vegetation type—all of which also affect decomposition and NPP directly.)

If changes in environmental conditions alter the disturbance regime, they will also change the NEP balance and eventually the soil organic matter content. For example, if a change in macroclimatic patterns increases the frequency of severe droughts, fire frequency will probably also increase and soil organic matter content will decline. For all practical purposes, this is a new period of "primary succession" as the soil organic matter pool slowly adjusts to new environmental conditions. If human activities alter the disturbance regime, similar changes in organic matter pools are to be expected. Perhaps the best-known example of this is the decrease in the organic matter of grassland soils when converted to farmland. Voroney *et al.* (1981) have modeled this process and found that even for this drastic annual disturbance it takes 60–100 years for soil organic matter to reach a new equilibrium level. For natural forest and shrubland ecosystems, with their much lower disturbance frequencies, the response time should be much longer. Since disturbance regimes are partly controlled by climate, which seems to change moderately every few hundred years and drastically every few thousand (Brubaker, 1981; Davis, 1981), soil organic matter in many ecosystems may never come into equilibrium with natural disturbance. It appears that one need not look at the vast time scales discussed by Stark (1978) and Walker *et al.* (1981) to conclude that primary succession is a process that never ends.

III. CONCLUSION—DISTURBANCE, ENERGETICS, AND RESOURCE MANAGEMENT

A landscape under a stable natural disturbance regime should (at least hypothetically) develop toward a dynamically stable mosaic of patches of different ages, each resulting from a single disturbance at some time in the past [see Cooper (1913) and many others since]. Whether or not such a stable mosaic is actually realized in any particular region depends on the size of the average disturbance. Where individual disturbances involve a few hundred acres, a true equilibrium mosaic may be seen (see, e.g., Sprugel, 1976; Sprugel and Bormann, 1981). However, where the typical disturbance covers hundreds or thousands of square kilometers (see, e.g., Franklin and Hemstrom, 1981), an equilibrium mosaic can be only a statistical concept. Development of a stable mosaic may also be prevented by inherent variability in the disturbance regime; for example, fire frequencies may exhibit long-period variation driven by slow climatic changes (Romme, 1982). Despite these restrictions, the equilibrium mosaic is a useful concept for considering the energetic implications and consequences of disturbance regimes and resource management practices.

The productivity and energy flow patterns of a mosaic landscape depend on both the frequency of disturbance and the trends of energetic parameters during the course of stand development. For example, the mean NPP integrated over an entire mosaic will be determined by the number of patches in each developmental stage multiplied by the productivity of that stage. Since NPP in most forested ecosystems is relatively low during the stand reinitiation stage, highest in the stem exclusion stage, and lower again in the old-growth stage (Section II,C), the NPP of the landscape mosaic will be highest under a disturbance regime in which most patches spend most of their time in the stem exclusion stage. Higher disturbance frequencies will result in lower mean productivity, because at any given time there will be more patches in the less productive stand reinitiation stage. Lower disturbance frequencies will also decrease the mean productivity of the mosaic because a high percentage of patches will reach the less productive old-growth stage.

Similar reasoning can be applied to other energy flow parameters. Even though primary productivity is usually low in the stand reinitiation stage, secondary productivity is often high, because much of the biomass produced is accessible to ground-dwelling herbivores such as mice, deer, and elk (Kirkland, 1977; Andersen *et al.*, 1980; Romme and Knight, 1982). Thus, secondary productivity of the entire landscape mosaic would be greatest under a high disturbance frequency, which would maintain most of the landscape in the early stages of regeneration. Woody biomass increment, on the other hand, is greatest in the early part of the stem exclusion stage and remains high for some time thereafter; thus, forests managed for maximum wood production will be harvested at a rate that will keep most patches in this stage most of the time. An understanding of energetic trends during stand development is obviously an important tool in developing resource management strategies.

Altering disturbance frequencies to serve management goals can also have unintended consequences. Since NEP is typically negative in the early stages of secondary succession, imposing an excessively high disturbance frequency over a long period of time will result in a decline in the organic matter content of the soil, accompanied by a loss of the nutrients formerly contained in the organic matter (Voroney *et al.*, 1981). While in some ecosystems this might not be a cause for serious concern, in others (e.g., those on excessively drained sandy soils) such losses could reverse decades or centuries of nutrient and organic matter accumulation. In the long run, the organic matter loss may be more serious than the loss of nutrients, because if the organic matter content declines too far, the soil may not be able to retain nutrients even if they are replaced by precipitation, nitrogen fixation, and weathering. Thus, prolonged periods of excessively frequent disturbance (e.g., by harvesting or man-caused fires) could result in long-term reductions of site productivity that might take millennia to recover.

Reducing disturbance frequencies (e.g., by preventing fires and insect outbreaks) can also have serious consequences, especially in ecosystems where disturbance is normally common. In many ecosystems, primary productivity declines after long disturbance-free periods due to sequestration of nutrients in organic matter. Secondary productivity also drops off (Andersen *et al.*, 1980), partly because of the decline in primary productivity and partly because little of the production in old-growth ecosystems is accessible to ground-dwelling herbivores such as mice, deer, and elk (Kirkland, 1977; Romme, 1982). Reduced NPP may also increase susceptibility to catastrophic disturbance, since senescing trees lose their resistance to predator attack (Loope and Gruell, 1973) and become more susceptible to fire (Romme, 1982). Thus, in addition to decreasing overall NPP, reducing disturbance frequency may increase the danger of catastrophic large-scale disturbances that destroy the very resources managers are trying to protect.

There have been several excellent statements of the need for an appropriate amount of disturbance to maintain habitat diversity in nature preserves (see, e.g., Pickett and Thompson, 1978). Ecosystem energetics provides additional powerful arguments for the importance of striking a proper balance between disturbance and recovery in any kind of ecosystem management.

IV. SUMMARY

1. Postdisturbance stand development in most forests and shrublands exhibits the following recognizable stages: (a) stand reinitiation; (b) stem exclusion; (c) understory reinitiation; and (d) old growth.

2. Biomass accumulation during stand development follows a sigmoid pattern, with total biomass increasing slowly at first, more rapidly after crown closure, then more slowly again in older stands. Foliage biomass initially increases rapidly, but then reaches a plateau and thereafter remains relatively constant. Bole and branch biomass continues to increase until the old-growth stage.

3. Net primary productivity is low immediately after disturbance, but then increases and reaches a peak early in the stem exclusion stage. Thereafter it remains relatively constant, at least until the beginning of the understory reinitiation stage. In most ecosystems that have been studied, NPP declines in the old-growth stage, apparently because of nutrient sequestration in the forest floor or soil organic matter.

4. Autotrophic respiration has not been well documented, but is apparently low in the stand reinitiation stage and then higher in the stem exclusion and understory reinitiation stages. Dogma holds that it continues to increase throughout the course of stand development, but this is not supported by evidence.

5. Net ecosystem productivity (NEP) is apparently negative in the stand reinitiation stage, then positive in the stem exclusion and understory reinitiation stages, possibly dropping to zero in the old-growth stage. Since NEP should average out to zero through an entire disturbance cycle, there may be measurable differences in NEP trends following fire (which consumes biomass) and windthrow (which does not).

6. Regardless of the goal toward which an ecosystem is being managed, management plans must give careful consideration to the long-term consequences of altering natural disturbance frequencies.

ACKNOWLEDGMENTS

M. D. Abrams and P. M. Vitousek read an earlier version of this chapter and made many useful suggestions. I appreciate their help.

RECOMMENDED READINGS

O'Neill, R. V., and DeAngelis, D. L. (1981). Comparative productivity and biomass relations of forest ecosystems. *In* "Dynamic Properties of Forest Ecosystems" (D. E. Reichle, ed.), pp. 411–449. Cambridge Univ. Press, London and New York.

Romme, W. H. (1982). Fire and landscape diversity in subalpine forests of Yellowstone National Park. *Ecol. Monogr.* **52,** 199–221.

Romme, W. H., and Knight, D. H. (1982). Landscape diversity: the concept applied to Yellowstone Park. *BioScience* **32,** 664–670.

Schlesinger, W. H., Gray, J. T., Gill, D. S., and Mahall, B. E. (1982). *Ceanothus megacarpus* chaparral: a synthesis of ecosystem processes during development and annual growth. *Bot. Rev.* **48,** 71–117.

Sprugel, D. G. (1984). Density, biomass, productivity, and nutrient cycling changes during stand development in wave-regenerated balsam fir forests. *Ecol. Monogr.* **54,** 165–186.

Chapter **20**

Modeling Forest Landscapes and the Role of Disturbance in Ecosystems and Communities

H. H. SHUGART[1]

Environmental Sciences Division
Oak Ridge National Laboratory
Oak Ridge, Tennessee

S. W. SEAGLE

Graduate Program in Ecology
University of Tennessee
Knoxville, Tennessee

I. INTRODUCTION

Over the last few decades, two paradigms have come to dominate the mathematical modeling of the dynamics of ecological systems. One uses a set of general, nonlinear differential equations that have come to be known in ecology as the

---

[1]Present address: Department of Environmental Sciences, University of Virginia, Charlottesville, Virginia 22903.

THE ECOLOGY
OF NATURAL DISTURBANCE
AND PATCH DYNAMICS

"Lotka-Volterra equations." These equations are autonomous, meaning that the effects of things outside the system on the components of the system are not included in any explicit way in the formulation. The Lotka-Volterra equations have formed a basis for a still developing theory of community interactions (see, e.g., Vandermeer, 1970; May, 1973; Davidson *et al.*, 1980) that emphasizes the internal interactions of populations. Attempts to use these equations to gain insight into the performance of ecological systems under the influence of disturbance have centered on determining the conditions needed for stability (see, e.g., May, 1973). The rationale underlying such investigations is that an unstable system (in an environment that frequently disturbs the system) will not persist. A stable system tends to return toward equilibrium following a disturbance. Much of modern ecological theoretical work derived from the Lotka-Volterra equations has emphasized parametric or structural conditions that confer stability. The resultant body of theory is essentially a qualitative theory.

The second, codominant modeling paradigm in ecology uses forced linear systems of differential equations as ecosystem models. These equations (sometimes called "compartment models") do have inputs from and losses to the environment outside the system. Unlike the Lotka-Volterra equations, compartment models are strongly constrained to conserve matter. The equations (Olson, 1963; Webster *et al.*, 1974) are generally taken to represent the dynamics of the modeled system in the vicinity of an equilibrium state. Investigations of the performance of the equations in response to disturbance usually consist of analyzing the effects of small changes in the input terms to the equations.

Both of the aforementioned modeling approaches tend to emphasize the dynamics of ecological systems at equilibrium (or near an equilibrium point). Neither approach typically includes the spatial heterogeneity that could alter ecosystem behavior, nor does either have a high degree of realism in representing a particular ecosystem as an important objective.

This chapter provides two examples of alternative approaches to modeling disturbance effects on forested landscapes. The modeling does include a degree of realism and incorporate spatial heterogeneity. Because of this latter emphasis, the ecosystems modeled can be thought of as landscapes (Forman and Gordon, 1981).

Weinstein and Shugart (1983) have pointed out that a basic consideration in modeling landscape systems is whether the landscape can reasonably be divided into a number of discrete, smaller elements. When this is the case, one can simulate the dynamics of the landscape by simulating the change in each of the landscape elements. The responses of these elements can then be combined statistically to obtain the landscape response. The less the dynamics of the landscape are a consequence of contagion among the mosaic elements, the more easily the computations of the model are performed. If one cannot reasonably divide the landscape into mosaic elements, then the most appropriate mathematics for landscape simulation would likely involve partial differential equations. While there are examples of such applications (see, e.g., Smith, 1980), the parameter estimation difficulties and numerical difficulties associated with solving such equations have greatly limited their use in ecological problems.

Landscape elements might be defined according to several different criteria. For example, the scale could be determined by sampling considerations (e.g., the pixel sampled by a LANDSAT satellite) or by the scale of the natural pattern of disturbance in the landscape (Shugart and West, 1980). Regardless of the logistics of such criteria, landscape elements should be defined so that they are reasonably homogeneous, so that elements of the same sort could be expected to behave similarly over time, and so that one can apply the same set of rules to predict landscape elements.

Simulation of landscape dynamics in response to disturbance is a problem in formulating the "rules" for how mosaic elements of the landscape change their particular attributes over time. Categories of landscape models can be devised by considering the time (either discrete or continuous) and scale (either a large or a small number of elements comprising the total landscape). Weinstein and Shugart

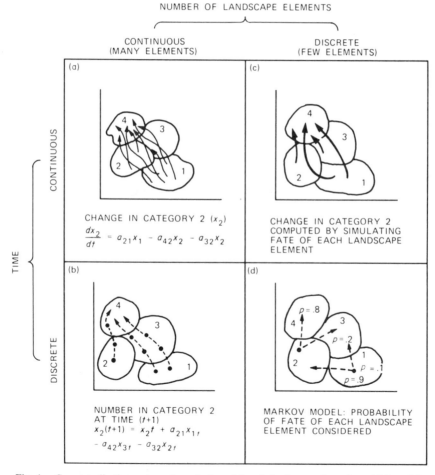

Fig. 1. Conceptualization and mathematical models of landscape dynamics. (Adapted from Weinstein and Shugart, 1983.)

(1983) pictured an abstract model of landscape change as a change of points in an abstract n-dimensional space (a state-space) where the axes were state-variables for the mosaic elements (Fig. 1). Hypervolumes in this state-space correspond to, for example, forest types. The state of a forested landscape is the number of mosaic elements in each of the hypervolume forest types at a given time. When the number of elements is large and time is continuous (Fig. 1a) then the appropriate model paradigm to use is ordinary differential equations. The use of such models has been outlined in Shugart *et al.* (1973) and Johnson and Sharpe (1976) for landscape models of northern lower Michigan and the Georgia Piedmont (respectively). If time is discrete (such as is the case if the mosaic elements are sampled periodically), then it may be appropriate to use a difference equation approach (Fig. 1b).

If time is discrete and the elements can be thought of as having different probabilities of being in one state or another as a function of their state at a prior time, then a Markov modeling approach may be used (Fig. 1c). There are abundant examples of the use of Markov models (and related approaches) as a landscape-modeling paradigm in both applied (Hool, 1966) and basic (Waggoner and Stephens, 1970; Horn, 1976; Wilkins, 1977) contexts. One good example, which includes discussions of parameter estimates and which is explicitly oriented toward landscapes, is found in the succession modeling approach used by Kessell and colleagues (see, e.g., Kessell, 1976, 1979; Cattelino *et al.,* 1979).

The approaches used to model the landscape dynamics discussed above are related in that the same data could be used to compute model parameters for any of the three modeling approaches. In cases in which there are relatively few elements considered over continuous time (Fig. 1d), fairly detailed computer simulation models are often used to capture more of the detail of the system dynamics. In the remainder of this chapter, forest models of this sort will be discussed. We will also provide an example of a Markov landscape simulator used to inspect the responses of animal populations in dynamic mosaic landscapes.

II. FOREST DYNAMICS MODELS

The past decade has seen the development of several detailed models of forest change based on the growth of individual trees in a simulated forest stand. These models have been developed more or less independently by foresters and ecologists (Munro, 1974; Shugart and West, 1980). The increased availability of large, high-speed digital computers has undoubtedly been a major contributing factor in the development of these models. Perhaps because these models evolved in the computer age, they are often formulated so that it is difficult (or impossible) to solve them analytically. Thus, results from these models are usually obtained from digital computer simulations.

One of the principal considerations in building forest growth models is whether the exact position in space of each tree is considered explicitly in the formulation. When tree spacing is explicit, highly detailed considerations of tree geometry can be used to effect competition among individual trees. Mitchell (1975), for example, models branch pruning among individual trees to assess competitive interaction in Douglas fir (*Pseudotsuga menzesii*) forests. As another example of the level of detail that is available in spatially explicit tree-based forest models, Ek and Monserud (1974) compute a "seed rain" varying in magnitude as a function of the distance from each seed-bearing tree in the simulated stand. As appealing as the high level of detail found in the spatial models may seem, this detail requires a very thorough study of the tree species used in the model so that the parameters of the equations can be quantified accurately. This high information requirement of spatially explicit models has tended to limit their development to well-studied forests of low diversity [the Wisconsin forest simulated by the Ek and Monserud (1974) FOREST model] and, more typically, to monospecies, commercial forests (Mitchell, 1975), or plantations (Hegyi, 1974).

An alternative approach has been developed for a number of diverse natural forest ecosystems that employs a vertical (one-dimensional) formulation for shading competition rather than a two- or three-dimensional formulation for spatial competition. This approach is most appropriate at small spatial scales and when the vertical dimension (shading) dominates competitive interactions. Such models have been called "gap models" (Shugart and West, 1980) because their structure seems particularly well suited for simulating some of the phenomena associated with the gap-phase mechanistic interpretation of forest regeneration by Watt (1947). In this chapter, we will outline the function of gap models and discuss their relation to other computer models of forested systems. Gap models are used as landscape models if one believes that a landscape may be considered a dynamic mosaic. This larger-scale forest response of landscapes as simulated by gap models will also be discussed.

III. GAP MODELS

The detailed functioning of gap models has been described in several papers (Botkin *et al.*, 1972a,b; Shugart and West, 1977, 1980, 1981). In general, these models conceptualize forest stand dynamics as a process of competition (for light and other limiting resources) among the trees living on a small patch of land (Fig. 2). Each tree's diameter increment is computed annually as a function both of species attributes (e.g., geometry, maximum growth rate) and of the individual tree's environment (e.g., crowding, shading by other trees). Each tree is subjected to a probability of annual mortality that is determined by the species' longevity. If the tree is growing slowly, then the intrinsic mortality rate is increased. Regeneration is determined as a stochastic outcome of both sprouting (for appropriate species) and seed regeneration. The model output resembles a stand tally sheet, listing the species and diameter of each tree on the modeled stand.

Gap models have been tested against independent data in a variety of forests under a reasonably wide spectrum of test conditions (Table 1). These tests are

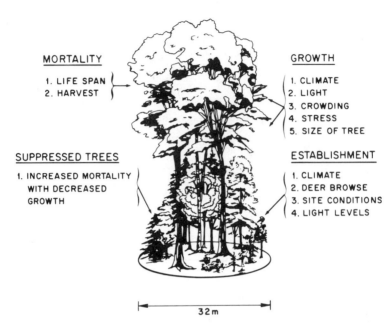

MORTALITY

1. LIFE SPAN
2. HARVEST

GROWTH

1. CLIMATE
2. LIGHT
3. CROWDING
4. STRESS
5. SIZE OF TREE

SUPPRESSED TREES

1. INCREASED MORTALITY
   WITH DECREASED
   GROWTH

ESTABLISHMENT

1. CLIMATE
2. DEER BROWSE
3. SITE CONDITIONS
4. LIGHT LEVELS

32 m

Fig. 2. Conceptualization of a forest gap as represented in a typical gap model, FORET. (Shugart and West, 1977.)

limited to cases in which gap models might be reasonably expected to perform well. For example, a gap model could not be expected to simulate accurately variation in spatial patterns of trees because the models do not have spatial location of trees as a simulated feature. Or the models have not been used to inspect the effect of seed source on regeneration, since there is no consideration (at present) of interplot contagion. Nonetheless, gap models do project a wide array of forest ecosystem dynamics (Table 1) and provide a method of projecting the consequences of species attributes to the forest ecosystem dynamics.

One gap model that has been applied in the context of simulating landscape-level responses is the BRIND model (Shugart and Noble, 1981). This model was designed to mimic the long-term behavior of the *Eucalyptus*-dominated forests of the upper-elevational Brindabella Range near Canberra, Australia.* The BRIND model is largely derived from the FORET model (Shugart and West, 1977) by the addition of explicit consideration of several phenomena that are important in the Australian system (e.g., seed source availability, seed storage and the effects of fire on mortality, reproduction and growth of trees). One of the validation tests of the BRIND

---

*The ecosystems, hydrology, and land use of a section of the Brindabella Ranges have been described in detail in Anonymous (1973). Lang (1970) developed a classification of the Brindabellas, and a reanalysis of this data is available in Dale and Quadraccia (1973). Pryor and Moore (1954) also described the plant communities in parts of the Brindabellas. Costin's (1954) description of the ecosystems of the neighboring Snowy Mountains region is generally applicable to the Brindabella Range.

model was an inspection of its ability to simulate the pattern of forest altitudinal zonation on the southeastern slopes of the Brindabella Range under different probabilities of wildfire. This was done by running 10 replicate simulations for each cell of a matrix, composed of four wildfire probabilities by four growing-degree-days (160 replicates *in toto;* Fig. 3). Each replicate was simulated for 500 years. Species dominance was calculated each year based on the biomass contribution of each tree species to the total biomass of the stand. If 90% of the total biomass of a plot could be attributed to a single species, the plot was given the type name of that species. Otherwise, if two species accounted for 90% of the total biomass, then the plot was given the type name of both species. The BRIND model's predictions of forest types associated with altitudes (as expressed by differences in the degree-day heat sum) and wildfire probabilities are summarized as forest-type constellations (Fig. 3).

Results of the BRIND simulations include the following:

1. At 850 m, the model simulates stands with mixtures of *Eucalyptus robertsonii, E. rubida, E. viminalis* and, rarely, a low-altitude form of *E. pauciflora.* Of these species, only *E. robertsonii* is able to show clear dominance (Fig. 2). These model predictions are consistent with the pattern of this landscape as described by others (Anonymous, 1973; Costin, 1954), although *E. viminalis* may be underestimated.

2. At 1050 m (Fig. 2), both *E. robertsonii* and *E. fastigata* follow a pattern of increasing dominance with decreasing fire frequency. The simulations seem to be appropriate for both the Brindabella Range (Anonymous, 1973) and nearby ranges (Costin, 1954). *Eucalyptus rubida,* although it is often present in mixtures with *E. robertsonii,* is never sufficiently dominant to account for 90% of the biomass on the stand. The inability of *E. rubida* to show such dominance (as well as the rarity of *E. dives* in the simulations) is a desirable result that is consistent with this forest ecosystem (Anonymous, 1973).

3. At elevations of c. 1300 m, the model frequently simulates the occurrence of pure stands of *E. delegatensis* and, less frequently, stands dominated by either *E. dalrympleana* and *E. pauciflora.* Forest dynamics in this elevational zone apparently favor the rapidly growing *E. delegatensis.*

4. At the highest elevations (c. 1500 m; Fig. 3) the model correctly (Anonymous, 1973; Costin, 1954; H. H. Shugart and S. W. Seagle, personal observation) simulates stands dominated by *E. pauciflora* either in pure stands or in mixtures with *Bouksia marginata* and *Acacia dealbata.*

The use of this model to simulate the larger-scale pattern of a landscape has analogous applications in the cases of other gap models. Botkin *et al.* (1972a) tested their model of northern hardwood forests in New Hampshire by determining its ability to predict the location of the deciduous–coniferous transition on a simulated mountain gradient. Weinstein *et al.* (1982) similarly used the FORNUT model to predict the nature of forests in various locations within watersheds of the Great Smoky Mountains and the Ridge and Valley Province in eastern Tennessee. Sol-

TABLE 1

Tests That Have Been Performed on Various Gap Models[a,b]

Type of model test	Structural response	Functional response
Verification Model can be made to predict a known feature of a forest	Consistent with the structure and composition of forests in New Hampshire (JABOWA), Tennessee (FORET), Puerto Rico (FORICO), and floodplain of the Mississippi River (FORMIS) Compares to forest of known age in subtropical rain forests (KIAMBRAM) Predicts Arkansas upland forests based on 1859 reconnaissance (FORAR)	Predicts forestry yield tables for loblolly pine (*Pinus taeda*) in Arkansas (FORAR) Predicts relations of forest types in succession in the middle altitudinal zone of the Australian Alps (BRIND) Predicts response to clearcutting in Arkansas wetlands (SWAMP) Predicts changes in forest types as a function of flood frequency in the Arkansas wetlands (SWAMP) and the Mississippi floodplain (FORMIS)
Validation Model independently predicts a known feature of a forest	Predicts frequency of trees of various diameters in rain forests of Puerto Rico (FORICO) and uplands of Arkansas (FORAR) Predicts vegetation change in response to elevation in New Hampshire (JABOWA) and in the Australian Alps (BRIND) Determines effects of hurricanes on the diversity of Puerto Rican rain forest (FORICO)	Predicts response of *Eucalyptus* forests to fire (BRIND) Assesses effects of the chestnut blight on forest dynamics in southern Appalachian forests (FORET) Predicts forestry yield tables for alpine ash (*Eucalyptus delegatensis*) in New South Wales (BRIND)

Application		
Model is used to predict a response of a forest to changed conditions	Predicts changes in a 16,000-year pollen chronology from east Tennessee in response to climate change (FORET)	Predicts response of the northern hardwood forest to increased levels of CO_2 in the atmosphere (JABOWA)
	Assesses habitat management schemes for endangered species (FORAR), non-game bird species (FORET), and ducks (SWAMP)	Predicts response of the southern Appalachian hardwood forest to decreased growth due to air pollutants (FORET)
		Predicts response of the northern hardwood forest (JABOWA), southern Appalachian forest (FORET), Arkansas upland forests (FORAR), Arkansas wetlands (SWAMP), and Australian subtropical rain forest (KIAMBRAM) to various timber management schemes

[a] From Shugart and West (1981).

[b] The mnemonics in parentheses indicate the following models: BRIND, a model of Australian *Eucalyptus* forests (Shugart and Noble, 1981); FORAR, a model of Arkansas mixed pine–oak forests (Mielke *et al.*, 1978); FORET, a model of Tennessee Appalachian hardwood forests (Shugart and West, 1977); FORICO, a model of Puerto Rican Tabonuco montane rain forest (Doyle, 1981); FORMIS, a model of Mississippi River floodplain deciduous forest (Tharp, 1978); JABOWA, a model of northern hardwood forest (Botkin *et al.*, 1972a); KIAMBRAM, a model of Australian subtropical rain forest (Shugart *et al.*, 1981); SWAMP, a model of Arkansas wetlands forest (Phipps, 1979).

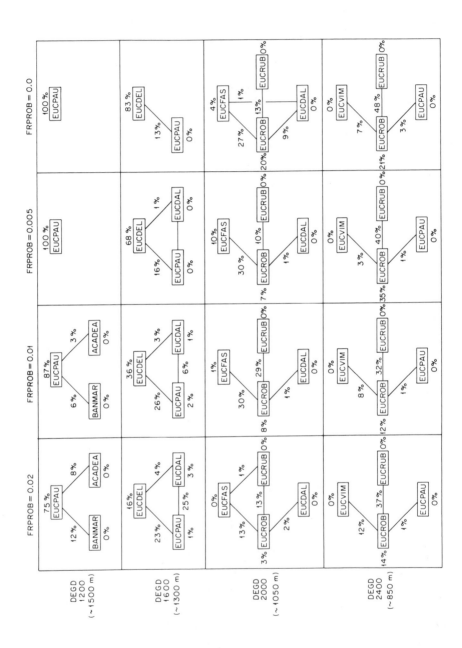

omon and Shugart (1984) have used a modified version of the FORET model to simulate changes in landscapes in Minnesota since the retreat of the Wisconsin glaciers. In these applications, the gap models have been used to simulate the dynamics of parts of the landscape (~0.1-ha plots), and the total landscape has been viewed as a statistical average of these plots.

A landscape viewed in the gap-phase replacement context is a mosaic of changing patches of trees. The detailed output of gap models allows statistical averaging of these patch characteristics to provide descriptions not only of forest landscapes but also of physiognomic factors that influence the distributions of animal species. Thus, by stratifying the replicate simulations in a manner resembling the pattern of a landscape, and interpreting the results as stratified plot survey data, predictions can be made of animal habitat availability. An example of this treatment is the FOR-HAB simulation of Smith et al. (1981), in which forest inventory quadrats were loaded into a gap model as initial conditions and future bird habitat changes were predicted. As indicated by this model, forest patches may correspond to animal habitat patches.

Since landscapes can be viewed as a shifting mosaic of habitats, the effects of landscape dynamics on animal communities and species distributions provide a potentially enlightening yet unexplored area for study. One aspect of species distributions, the species–area relationship, has a broad empirical basis and is an appropriate phenomenon for study at the landscape scale. In the following discussion, the effect of landscape dynamics on the development of the species–area curve will be explored by using a Markov model to simulate habitat dynamics.

IV. A MODEL OF ANIMAL RESPONSE TO DISTURBANCE

Unlike the more detailed gap models discussed above, the Markov model omitted the detailed interaction of plant species. Rather, emphasis was placed on the change in the landscape that results from concurrent succession and disturbance and on the emergent properties of the animal communities that inhabit such a dynamic landscape. The computational aspects of this model appear in the flow diagram (Fig. 4). The pattern of landscape change was based on a Markov model of vegetation dynamics of a Tasmanian wet sclerophyll-rain forest ecosystem (Noble and Slatyer, 1980). Ten habitat types were recognized, and the annual probability of change

---

Fig. 3. Summary forest-type constellations for one- and two-species-dominated forest types simulated by the BRIND model at four different altitudes (expressed by the heat sum, DEGD) by four different wildfire probabilities (FRPROB). The value in a box is the percentage of instances in which the indicated species contributed 90% or more to the stand's total biomass. The values on the lines between boxes are the percentage of instances in which the two named species contributed 90% or more of a stand's total biomass. Species mnemonics are taken from the first three letters of the scientific bionomial: ACADEA, *Acacia dealbata;* BANMAR, *Banksia marginata;* EUCDAL, *Eucalyptus dalrympleana;* EUCDEL, *E. delegatensis;* EUCFAS, *E. fastigata;* EUCPAU, *E. pauciflora;* EUCROB, *E. robertsonii;* EUCRUB, *E. rubida;* EUCVIM, *E. viminalis.*

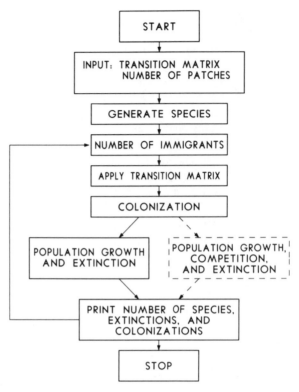

Fig. 4. Flow chart of operations in a model of animal responses to dynamically changing landscapes of different sizes.

from one habitat type to all others was calculated as the reciprocal of the successional time between habitat types. The probabilities were tabulated as a 10×10 matrix of transition probabilities so that the model characterizes the landscape dynamics as a first-order Markov process. During each year or iteration of the model, the appropriate elements of this matrix were used to determine whether each patch of habitat succeeded to another habitat type or simply aged 1 year. Three hundred iterations comprised each model run, and the results of the patch dynamics were tabulated in a 10×300 habitat–age matrix. The landscape area was represented by the number of habitat patches simulated in the model. Starting conditions for each model run were uniform, and consisted of the equilibrium configuration of habitat patches expected for the landscape and determined during model development.

A pool of 150 hypothetical animal species was generated from uniform statistical distributions for different species-level characteristics. These characteristics included range of habitats utilized ($x = 2$, SD $= 0.5$), range of habitat ages utilized ($x = 15$, SD $= 2.0$), intrinsic rate of population increase in individuals per year ($x = 3$, SD $= 0.05$), home range in number of habitat patches ($x = 20$, SD $= 5.0$), and efficiency of habitat utilization ($x = 5$, SD $= 1.5$). The ranges of habitat and age defined each species niche and were positioned randomly within the habitat-age matrix. Intrinsic rate of increase was used in calculations of population growth and home range size in determining colonization and carrying capacity. Efficiency of habitat utilization was used in determining competition coefficients.

Each model run began with no species occupying the landscape, so that colonization was similar to invasion of a defaunated island. The number of immigrant species during each model iteration was chosen from a uniform distribution with a mean of 12 and a standard deviation of 2.0. This number was chosen from the same distribution regardless of the landscape area under consideration. Thus, no target effect (MacArthur and Wilson, 1967) was included in the model. This aspect of approximately equal immigration to both large and small landscapes actually biases the results against a finding of more species on larger landscapes. Colonization of each immigrant species was possible only if the landscape had twice the number of patches per home range within the designated niche space of a species. If this criterion for colonization was met, a propagule of two individuals was established on the landscape. Established species chosen as immigrants in subsequent years were discarded. Thus, no "rescue effect" (Brown and Kodric-Brown, 1977) was incorporated, although a species could recolonize after extinction.

During each year of a model run, all established species were allowed population growth using the logistic growth model as a difference equation:

$$N_{t+1} = N_t + rN_t (1 - N_t/K)$$

In this equation, N_t is the population size at time t, N_{t+1} is the population size at time $t + 1$, r is the species-specific intrinsic rate of increase, and K is the species-specific carrying capacity of the landscape. Carrying capacity was calculated by

$$K = \frac{\sum\limits_{i \, = \, \text{MINHAB}}^{\text{MAXHAB}} \sum\limits_{j \, = \, \text{MINAGE}}^{\text{MAXAGE}} (P_{ij})}{\text{HR}}$$

where HR is the number of habitat patches required by an individual, P is the number of patches within one element of the habitat-age matrix, and the subscripts on the summation symbols denote the position of a species niche within the habitat-age matrix. In the version of the model in which species are allowed to compete, the following equation was used:

$$N_{i,t+1} = N_{i,t} + r_i N_{i,t} 1 - \frac{N_{i,t}}{K} - \frac{\sum\limits_{j=1}^{n} \alpha_j N_j}{K}$$

where α_j is the product of the proportion of patches species j shares with species i and the ratio of their resource utilization efficiencies, N_j is the population size of species j, and n is the number of species established on the landscape.

A single criterion was also used for species extinction based on population size. The expression $1/2^{N-1}$, which gives the probability of all offspring in a single generation being of the same sex, was compared to a randomly generated probability to determine whether extinction occurred. This probability rapidly decreases as population size increases and therefore reflects the assumptions of MacArthur and Wilson (1967) and the empirical results of Schoener and Schoener (1983) concerning population size and extinction. For extinction to occur, then, population sizes must reach a low level. Depending on the population growth equation utilized, such low population sizes may result from habitat loss or habitat loss combined with competition.

Both habitat diversity and its constancy through time may affect species richness on a landscape of a given size. Figure 5 shows how these two factors may be affected by a single disturbance regime interacting with landscapes of different sizes. In this figure, habitat diversity (calculated by the Shannon-Weiner Index) is plotted against time for three landscape areas that vary by two orders of magnitude in the number of patches. The average diversity does not vary greatly for the three landscape areas; however, the constancy of this diversity measure increases greatly as the landscape area increases. Larger landscapes thus buffer the effect of disturbance on habitat diversity. These results indicate that differences in species richness on landscapes of various sizes are due to temporal variability in habitat diversity.

Figure 6 illustrates the effect of increasing landscape area on the number of established species both with and without competitive interactions among them. In each case, the resulting curve displays a rapid increase in the number of species per unit area followed by a gradual decrease in the slope of the curve. With no competition, the species–area curve equilibrates at approximately 23 species, implying a

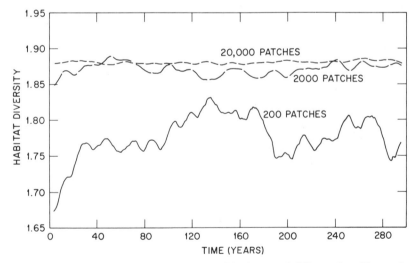

Fig. 5. Changes in habitat diversity for simulated landscapes of different sizes. The number of patches in the landscape is used to indicate the area.

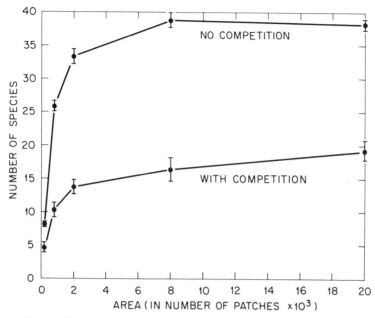

Fig. 6. Number of species at equilibrium for simulated islands of different sizes with either competing or noncompeting colonizing animal species.

balance between immigration and extinction due to habitat loss from landscape change. The steep slope at small areas and the equal rate of immigration for different landscape areas show that interaction between the disturbance regime embodied by the transition matrix and the landscape area produce different patch dynamics. The result is that smaller landscapes support small populations that are sensitive to landscape changes that lower carrying capacities and increase extinction probabilities.

When colonizing species are allowed to compete, the resulting species–area curve is qualitatively similar to the noncompetitive case (Fig. 6). The distinguishing feature of this curve is its equilibration at approximately eight species. The lower equilibrium number of species in the competitive case is not unexpected. However, the graphic similarity of the two curves suggests that the general shape of the species–area curve is determined by the interaction of species habitat requirements and landscape dynamics.

The power function model, $S = CA^z$, has long been used to describe the relationship between area and number of species (Preston, 1962; MacArthur and Wilson, 1967; Connor and McCoy, 1979). In this equation, S is the number of species and A is the area, with C and z fitted constants. While the interpretability of the constant z is debatable (Connor and Simberloff, 1979; Sugihara, 1981), its value is remarkably constant over many empirical studies (MacArthur and Wilson, 1967; Connor and McCoy, 1979). We estimated z by linear regression from $\log S = C +$

$z \log A$, the log–log transformation of the power function model, in order to compare the model's results with those of empirical studies. For the case with no competition, z was 0.31, with a correlation between $\log S$ and $\log A$ of 0.86. Addition of competition to the model lowered the value of z to 0.29 and raised the correlation to 0.94. These z values fall within the 0.20–0.35 range suggested by MacArthur and Wilson (1967).

V. CONCLUDING REMARKS

The landscape models discussed in this chapter are intended to demonstrate the potential utility of computer simulation models in landscape/disturbance problems, rather than to be an exhaustive compilation. These models are available as tools to aid in the development and evaluation of theory in landscape ecology. The dynamic nature of landscape ecosystems and the potentially long time scales of the response of ecosystems to infrequent disturbance make the study of disturbance as an ecological phenomenon difficult. It is hard to design a study to assess the response of the floodplain forest of the Mississippi River to the 100-year flood cycle, for example. Since we have altered the pattern of this 100-year cycle (due to a variety of public works projects), we may not be able to see the floodplain ecosystem operating in its natural state (if such a single state ever existed) even if we could monitor the dynamics of the ecosystem over several centuries. This example is, perhaps, contrived for purpose of illustration, but it is probably illustrative of conditions found in many ecosystems. The human life span does not provide sufficient time to watch the longer-term dynamics of forested landscapes. Further, many landscapes have been altered from their natural operating state. These situations indicate the need for tools that can synthesize what we know of ecological processes at smaller spatial and temporal scales into an understanding of the larger-scale function of ecological systems. Forest models are tools that have been tested in their predictive ability and are now being used as devices for synthesis. We believe that this will continue to be an important use for these models in the future.

ACKNOWLEDGMENTS

Research supported by the National Science Foundation's Ecosystem Studies Program under Interagency Agreement No. DEB 80-21024 with the U.S. Department of Energy, under contract W-7405-eng-26 with Union Carbide Corporation. Publication No. 2450, Environmental Sciences Division, ORNL.

Part V

SYNTHESIS

Chapter **21**

Patch Dynamics:
A Synthesis

S. T. A. PICKETT

Department of Biological Sciences
and Bureau of Biological Research
Rutgers University
New Brunswick, New Jersey

P. S. WHITE

Uplands Field Research Laboratory
Great Smoky Mountains National Park
Twin Creeks Area
Gatlinburg, Tennessee
and
Graduate Program in Ecology
University of Tennessee
Knoxville, Tennessee

I. INTRODUCTION

There can be no doubt that disturbance is an important and widespread phenomenon in nature, as attested to by the contributors to this volume (West *et al.,* 1981; White, 1979). Indeed, references to disturbance appear in some of the earliest major works in ecology (Clements, 1916, p. 4):

THE ECOLOGY
OF NATURAL DISTURBANCE
AND PATCH DYNAMICS

> Even the most stable association is never in complete equilibrium, nor is it free from disturbed
> areas in which secondary succession is evident.

However, this early reference saw disturbance only as a mechanism resetting the inexorable march toward equilibrium. Disturbance per se had no place in the theory of community and ecosystem dynamics until relatively recently. Early workers did observe the effects of disturbance in various communities (see, e.g., Cooper, 1913; Aubréville, 1938; Jones, 1945; Watt, 1947; van Steenis, 1958). It may be significant that these workers were less under the influence of Clements than others, either because they wrote before his major publication or because they worked in Europe. Raup (1957), Egler (1977), and Heinselman (1973), who are known as independent thinkers or who worked in particularly dynamic systems, opened the way for the study of disturbance itself as an ecologically significant phenomenon. Those who have followed their lead have often concluded that disturbance is important. This chapter will first state what is important about disturbance; second, it will examine the mechanistic predictions we can or should make about disturbance. Finally, it will examine the context in which our predictions must be couched and the factors that control the role of disturbance in particular situations.

II. THE IMPORTANCE OF DISTURBANCE

The contributors to this volume have demonstrated that natural disturbance occurs in a wide variety of biomes, affecting both animals and plants (see also Bazzaz, 1983; MacMahon, 1980, 1981). Among coniferous forests, it affects boreal (Sprugel, Chapter 19, this volume), montane (Romme and Knight, 1981), and coastal plain types (Forman and Boerner, 1981). Deciduous forests in all North American associations, from northern hardwoods to mixed mesophytic to oak (Runkel, Chapter 2, this volume; Oliver, 1981), are affected. Studies on tropical dry season-deciduous forests are lacking, although these forests are composed of tree species capable of resprouting after fire (Ewel, 1980). Evergreen forests in temperate areas (Veblen, Chapter 3; Shugart and Seagle, Chapter 20, this volume) and throughout the tropics (Brokaw, Chapter 4, this volume) are subject to disturbance.

Not only do forest systems undergo disturbance, but grasslands (Loucks *et al.*, Chapter 5, this volume) and shrublands (Christensen, Chapter 6, this volume; Rundel, 1981a) do as well. Other systems have received less attention, but tundra and deserts, among others, occasionally experience natural disturbance (Pickett and Thompson, 1978).

Disturbances in nature are not restricted to primary producers or to vegetation-dominated systems. Sessile subtidal and intertidal animal communities are naturally disturbed at intervals (Connell and Keough, Chapter 8; Sousa, Chapter 7; this volume). Freshwater fish, too, experience natural disturbance (Vrijenhoek, Chapter 15; Karr and Freemark, Chapter 9; Denslow, Chapter 17; this volume). Insects (Schowalter, Chpater 13, this volume), birds, and mammals (Wiens, Chapter 10,

Karr and Freemark, Chapter 9; Shugart and Seagle, Chapter 21; this volume) are subject to the effects of disturbance in vegetation, but they themselves may also cause disturbance or enhance abiotic disturbance to vegetation or other trophic levels (Wiens, Chapter 10; Schowalter, Chapter 13; Thompson, Chapter 14; Denslow, Chapter 17; this volume).

Not only does disturbance affect different biomes, taxa, and trophic levels, but its impacts are also observable at all levels of ecological organization. The node at which many of these effects are focused is the individual level of organization (MacMahon et al., 1978; Bazzaz, 1983). Growth, architecture, reproduction, dispersal, and the capacity for antagonistic interactions are features of individual organisms affected by disturbance (Canham and Marks, Chapter 11; Collins et al., Chapter 12; Thompson, Chapter 14; Schowalter, Chapter 13; Wiens, Chapter 10; Connell and Keough, Chapter 8; Sousa, Chapter 7; this volume). Through the collective effect on individuals, populations are also affected by disturbance. Age, size (Veblen, Chapter 3, this volume), and genetic structures (Vrijenhoek, Chapter 15; Thompson, Chapter 14; Rice and Jain, Chapter 16; this volume) of populations can alter as a result of disturbance, but much more work is needed here. Indeed, life history attributes interlock with the disturbance regime and may evolve in response to disturbance (Collins et al., Chapter 12; Canham and Marks, Chapter 11; Thompson, Chapter 14; this volume; Bazzaz, 1983). Dispersion within populations may reflect disturbance as well (Thompson, Chapter 14; Schowalter, Chapter 13; this volume).

Disturbance has demonstrated effects on community characteristics, including richness (Denslow, Chapter 17, this volume), dominance, and structure (see, e.g., Brokaw, Chapter 4, this volume). The functional attributes of ecosystems are also governed to some extent by disturbance. Nutrient cycling and energetics respond to disturbance as well as to biotic and abiotic opportunities and limiting factors (Sprugel, Chapter 19; Vitousek, Chapter 18; this volume; Peet, 1981; Reiners, 1983). Most information on ecosystem response to disturbance is at a fairly coarse scale of patch resolution, leaving much opportunity for fine-scale work.

The most obvious role that disturbance plays in ecosystems is in the deflection of a community from some otherwise predictable successional path. If we measure the "instantaneous" rate of succession at any point in a stand, which is a function of local conditions and varies within every stand, we find that disturbance and environmental fluctuation prevent this path from being followed for any effective length of time. Stated another way, relatively infrequent events have a large role in shaping community structure—particularly when establishment of new individuals is rare relative to the life span of dominant organisms.

It is clear from this brief summary of the contributions to this volume that disturbance is important. Specifically, (a) disturbance occurs in a wide variety of biotic assemblages; (b) it occurs at all ecological levels of organization; and (c) the effects of disturbance merge with secular environmental changes over various spatial scales and time frames (Karr and Freemark, Chapter 9, this volume; MacMahon, 1980; Delcourt et al., 1983; Delcourt and Delcourt, 1983). It may be argued, though, that glacial cycles and alterations of global rainfall and wind patterns are not "events" in the sense used to define disturbance here (White and Pickett, Chapter 1, this volume). However, those coarse-scale changes affect system structure and resource availability as well, and as a result profoundly influence the attributes of individuals, populations, communities, and ecosystems. Wiens (Chapter 10, this volume) discusses such diffuse causes of patch dynamics at fine scales. Thus, disturbance is common to many systems, spatial and temporal scales, and is continuous over all ecological levels of organization. That is the essence of its importance.

Failure to recognize the importance of disturbance has led to two kinds of frequent misinterpretation in field ecology: (a) extrapolation of events measured during disturbance-free years to predict future system states, and (b) use of a plot scale that integrates different kinds of patches. In fact, variance generated by patch dynamics is likely to be one of the most important constraints on nonexperimental sampling strategies, although this factor has seldom been considered in designing or carrying out a field sampling project.

III. TOWARD A THEORY OF DISTURBANCE

The commonality, over various levels of organization, and continuity, over different scales, of the process and effects of disturbance suggest the need for a theory of disturbance. Such a theory is currently lacking. However, it may now be possible to erect a conceptual framework in which to view disturbance, one that may provide a prolegomena to disturbance theory. Without such a framework, observations on disturbance are likely to be confused and discussion diffuse. As a contribution toward this theory, we will examine the existing conceptualization of disturbance and explore some extensions. These extensions will be based on a consideration of the parameters about which predictions should be made, and will enumerate what must be known in order to make such predictions. This discussion also affords an opportunity to point out emergent generalizations about disturbance beyond those presented in Section II.

A. Existing Concepts

The most important concepts of disturbance are embodied in the definition (White and Pickett, Chapter 1, this volume). There are several additional features of disturbance that must be taken into account in understanding it and comparing its role in a variety of situations.

Disturbance is often patchy. It may create discrete openings of gaps in either or both of the above- and below-substrate components of a system. Patches themselves are characterized by size, shape, dispersion, and internal heterogeneity (Denslow, Chapter 17; Collins *et al.,* Chapter 12; this volume). It is particularly important to state how patches or gaps are defined, i.e., what constitutes the lateral border; what constitutes the depth; what minimum (and maximum) size is recognized? Without such explicit definition, comparison between systems and times will be unproductive.

The nature of patchiness affects the level of resource availability in the disturbed patches (often increasing availability with gap size), the survival of residual organisms in the patch, and the rate of invasion and success of establishment of new organisms. Not all disturbances create discrete patches within an undisturbed matrix. Some disturbances may thin a canopy of plants or sessile animals, or a root mat, and leave relatively sparse survivors (Loucks *et al.*, Chapter 5; Sousa, Chapter 7; this volume). At a particular spatial scale, such thinning or diffuse disturbance delimits a continuum with discrete disturbance. (On a coarser spatial scale, disturbed patches versus the undisturbed matrix may still be recognizable.) Drought, epidemics, or climatic shifts may, on the appropriate temporal and spatial scales (Delcourt *et al.*, i983), exemplify diffuse disturbance.

Disturbance is distributed in time. The frequency (events per year) of disturbance may be assessed as recurrence at a point, irrespective of the size of the patch involved. The inverse of the frequency is the recurrence interval, given in years per event. These two parameters, for some organisms, may be the most biologically meaningful assessment, since they indicate how often a genet [*sensu* Harper (1977)] is exposed to altered resource levels. A different assessment, the rotation period, determines how often an area equivalent to the entire study area is disturbed. Not all of the actual surface area may have been visited by disturbance during that time. It seems prudent to prevent blurring of these parameters, since their biological meanings are so different. The different parameters may help explain different properties of species or assemblages. Disturbance rotation period may explain the distribution and abundance of a taxon in a landscape because long rotations probably permit the existence of more extensive refugia or reinvasion sources than short rotations. In contrast, the disturbance recurrence interval may be more useful in explaining the persistence or performance of poor competitors within an assemblage or the genetic or age structure of populations in communities.

Recognizing the difference between return time and rotation time also focuses attention on the existence of a spectrum of patch types in a landscape. At one extreme are small, frequent disturbance gaps, while at the other are large, rare patches. The rare extreme is little investigated in most systems for obvious reasons. However, because the size of a disturbed patch is often correlated with both resource level and environmental stress, it would be useful to construct the actual spectrum of disturbance sizes and calculate the probabilities of occurrence of each size class (Veblen, Chapter 3, this volume). This may be, for some purposes, more valuable than simple quantification of the return interval and rotation time. Neglect of the rare disturbances in a system may prevent an understanding of species persistence; evolution of life history strategy; and aspects of energetics, soil, and nutrient relations (Franklin and Hemstrom, 1981).

Disturbances vary in impact or magnitude. The literature on fire recognizes a difference between the physical intensity of the disturbance, as heat flux or temperature, and the biological effect, the severity. The amount of biomass destroyed and the degree of substrate disruption can be used to quantify the intensity of disturbance. Exactly how much emphasis is given to each of these parameters, or indeed whether other parameters should be included, depends on the specific system of interest. In some systems, especially those where substrate disruption is not a factor, severity may be equivalent to size. However, in the interest of generality, it is important not to confuse these two characteristics of disturbance.

Two agents of disturbance may act synergistically, as well as interact with stress. Schowalter (Chapter 13, this volume) shows that stress may predispose a tree to fungus attack and that tree stress, insect outbreaks, windfall, drought, and fire may influence each other synergistically. This is likely to be true in other cases as well, e.g., the combined influences of physical disturbance and predators on marine systems (Connell and Keough, Chapter 8, this volume).

Conservation decisions must consider the role of disturbance (Karr and Freemark, Chapter 9; Wiens, Chapter 10; Sprugel, Chapter 19; this volume). An essential paradox of wilderness conservation is that we seek to preserve what must change (White and Bratton, 1980). Denslow (Chapter 17, this volume) has stressed the contrast between natural and anthropogenic disturbances (see also Harmon *et al.*, 1984). Pickett and Thompson (1978) have noted that patch dynamics considerations are essential to nature preservation. Dolan *et al.* (1978) have reviewed the conservation issues for coastal erosion.

The features of disturbance enumerated above can be summarized in two concepts. First is the "disturbance regime," which has come to refer to the temporal and spatial pattern of creation of open or altered patches. This is the demography and dispersion of gaps. Basic information on the disturbance regime is accumulating in many systems, but more study is required. The apparent constancy of rates of gap creation and disappearance in mesic forests in tropical and temperate zones (Runkle, Chapter 2; Brokaw, Chapter 4; this volume; Lang and Knight, 1983) is an example of an emerging generalization that needs testing (Romme and Martin, 1982) and further comparative study. Observations of permanent plots, attention to rare disturbances, relation to climatic change and topographic heterogeneity, and assessment of variance need emphasis.

Variance of disturbance is an important component of the disturbance regime that is underinvestigated (Shugart and Seagle, Chapter 20, this volume). There are many levels on which disturbances have been shown to be variable. Beyond the superficial recognition that one kind of disturbance is unlike another (fire versus wind) is the observation that for even a given disturbance (wind), the actual physical effects at the site can be different (uprooting versus trunk snap; Runkle, Chapter 2, this volume; Veblen, Chapter 3, this volume; Brokaw, Chapter 4, this volume). Several contributors to this volume note that no two disturbance events are alike. Further, for a given event, a mosaic of effects is present (Broakw, Chapter 4, this volume) and patches are internally heterogeneous (Runkle, Chapter 2; Connell and Keough, Chapter 8; this volume). Even in the case of small-scale patches for a monocarpic perennial plant, species interactions and gene flow vary with internal patch characteristics (Thompson, Chapter 14, this volume). Disturbance also affects different trophic levels differently (Connell and Keough, Chapter 8; Karr and Freemark, Chapter 9; Wiens, Chapter 10; this volume).

The second major concept offering a broad comparative framework for study and theory of disturbance is "patch dynamics" (Pickett and Thompson, 1978). This includes disturbance regimes but also phenomena such as the filling and changes within patches; the relation of patches to one another and the matrix (Forman and Godron, 1981); and the flows of organisms, materials, and energy among patches (Peet, 1981; Reiners, 1983; Vitousek, Chapter 18, this volume). This idea emphasizes the temporal and spatial changes in these relationships.

B. Predictions about Disturbance

1. Parameters of Interest

The development of theory requires the ability to make predictions. Predictions from first principles, which treat all taxa, ecological systems, or situations as equivalent particles without an accrued history, such as are possible in physics, are not likely to be productive in the study of disturbance. Organisms, communities, populations, and ecosystems vary, have a history, and are subject to the vagaries of climatic and geological background. Thus, the sorts of predictions we can make about disturbance are mechanistic—those that take into account the peculiarities of a particular system and situation. In that context, and considering the evolved or acquired characteristics of the entities involved, meaningful ecological predictions can be made. For example, such predictions can explain a particular situation or suggest the impact of an altered disturbance regime.

Because of disturbance variation, the importance of community history, the probabilistic nature of dispersal, and the existence of different kinds of competitive hierarchies, disturbance has been shown to result in unpredictable species composition within gaps (Brokaw, Chapter 4, this volume). The predictability of disturbance response may itself be correlated with disturbance scale: Connell and Keough (Chapter 8, this volume) have suggested that the larger and more infrequent the disturbance, the less predictable the postdisturbance composition. MacMahon (1981) has noted that the usefulness of deterministic community concepts is related to gradients in disturbance frequency and patch size.

It is necessary not only to limit prediction to the mechanistic sort but also to define clearly the parameters about which predictions are desired. Perhaps the most commonly encountered parameter of interest in the literature is diversity. This may be subdivided into species richness, distribution of dominance, community structure, and genetic diversity. Of these, richness has received the greatest attention, while little thought has been devoted to the others. Not all are likely to be equally affected or to vary in the same way with disturbance. The potential differences deserve serious study. For example, though composition may be unpredictable, ecosystem structural and functional characteristics may be predictable. Total leaf area (all species summed) returns to predisturbance levels relatively quickly (Sprugel, Chapter 19, this volume); total leaf area may also be relatively predictable at any stand age.

The other major attribute affected by disturbance is resource status. Here, the level of nutrients in soil, litter, and biomass, and the ability of entire landscapes or patches within them to sequester, retain, or disperse nutrients, are of interest (Vitousek, Chapter 18, this volume). The well-known phenomena of stability and resilience intersect with disturbance in considering these parameters (Denslow, Chapter 17, this volume). Likewise, the energetics of ecosystems may be affected by disturbance (Sprugel, Chapter 19, this volume). There is currently little basis for calculating the impact of disturbance on ecosystem nutrition, on either the fine scale or the long term. Again, the role of disturbance-generated heterogeneity in governing within-ecosystem energetic relations is unknown.

2. Current Hypotheses

Two "hypotheses" are current in the literature on disturbance. Their various incarnations and the relationships are thoroughly discussed by Peet *et al.* (1983). The first, and more widely cited, is the intermediate disturbance hypothesis, which is attributed to Connell (1978), although others prefigured it or present it in other terms (Grime, 1979). The hypothesis states that species richness will be greatest in communities experiencing some intermediate level of disturbance. Many observations presented in this volume support this generalization. However, the statement leaves much unspecified. First, which community and ecosystem parameters will behave in the expected way? For example, will intermediate disturbance enhance nutrient retention or productivity as well as richness?

The second problem concerns the maximum level of disturbance deemed possible. It is possible both to set the maximum absurdly high and to inadvertently omit reasonable or historically important events. Thus, explicit statements of the maximum and its justification are required. For example, the goals of a study, whether natural or human disturbances are the focus, and whether the rare component of a natural regime is included will determine the range of the gradient of disturbance along which impacts are compared. Some disturbances may be so severe as to decrease resource availability or to obliterate the system completely. The specific research questions will determine whether such disturbances should be included or omitted from consideration. Application of the intermediate disturbance hypothesis to management problems such as maintenance of diversity or sustainable yield may force ecologists to clarify these issues.

The final problem to be clarified in applying the intermediate disturbance hypothesis is that of quantification. How should the impact of disturbance be measured? Destruction of biomass is one reasonable parameter, but structural disruption of the community, substrate disruption, or in some cases the area of the disturbed patch may be appropriate measures. The frequency of the disturbance may be a plausible measure of impact as well. Disturbance impact is the issue, and it may be sub-divided into components presented earlier: magnitude, frequency, and size. Inter-mediacy in each of these variables of disturbance may mean different things. Such considerations have been neglected in the literature on disturbance to date (Peet *et al.*, 1983; Armesto and Pickett, 1984). Without an explicit statement of which parameters of response to disturbance are of interest, the extremes of disturbance included, and the component(s) of disturbance impact allowed to vary, the ideas of intermediate disturbance are best considered heuristic conceptual generalizations rather than testable hypotheses.

A second major generalization relates disturbance frequency to species richness. Where disturbance recurs more frequently than the time required for competitive exclusion, richness should be maintained (Huston, 1979). This is an important generalization with clear implications for the intermediate disturbance idea (Peet *et al.*, 1983). However, it suffers from much of the same uncertainty as the intermedi-ate disturbance hypothesis.

Refinement of the hypotheses presented above and elaboration of new ones, which together may constitute a theory of disturbance, require more than careful selection of relevant parameters and an explicit statement of independent variables. Successful prediction about disturbance also requires a knowledge of what may constrain or enhance the impact of disturbance in a given situation. The following section discusses some of the possible contexts of disturbance and their implications.

C. Contexts and Implications of Disturbance

1. System Structure

"System structure" refers to the amount and disposition of biomass relative to the substrate and the degree of connectedness of the biomass to the substrate. These

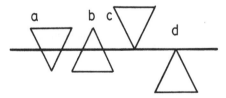

Fig. 1. Diagram of four contrasting sorts of community structure showing disposition of biomass relative to the substrate and degree of attachment of the organisms to the substrate. (a) Shoot-biased community having the preponderance of biomass above the substrate. Resource levels and community attributes are altered most by disturbances that disrupt the above ground portion of the community. (b) A root-biased community having the majority of biomass arrayed within the substrate. Disturbance to the within-substrate component will influence resources and community structure most significantly. (c) A surface-attached community having all biomass above the substrate and superficial attachment to the substrate. Although the terminology relies on plant structure, animal communities can have analogous structures; indeed, surface-attached communities are common among invertebrate animals. Communities of burrowing animals would form a fourth type, (d) substrate contained.

features of a community or ecosystem can determine how a particular disturbance event might alter a system. Four classes of structure can be abstracted from the systems surveyed in this book (Fig. 1). Both sessile animals and plant communities should be assignable to one of these types.

The structure of a system will determine (a) what sorts of disturbance may have an impact, (b) the threshold of intensity that is effective and (c) the dependence of species coexistence on disturbance. For example, above-substrate disturbances will alter resource availability most in "shoot-biased" systems but little in "root-biased" systems. Furthermore, disturbance of insufficient intensity to open the root mat in root-biased systems will have little impact on species coexistence in those systems.

2. Resource Base

The amount of resources available to the organisms at a particular site affect many system and population attributes and thus the sensitivity of the system to disturbance. The stature of the community, mass, and placement of storage organs may all be influenced by resource availability. For example, the structure of terrestrial systems in resource-limited environments may tend to be root biased. Another feature of some systems that may be affected by resource flux is fuel accumulation.

The rate of regrowth after disturbances is determined by the resource base. The effect of disturbance of a given intensity and frequency may be different in resource-limited systems than in resource-rich sites. Species richness may be reduced in systems experiencing a severe disturbance regime but having low resource bases. Since environmental stress may, by affecting resource uptake and productivity, also govern system structure, its role may be very similar to that of the resource base.

A particularly telling example of the role of the resource base in determining the impact of disturbance appears in the ancient primary succession on sand dune systems in eastern Australia (Walker *et al.,* 1981). During this long succession, nutrients are reduced overall and are leached below the rooting depth. Early in the succession, fire and other disturbances might cause a pulse of available nutrients, while on the very oldest dunes, fire might cause a net loss of nutrients and the biomass of the system vegetation might not recover to predisturbance levels (Walker *et al.,* 1981).

3. Life History

The life history strategies and assimilative capacities of species present or potentially available to a site can influence the role of disturbance in a system. "Life history strategy" refers to the genetically determined rate of growth, allocation of assimilate, structure, and timing of the life cycle events of organisms. Such features of organisms determine the timing and magnitude of their resource demands and the probability of success in reproduction, survival, and growth. Whether species grow rapidly, or store resources or not, will have much to do with how they respond to disturbance. Furthermore, their reproductive characteristics, including time needed to mature, vegetative versus sexual reproductive effort and output, and dispersability and persistence of propagules, all influence the response to disturbance. Not only may all of these features determine the capacity of a species to resist a potential disturbance or respond after disturbance, they may evolve in response to particular disturbance regimes.

Growth rates, longevity, dispersal, and other life history characteristics are important in determining disturbance as well as responding to it. All else being equal, the faster the growth rate, the greater the disturbance frequency in marine invertebrate communities (Connell and Keough, Chapter 8, this volume). Since these parameters (e.g., longevity) are also under environmental or biological constraint, biological characteristics per se may control or influence the disturbance rate. If life history traits are under selective control, the possibility of feedback control of disturbance frequency is stronger. The importance of biological factors then blurs the distinction, admittedly semantic, between outside and within-community factors in disturbance.

4. Competitive Abilities

One of the most common tacit assumptions in the disturbance literature is that competitive hierarches exist. Without the existence of linear competitive hierarchies, disturbance is unlikely to enhance or maintain species richness. Whether competitive hierarchies are absent, strictly transitive, cyclic or not will greatly determine the outcome of disturbance in a system. In systems where there are no clear linear competitive hierarchies, disturbance may influence community composition and structure but have no consistent effect on coexistence (Connell and Keough, Chapter 8, this volume).

Connell and Keough (Chapter 8, this volume) rightly stress that the structure of competitive hierarchies and the symmetry of competitive outcomes is fundamental to the role of patch dynamics in the structure of communities. Traditional plant ecology assumes a transitive competitive hierarchy and predictable asymmetrical outcomes. On the other hand, marine invertebrate competitive relations do not follow such a simple and predictable path. The result is that the mix of species may not be the result of competition-mediated niche differences.

The view of species as interchangeable in their role in community dynamics after disturbance has been espoused for terrestrial plant communities. Raup (1975) has suggested that shoreline species in his study area are essentially randomly mixed, with no repeating associations or relation to microsite. Brokaw (Chapter 4, this volume) and Denslow (Chapter 17, this volume) comment on this problem for diverse tropical forest. The issue is whether species have discrete regeneration niches (Grubb, 1977) or are randomly and interchangeably mixed after disturbance. For example, in a diverse tropical forest stand with small scale gap turnover of the canopy, are the species adapted to gap regeneration or is reproduction haphazard with regard to gap conditions?

5. Landscape Characteristics

How a particular disturbance event may alter a system or affect the distribution of resources, and the distribution and coexistence of species in an area, will be greatly influenced by the nature of the landscape. Landscape may be divided into elements generated by topography, substrate conditions, organisms (including people), or disturbance (Forman and Godron, 1981). To what extent a particular event will alter the resource base and environmental conditions, and affect organism survival and migration, will be determined in large part by the identity, size, isolation, and composition of the landscape elements present. If landscape elements are isolated in a biotically or abiotically very different matrix, the result of a particular disturbance regime may be very different from that in a landscape without such barriers. Likewise, the availability of colonizers after disturbance may be affected by landscape configuration (McDonnell and Stiles, 1983).

The topographic heterogeneity of a landscape may have major significance in defining disturbance regimes. Different topographic positions may have very different disturbance regimes (Runkle, Chapter 2; Shugart and Seagle, Chapter 20; this volume). Topography may also affect the resource base, species composition, population persistence, and migration of individuals, and thus control the outcome of disturbance.

Other environmental gradients also influence the impact of disturbance. Connell and Keough (Chapter 8, this volume) and Sousa (Chapter 7, this volume) have shown that the effects of a given disturbance vary with water depth. Harmon et al. (1983) have shown that disturbances can be differentiated on Whittaker-type (Whittaker, 1956) mosaic charts. It is likely that applications of direct gradient analysis that omit consideration of changing disturbance regimes will be unable to explain community continuance completely.

Whether any landscapes are or have been in equilibrium with a disturbance regime is unknown. Zedler and Goff (1973) demonstrated that the population structure of an early successional tree was stable if large enough blocks of land were analyzed. Large areas included a mosaic of recently disturbed and undisturbed stands. To generalize this situation, if patch birth and death rates are in balance, landscapes may be in an overall steady state (Paine and Levin, 1981). The disturbance regime of such a landscape is stable. Change on any one site is, in the clearest cases, cyclic and is mediated by recurrent disturbance.

On the other hand, where the spatial scale of disturbances is large relative to any consistent unit of landscape or vegetation (Caine and Swanson, 1983; Romme, 1982; Romme and Knight, 1981), equilibrium in landscapes may be prevented. This is exemplified by forests in western North American mountains. Further examples in which disturbance does not produce cyclic, steady-state change include the interaction of bamboo demography and tree regeneration (Veblen, Chapter 3, this volume) and disturbances that result in permanent directional trends in site resources (Walker et al., 1981). Climatic shifts, which influence disturbance regimes, occur on the same time scales as some disturbances themselves (Neilson and Wullstein, 1983) and may prevent equilibration with the disturbance regime.

The greatest likelihood of a steady state in the shifting mosaic occurs in systems in which disturbance is frequent and small in scale relative to an otherwise homogeneous area of habitat, and in which feedback mechanisms influence disturbance frequency. An example of such a feedback mechanism is the increasing flammability of forest stands with time since the last fire (Mutch, 1970) and the increasing vulnerability to disturbance that organisms experience as they age and increase in size (e.g., the stress on algae holdfasts increases with increasing algae biomass; Sousa, Chapter 7, this volume).

The various contexts and implications of disturbance outlined above are some of the more obvious ones. There may be others that act in particular systems, and in any event, the combination of these various attributes is likely to be system dependent. It is clear that there can be interactions and feedbacks among them. Thus, consideration of contexts of disturbance must be multivariate. These specific suites of features must be considered in making sound mechanistic predictions about the role of disturbance. Statements of hypotheses without specification of at least the important dimensions of the context of disturbance will not permit meaningful comparisons among systems and disturbance regimes or lead to a viable theory of disturbance.

IV. CONCLUSION

The contents of this volume and the literature upon which it draws lead to these conclusions:

1. Disturbance is common to many different systems. It functions or has functioned at all temporal and spatial scales and levels of organization of ecological and evolutionary interest.

2. The key processes common to all disturbances are alterations of resource availability and system structure.

3. Although an understanding of disturbance is of crucial importance in ecology, no coherent theory exists to further its study. Two major generalizations, one concerning intermediate disturbance intensity and the other the rate of competitive exclusion relative to disturbance frequency, are basic to the embryonic theory. Concepts of disturbance regime and of patch dynamics form a basic framework in which comparative and quantitative studies of disturbance should be couched.

4. In order to develop a theory of disturbance composed of unambiguous, testable hypotheses and capable of making sound mechanistic predictions, the relevant variables of disturbance must be established. These include at least magnitude, frequency, size, and dispersion.

5. The development of disturbance theory also requires an explicit statement of the parameters of systems that can respond to disturbance. Predictions must be made in terms of the variables of disturbance and the response parameters.

6 Predictions about disturbance must also recognize that disturbance will have a particular context that may enhance or constrain its impact in a given situation. At

least the following factors are required to define the multivariate context of disturbance: (a) system structure, (b) resource base, (c) life history characteristics, (d) nature of the competitive hierarchy, and (e) landscape composition and configuration.

An explicit statement of the parameters that respond to disturbance, the variables that determine the impact of disturbance, and consideration of the context and constraints of disturbance, can form the basis of a theory of disturbance. Placing studies of disturbance in various systems, particularly understudied ones, in this framework will further our ability to generalize and make appropriate predictions about disturbance.

ACKNOWLEDGMENTS

Many of the ideas expressed in this chapter were developed in the course of research supported by the NSF through grant DEB 80-03504 to S.T.A.P.

Bibliography

Abbott, D. T., and Crossley, D. A., Jr. (1982). Woody litter decomposition following clear-cutting. *Ecology* **63,** 35–42.

Abbott, M. B. (1973). Seasonal diversity and density in bryozoan populations of Block Island Sound (New York, U.S.A.). *In* "Living and Fossil Bryozoa" (G. P. Larwood, ed.), pp. 37–52. Academic Press, New York.

Aber, J. D., Melillo, J. M., and Federer, C. A. (1982). Predicting the effects of rotation length, harvest intensity, and fertilization on fiber yield from northern hardwood forests in New England. *For. Sci.* **28,** 31–45.

Abrams, M. D., Sprugel, D. G., and Dickmann, D. I. (1984). Multiple pathways in early successional development following disturbance to jack pine sites in northern Lower Michigan. *For. Ecol. Manage.* (in press).

Abramsky, Z. (1978). Small mammal community ecology. Changes in species diversity in response to manipulated productivity. *Oecologia* **34,** 113–123.

Abrell, D. B., and Jackson, M. T. (1977). A decade of change in an oldgrowth beech–maple forest in Indiana. *Am. Midl. Nat.* **98,** 22–32.

Abugov, R. (1982). Species diversity and phasing of disturbance. *Ecology* **63,** 289–293.

Acevedo, M. (1981). On Horn's model of forest dynamics with particular reference to tropical forests. *Theor. Popul. Biol.* **19,** 230–250.

Acevedo, M., and Marquis, R. (1978). "A Survey of the Light Gaps of the Tropical Rain Forest at Llorona, Peninsula de Osa," Book 78.3. Organ. Trop. Stud., San Jose, Costa Rica.

Adams, F., Ewing, P. A., and Huberty, M. R. (1947). "Hydrologic Aspects of Burning Brush and Woodland Grass Ranges in California." Calif. Div. For., Sacramento.

Adams, S., Strain, B. R., and Adams, M. S. (1970). Water-repellent soils, fire, and annual plant cover in a desert scrub community at southeastern California. *Ecology* **51,** 696–700.

Adams, S. N., Jack, W. A., and Dickson, D. A. (1970). The growth of Sitka spruce on poorly drained sites in Northern Ireland. *Forestry* **43,** 125–133.

Addicott, J. F. (1978a). Niche relationships among species of aphids feeding on fireweed. *Can. J. Zool.* **56,** 1837–1841.

Addicott, J. F. (1978b). The population dynamics of aphids on fireweed: a comparison of local populations and metapopulations. *Can. J. Zool.* **56,** 2554–2564.

Ahlgren, C. E. (1974). Effects of fires on temperate forests: North Central United States. "Fire and Ecosystems" (T. T. Kozlowski and C. E. Ahlgren, eds.), pp. 195–223. Academic Press, New York.

Ahlgren, I. F., and Ahlgren, C. E. (1960). Ecological effects of forest fires. *Bot. Rev.* **46,** 304–310.

Alagarswami, K., and Chellam, A. (1976). On fouling and boring organisms and mortality of pearl oysters in the farm at Veppalodai, Gulf of Mannar, India. *Indian J. Fish.* **23,** 10–22.

Albini, F. A., and Anderson, E. B. (1982). Predicting fire behavior in U.S. Mediterranean ecosystems. *In* "Dynamics and Management of Mediterranean-Type Ecosystems" (C. E. Conrad and W. C. Oechel, eds.), *USDA For. Serv. Gen. Tech. Rep. PSW* **PSW-58,** pp. 483–489.

Aldous, S. E. (1941). Deer management suggestions from northern white cedar types. *J. Wildl. Manage.* **5,** 90–94.

Aldous, S. E. (1952). Deer browse clipping study in the Lake States Region. *J. Wildl. Manage.* **16,** 401–409.

Alexandre, D.-Y. (1977). Régénération naturelle d'un arbre caractéristique de la forêt équatoriale de Côte d'Ivoire, *Turreanthus africana* Pellegr. *Oecol. Plant.* **12,** 241–262.

Alexandre, D.-Y. (1978). Observations sur l'écologie de *Trema guineensis* en basse Côte d'Ivoire. *Cah. ORSTOM,* Ser. Biol. **13,** 261–266.

Alexandre, D.-Y. (1982a). Aspects de la régénération naturelle en forêt dense de Côte d'Ivoire. *Candollea* **37,** 579–588.

Alexandre, D.-Y. (1982b). Pénétration de la lumière au niveau du sous-bois d'une forêt dense tropicale. *Ann. Sci. For.* **39,** 419–438.

Al-Hussaini, A. H. (1947). The feeding habits and the morphology of the alimentary tract of some teleosts living in the neighborhood of the Marine Biological Station, Ghardaqa, Red Sea. *Publ. Mar. Biol. Stn., Ghardaqa, Red Sea* **5,** 1–61.

Allard, R. W., Kahler, A. L., and Clegg, M. T. (1977). Estimation of mating cycle components of selection in plants. "Measuring Selection in Natural Populations" (F. B. Christiansen and T. M. Fenchel, eds.), pp. 1–20. Springer-Verlag, Berlin and New York.

Allee, W. C., Emerson, A. E., Park, O., and Schmidt, K. P. (1949). "Principles of Animal Ecology." Saunders, Philadelphia, Pennsylvania.

Allen, S. E. (1971). Chemical aspects of heather burning. *J. Appl. Ecol.* **1,** 347–367.

Allen, T. F. H., and Starr, T. B. (1982). "Hierarchy: Perspectives for Ecological Complexity." Univ. of Chicago Press, Chicago, Illinois.

Allen, T. F. H., Bartell, S. M., and Koonce, J. F. (1977). Multiple stable configurations in ordination of phytoplankton community change rates. *Ecology* **58,** 1076–1084.

Almeyda A., E., and Saez S., F. (1958). "Recopilación de Datos Climáticos de Chile y Mapas Sinópticos Respectivos." Minist. Agric., Santiago, Chile.

Anaya L., A. L. (1976). Consideraciones sobre el potencial alelopático de la vegetación secundaria. *In* Regeneración des Selvas" (A. Gómez-Pompa, C. Vázquez-Yanes, S. del Amo C. and A. Butanda C., eds.), pp. 428–446. Compañia Editorial Continental, Mexico City, Mexico.

Andersen, D. C., MacMahon, J. A., and Wolfe, M. L. (1980). Herbivorous mammals along a montane sere: community structure and energetics. *J. Mammal.* **61,** 500–519.

Anderson, G. D., and Walker, B. H. (1974). Vegetation composition and elephant damage in the Sengwa wildlife research area, Rhodesia. *J. S. Afr. Wildl. Manage. Assoc.* **4,** 1–14.

Anderson, G. W., and Anderson, R. L. (1963). The rate of spread of oak wilt in the Lake States. *J. For.* **63,** 823–825.

Anderson, J. A. R. (1964). Observations on climatic damage in peat swamp forests in Sarawak. *Emp. For. Rev.* **43,** 145–158.

Anderson, J. M., and Swift, M. J. (1983). Decomposition in tropical forests. *In* "The Tropical Rain Forest" (S. L. Sutton, T. C. Whitmore, and A. C. Chadwick, eds.), pp. 287–309. Blackwell, Oxford.

Anderson, R. M., Turner, B. D., and Taylor, L. R., eds. (1979). "Population Dynamics." *Symp. Br. Ecol. Soc.* **20.**

Anderson, R. M., Gordon, D. M., Crawley, M. J., and Hassell, M. P. (1982). Variability in the abundance of animal and plant species. *Nature (London)* **296,** 245–248.

Andersson, M. (1978). Optimal foraging area: size and allocation of search effort. *Theor. Popul. Biol.* **13,** 397–409.

Andrewartha, H. G., and Birch, L. C. (1954). "The Distribution and Abundance of Animals." Univ. of Chicago Press, Chicago, Illinois.

Andrews, R. M., and Rand, A. S. (1982). Seasonal breeding and long-term fluctuations in the lizard *Anolis limifrons*. *In* "The Ecology of a Tropical Forest: Seasonal Rhythms and Long-term Changes" (E. G. Leigh, Jr., A. S. Rand, and D. M. Windsor, eds.), pp. 405–412. Smithsonian Inst. Press, Washington, D.C.

Andrews, R. M., Rand, A. S., and Guerrero, S. (1983). Seasonal and spatial variation in the annual cycle of a tropical lizard. *In* "Advances in Herpetology and Evolutionary Biology: Essays in Honor of Ernest E. Williams" (A. G. J. Rhodin and K. Miyata, eds.), pp. 441–454. Mus. Comp. Zool., Cambridge, Massachusetts.

Angermeier, P. L., and Karr, J. R. (1983). Fish communities along environmental gradients in a system of tropical streams. *Environ. Biol. Fish* **9,** 117–135.

Angus, R. A., and Schultz, R. J. (1983). Geographical dispersal and clonal diversity in unisexual fish populations. *Am. Nat.* **115,** 531–550.

Annest, R. A., and Templeton, A. R. (1978). Genetic recombination and clonal selection in *Drosophila mercatorum*. *Genetics* **89,** 193–210.

Anonymous (1973). A resource and management survey of the Cotter River catchment. Report for For. Branch, Dep. Capital Territ. by Consultancy Group, Dep. For., Aust. Natl. Univ., Canberra.

Antonovics, J. (1976). The population genetics of mixtures. *In* "Plant Relations in Pastures" (J. Wilson, ed.), pp. 233–252. CSIRO, Brisbane, Australia.

Antonovics, J., and Levin, D. A. (1981). The ecological and genetic consequences of density-dependent regulation in plants. *Annu. Rev. Ecol. Syst.* **11,** 411–452.

Arianoutsou, M., and Margaris, N. S. (1981a). Early stages of regeneration after fire in a phryganic ecosystem (east Mediterranean). I. Regeneration by seed germination. *Biol. Ecol. Mediterr.* **8,** 119–128.

Arianoutsou, M., and Margaris, N. S. (1981b). Producers and the fire cycle in a phryganic ecosystem. *In* "Components of Productivity of Mediterranean-Climate Regions—Basic and Applied" (N. S. Margaris and H. A. Mooney, eds.), pp. 181–190. Junk, The Hague.

Arianoutsou, M., and Margaris, N. S. (1982). Phryganic (east Mediterranean) ecosystems and fire. *Ecol. Mediterr.* **8,** 473–480.

Armesto, J. J., and Pickett, S. T. A. (1984). Experiments on disturbance in old-field plant communities: Impact on species richness and abundance. *Ecology* **65,** in press.

Armson, K. A., and Fessenden, R. J. (1973). Forest windthrows and their influence on soil morphology. *Soil Sci. Soc. Am. Proc.* **37,** 781–783.

Armstrong, R. A. (1976). Fugitive species: experiments with fungi and theoretical considerations. *Ecology* **57,** 953–963.

Ash, J. E., and Barkham, J. P. (1976). Changes and variability in the field layer of a coppiced woodland in Norfolk, England. *J. Ecol.* **64,** 697–712.

Ashton, D. H. (1981). Tall open forests. *In* "Australian Vegetation" (R. H. Groves, ed.), pp. 121–151. Cambridge Univ. Press, London and New York.

Ashton, P. S. (1969). Speciation among tropical forest trees: some deductions in light of recent evidence. *Biol. J. Linn. Soc.* **1,** 155–196.

Ashton, P. S. (1978). Crown characteristics of tropical trees. *In* "Tropical Trees as Living Systems" (P. B. Tomlinson and M. H. Zimmerman, eds.), pp. 591–615. Cambridge Univ. Press, London and New York.

Ashton, P. S. (1981). The need for information regarding tree age and growth in tropical trees. *In* "Age and Growth Rate of Tropical Trees'' (F. H. Bormann and G. Berlyn, eds.), *Bull. Yale Univ., Sch. For. Environ. Stud.* No. 94, pp. 3–6.

Ashton, P. S., Riswan, S., and Kenworthy, J. B. (1984). Rainforests of the Far East and Latin America: some comparisons. Manuscript.

Aston, A. R., and Gill, A. M. (1976). Coupled soil moisture, heat and water vapour transfers under simulated fire conditions. *Aust. J. Soil Sci.* **14**, 55–66.

Aubréville, A. (1938). La forêt coloniale: Les forêts de l'Afrique occidentale française. *Ann. Acad. Sci. Colon. Paris* **9.**

Aubréville, A. (1971). Regeneration patterns in the closed forest of Ivory Coast. *In* ''World Vegetation Types'' (S. R. Eyre, ed.), pp. 41–55. Columbia Univ. Press, New York.

Auchmoody, L. R. (1979). Nitrogen fertilization stimulates germination of dormant pin cherry seed. *Can. J. For. Res.* **9**, 514–516.

Auclair, A. N. D. (1975). Sprouting response in *Prunus serotina* Erhr: Multivariate analysis of site forest structure and growth rate relationships. *Am. Midl. Nat.* **94**, 72–87.

Auclair, A., and Goff, F. G. (1971). Diversity relations of upland forests in the Western Great Lakes area. *Am. Nat.* **105**, 499–528.

Augspurger, C. K. (1983). Seed dispersal of the tropical tree *Platypodium elegans,* and the escape of its seedlings from fungal pathogens. *J. Ecol.* **71**, 759–771.

Ayling, A. L. (1983a). Factors affecting the spatial distributions of thinly encrusting sponges from temperate waters. *Oecologia* **60**, 412–418.

Ayling, A. L. (1983b). Growth and regeneration rates in thinly encrusting demospongiae from temperate waters. *Biol. Bull. (Woods Hole, Mass.)* **165**, 343–352.

Ayling, A. M. (1981). The role of biological disturbance in temperate subtidal encrusting communities. *Ecology* **62**, 830–847.

Ayyad, M. A. (1981). Soil–vegetation–atmosphere interactions. *In* ''Arid-land Ecosystems: Structure, Functioning and Management'' (D. W. Goodall, R. A. Perry, and K. M. W. Howes, eds.), Vol. 2, pp. 9–31. Cambridge Univ. Press, London and New York.

Bak, R. P. M., Brouns, J. J. W. M., and Heys, F. M. L. (1977). Regeneration and aspects of spatial competition in the scleractinian corals *Agaricia agaricites* and *Montastrea annularis. Proc.—Int. Coral Reef Symp., 3rd, Miami, Fla.* **1**, 143–149.

Bak, R. P. M., Sybesma, J., and Van Duyl, F. C. (1981). The ecology of the tropical compound ascidian *Trididemnum solidum.* II. Abundance, growth and survival. *Mar. Ecol.: Prog. Ser.* **6**, 43–52.

Baker, H. G. (1965). Characteristics and modes of origin of weeds. *In* ''Genetics of Colonizing Species'' (H. G. Baker and G. L. Stebbins, eds.), pp. 147–172. Academic Press, New York.

Baker, H. G. (1974). The evolution of weeds. *Annu. Rev. Ecol. Syst.* **5**, 1–24.

Baker, H. G., and Stebbins, G. L., eds. (1965). ''The Genetics of Colonizing Species.'' Academic Press, New York.

Bannister, B. A. (1970). Ecological life cycle of *Euterpe globosa. In* ''A Tropical Rainforest'' (H. Odum and R. F. Pigeon, eds.), pp. B299–B314. USAEC, Washington, D.C.

Barbour, M. A., and DeJong, T. M. (1977). Response of West Coast beach taxa to salt spray, seawater inundation, and soil salinity. *Bull. Torrey Bot. Club* **104**, 29–34.

Barclay-Estrup, P. (1970). The description and interpretation of cyclical processes in a heath community. II. Changes in biomass and shoot production during the *Calluna* cycle. *J. Ecol.* **58**, 243–249.

Barclay-Estrup, P., and Gimingham, C. H. (1969). The description and interpretation at cyclical processes in the heath community. I. Vegetational change in relation to the *Calluna* cycle. *J. Ecol.* **57**, 737–758.

Bard, G. E. (1952). Secondary succession on the Piedmont of New Jersey. *Ecol. Monogr.* **22**, 195–215.

Barden, L. S. (1979). Tree replacement in small canopy gaps of a *Tsuga* canadensis forest in the southern Appalachians, Tennessee. *Oecologia* **44**, 141–142.

Barden, L. S. (1980). Tree replacement in a cove hardwood forest of the southern Appalachians. *Oikos* **35**, 16–19.

Barden, L. S. (1981). Forest development in canopy gaps of a diverse hardwood forest of the southern Appalachian Mountains. *Oikos* **37**, 205–209.

Barden, L. S., and Woods, F. W. (1976). Effects of fire on pine and pine-hardwood forests in the southern Appalachians. *For. Sci.* **22**, 399–403.

Barkham, J. P. (1980a). Population dynamics of the wild daffodil (*Narcissus pseudonarcissus*). I. Clonal growth, seed reproduction, mortality and the effects of density. *J. Ecol.* **68**, 607–633.

Barkham, J. P. (1980b). Population dynamics of the wild daffodil (*Narcissus pseudonarcissus*). II. Changes in number of shoots and flowers, and the effect of bulb depth on growth and reproduction. *J. Ecol.* **68**, 635–664.

Barnes, H., and Powell, H. T. (1950). The development, general morphology and subsequent elimination of barnacle populations, *Balanus crenatus* and *B. balanoides,* after a heavy initial settlement. *J. Anim. Ecol.* **19**, 175–179.

Barnes, H., and Topinka, J. A. (1969). Effect of the nature of the substratum on the force required to detach a common littoral alga. *Am. Zool.* **9**, 753–758.

Bartholomew, B. (1970). Bare zones between California shrub and grassland communities: the role of animals. *Science (Washington, D.C.)* **170**, 1210–1212.

Barton, A. M. (1984). The partitioning of treefall gaps between tropical pioneer and shade tolerant tree species. Manuscript.

Bauer, H. L. (1936). Moisture relations in the chaparral of the Santa Monica Mountains, California. *Ecol. Monogr.* **6**, 409–454.

Baur, G. N. (1964). "The Ecological Basis of Rainforest Management." For. Comm., Sydney, New South Wales, Australia.

Baxter, F. P., and Hole, F. D. (1967). Ant (*Formicia cinerea*) pedoturbation in a prairie soil. *Soil Sci. Soc. Am. Proc.* **31**, 425–428.

Bazzaz, F. A. (1975). Plant species diversity in old field ecosystems in southern Illinois. *Ecology* **56**, 485–488.

Bazzaz, F. A. (1979). The physiological ecology of plant succession. *Annu. Rev. Ecol. Syst.* **10**, 351–371.

Bazzaz, F. A. (1983). Characteristics of populations in relation to disturbance in natural and man-modified ecosystems. *In* "Disturbance and Ecosystems" (H. A. Mooney and M. Godron, eds.), pp. 259–277. Springer-Verlag, Berlin and New York.

Bazzaz, F. A., and Carlson, R. W. (1982). Photosynthetic acclimation to variability in light environment of early and late successional plants. *Oecologia* **54**, 313–316.

Bazzaz, F. A., and Pickett, S. T. A. (1980). The physiological ecology of tropical succession: a comparative review. *Annu. Rev.Ecol. Syst.* **11**, 287–310.

Bazzaz, F. A., Levin, D. A., and Schmierbach, M. R. (1982). Differential survival of genetic variants in crowded populations in *Phlox. J. Appl. Ecol.* **19**, 891–900.

Beals, E. W. (1969). Vegetational change along altitudinal gradients. *Science (Washington, D.C.)* **165**, 981–985.

Beard, J. S. (1945). The progress of plant succession on the Soufriere of St. Vincent. *J. Ecol.* **33**, 1–9.

Beattie, A. J. (1971). Itinerant pollinators in a forest. *Madrono* **21**, 120–124.

Beattie, A. J., and Lyons, N. (1975). Seed dispersal in *Viola* (Violaceae): Adaptations or strategies. *Am. J. Bot.* **62**, 714–722.

Beck, D. E., and Della-Bianca, L. (1981). Yellow-poplar: characteristics and management. *U.S. Dep. Agric., Agric. Handb.* No. 583.

Bell, C. R. (1970). Seed distribution and germination experiment. *In* "A Tropical Rainforest" (H. Odum and R. F. Pigeon, eds.), pp. D177–D182. USAEC, Washington, D.C.

Bell, D. T., and del Moral, R. (1977). Vegetation gradients in the stream-side forest of Hickory Creek, Will County, Illinois. *Bull. Torrey Bot. Club* **104**, 127–135.

Bell, R. H. V. (1970). The use of the herb layer by grazing ungulates in the Serengeti. *In* "Animal Populations in Relation to Their Food Resources" (A. Watson, ed.), pp. 111–123. Blackwell, Oxford.

Bender, E. A., Case, T. J., and Gilpin, M. E. (1984). Perturbation experiments in community ecology: theory and practice. *Ecology* **65**, 1–13.

Benson, J. A. (1983). Physical disturbance and the community structure of epibionts on *Cystoseira neglecta*. Ph.D. Thesis, Dep. Biol., Univ. of California at Los Angeles.

Bentley, J. R., and Fenner, R. L. (1958). Soil temperatures during burning related to post fire seed beds on woodland range. *J. For.* **56**, 737–740.

Berger, E. (1976). Heterosis and the maintenance of enzyme polymorphism. *Am. Nat.* **110**, 823–839.

Bernstein, B. B., and Jung, N. (1979). Selective pressures and coevolution in a kelp canopy community in southern California, U.S.A. *Ecol. Monogr.* **49**, 335–355.

Bernstein, F. (1931). Die geographische Verteilung der Blutgruppen und ihre autropolögische Bedeutung. *In* "Comitato Italiano per lo Studio dei Problemi della Populazione." Instituto Poligrafico dello Stato, Rome.

Berry, A. B. (1964). Effect of strip width on proportion of daily light reaching the ground. *For. Chron.* **40**, 130–131.

Berryman, A. A. (1981). "Population Systems: A General Introduction." Plenum, New York.

Berryman, A. A., and Wright, L. C. (1978). Defoliation, tree condition, and bark bettles. *In* "The Douglas-fir Tussock Moth: A Synthesis" (M. H. Brookes, R. W. Stark, and R. W. Campbell, eds.), *U.S. For. Serv., Pac. Northwest For. Range Exp. Stn., Tech. Bull.* No. 1585, pp. 81–87.

Bertram, B. C. R. (1979). Serengeti predators and their social systems. *In* "Serengeti: Dynamics of an Ecosystem" (A.R.E. Sinclair and M. Norton-Griffiths, eds.), pp. 221–248. Univ. of Chicago Press, Chicago, Illinois.

Bierzychudek, P. (1982). The demography of jack-in-the-pulpit, a forest perennial that changes sex. *Ecol. Monogr.* **52**, 335–351.

Billings, W. D., and Peterson, K. M. (1980). Vegetational change and ice-wedge polygons through the thaw-lake cycle in Arctic Alaska. *Arct. Alp. Res.* **12**, 413–432.

Birk, E. M. (1983). Nitrogen availability, N cycling and N use efficiency on the Savannah River Plant. *Bull. Ecol. Soc. Am.* **64**, 93.

Birkeland, C. (1982). Terrestrial runoff as a cause of outbreaks of *Acanthaster planci* (Echinodermata: asteroidea). *Mar. Biol. (Berlin)* **69**, 175–185.

Bjorkbom, J. C., and Larson, R. G. (1977). The Tionesta scenic and research natural areas. *USDA For. Ser. Cen. Tech. Rep. NE* **NE-31.**

Bjorkman, O. (1968). Further studies on differentiation of photosynthetic properties in sun and shade ecotypes of *Solidago virgaurea*. *Physiol. Plant.* **21**, 84–99.

Bjorkman, O., and Holmgren, P. (1966). Photosynthetic adaptation to light intensity in plants native to shade and exposed habitats. *Physiol. Plant.* **19**, 854–889.

Black, J. N. (1957). The early vegetative growth of three strains of subterranean clover (*Trifolium subterraneum* L.) in relation to size of seed. *Aust. J. Agric. Res.* **8**, 1–14.

Black, R. (1976). The effects of grazing by the limpet, *Acmaea insessa,* on the kelp, *Egregia laevigata,* in the intertidal zone. *Ecology* **57**, 265–277.

Blackman, G. E., and Rutter, A. J. (1949). Physiological and ecological studies in the analysis of plant environment. IV. The interaction between light intensity and mineral nutrient on the uptake of nutrients by the bluebell (*Scilla non-scripta*). *Ann. Bot. (London)* **13**, 453–489.

Blackman, G. E., and Rutter, A. J. (1950). Physiological and ecological studies in the analysis of plant environment. V. An assessment of the factors controlling the distribution of the bluebell (*Scilla non-scripta*) in different communities. *Ann. Bot. (London)* **14**, 407–420.

Blais, J. R. (1954). The recurrence of spruce budworm infestations in the past century in the Lac Seul area of northwestern Ontario. *Ecology* **35**, 62–71.

Blaisdell, J. P. (1953). Ecological effects of planned burning of sagebrush-grass range on the upper Snake River Plains, *U.S. Dep. Agric., Tech. Bull.* No. 1075.

Blank, J. L., Olson, R. K., and Vitousek, P. M. (1980). Nutrient uptake by a diverse spring ephemeral community. *Oecologia* **47**, 96–98.

Bledsoe, L. J., and Van Dyne, G. M. (1971). A compartment model simulation of secondary succession.

In "Systems Analysis and Simulation in Ecology" (B. M. Patten, ed.), pp. 479–511. Academic Press, New York.

Bliss, L. C. (1979). Arctic heathlands. *In* "Ecosystems of the World. Vol. 9A: Heathlands and Related Shrublands" (R. L. Specht, ed.), pp. 415–424. Elsevier, Amsterdam.

Bloom, S. A. (1975). The motile escape response of a sessile prey: a sponge-scallop mutualism. *J. Exp. Mar. Biol. Ecol.* **17,** 311–321.

Blum, K. E. (1968). Contributions toward an understanding of vegetational development in the Pacific lowlands of Panama. Ph.D. Thesis, Florida State Univ., Tallahassee.

Blum, M. S. (1980). Arthropods and ecomones: better fitness through ecological chemistry. *In* "Animals and Environmental Fitness" (R. Gilles, ed.), pp. 207–222. Pergamon, New York.

Boaden, P. J. S., O'Connor, R. J., and Seed, R. (1976). The fauna of a *Fucus serratus* L. community: ecological isolation in sponges and tunicates. *J. Exp. Mar. Biol. Ecol.* **21,** 249–267.

Boag, P. T., and Grant, P. R. (1981). Intense natural selection in a population of Darwin's finches (Geospizinae) in the Galapagos. *Science (Washington, D.C.)* **214,** 82–84.

Boal, J. (1980). Pacific harbor seal (*Phoca vitulina richardii*). Haul out impact on the rocky midtidal zone. *Mar. Ecol.: Prog. Ser.* **2,** 265–269.

Boardman, N. K. (1977). Comparative photosynthesis of sun and shade plants. *Annu. Rev. Plant Physiol.* **28,** 355–377.

Bock, C. E., and Bock, J. H. (1978). Response of birds, small mammals, and vegetation to burning sacaton grasslands in southeastern Arizona. *J. Range Manage.* **31,** 296–300.

Boerner, R. E. J. (1982). Fire and nutrient cycling in temperate ecosystems. *BioScience* **32,** 187–192.

Boone, R. D. (1982). Patterns of soil organic matter and microclimate accompanying the death and regeneration of a mountain hemlock forest. M.D. Thesis, Oregon State Univ., Corvallis.

Borchert, R. (1976). Size and shoot growth patterns in broadleaved trees. *Cent. Hardwoods For. Conf.* **1,** 221–230.

Borchert, R., and Slade, N. A. (1981). Bifurcation ratios and the adaptive geometry of trees. *Bot. Gaz.* **142,** 394–401.

Boring, L. R., Monk, C. D., and Swank, W. T. (1981). Early regeneration of a clear-cut southern Appalachian forest. *Ecology* **62,** 1244–1253.

Bormann, F. H., and Likens, G. E. (1979). "Pattern and Process in a Forested Ecosystem." Springer-Verlag, Berlin and New York.

Bormann, R. E. (1982). Agricultural disturbance and forest recovery at Mt. Cilley. Ph.D. Thesis, Yale Univ., New Haven, Connecticut.

Bothwell, A. (1983). Fragmentation, a means of asexual reproduction and dispersal in the coral genus *Acropora* (Scleractinia: Astrocoeniida: Acroporidae)—a preliminary report. *Proc.—Int. Coral Reef Symp., 4th, Manila, 1981* **2,** 137–144.

Botkin, D. B. (1981). Causality and succession. *In* "Forest Succession: Concepts and Application" (D. C. West, H. H. Shugart, and D. B. Botkin, eds.), pp. 36–55. Springer-Verlag, Berlin and New York.

Botkin, D. B., Janak, J. F., and Wallis, J. R. (1972a). Some ecological consequences of a computer model of forest growth. *J. Ecol.* **60,** 849–873.

Botkin, D. B., Janak, J. F., and Wallis, J. R. (1972b). Rationale, limitations and assumptions of a northeastern forest growth simulator. *IBM J. Res. Dev.* **16,** 101–116.

Bourdo, E. A. (1956). A review of the General Land Office survey and of its use in quantitative studies of former forests. *Ecology* **37,** 754–768.

Bowers, M. A. (1982). Foraging behavior of heteromyid rodents: field evidence of resource partitioning. *J. Mammal.* **63,** 361–367.

Bowker, R. G., and Johnson, O. W. (1980). Thermoregulatory precision in three species of whiptail lizards (Lacertilia: Teiidae). *Physiol. Zool.* **53,** 176–185.

Bowman, R. S., and Lewis, J. R. (1977). Annual fluctuations in the recruitment of *Patella vulgata* L. *J. Mar. Biol. Assoc. U.K.* **57,** 793–815.

Bradshaw, A. D. (1972). Some of the evolutionary consequences of being a plant. *Evol. Biol.* **5,** 25–47.

Bradstock, R. A., and Myerscough, P. J. (1981). Fire effects on seed release and the emergence and establishment of seedlings in *Banksia ericifolia* L.f. *Aust. J. Bot.* **29,** 521–531.

Branch, G. M. (1975). Ecology of *Patella* species from the Cape Peninsula, South Africa. 4. Desiccation. *Mar. Biol. (Berlin)* **32,** 179–188.

Brandani, A. (1984). Report on field work done at San Carlos de Rio Negro, Venezuela. Manuscript.

Bratton, S. P. (1974). The effect of the European wild boar (*Sus scrofa*) on the high-elevation vernal flora in the Great Smoky Mountains National Park. *Bull. Torrey Bot. Club* **101,** 198–206.

Bratton, S. P. (1975). The effect of the European wild boar (*Sus scrofa*) on the gray beech forest in the Great Smoky Mountains. *Ecology* **56,** 1356–1366.

Braun, E. L. (1950). "Deciduous Forests of Eastern North America." Hafner, New York.

Bray, J. R. (1956). Gap phase replacement in a maple–basewood forest. *Ecology* **37,** 598–600.

Breitburg, D. L. (1984). Residual effects of grazing: inhibition of competitor recruitment by encrusting colonial algae. *Ecology* **65,** 1136–1143.

Brokaw, N. V. L. (1980). Gap phase regeneration in a neotropical forest. Ph.D. Thesis, Univ. of Chicago, Chicago, Illinois.

Brokaw, N. V. L. (1982a). Treefalls, frequency, timing and consequences. *In* "Seasonal Rhythms in a Tropical Forest" (E. G. Leigh, A. S. Rand, and D. M. Windsor, eds.), pp. 101–108. Smithsonian Inst. Press, Washington, D.C.

Brokaw, N. V. L. (1982b). The definition of treefall gap and its effect on measures of forest dynamics. *Biotropica* **14,** 158–160.

Brokaw, N. V. L. (1984). Ground layer dominance and apparent inhibition of tree regeneration by *Aechmea magdalenae* (Bromeliaceae) in a tropical forest. *Trop. Ecol.* (in press).

Brokaw, N. V. L. (1985). Gap-phase regeneration in a tropical forest. *Ecology* (in press).

Brooks, J. L., and Dodson, S. I. (1965). Predation, body size, and composition of plankton. *Science (Washington, D.C.)* **150,** 28–35.

Brown, A. H. D. (1979). Enzyme polymorphisms in plant populations. *Theor. Popul. Biol.* **15,** 1–42.

Brown, A. H. D., and Marshall, D. R. (1981). Evolutionary changes accompanying colonization in plants. *Evol. Today, Proc. Int. Congr. Syst. Evol. Biol., 2nd, Vancouver, B.C., 1980* pp. 351–163.

Brown, A. H. D., Feldman, M. W., and Nevo, E. (1980). Multilocus structure of natural populations of *Hordeum spontaneum. Genetics* **96,** 523–536.

Brown, J. H. (1975). Geographical ecology of desert rodents. *In* "Ecology and Evolution of Communities" (M. L. Cody and J. M. Diamond, eds.), pp. 315–341. Belknap Press, Cambridge, Massachusetts.

Brown, J. H., and Kodric-Brown, A. (1977). Turnover rates in insular biogeography: Effect of immigration on extinction. *Ecology* **58,** 445–449.

Brown, J. H., and Lieberman, G. A. (1973). Resource utilization and coexistence of seed-eating desert rodents in sand dune habitats. *Ecology* **54,** 788–797.

Brown, J. H., Reichman, O. J., and Davidson, D. W. (1979). Granivory in desert ecosystems. *Annu. Rev. Ecol. Syst.* **10,** 201–227.

Brown, W. H., and Mathews, D. M. (1914). Philippine dipterocarp forests. *Philipp. J. Sci., Sect. A* **9,** 413–561.

Brubaker, L. B. (1981). Long-term forest dynamics. *In* "Forest Succession: Concepts and Application" (D. C. West, H. H. Shugart, and D. B. Botkin, eds.), pp. 95–106. Springer-Verlag, Berlin and New York.

Brun, R. (1975). Estructura y potencialidad de distintos tipos de bosque nativo en el sur de Chile. *Bosque* **1,** 6–17.

Brunig, E. F. (1964). A study of damage attributed to lightning in two areas of *Shorea albida* forest in Sarawak. *Emp. For. Rev.* **43,** 134–144.

Buckley, R. C. (1981). Scale-dependent equilibrium in highly heterogeneous islands: plant geography of the northern Great Barrier Reef sand cays and shingle islets. *Aust. J. Ecol.* **6**, 143–147.

Budowski, G. (1965). Distribution of tropical American rain forest species in the light of successional processes. *Turrialba* **15**, 40–42.

Buell, M. F., and Buell, H. F. (1975). Moat bogs in the Itasca Park area, Minnesota. *Bull. Torrey Bot. Club* **102**, 6–9.

Bulger, A. J., and Schultz, R. J. (1979). Heterosis and interclonal variation in thermal tolerance in unisexual fish. *Evolution* **33**, 848–859.

Bulger, A. J., and Schultz, R. J. (1982). Origins of thermal adaptation in northern vs. southern populations of a unisexual hybrid fish. *Evolution* **36**, 1041–1050.

Burschel N., P., Gallegos G., C., Martínez M., O., and Moll, W. (1976). Composición y dinámica regenerativa de un bosque virgen mixto de raulí y coigue. *Bosque* **1**, 55–74.

Burton, D. H., Anderson, H. W., and Wiley, L. F. (1969). Natural regeneration of yellow birch in Canada. *In* "Birch Symposium Proceedings" (E. H. Larson, ed.), pp. 55–73. Northeast. For. Exp. Stn., Upper Darby, Pennsylvania.

Buss, L. W. (1980). Competitive intransitivity and size frequency distributions of interacting populations. *Proc. Natl. Acad. Sci. U.S.A.* **77**, 5355–5359.

Buss, L. W. (1981). Group living and the evolution of cooperation in a sessile invertebrate, *Bugula turrita*. *Science (Washington, D.C.)* **213**, 1012–1014.

Buss, L. W., and Jackson, J. B. C. (1979). Competitive networks: non-transitive competitive relationships in cryptic coral reef environments. *Am. Nat.* **113**, 223–234.

Butler, A. J., and Brewster, F. J. (1979). Size distributions and growth of the fanshell *Pinna bicolor* Gmelin (Mollusca: Eulamellibranchia) in South Australia. *Aust. J. Mar. Freshwater Res.* **30**, 25–39.

Butler, A. J., and Keough, M. J. (1981). Distribution of *Pinna bicolor* Gmelin (Mollusca: Bivalvia) in South Australia, with observations on recruitment. *Trans. R. Soc. South Aust.* **105**, 29–39.

Byram, G. M. (1959). Combustion of forest fuels. *In* "Forest Fire: Control and Use" (K. P. Davis, ed.), pp. 61–89. McGraw-Hill, New York.

Byrne, R., Michaelsen, J., and Soutar, A. (1977). Fossil charcoal as a measure of wildfire frequency in southern California: A preliminary analysis. *In* "Environmental Consequences of Fire and Fuel Management in Mediterranean Ecosystems" (H. A. Mooney and C. E. Conrad, eds.), *Gen. Tech. Rep. WO—U.S. For. Serv. [Wash. Off.]* **GTR-WD-3**, pp. 361–367.

Caffey, H. M. (1982). No effect of naturally-occurring rock types on settlement or survival in the intertidal barnacle, *Tesseropera rosea* (Krauss). *J. Exp. Mar. Biol. Ecol.* **63**, 119–132.

Caine, N., and Swanson, F., eds. (1983). "Report of Long-term Ecological Research Workshop on Disturbance Regimes of Ecosystems," Long-term Ecol. Res. Work. Pap. No. 83/6.

Caldwell, M. M. (1979). Root structure: the considerable cost of belowground function. *In* "Topics in Plant Population Biology" (O. T. Solbrig, S. Jain, G. B. Johnson, and P. H. Raven, eds.), pp. 408–427. Columbia Univ. Press, New York.

Calow, P., and Townsend, C. R. (1981). Energetics, ecology, and evolution. *In* "Physiological Ecology: An Evolutionary Approach to Resource Use" (C. R. Townsend and P. Calow, eds.), pp. 3–19. Sinauer, Sunderland, Massachusetts.

Canham, C. D. (1981). Canopy architecture, leaf display and growth patterns of understory saplings of sugar maple (*Acer saccharum* Marsh.). *Bull. Ecol. Soc. Am.* **62**, 160.

Canham, C. D. (1984). Canopy recruitment in shade tolerant trees: The response of *Acer saccharum* and *Fagus grandifolia* to canopy openings. Ph. D. Thesis, Cornell University, Ithaca, New York.

Canham, C. D., and Loucks, O. L. (1984). Catastrophic windthrow in the presettlement forests of Wisconsin. *Ecology* **65**, 803–809.

Cannell, M. G. R., and Willett, S. C. (1976). Shoot growth phenology, dry matter distribution and root:shoot ratios of provenances of *Populus trichocarpa*, *P. sitchensis* and *P. contorta* growing in Scotland. *Silvae Genet.* **25**, 49–59.

Carosella, T. L. (1978). Population responses of *Opuntia compressa* (Salisb.) Macbr. in a southern Wisconsin sand prairie. M.S. Thesis, Univ. of Wisconsin, Madison.

Carpenter, F. L., and Recher, H. F. (1979). Pollination, reproduction, and fire. *Am. Nat.* **113**, 871–879.

Carson, H. (1975). The genetics of speciation at the diploid level. *Am. Nat.* **109**, 73–92.

Casertano, L. (1963). General characteristics of active Andean volcanoes and a summary of their activities during recent centuries. *Bull. Seismol. Soc. Am.* **53**, 1415–1433.

Castric-Fey, A., Girard-Descatoire, A., and Lafargue, F. (1973). Les peuplements sessile de l'Archipele de Glenan repartition de la faune dans les differents horizons. *Vie Milieu* **28**, 51–67.

Castro A., R., and Guevara S., S. (1976). Viabilidad de semillas en muestras de suelo almacenado de "Los Tuxtlas", Veracruz. *In* "Regeneración des Selvas" (A. Gómez-Pompa, C. Vázquez-Yanes, S. del Amo R., and A. Butanda C., eds.). pp. 233–249. Compañia Editorial Continental, Mexico City, Mexico.

Caswell, H. (1978). Predator-mediated coexistence:a non-equilibrium model. *Am. Nat.* **112**, 127–154.

Cattelino, P. J., Noble, I. R., Slatyer, R. O., and Kessell, S. R. (1979). Predicting the multiple pathways of plant succession. *Environ. Manage. (N.Y.)* **3**, 41–50.

Caughley, G. (1976). Plant-herbivore systems. *In* "Theoretical Ecology" (R. May, ed.), pp. 94–113. Saunders, Philadelphia, Pennsylvania.

Causton, D. R., and Venus, J. C. (1981). "The Biometry of Plant Growth." Edward Arnold, London.

Cavalli-Sforza, L. L. (1966). Population structure and evolution. *Proc. R. Soc. London, Ser. B* **164**, 362–379.

Chabot, B. F. (1978). Environmental influences on photosynthesis and growth in *Fragaria vesca. New Phytol.* **80**, 87–98.

Chabot, B. F., and Chabot, J. F. (1977). Effects of light and temperature on leaf anatomy and photosynthesis in *Fragaria vesca. Oecolgia* **26**, 363–377.

Chabrek, R. H., and Palmisano, A. W. (1973). The effects of Hurricane Camille on the marshes of the Mississippi River delta. *Ecology* **54**, 1118–1123.

Chapin, F. S., III (1981). The mineral nutrition of wild plants. *Annu. Rev. Ecol. Syst.* **11**, 233–260.

Chapin, F. S., III, and Van Cleve, K. (1981). Plant nutrient absorption and retention under differing fire regimes. *In* "Fire Regimes and Ecosystem Properties" (H. A. Mooney, T. M. Bonnicksen, N. L. Christensen, J. E. Lotan, and W. A. Reiners, eds.), *Gen. Tech. Rep. WO—U.S. For. Serv.* [*Wash. Off.*] **GTR-WO-26**, pp. 301–321.

Charlesworth, B. (1980). "Evolution in Age Structured Populations." Cambridge Univ. Press, London and New York.

Charlesworth, B., and Charlesworth, D. (1979). The evolutionary genetics of sexual systems in flowering plants. *Proc. R. Soc. London, Ser. B* **205**, 513–530.

Charley, J. L., and West, N. E. (1975). Plant-induced soil chemistry patterns in some shrub-dominated semi-desert ecosystems of Utah. *J. Ecol.* **63**, 945–964.

Charnov, E. L. (1976). Optimal foraging: the marginal value theorem. *Theor. Popul. Biol.* **9**, 129–136.

Chazdon, R. L., and Fetcher, N. (1984). Photosynthetic light environments in a lowland tropical rainforest in Costa Rica. *J. Ecol.* **72**, 553–564.

Cheke, A. S., Nanakorn, W., and Yankoses, C. (1979). Dormancy and dispersal of seeds of secondary forest species under the canopy of a primary tropical rain forest in northern Thailand. *Biotropica* **11**, 88–95.

Chesson, P. L. (1978). Predator–prey theory and variability. *Annu. Rev. Ecol. Syst.* **9**, 323–347.

Chesson, P. L. (1981). Models for spatially distributed populations: the effect of within-patch variability. *Theor. Popul. Biol.* **19**, 288–325.

Chew, R. M., and Chew, A. E. (1965). Primary productivity of a desert shrub (*Larrea trideubata*) community. *Ecol. Monogr.* **35**, 355–375.

Choat, J. H. (1977). The influence of sessile organisms on the biology of three species of acmaeid limpets. *J. Exp. Mar. Biol. Ecol.* **26**, 1–26.

Chou, C. H., and Muller, C. H. (1972). Allelopathic mechanisms of *Arctostaphylos glandulosa* var. *zacaensis. Am. Midl. Nat.* **88**, 324–347.

Christensen, N. L. (1973). Fire and the nitrogen cycle in chaparral. *Science (Washington, D.C.)* **181,** 66–68.

Christensen, N. L. (1977). Fire and soil–plant nutrient relations in a pine–wiregrass savanna on the coastal plain of North Carolina. *Oecologia* **31,** 27–44.

Christensen, N. L. (1981). Fire regimes in sougheastern ecosystems. *In* "Fire Regimes and Ecosystem Properties" (H. A. Mooney, T. M. Bonnicksen, N. L. Christensen, J. E. Lotan, and W. A. Reiners, eds.), *Gen. Tech. Rep. WO—U.S. For. Serv. (Wash. Off.)* **GTR-WO-26,** pp. 112– 136.

Christensen, N. L., and Muller, C. H. (1975a). Effects of fire on factors controlling plant growth in *Adenostoma* chaparral. *Ecol. Monogr.* **45,** 29–55.

Christensen, N. L., and Muller, C. H. (1975b). Relative importance of factors controlling germination and seedling survival in *Adenostoma* chaparral. *Am. Midl. Nat.* **93,** 71–78.

Christensen, N. L., and Peet, R. K. (1981). Secondary forest succession on the North Carolina Piedmont. *In* "Forest Succession: Concept and Application" (D. C. West, H. H. Shugart, and D. B. Botkin, eds.), pp. 230–245. Springer-Verlag, Berlin and New York.

Christensen, N. L., and Peet, R. K. (1985). Convergence during secondary forest succession. *J. Ecol.* **72,** 25–36.

Christensen, N. L., and Wilbur, R. B. (1984). Vegetation changes associated with burning and clearing in a North Carolina Coastal Plain pocosin. Manuscript.

Christensen, N. L., Burchell, R. B., Liggett, A., and Simms, E. L. (1981). The structure and development of pocosin vegetation. *In* "Pocosin Wetlands" (C. J. Richardson, ed.), pp. 43–61. Dowden, Hutchinson & Ross, Stroudsburg, Pennsylvania.

Christian, D. P. (1980). Vegetative cover, water resources, and microdistributional patterns in a desert rodent community. *J. Anim. Ecol.* **49,** 807–816.

Christian, K., Tracy, C. R., and Porter, W. P. (1983). Seasonal shifts in body temperature and use of microhabitats by Galapagos land iguanas (*Conolophus pallidus*). *Ecology* **64,** 463–468.

Churchill, E. C., and Hanson, H. C. (1958). The concept of climax in arctic and alpine vegetation. *Bot. Rev.* **24,** 127–191.

Clark, F. B., and Boyce, S. G. (1964). Yellow-poplar seed remains viable in the forest litter. *J. For.* **62,** 564–567.

Clark, W. C. (1979). Spatial structure relationship in a forest insect system: simulation models and analysis. *Mitt. Schwiez. Entomol. Ges.* **52,** 235–257.

Clegg, M. T., and Allard, R. W. (1972). Patterns of genetic differentiation in the slender wild oat species, *Avena barbata. Proc. Natl. Acad. Sci. U.S.A.* **69,** 1820–1824.

Clegg, M. T., and Allard, R. W. (1973). Viability versus fecundity selection in the slender Wild Oat, *Avena barbata* L. *Science (Washington, D.C.)* **181,** 667–668.

Clegg, M. T., Kahler, A. L., and Allard, R. W. (1978). Genetic demography of plant populations. *In* "Ecological Genetics: The Interface" (P. F. Brussard, ed.), pp. 173–188. Springer-Verlag, Berlin and New York.

Clements, F. E. (1916). Plant succession: An analysis of the development of vegetation. *Carnegie Inst. Washington Publ.* No. 242.

Clements, F. E., and Shelford, V. E. (1939). "Bioecology." Wiley, New York.

Clough, J. M., Alberte, R. S., and Teeri, J. A. (1979). Photosynthetic adaptation of *Solanum dulcamara* L. to sun and shade environments. II. Physiological characterization of phenotypic response to the environment. *Plant Physiol.* **64,** 25–30.

Cody, M. L. (1975). Towards a theory of continental species diversities. *In* "Ecology and Evolution of Communities" (M. L. Cody and J. M. Diamond, eds.), pp. 214–257. Harvard Univ. Press, Cambridge, Massachusetts.

Cogbill, C. V. (1980). Forest history and tree growth dynamics in the northern hardwood forest at Hubbard Brook, New Hampshire. *Bull. Ecol. Soc. Am.* **61,** 91.

Cole, B. J. (1981). Colonizing abilities, island size, and the number of species on archipelagoes. *Am. Nat.* **117,** 629–638.

Cole, S., Hainsworth, F. R., Kamil, A. C., Mercier, T., and Wolf, L. L. (1982). Spatial learning as an adaptation in hummingbirds. *Science (Washington, D.C.)* **217**, 655–657.

Cole, W. E., and Amman, G. D. (1980). Mountain pine beetle dynamics in lodgepole pine forests. Part 1: Course of an infestation. *USDA For. Serv. Gen. Tech. Rep. INT* **INT-89.**

Coley, P. D. (1982). Rates of herbivory on different tropical trees. *In* "The Ecology of a Tropical Forest" (E. G. Leigh, Jr., A. S. Rand, and D. M. Windsor, eds.), pp. 123–132. Smithsonian Institution Press, Washington, D.C.

Coley, P. D. (1983). Herbivory and defensive characteristics of tree species in a lowland tropical forest. *Ecol. Monogr.* **53**, 209–233.

Colgan, M. W. (1983). Succession and recovery of a coral reef after predation by *Acanthaster planci* (L.). *Proc—Int. Coral Reef Symp., 4th, Manila, 1981* **2**, 333–338.

Colwell, R. K. (1974). Predictability, constancy, and contingency in periodic phenomena. *Ecology* **55**, 1148–1153.

Comstock, R. E. (1977). Quantitative genetics and design of breeding programs. *In* "Proceedings of the International Conference on Quantitative Genetics" (E. Pollack, O. Kempthorne, and T. B. Bailey, eds.), pp. 705–18. Iowa State Univ. Press, Ames.

Connell, J. H. (1961a). The influence of interspecific competition and other factors on the distribution of the barnacle *Chthamalus stellatus. Ecology* **42**, 710–723.

Connell, J. H. (1961b). Effects of competion, predation by *Thais lapillus,* and other factors on natural populations of the barnacle *Balanus balanoides. Ecol. Monogr.* **31**, 61–104.

Connell, J. H. (1970). A predator–prey system in the marine intertidal region. 1. *Balanus glandula* and several predatory species of *Thais. Ecol. Monogr.* **40**, 49–78.

Connell, J. H. (1971). On the role of natural enemies in preventing competitive exclusion in some marine animals and in rain forest trees. *In* "Dynamics of Populations" (P. J. denBoer and G. R. Gradwell, eds.), Proceedings of the Advanced Study Institute on Dynamics of Numbers in Populations, pp. 298–312. Cent. Agric. Publ Doc., Wageningen.

Connell, J. H. (1972). Community interactions on marine rocky intertidal shores. *Annu. Rev. Ecol. Syst.* **3**, 169–192.

Connell, J. H. (1973). Population ecology of reef-building corals. *In* "Biology and Geology of Coral Reefs. Vol. 2: Biology 1" (O. A. Jones and R. Endean, eds.), pp. 205–245. Academic Press, New York.

Connell, J. H. (1976). Competitive interactions and the species diversity of corals. *In* "Coelenterate Ecology and Behavior" (G. O. Mackie, ed.), pp. 51–58. Plenum, New York.

Connell, J. H. (1978). Diversity in tropical rain forests and coral reefs. *Science (Washington, D.C.)* **199**, 1302–1310.

Connell, J. H. (1979). Tropical rain forests and coral reefs as open non-equilibrium systems. *In* "Population Dynamics" (R. Anderson, B. Turner, and L. Taylor, eds.), pp. 141–163. Blackwell, Oxford.

Connell, J. H. (1980). Diversity and the coevolution of competitors, or the ghost of competition past. *Oikos* **35**, 131–138.

Connell, J. H. (1983). On the prevalence and relative importance of interspecific competition: evidence from field experiments. *Am. Nat.* **122**, 661–696.

Connell, J. H. (1984). Disturbance and the maintenance of diversity of corals. In preparation.

Connell, J. H., and Orias, E. (1964). The ecological regulation of species diversity. *Am. Nat.* **98**, 399–414.

Connell, J. H., and Slatyer, R. O. (1977). Mechanisms of succession in natural communities and their role in community stability and organization. *Am. Nat.* **111**, 1119–1144.

Connell, J. H., and Sousa, W. P. (1983). On the evidence needed to judge ecological stability or persistence. *Am. Nat.* **121**, 789–824.

Connor, E. F., and McCoy, E. D. (1979). The statistics and biology of the species–area relationship. *Am. Nat.* **113**, 791–833.

Connor, E. F., and Simberloff, D. (1979). The assembly of species communities: chance or competition? *Ecology* **60**, 1132–1140.

Continho, L. M. (1977). Ecological aspects of fire in the Cerrado. II—Fire and seed dispersion in some anemochoric species of the herbaceous layer. *Bol. Bot., Univ. Sao Paulo* **5**, 57–64.

Cook, R. E. (1980). The biology of seeds in the soil. *In* "Demography and Evolution in Plant Populations" (O. T. Solbrig, ed.), Botanical Monographs, Vol. 15, pp. 107–129. Blackwell, Oxford.

Cook, R. E., and Lyons, E. E. (1983). The biology of *Viola fimbriatula* in a natural disturbance. *Ecology* **64**, 654–660.

Coombe, D. E. (1960). An analysis of the growth of *Trema guineensis*. *J. Ecol.* **48**, 219–231.

Coombe, D. E., and Hadfield, W. (1962). An analysis of the growth of *Macaranga cecropiodes*. *J. Ecol.* **50**, 221–234.

Cooper, A. W. (1981). Above-ground biomass accumulation and net primary production during the first 70 years of succession in *Populus grandidentata* stands on poor sites in northern Lower Michigan. *In* "Forest Succession: Concepts and Application" (D. C. West, H. H. Shugart, and D. B. Botkin, eds.), pp. 339–360. Springer-Verlag, Berlin and New York.

Cooper, C. F. (1961). Changes in vegetation, structure, and growth of southwestern pine forests since white establishment. *Ecol. Monogr.* **30**, 129–164.

Cooper, W. E., and Crowder, L. B. (1979). Patterns of predation in simple and complex environments. *In* "Predator–Prey Systems in Fisheries Management" (R. H. Stroud and H. Clepper, eds.), pp. 257–268. Sport Fish. Inst., Washington, D.C.

Cooper, W. S. (1913). The climax forest of Isle Royale, Lake Superior and its development. *Bot. Gaz.* **15**, 1–44, 115–140, 189–235.

Cornell, H. V. (1982). The notion of minimum distance or Why rare species are clumped. *Oecologia* **52**, 278–280.

Costin, A. B. (1954). "A Study of the Ecosystems of the Monaro Region of New South Wales." Gov. Printer, Sydney.

Coulson, R. N. (1979). Population dynamics of bark beetles. *Annu. Rev. Entomol.* **24**, 417–447.

Countryman, C. M., and Philpot, C. W. (1970). Physical properties of chamise as a wildland fuel. *U.S. For. Serv., Res. Pap. PSW* No. 66.

Covington, W. W. (1981). Changes in forest floor organic matter and nutrient content following clear cutting in northern hardwoods. *Ecology* **62**, 41–48.

Craddock, G. W. (1929). The successional influence of fire on the chaparral type. M.S. Thesis, Univ. of California. Berkeley.

Crawford, C. S., and Gosz, J. R. (1982). Desert ecosystems: their resources in space and time. *Environ. Conserv.* **9**, 181–195.

Crawford, H. S. (1976). Relationships between forest cutting and understory vegetation. An overview of eastern hardwood stands. *USDA For. Serv. Res. Pap. NE* **NE-349.**

Creese, R. G. (1982). Distribution and abundance of the acmaeid limpet, *Patelloida latistrigata,* and its interaction with barnacles. *Oecologia* **52**, 85–96.

Crisp, D. J. (1964). The effects of the severe winter of 1962–63 on marine life in Britain. *J. Anim. Ecol.* **33**, 165–210.

Crisp, D. J. (1974). Factors influencing the settlement of marine invertebrate larvae. *In* "Chemoreception in Marine Organisms" (P. T. Grant and A. M. Mackie, eds.), pp. 177–265. Academic Press, New York.

Crisp, D. J. (1979). Dispersal and re-aggregation in sessile marine invertebrates, particularly barnacles. *In* "Biology and Systematics of Colonial Organisms" (L. Larwood and B. R. Rosen, eds.), pp. 319–327. Academic Press, New York.

Crisp, M. D. (1978). Demography and survival under grazing of three Australian semi-desert shrubs. *Oikos* **30**, 520–528.

Croat, T. B. (1975). Phenological behavior of habit and habitat classes on Barro Colorado Island (Panama Canal Zone). *Biotropica* **7**, 270–277.

Crook, J. H. (1965). The adaptive significance of avian social organizations. *Symp. Zool. Soc. London* No. 14, 181–218.

Crow, J. F., and Kimura, M. (1970). "An Introduction to Population Genetics Theory." Harper & Row, New York.

Crow, J. F., and Morton, N. E. (1955). Measurement of gene frequency drift in small populations. *Evolution* **9**, 202–214.

Crow, T. R. (1980). A rainforest chronicle: a 30 year record of change in structure and composition at El Verde, Puerto Rico. *Biotropica* **12**, 42–55.

Croze, H. (1970). Searching image in carrion crows. *Z. Tierpsychol., Suppl.* **5**, 1–85.

Cubit, J. D. (1984). Herbivory and the seasonal abundance of algae on a high intertidal rocky shore. *Ecology* **65**, 1904–1917.

Culver, D. C., and Beattie, A. J. (1983). Effects of ant mounds on soil chemistry and vegetation patterns in a Colorado montane meadow. *Ecology* **64**, 485–492.

Curtis, J. T. (1959). "The Vegetation of Wisconsin." Univ. of Wisconsin Press, Madison.

Cypher, J., and Boucher, D. H. (1982). Beech–maple coexistence and seedling growth rates at Mont Saint Hilaire, Quebec. *Can. J. Bot.* **60**, 1279–1281.

Dale, M. B., and Quadraccia, L. (1973). Computer assisted tabular sorting of phytosociological data. *Vegetatio* **28**, 57–73.

Daly, J. C. (1981). Effects of social organization and environmental diversity on determining the genetic structure of a population of the wild rabbit, *Oryctolagus cuniculus. Evolution* **35**, 689–706.

Daly, M. A., and Mathieson, A. C. (1977). The effects of sand movement on intertidal seaweeds and selected invertebrates at Bound Rock, New Hampshire, USA. *Mar. Biol. (Berlin)* **43**, 45–55.

Daniel, T. W., Helms, J. A., and Baker, F. S. (1979). "Principles of Silviculture." McGraw-Hill, New York.

Daubenmire, R. F. (1968). Ecology of fire in grasslands. *Adv. Ecol. Res.* **5**, 209–266.

Davidson, D. W., Brown, J. H., and Inouye, R. S. (1980). Competition and the structure of granivore communities. *BioScience* **30**, 233–238.

Davies, R. I., Coulson, C. B., and Lewis, D. A. (1964). Polyphenols in plant, humus, and soil. II. Factors leading to increase in biosynthesis of polyphenol in leaves and their relationship to mull and mor formation. *J. Soil. Sci.* **15**, 310–318.

Davis, G. E. (1982). A century of natural change in coral distribution at the Dry Tortugas: a comparison of reef maps from 1881 and 1976. *Bull. Mar. Sci.* **32**, 608–623.

Davis, M. B. (1981). Quaternary history and the stability of plant communities. *In* "Forest Succession: Concepts and Application" (D. C. West, H. H. Shugart, and D. B. Botkin, eds.), pp. 132–153. Springer-Verlag, Berlin and New York.

Davis, S. N., and Karzulovíc, J. (1963). Landslides at Lago Rinihue, Chile. *Bull. Seismol. Soc. Am.* **53**, 1403–1414.

Dawkins, H. C. (1966). The time dimension of tropical trees. *J. Ecol.* **53**, Suppl., 837–838.

Day, R. W. (1977). The ecology of settling organisms on the coral reef at Heron Island, Queensland. Ph.D. Thesis, Univ. of Sydney.

Dayton, P. K. (1971). Competition, disturbance, and community organization: The provision and subsequent utilization of space in a rocky intertidal community. *Ecol. Monogr.* **41**, 351–389.

Dayton, P. K. (1973). Dispersion, dispersal, and persistence of the annual intertidal alga, *Postelsia palmaeformis* Ruprecht. *Ecology* **54**, 433–438.

Dayton, P. K. (1975). Experimental evaluation of ecological dominance in a rocky intertidal algal community. *Ecol. Monogr.* **45**, 137–159.

Dayton, P. K., Robilliard, G. A., and Paine, R. T. (1970). Benthic faunal zonation as a result of anchor ice at McMurdo Sound, Antartica. *In* "Antarctic Ecology" (M. W. Holdgate, ed.), Vol. 1, pp. 244–258. Academic Press, New York.

DeAngelis, D. L., Travis, C. C., and Post, W. M. (1979). Persistence and stability of seed-dispersed species in a patchy environment. *Theor. Popul. Biol.* **16**, 107–125.

Debano, L. F., and Conrad, C. E. (1976). Nutrients lost in debris and run off water from a burned chaparral watershed. *Proc. Fed. Inter-Agency Sediment. Conf., 3rd, Denver, Colo.* pp. 3/13–3/27.

Debano, L. F., and Conrad, C. E. (1978). The effect of fire on nutrients in a chaparral ecosystem. *Ecology* 59, 489–497.

Debano, L. F., Osborn, J. F., Krammes, J. S., and Letey, J., Jr. (1967). Soil wettability and wetting agents . . . our current knowledge of the problem. *U.S. For. Serv., Res. Pap. PSW* No. 43.

de Jong, G. (1979). The influence of the distribution of juveniles over patches of food on the dynamics of a population. *Neth. J. Zool.* 29, 33–51.

del Amo R., S. and Gómez-Pompa, A. (1976). Crecimiento de estados juveniles de plantas en selva tropical alta perennifolia. *In* "Regeneración des Selvas" (A. Gómez-Pompa, C. Vázquez-Yanes, S. del Amo R. and A. Butanda C., eds.), pp. 549–565. Compañia Editorial Continental, Mexico City, Mexico.

del Amo R., S., and Nieto de Pascual, J. (1981). Applications of models and mathematical equations to evaluate growth rates and age determination of tropical trees. *In* "Age and Growth Rate of Tropical Trees" (F. H. Bormann and G. Berlyn, eds.), *Bull.—Yale Univ., Sch. For. Environ. Stud.* No. 94, 128–133.

Delcourt, H. R., and Harris, W. F. (1980). Carbon budget of the southeastern United States biota: analysis of historical change in trend from source to sink. *Science (Washington, D.C.)* 210, 321–323.

Delcourt, H. R., Delcourt, P. A., and Webb, T. (1983). Dynamic plant ecology: The spectrum of vegetational change in space and time. *Quat. Sci. Rev.* 1, 153–175.

Delcourt, P. A., and Delcourt, H. R. (1983). Late quaternary vegetational dynamics and community stability reconsidered. *Quat. Res. (N.Y.)* 19, 265–271.

del Moral, R. (1983). Competition as a control mechanism in subalpine meadows. *Am. J. Bot.* 70, 232–245.

Den Boer, P. J. (1968). Spreading of risk and stabilization of animal numbers. *Acta Biotheor.* 18, 165–194.

Den Boer, P. J. (1981). On the survival of populations in a heterogeneous and variable environment. *Oecologia* 50, 39–53.

den Hartog, C. (1972). Substratum: multicellular plants. *In* "Marine Ecology" (O. Kinne, ed.), Vol. 1, pp. 1277–1289. Wiley (Interscience), New York.

Denley, E. J., and Underwood, A. J. (1979). Experiments on factors influencing settlement, survival, and growth of two species of barnacles in New South Wales. *J. Exp. Mar. Biol. Ecol.* 36, 269–293.

Denno, R. F., and McClure, M. S., eds. (1983). "Variable Plants and Herbivores in Natural and Managed Systems." Academic Press, New York.

Denny, M. W., Daniel, T. L., and Koehl, M. A. R. (1985). Mechanical limits to size in wave-swept *Monogr.* 55 (in press).

Denslow, J. S. (1978). Secondary succession in a Colombian rainforest: strategies of species response across a disturbance gradient. Ph.D. Thesis, Univ. of Wisconsin, Madison.

Denslow, J. S. (1980a). Gap partitioning among tropical rainforest trees. *Biotropica* 12, Suppl., 47–55.

Denslow, J. S. (1980b). Patterns of plant species diversity during succession under different disturbance regimes. *Oecologia* 46, 18–21.

Dethier, M. N. (1984). Disturbance and recovery in intertidal pools: Maintenance of mosaic patterns. *Ecol. Monogr.* 54, 99–118.

Diamond, J. M. (1972). Biogeographic kinetics: estimation of relaxation times for avifaunas of southwest Pacific islands. *Proc. Natl. Acad. Sci. U.S.A.* 69, 3199–3203.

Diamond, J. M. (1975). The island dilemma: lessons of modern biogeographic studies for the design of natural reserves. *Biol. Conserv.* 7, 129–146.

Diamond, J. M., and May, R. M. (1976). Island biogeography and the design of natural reserves. *In* "Theoretical Ecology" (R. M. May, ed.), pp. 163–186. Saunders, Philadelphia, Pennsylvania.

Di Castri, F. (1981). Mediterranean-type shrublands of the World. *In* "Mediterranean-Type

Shrublands'' (F. di Castri, D. W. Goodall, and R. L. Specht, eds.), pp. 1–52. Elsevier, Amsterdam.

Dickinson, H., and Antonovics, J. (1973). The effects of environmental heterogeneity on the genetics of finite populations. *Genetics* **73**, 713–735.

Dixon, J. D. (1978). Determinants of the local distribution of four closely-related species of herbivorous marine snails. Ph.D. Thesis, Univ. of California, Santa Barbara.

Dixon, J. D., Schroeter, S. C., and Kastendiek, J. (1982). Effects of the encrusting bryozoan *Membranipora membranacea* on the loss of blades and fronds by the Giant Kelp *Macrocystis pyrifera* (Laminariales). *J. Phycol.* **17**, 341–345.

Dobzhansky, T. (1950). Genetics of natural populations. XIX. Origin of heterosis through natural selection in populations of *Drosophila pseudoobscura*. *Genetics* **35**, 588–602.

Dobzhansky, T., and Pavlovsky, O. (1953). An experimental study of interaction between genetic drift and natural selection. *Evolution* **7**, 198–210.

Doherty, P. J. (1983). Tropical territorial damselfishes: is density limited by aggression or recruitment? *Ecology* **64**, 176–190.

Dolan, R., Hayden, B. P., and Soucie, G. (1978). Environmental dynamics and resource management in the U.S. National Parks. *Environ. Manage.* **2**, 249–258.

Dollar, S. J. (1982). Wave stress and coral community structure in Hawaii. *Coral Reefs* **1**, 71–81.

Donoso Z., C. (1981). ''Tipos Forestales de Los Bosques Nativos de Chile.'' Corp. Nac. For., Santiago, Chile.

Downs, A. A. (1946). Response to release of sugar maple, white oak, and yellow poplar. *J. For.* **44**, 22–27.

Doyle, T. W. (1981). The role of disturbance in the gap dynamics of a montane rain forest: An application of a tropical forest succession model. *In* ''Forest Succession: Concepts and Application'' (D. C. West, H. H. Shugart, and D. B. Botkin, eds.), pp. 56–73. Springer-Verlag, Berlin and New York.

Driscoll, E. G. (1968). Sublittoral attached epifaunal development in Buzzard's Bay, Massachusetts. *Hydrobiologia* **32**, 27–32.

Drury, W. H. (1956). Bog flats and physiographic processes in the upper Kuskokwim River region, Alaska. *Contrib. Gray Herbarium* **178**, 1–30.

Drury, W. H., and Nisbet, I. C. T. (1971). Inter-relations between developmental models in geomorphology, plant ecology, and animal ecology. *Gen. Syst.* **16**, 57–68.

Drury, W. H., and Nisbet, I. C. (1973). Succession. *J. Arnold Arbor. Harv. Univ.* **54**, 331–368.

Dustan, P. (1975). Genecological differentiation in the reef-building coral *Montastrea annularis*. Ph.D. Thesis, State Univ. of New York, Stony Brook.

Eanes, W. F. (1978). Morphological variance and enzyme heterozygosity in the monarch butterfly. *Nature (London)* **276**, 263–264.

Edmisten, J. (1970). Some autecological studies of *Ormosia kruggi*. *In* ''A Tropical Rainforest'' (H. Odum and R. F. Pigeon, eds.), pp. B291–298. USAEC, Washington, D.C.

Edmonds, R. L. (1979). Decomposition and nutrient release in Douglas-fir needle litter in relationship to stand development. *Can. J. For. Res.* **9**, 132–140.

Edmunds, G. G., Jr., and Alstad, D. N. (1978). Coevolution in insect herbivores and conifers. *Science (Washington, D.C.)* **199**, 941–945.

Eggler, W. A. (1971). Quantitative studies of vegetation on sixteen young lava flows on the island of Hawaii. *Trop. Ecol.* **12**, 66–100.

Eggleston, D. (1972). Factors influencing the distribution of sub-littoral ectoprocts off the south of the Isle of Man (Irish Sea). *J. Nat. Hist.* **6**, 247–260.

Egler, F. E. (1947). Arid southeast Oahu vegetation. Hawaii. *Ecol. Monogr.* **17**, 383–435.

Egler, F. E. (1954). Vegetation science concepts. I. Initial floristic composition—a factor in old-field vegetation development. *Vegetatio* **4**, 412–417.

Egler, F. E. (1977). ''The Nature of Vegetation.'' Connecticut Conserv. Assoc., Bridgewater.

Ehrenfeld, J. G. (1980). Understory response to canopy gaps of varying size in a mature oak forest. *Bull. Torrey Bot. Club* **107**, 29–41.

Ehrlich, P., and Ehrlich, A. (1981). "Extinction: The Causes and Consequences of the Disappearance of Species." Random House, New York.

Eisenbrey, A. B., and Moore, W. S. (1981). Evolution of histocompatibility diversity in an asexual vertebrate, *Poeciliopsis 2 monacha-lucida* (Pisces: Poeciliidae). *Evolution* **35**, 1180–1191.

Ek, A. R., and Monserud, R. A. (1974). Trials with program FOREST: Growth and reproduction simulation for mixed species even or uneven-aged forest stands. *In* "Growth Models and Forest Stand Simulation" (J. Fries, ed.), Research Notes No. 30, pp. 56–73, Dep. For. Yield Res., R. Coll. For., Stockholm.

Elias, P. (1978). Water deficit of plants in an oak–hornbeam forest. *Preslia* **50**, 173–188.

Elias, P. (1981). Some ecophysiological leaf-characteristics of components of spring synuzium in temperate deciduous forests. *Biologia (Bratislava)* **36**, 841–849.

Emerson, S. E., and Zedler, J. B. (1978). Recolonization of intertidal algae: an experimental study. *Mar. Biol. (Berlin)* **44**, 315–324.

Encina, F. A. (1954). "Resumen de la Historia de Chile," Vol. 1. Zig-Zag, Santiago, Chile.

Endler, J. A. (1977). "Geographic Variation, Speciation and Clines." Princeton Univ. Press, Princeton, New Jersey.

Enright, N. J., and Hartshorn, G. S. (1981). The demography of tree species in undisturbed tropical rainforest. *In* "Age and Growth Rate of Tropical Trees" (F. H. Bormann and G. Berlyn, eds.), *Bull.—Yale Univ., Sch. For. Environ. Stud.* No. 94, 107–121.

Environmental Data Service (1975). "Storm Data," Vol. 17. U.S. Dep. Commer., Nat. Oceanic Atmos. Adm., Washington, D.C.

Erdmann, G. G., Godman, R. M., and Oberg, R. R. (1975). Crown release accelerates diameter growth and crown development of yellow birch samplings. *U.S. For. Serv., Res. Pap. NC* **NC-117.**

Ernst, W. H. O. (1979). Population biology of *Allium ursinum* in northern Germany. *J. Ecol.* **67**, 347–362.

Errington, P. L. (1963). "Muskrat Populations." Iowa State Univ. Press. Ames.

Eussen, J. H. H., and Wirjahardja, S. (1973). Studies on an alang-alang (*Imperata cylindrica* (L.) Beauv.) vegetation. *Biotrop Bull.* No. 6, 1–24.

Evans, C. C., and Allen, S. E. (1971). Nutrient losses in smoke produced during heather burning. *Oikos* **22**, 149–154.

Evans, G. C. (1972). "The Quantitative Analysis of Plant Growth." Univ. of California Press, Berkeley.

Evans, H. L. (1968). "Laminar Boundary-Layer Theory." Addison-Wesley, Reading, Massachusetts.

Evans, L. T. (1975). Beyond photosynthesis—the role of respiration, translocation and growth potential in determining productivity. *In* "Photosynthesis and Production in Different Environments" (J. P. Cooper, ed.), pp. 501–507. Cambridge Univ. Press, London and New York.

Ewel, J. J. (1980). Tropical succession: Manifold routes to maturity. *Biotropica* **12**, Suppl., 2–7.

Ewel, J., Barish, C., Brown, B., Price, N., and Raich, J. (1981). Slash and burn impacts on a Costa Rican wet forest site. *Ecology* **62**, 816–829.

Ewel, J. J., Ojima, D. S., Karl, D. A., and DeBusk, W. F. (1982). "Schinus in Successional Ecosystems of Everglades National Park," Rep. T-676. South Florida Res. Cent., Homestead, Florida.

Falconer, D. S. (1960). "Introduction to Quantitative Genetics." Ronald Press, New York.

Falconer, D. S. (1977). Some results of the Edinburgh selection experiments with mice. *In* "Proceedings of the International Conference on Quantitative Genetics" (E. Pollack, O. Kempthorne, and T. B. Bailey, eds.), pp. 101–116. Iowa State Univ. Press, Ames.

Fausch, K. D., Karr, J. R., and Yant, P. R. (1984). Regional application of an index of biotic integrity based on stream fish communities. *Trans. Am. Fish. Soc.* (in press).

Federer, C. A. (1973). Annual cycles of soil and water temperatures at Hubbard Brook. *U.S. For. Serv., Northeast. For. Exp. Stn., Res. Note NE* **NE-167.**

Fell, P. E. (1974). Porifera. *In* "Reproduction in Marine Invertebrates" (A. C. Giese and J. S. Pearse, eds.), pp. 51–132. Academic Press, New York.

Felsenstein, J. (1971). Inbreeding and variance effective numbers in populations with overlapping generations. *Genetics* **68**, 581–597.

Felsenstein, J. (1974). The evolutionary advantage of recombination. *Genetics* **78**, 737–756.

Felsenstein, J. (1976). The theoretical population genetics of variable selection and migration. *Annu. Rev. Genet.* **10**, 253–286.

Fernald, M. L. (1950). "Gray's Manual of Botany," 8th Ed. American Book Co., New York.

Fetcher, N., Strain, B. R., and Oberbauer, S. F. (1983). Effects of light regime on the growth, leaf morphology, and water relations of seedlings of two species of tropical trees. *Oecologia* **58**, 314–319.

Fishelson, L. (1973). Ecological and biological phenomena influencing coral-species composition on the reef tables at Eilat (Gulf of Aquaba, Red Sea). *Mar. Biol. (Berlin)* **19**, 183–196.

Fisher, J. B., and Honda, H. (1979). Branch geometry and effective leaf area: a study of *Terminalia* branching pattern. 1. Theoretical trees. *Am. J. Bot.* **66**, 633–644.

Fisher, R. A. (1930). "The Genetical Theory of Natural Selection." Oxford Univ. Press, London and New York.

Fisher, R. F., Gholz, H. L., and Pritchett, W. L. (1982). Nutrient limits to productivity after plantation establishment on the Coastal Plain region of the Southeastern U.S. *Bull. Ecol. Soc. Am.* **63**, 181–182.

Fisher, S. G., Gray, L. J., Grimm, N. B., and Busch, D. E. (1982). Temporal succession in a desert stream ecosystem following flash flooding. *Ecol. Monogr.* **52**, 93–110.

Fitter, A. H., and Ashmore, C. J. (1974). Response of two *Veronica* species to a simulated woodland light climate. *New Phytol.* **73**, 997–1001.

Flaccus, E. (1959). Revegetation of landslides in the White Mountains of New Hampshire. *Ecology* **40**, 692–703.

Fleming, T. H., and Heithaus, E. R. (1981). Frugivorous bats, seed shadows, and the structure of tropical forests. *Biotropica* **13**, Suppl., 45–53.

Fleming, T. H., Hooper, E. T., and Wilson, D. E. (1972). Three Central American bat communities: Structure, reproductive cycles, and movement patterns. *Ecology* **53**, 555–569.

Flieger, B. W. (1970). Forest fire and insects: the relation of fire to insect outbreak. *Proc. Annu. Tall Timbers Fire Ecol. Conf.* **10**, 107–120.

Florence, J. (1981). Chablis et sylvigenèse dans une forêt dense humide sempervirente du Gabon. Ph.D. Thesis, Univ. Louis Pasteur de Strasbourg, Strasbourg, France.

Florence, R. G. (1965). Decline of old-growth redwood forests in relation to some soil microbiological processes. *Ecology* **46**, 52–64.

Fogden, M. P. L. (1972). The seasonality and population dynamics of equatorial forest birds in Sarawak. *Ibis* **114**, 307–343.

Foin, T. C., and Jain, S. K. (1977). Ecosystems analysis and population biology: Lessons for the development of community biology. *BioScience* **27**, 532–538.

Fonck, F. (1896). "Viajes de Fray Francisco Menéndez a la Cordillera." Niemeyer, Valparaiso, Chile.

Force, D. C. (1981). Postfire insect succession in southern California chaparral. *Am. Nat.* **117**, 575–582.

Force, D. C. (1982). Postburn insect fauna in southern California chaparral. *In* "Dynamics and Management of Mediterranean-Type Ecosystems" (C. E. Conrad and W. C. Oechel, eds.), *USDA For. Ser. Gen. Tech. Rep. PSW* **PSW-58**, pp. 234–240.

Forcier, L. K. (1975). Reproductive strategies and the cooccurence of climax tree species. *Science (Washington, D.C.)* **189**, 808–810.

Ford, E. D. (1975). Competition and stand structure in some even-aged plant monocultures. *J. Ecol.* **63**, 311–333.

Ford, S. E., and Haskin, H. H. (1982). History and epizootiology of *Haplosporidium nelsoni* (MSX) an oyster pathogen in Delaware Bay, 1957–1980. *J. Invertebr. Pathol.* **40**, 118–141.

Forman, R. T. T. (1983). An ecology of the landscape. *BioScience* **33**, 535.

Forman, R. T. T., and Boerner, R. E. J. (1981). Fire frequency and the Pine Barrens of New Jersey. *Bull. Torrey Bot. Club* **108**, 34–50.

Forman, R. T. T., and Godron, M. (1981). Patches and structural components for a landscape ecology. *BioScience* **31**, 733–740.

Forster, G. R. (1958). Underwater observations on the fauna of shallow rocky areas in the neighbourhood of Plymouth. *J. Mar. Biol. Assoc. U.K.* **37**, 473–482.

Foster, M. A., and Stubbendieck, J. (1980). Effects of the plains pocket gopher (*Geomys bursarius*) on rangeland. *J. Range Manage.* **33**, 74–78.

Foster, M. S. (1975). Regulation of algal community development in a *Macrocystis pyrifera* forest. *Mar. Biol. (Berlin)* **32**, 331–342.

Foster, R. B. (1977). *Tachigalia versicolor* is a suicidal neotropical tree. *Nature (London)* **268**, 624–626.

Foster, R. B. (1980). Heterogeneity and disturbance in tropical vegetation. *In* "Conservation Biology: An Evolutionary–Ecological Perspective" (M. E. Soule and B. A. Wilcox, eds.), pp. 75–92. Sinauer, Sunderland, Massachusetts.

Foster, R. B. (1982). Famine on Barro Colorado Island. *In* "The Ecology of a Tropical Forest" (E. Leigh, Jr., A. S. Rand, and D. M. Windsor, eds.), pp. 201–212. Smithsonian Inst. Press, Washington, D.C.

Foster, R. B., and Brokaw, N. V. L. (1982). Structure and history of the vegetation of Barro Colorado Island. *In* "The Ecology of a Tropical Forest" (E. G. Leigh, Jr., A. S. Rand, and D. M. Windsor, eds.), pp. 67–81. Smithsonian Inst. Press, Washington, D.C.

Fowells, H. A. (1965). Silvics of forest trees of the United States. *U.S. Dep. Agric., Agric. Handb.* No. 271.

Fowler, N. (1982). Competition and coexistence in a North Carolina grassland. III. Mixtures of component species. *J. Ecol.* **70**, 77–92.

Fox, J. F. (1977). Alternation and coexistence of tree species. *Am. Nat.* **111**, 69–89.

Fox, L. R., and Macauley, B. J. (1977). Insect grazing on *Eucalyptus* in response to variation in leaf tannins and nitrogen. *Oecologia* **29**, 145–162.

Frank, P. W. (1965). The biodemography of an intertidal snail population. *Ecology* **46**, 831–844.

Frankel, O. H., and Soulé, M. E. (1981). "Conservation and Evolution." Cambridge Univ. Press, London and New York.

Frankie, G. W., Baker, H. G., and Opler, P. A. (1974). Comparative phenological studies of trees in tropical wet and dry forests in the lowlands of Costa Rica. *J. Ecol.* **62**, 881–919.

Franklin, I. R. (1980). Evolutionary change in small populations. *In* "Conservation Biology: An Evolutionary–Ecological Perspective" (M. E. Soulé and B. A. Wilcox, eds.), pp. 135–149. Sinauer, Sunderland, Massachusetts.

Franklin, J. F., and Hemstrom, M. A. (1981). Aspects of succession in the coniferous forests of the Pacific Northwest. *In* "Forest Succession: Concepts and Application" (D. C. West, H. H. Shugart, and D. B. Botkin, eds.), pp. 212–239. Springer-Verlag, Berlin and New York.

Fretwell, S. D. (1969). Dominance behavior and winter habitat distribution in juncos (*Junco hyemalis*). *Bird Banding* **40**, 1–25.

Fretwell, S. D., and Lucas, H. L. (1969). On territorial behavior and other factors influencing habitat distribution in birds. I. Theoretical development. *Acta Biotheor.* **19**, 16–36.

Frissell, S. S. (1973). The importance of fire as a natural ecological factor in Itasca State Park, Minnesota. *Quat. Res. (N.Y.)* **3**, 397–407.

Fritz, R. S. (1980). Consequences of insular population structure: distribution and extinction of Spruce Grouse populations in the Adirondack Mountains. *Acta Congr. Int. Ornithol.* **17**, 757–763.

Frome, M. (1966). "Strangers in High Places." Doubleday, Garden City, New York.

Frye, R. J., and Rosenzweig, M. L. (1980). Clump size selection: a field test with two species of *Dipodomys*. *Oecologia* **47**, 323–327.

Fuentes, E. R., and Cancino, J. (1979). Rock-ground patchiness in a simple *Liolaemus* lizard community (Reptilia, Lacertilia, Iguanidae). *J. Herpetol.* **13,** 343–350.

Fuentes, E. R., Etchegaray, J., Aljaro, M. E., and Montenegro, G. (1981). Shrub defoliation by insects. *In* "Mediterranean-Type Shrublands" (F. di Castri, D. W. Goodall, and R. L. Specht, eds.), pp. 345–359. Elsevier, Amsterdam.

Fujino, K., and Kang, T. (1968). Transferrin groups of tunas. *Genetics* **59,** 79–91.

Fujita, T. T. (1976). "Tornado Map." Univ. of Chicago, Chicago, Illinois.

Fulton, R. E., and Carpenter, F. L. (1979). Pollination, reproduction, and fire in *Arctostaphylos*. *Oecologia* **46,** 322–329.

Furniss, R. L., and Carolin, V. M. (1977). Western forest insects. *Misc. Publ.—U.S. Dep. Agric.* No. 1339.

Futuyma, D., and Slatkin, M., eds. (1983). "Coevolution." Sinauer, Sunderland, Massachusetts.

Gadgil, M. (1971). Dispersal: Population consequences and evolution. *Ecology* **52,** 253–261.

Gadgil, M., and Bossert, W. H. (1970). Life historical consequences of natural selection. *Am. Nat.* **104,** 1–24.

Gadgil, M., and Solbrig, O. T. (1972). The concept of r- and K-selection: evidence from wild flowers and some theoretical considerations. *Am. Nat.* **106,** 14–31.

Gaines, M. R., McClenaghan, L. R., Jr., and Rose, K. K. (1978). Temporal patterns of allozymic variation in fluctuating populations of *Microtus ochrogaster*. *Evolution* **32,** 723–739.

Gardner, C. A. (1957). The fire factor, in relation to the vegetation of Western Australia. *West. Aust. Nat.* **5,** 166–173.

Garten, C. T. (1976). Relationships between aggressive behavior and genic heterozygosity in the oldfield mouse, *Peromyscus polionotus*. *Evolution* **30,** 59–72.

Garwood, N. C. (1983). Seed germination in a seasonal tropical forest in Panama: a community study. *Ecol. Monogr.* **53,** 159–181.

Garwood, N. C., Janos, D. P., and Brokaw, N. (1979). Earthquake caused landslides: a major disturbance to tropical forests. *Science (Washington, D.C.)* **205,** 997–999.

Gates, D. M. (1980). "Biophysical Ecology." Springer-Verlag, Berlin and New York.

Gauhl, E. (1976). Photosynthetic response to varying light intensity in ecotypes of *Solanum dulcamara* L. from shaded and exposed habitats. *Oecologia* **22,** 275–276.

Gause, G. F., and Witt, A. A. (1935). Behavior of mixed populations and the problem of natural selection. *Am. Nat.* **69,** 596–609.

Geiger, R. (1965). "The Climate near the Ground." Harvard Univ. Press, Cambridge, Massachusetts. (Transl. of "Das Klima der Bodennahen Luftschicht," 4th Ed.)

Gerrodette, T. (1981). Dispersal of the solitary coral *Balanophyllia elegans* by demersal planular larvae. *Ecology* **62,** 611–619.

Ghiselin, M. (1974). "The Economy of Nature and the Evolution of Sex." Univ. of California Press, Berkeley.

Gholz, H. L. (1980). Production and the role of vegetation in element cycles for the first three years on an unburned clearcut watershed in western Oregon. *Bull. Ecol. Soc. Am.* **61,** 149.

Gholz, H. L. (1982). Environmental limits on aboveground net primary production, leaf area, and biomass in vegetation zones of the Pacific Northwest. *Ecology* **63,** 469–481.

Giesel, J. T. (1976). Reproductive strategies as adaptations to life in temporally heterogeneous environments. *Annu. Rev. Ecol. Syst.* **7,** 57–79.

Gilbert, L. E. (1980). Food web organization and the conservation of neotropical diversity. *In* "Conservation Biology: An Evolutionary–Ecological Perspective" (M. E. Soulé and B. A. Wilcox, eds.), pp. 11–33. Sinauer, Sunderland, Massachusetts.

Gill, A. M. (1975). Fire and the Australian flora. *Aust. For.* **38,** 4–25.

Gill, A. M. (1981). Fire adaptive traits of vascular plants. *In* "Fire Regimes and Ecosystems Properties" (H. A. Mooney, T. M. Bonnicksen, N. L. Christensen, J. E. Lotan, and W. A. Reiners, eds.), *Gen. Tech. Rep. WO—U.S. For. Serv. (Wash. Off.)* **GTR-WO-26,** pp. 208–230.

Gill, A. M., and Groves, R. H. (1981). Fire regimes in heathlands and their plant ecologic effects. *In* "Ecosystems of the World. Vol. 9B: Heathland and Related Shrublands" (R. L. Specht, ed.), pp. 61–84. Elsevier, Amsterdam.

Gill, A. M., and Ingwerson, F. (1976). Growth of *Xanthorrea australis* R.Br. in relation to fire. *J. Appl. Ecol.* **13**, 195–203.

Gill, F. B., and Wolf, L. L. (1977). Nonrandom foraging by sunbirds in a patchy environment. *Ecology* **58**, 1284–1296.

Gillespie, J. H. (1975). The role of migration in the genetic structure of populations in temporally and spatially varying environments. I. Conditions for polymorphism. *Am. Nat.* **109**, 127–135.

Gillespie, J. H. (1977). A general model to account for enzyme variation in natural populations. III. Multiple alleles. *Evolution* **31**, 85–90.

Gillespie, J. H. (1978). A general model to account for enzyme variation in natural populations. V. The SAS–CFF model. *Theor. Popul. Biol.* **14**, 1–45.

Gimingham, C. H. (1971). British heathland ecosystems: the outcome of many years of management by fire. *Proc. Annu. Tall Timbers Fire Ecol. Conf.* **10**, 293–321.

Gimingham, C. H. (1972). "Ecology of Heathlands." Chapman & Hall, London.

Gimingham, C. H., Chapman, S. B., and Webb, N. R. (1979). European heathlands. *In* "Ecosystems of the World. Vol. 9A: Heathlands and Related Shrublands" (R. L. Specht, ed.), pp. 365–413. Elsevier, Amsterdam.

Givnish, T. J. (1978). On the adaptive significance of compound leaves, with particular reference to tropical trees. *In* "Tropical Trees as Living Systems" (P. B. Tomlinson and M. H. Zimmerman, eds.), pp. 351–380. Cambridge Univ. Press, London and New York.

Givnish, T. J. (1979). On the adaptive significance of leaf form. *In* "Topics in Plant Population Biology" (O. T. Solbrig, S. Jain, G. B. Johnson, and P. H. Raven, eds.), pp. 375–407. Columbia Univ. Press, New York.

Givnish, T. J. (1982). On the adaptive significance of leaf height in forest herbs. *Am. Nat.* **120**, 353–381.

Givnish, T. J., and Vermeij, G. J. (1976). Sizes and shapes of liane leaves. *Am. Nat.* **100**, 743–778.

Glanz, W. E. (1982). The terrestrial mammal fauna of Barro Colorado Island: Censuses and long-term changes. *In* "The Ecology of Tropical Forest: Seasonal Rhythms and Long-term Changes" (E. G. Leigh, Jr., A. S. Rand, and D. M. Windsor, eds.), pp. 455–468. Smithsonian Inst. Press, Washington, D.C.

Glass, G. E., and Slade, N. A. (1980a). Population structure as a predictor of spatial association between *Sigmodon hispidus* and *Microtus ochrogaster*. *J. Mammal.* **61**, 473–485.

Glass, G. E., and Slade, N. A. (1980b). The effect of *Sigmodon hispidus* on spatial and temporal activity of *Microtus ochrogaster:* evidence for competition. *Ecology* **61**, 358–370.

Glavac, V., and Koenies, H. (1978). Mineral Stickstoff-Gehalte und N-Nettomineralisation in Boden eines Fichtenforstes und seines Kahlschlages während der Vegetationsperiode. *Oecol. Plant.* **13**, 207–218.

Gleason, H. A. (1927). Further views on the succession concept. *Ecology* **8**, 299–326.

Gloyne, R. W. (1968). The structure of the wind and its relevance to forestry. *For. Suppl.* **20**, 7–19.

Glynn, P. W. (1968). Mass mortalities of echinoids and other reef flat organisms coincident with midday, low water exposures in Puerto Rico. *Mar. Biol. (Berlin)* **1**, 226–243.

Glynn, P. W. (1976). Some physical and biological determinants of coral community structure in the eastern Pacific. *Ecol. Monogr.* **46**, 431–456.

Glynn, P. W., Stewart, R. H., and McCosker, J. E. (1972). Pacific coral reefs of Panama: structure, distribution and predators. *Geol. Rundsch.* **61**, 483–519.

Godman, R. M., and Mattson, G. A. (1976). Seed crops and regeneration problems of 19 species in northeastern Wisconsin. *U.S. For. Serv., Res. Pap. NC* **NC-123.**

Goheen, D. J., and Cobb, F. W., Jr. (1978). Occurrence of *Verticicladiella wagenerii* and its perfect state, *Ceratocystis wageneri* sp. nov., in insect galleries. *Phytopathology* **68**, 1192–1195.

Goheen, D. J., and Cobb, F. W., Jr. (1980). Infestation of *Ceratocystis wageneri*-infected ponderosa pines by bark beetles (Coleoptera: Scolytidae) in the central Sierra Nevada. *Can. Entomol.* **112,** 725–730.

Golden, M. S. (1981). An integrated multivariate analysis of forest communities of the central Great Smoky Mountains. *Am. Midl. Nat.* **106,** 37–53.

Gomez-Pompa, A., Vazquez-Yanes, C., and Guevara, S. (1972). The tropical rain forest: a nonrenewable resource. *Science (Washington, D.C.)* **177,** 762–765.

Good, R. E., Whigham, D. F., and Simpson, R. L., eds. (1978). "Freshwater Wetlands: Ecological Processes and Management Potential." Academic Press, New York.

Goodlett, J. C. (1954). Vegetation adjacent to the border of the Wisconsin drift in Potter County, Pennsylvania. *Harv. For. Bull.* **25.**

Gordon, D. P. (1972). Biological relationships of an intertidal bryozoan population. *J. Nat. Hist.* **6,** 503–514.

Gorham, E. (1961). Factors influencing supply of major ions to inland waters, with special reference to the atmosphere. *Geol. Soc. Am. Bull.* **72,** 795–840.

Gorham, E., Vitousek, P. M., and Reiners, W. A. (1979). The regulation of chemical budgets over the course of terrestrial ecosystem succession. *Annu. Rev. Ecol. Syst.* **10,** 53–84.

Goryshina, T. K., Zabotina, L. N., and Pruzhina, E. C. (1981). Mesostructure of the photosynthetic apparatus in the wood anemone (*Anemone nemorosa* L.) from different habitats. *Sov. J. Ecol. (Engl. Transl.)* **12,** 13–19.

Gosz, J. R. (1980a). Biomass distribution and production budget for a nonaggrading forest ecosystem. *Ecology* **61,** 507–514.

Gosz, J. R. (1980b). Nutrient budget studies for forests along an elevational gradient in New Mexico. *Ecology* **61,** 515–521.

Gowen, J. W. (1952). "Heterosis." Univ. of Iowa Press, Ames.

Graham, J. B., Kramer, D. L., and Pineda, E. (1977). Respiration of the air breathing fish *Piabucina festae. J. Comp. Physiol.* **122,** 295–310.

Graham, J. B., Kramer, D. L., and Pineda, E. (1978). Comparative respiration of an air breathing and a non-airbreathing characoid fish and the evolution of aerial respiration in characins. *Physiol. Zool.* **51,** 279–288.

Graham, S. A. (1958). Results of deer exclosure experiments in the Ottawa National forest. *Trans. North Am. Wildl. Nat. Resour. Conf.* **23,** 478–490.

Grant, W. E., Birney, E. C., French, N. R., and Swift, D. M. (1982). Structure and productivity of grassland small mammal communities related to grazing-induced changes in vegetative cover. *J. Mammal.* **63,** 248–260.

Grant, W. S. (1977). High intertidal community organization on a rocky headland in Maine, USA. *Mar. Biol. (Berlin)* **44,** 15–25.

Gray, J. T. (1982). Community structure and productivity in *Ceanothus* chaparral and coastal sage scrub of southern California. *Ecol. Monogr.* **52,** 415–435.

Gray, J. T., and Schlesinger, W. H. (1981). Nutrient cycling in Mediterranean type ecosystems. *In* "Resource Use by Chaparral and Matorral" (P. C. Miller, ed.), pp. 259–285. Springer-Verlag, Berlin and New York.

Green, L. R. (1982). Prescribed burning in the California Mediterranean Ecosystem. *In* "Dynamics and Management of Mediterranean-Type Ecosystems (C. E. Conrad and W. C. Oechel, eds.), *USDA For. Ser. Gen. Tech. Rep. PSW* **PSW-58,** pp. 464–471.

Greene, C. H., and Schoener, A. (1982). Succession on marine hard substrata: a fixed lottery. *Oecologia* **55,** 289–297.

Greig-Smith, P. (1979). Pattern in vegetation. *J. Ecol.* **67,** 755–779.

Grier, C. C. (1975). Wildfire effects on nutrient distribution and leaching in a coniferous ecosystem. *Can. J. For. Res.* **5,** 599–607.

Grier, C. C., and Logan, R. S. (1977). Old-growth *Pseudotsuga menziesii* communities of a western Oregon watershed: Biomass distribution and production budgets. *Ecol. Monogr.* **47,** 373–400.

Grier, C. C., and Running, S. W. (1977). Leaf area of mature northwestern coniferous forests: relation to site water balance. *Ecology* **58**, 893–899.

Grier, C. C., Vogt, K. A., Keyes, M. R., and Edmonds, R. L. (1981). Biomass distribution and above- and below-ground production in young and mature *Abies amabilis* zone ecosystems of the Washington Cascades. *Can. J. For. Res.* **11**, 155–167.

Grigg, R. W., and Maragos, J. E. (1974). Recolonization of hermatypic corals on submerged lava flows in Hawaii. *Ecology* **55**, 387–395.

Grime, J. P. (1965). Shade tolerance in flowering plants. *Nature (London)* **208**, 161–163.

Grime, J. P. (1966). Shade avoidance and shade tolerance in flowering plants. *In* "Light as an Ecological Factor" (R. Bainbridge, G. C. Evans, and O. Rackham, eds.), *Symp. Br. Ecol. Soc.* **6**, 187–207.

Grime, J. P. (1974). Vegetation classification by reference to strategies. *Nature (London)* **250**, 26–31.

Grime, J. P. (1977). Evidence for the existence of three primary strategies in plants and its relevance to ecological and evolutionary theory. *Am. Nat.* **111**, 1169–1194.

Grime, J. P. (1979). "Plant Strategies and Vegetation Processes." Wiley, New York.

Grime, J. P., and Hunt, R. (1975). Relative growth rate: its range and adaptive significance in a local flora. *J. Ecol.* **63**, 393–422.

Gross, K. L. (1980). Colonization by *Verbascum thapsus* (mullein) of an old field in Michigan: experiments on the effects of vegetation. *J. Ecol.* **68**, 919–927.

Gross, K. L., and Werner, P. A. (1982). Colonizing abilities of "biennial" plant species in relation to ground cover: implications for their distributions in a successional sere. *Ecology* **63**, 921–931.

Groves, R. H. (1981a). Heathland soils and their fertility status. *In* "Ecosystems and Related Shrublands" (R. L. Specht, ed.), pp. 143–150. Elsevier, Amsterdam.

Groves, R. H. (1981b). Nutrient cycling in heathlands. *In* "Ecosystems of the World. Vol. 9b: Heathlands and Related Shrublands" (R. L. Specht, ed.), pp. 151–163. Elsevier, Amsterdam.

Grubb, P. J. (1977). The maintenance of species-richness in plant communities: the importance of the regeneration niche. *Biol. Rev. Cambridge Philos. Soc.* **52**, 107–145.

Guarda, G., F. (1953). "Historia de Valdivia, 1552–1952." Imprenta Cultural, Santiago, Chile.

Guevara, S. S., and Gómez-Pompa, A. (1972). Seeds from surface soils in a tropical region of Veracruz, Mexico. *J. Arnold Arbor., Harv. Univ.* **53**, 312–335.

Gulliksen, B. (1980). The macrobenthic rocky-bottom fauna of Borgenfjorden, North-Trondelag, Norway. *Sarsia* **65**, 115–136.

Gulliksen, B., Haug, T., and Sandnes, O. K. (1980). Benthic macrofauna on new and old lava grounds at Jan Mayen. *Sarsia* **65**, 137–148.

Gunckel, L. H. (1948). La floración de la quila y del colihue en la Araucanía. *Cienc. Invest.* **4**, 91–95.

Guntenspergen, G., and Stearns, F. W. (1984). Temperature in a circular clearing in the forest. In preparation.

Guries, R. P., and Ledig, F. T. (1982). Genetic diversity and population structure in pitch pine (*Pinus rigida* Mill.). *Evolution* **36**, 387–402.

Guttman, S. I., Wood, T. K., and Karlin, A. A. (1981). Genetic differentiation along host plant lines in the sympatric *Echenopa binotata* Say complex (Homoptera: Membracidae). *Evolution* **35**, 205–217.

Gysel, L. W. (1951). Borders and openings of beech–maple woodlands in southern Michigan. *J. For.* **49**, 13–19.

Hadley, E. B. (1961). Influence of temperature and other factors on Ceanothus magecarpus seed germination. *Madrono* **16**, 132–138.

Haggett, P., Cliff, A. D., and Frey, A. (1977). "Locational Analysis in Human Geography." Arnold, London.

Haila, Y., Hanski, I., Järvinen, O., and Ranta, E. (1982). Insular biogeography: a northern European perspective. *Acta Oecol. [Ser.]: Oecol. Gen.* **3**, 303–318.

Halkka, O. (1978). Influence of spatial and host–plant population on polymorphism in *Philaenus spumarius. In* "Diversity of Insect Faunas" (L. A. Mound and N. Waloff, eds.), pp. 41–55. Blackwell, Oxford.

Halkka, O., Halkka, L., Hovinen, R., Raatikainen, M., and Vasarainen, A. (1975a). Genetics of *Philaenus* colour polymorphism: the 28 genotypes. *Hereditas* **79**, 308–310.

Halkka, O., Raatikainen, M., Halkka, L., and Hovinen, R. (1975b). The genetic composition of *Philaenus spurmarius* populations in island habitats variably effected by voles. *Evolution* **29**, 700–706.

Hall, J. B., and Swaine, M. D. (1976). Classification and ecology of a closed-canopy forest in Ghana. *J. Ecol.* **64**, 913–951.

Hall, J. B., and Swaine, M. D. (1980). Seed stocks in Ghanian forest soils. *Biotropica* **12**, 256–263.

Hall, J. B., and Swaine, M. D. (1981). "Distribution and Ecology of Vascular Plants in a Tropical Rain Forest: Forest Vegetation in Ghana." Junk, The Hague.

Hallé, F., Oldeman, R. A. A., and Tomlinson, P. B. (1978). "Tropical Trees and Forests: An Architectural Analysis." Springer-Verlag, Berlin and New York.

Halligan, J. P. (1973). Bare areas associated with shrub stands in grassland: the case of *Artemesia california. BioScience* **23**, 429–432.

Hamilton, W. D., and May, R. M. (1977). Dispersal in stable habitats. *Nature (London)* **269**, 578–581.

Hamrick, J. L. (1982). Plant population genetics and evolution. *Am. J. Bot.* **69**, 1685–1693.

Hamrick, J. L., Linhart, Y. B., and Mitton, J. B. (1979). Relationships between life history characteristics and electrophoretically detectable genetic variation in plants. *Annu. Rev. Ecol. Syst.* **10**, 175–200.

Hancock, J. F., and Bringhurst, R. S. (1978). Interpopulational differentiation and adaptation in the perennial, diploid species of *Fragaria vesca* L. *Am. J. Bot.* **65**, 795–803.

Handel, S. N. (1976). Dispersal ecology of *Carex pedunculata* (Cyperaceae), a new North American myrmecochore. *Am. J. Bot.* **63**, 1071–1079.

Hanes, T. L. (1971). Succession after fire in the chaparral of southern California. *Ecology* **48**, 259–264.

Hansen, E. A., and Baker, J. B. (1979). Biomass and nutrient removal in short-rotation intensively cultured plantations. *In* "Impact of Intensive Harvesting on Forest Nutrient Cycling" (A. L. Leaf, ed.), pp. 130–151. USDA For. Serv. Northeast. For. Exp. Stn., Broomall, Pennsylvania.

Hanski, I. (1980). Spatial patterns and movements in coprophagous beetles. *Oikos* **34**, 293–310.

Hanski, I. (1981). Coexistence of competitors in patchy environment with and without predation. *Oikos,* **37**, 306–312.

Hanski, I. (1982a). On patterns of temporal and spatial variation in animal populations. *Ann. Zool. Fenn.* **19**, 21–37.

Hanski, I. (1982b). Dynamics of regional distribution: the core and satellite species hypothesis. *Oikos* **38**, 210–221.

Hanski, I. (1983). Coexistence of competitors in patchy environment. *Ecology* **64**, 493–500.

Hansson, L. (1977). Landscape ecology and stability of populations. *Landscape Plann.* **4**, 85–93.

Hansson, L. (1979). On the importance of landscape heterogeneity in northern regions for the breeding population densities of homeotherms: a general hypothesis. *Oikos* **33**, 182–189.

Harcombe, P. A. (1977). Nutrient accumulation by vegetation during the first year of recovery of a tropical forest ecosystem. *In* "Recovery and Restoration of Damaged Ecosystems" (J. Cairns, K. L. Dickison, and E. E. Herricks, eds.), pp. 347–348. Univ. of Virginia Press, Charlottesville.

Harcombe, P. A. (1980). Soil nutrient loss as a factor in early tropical secondary succession. *Biotropica* **12**, Suppl., 8–15.

Harding, J. A., and Barnes, K. (1977) Genetics of *Lupinus*. X. Genetic variability, heterozygosity and outcrossing in colonial populations of *Lupinus succulentus. Evolution* **31**, 247–255.

Harger, J. R. E., and Landenberger, D. E. (1971). The effects of storms as a density dependent mortality factor on populations of sea mussels. *Veliger* **14**, 195–201.

Harlin, M. M., and Lindbergh, J. M. (1977). Selection of substrata by seaweeds: optimal surface relief. *Mar. Biol. (Berlin)* **40**, 33–40.

Harmon, M. E. (1982). Fire history of the westernmost portion of Great Smoky Mountains National Park. *Bull. Torry Bot Club* **109**, 74–79.

Harmon, M. E., Bratton, S. P., and White, P. S. (1983). Disturbance and vegetation response in relation to environmental gradients in the Great Smoky Mountains. *Vegetatio* **55**, 129–139.

Harper, J. L. (1969). The role of predation in vegetational diversity. *Brookhaven Symp. Biol.* No. 22, 48–62.

Harper, J. L. (1977). "Population Biology of Plants." Academic Press, New York.

Harrington, R. W., Jr. (1968). Delimitation of the thermolabile phenocritical period of sex determination and differentiation in the ontogeny of the normally hermaphroditic fish, *Rivulus marmoratus* Poey. *Physiol. Zool.* **41**, 447–460.

Harris, J. (1978). The ecology of marine epifaunal communities on the pilings of Portsea pier, Victoria. B.S. Thesis, Univ. of Melbourne.

Harris, W. F., Sollins, P., Edwards, N. T., Dinger, B. E., and Shugart, H. H. (1975). Analysis of carbon flow and productivity in a temperate forest ecosystem. *In* "Productivity of World Ecosystems" (D. E. Reichle, J. F. Franklin, and D. W. Goodall, eds.), pp. 116–122. Natl. Acad. Sci., Washington, D.C.

Harrison, P. L., Babcock, R. C., Bull, G. D., Oliver, J. K., Wallace, C. C., and Willis, B. L. (1984). Mass spawning in tropical reef corals. *Science (Washington, D.C.)* **223**, 1186–1189.

Hartshorn, G. S. (1978). Treefalls and tropical forest dynamics. *In* "Tropical Trees as Living Systems" (P. B. Tomlinson and M. H. Zimmermann, eds.), pp. 617–638. Cambridge Univ. Press, London and New York.

Hartshorn, G. S. (1980). Neotropical forest dynamics. *Biotropica* **12**, Suppl., 23–30.

Hassell, M. P. (1976). Arthropod predator–prey systems. *In* "Theoretical Ecology" (R. M. May, ed.), pp. 71–93. Saunders, Philadelphia, Pennsylvania.

Hastings, A. (1978). Spatial heterogeneity and the stability of predator–prey systems: predator-mediated coexistence. *Theor. Popul. Biol.* **14**, 380–395.

Hastings, A. (1980). Disturbance, coexistence, history and competition for space. *Theor. Popul. Biol.* **18**, 363–373.

Haven, S. B. (1971). Effects of land-level changes on intertidal invertebrates, with discussion of earthquake ecological succession. *In* "The Great Alaskan Earthquake of 1964: Biology," Publ. No. 1604, pp. 82–126. Committee on the Great Alaskan Earthquake of 1964, Natl. Res. Counc.– Natl. Acad. Sci., Washington, D.C.

Hawkins, S. J. (1981). The influence of season and barnacles on the algal colonization of *Patella vulgata* exclusion areas. *J. Mar. Biol. Assoc. U.K.* **61**, 1–15.

Hawkins, S. J., and Hartnoll, R. G. (1982). Settlement patterns of *Semibalanus balanoides* L. in the Isle of Man (1977–1981). *J. Exp. Mar. Biol. Ecol.* **62**, 271–283.

Hay, M. E. (1981). The functional morphology of turf-forming seaweeds: persistence in stressful marine habitats. *Ecology* **62**, 739–750.

Hay, M. E., and Fuller, P. J. (1981). Seed escape from heteromyid rodents: the importance of microhabitat and seed preference. *Ecology* **62**, 1395–1399.

Hayward, P. J. (1973). Preliminary observations on settlement and growth in populations of *Alcyonidium hirsutum* (Fleming). *In* "Living and Fossil Bryozoa" (G. P. Larwood, ed.), pp. 107–113. Academic Press, New York.

Hebert, P. D. N., Ward, R. D., and Gibson, J. B. (1972). Natural selection for enzyme variants among parthenogenetic *Daphnia magna*. *Genet. Res.* **19**, 173–176.

Hedgpeth, J. W. (1957). Sandy beaches. *In* "Treatise on Marine Ecology and Paleoecology. Vol. 1: Ecology" (J. W. Hedgpeth, ed.), Mem. No. 67, pp. 587–608. Geol. Soc. Am., Boulder, Colorado.

Hedlin, A. F., Miller, G. E., and Ruth, D. S. (1982). Induction of prolonged diapause in *Barbara colfaxiana* (Lepidoptera: Olethreutidae): correlations with cone crops and weather. *Can. Entomol.* **114**, 465–471.

Hedrick, P. W. (1981). The establishment of chromosomal variants. *Evolution* **35**, 322–332.

Hedrick, P. W. (1983). "Genetics of Populations." Van Nostrand-Reinhold, New York.

Hedrick, P. W., and Holden, L. (1979). Hitchhiking: an alternative to coadaptation for the barley and slender oat examples. *Heredity* **43**, 79–86.

Hedrick, P. W., Ginevan, M. E., and Ewing, E. P. (1976). Genetic polymorphism in heterogeneous environments. *Annu. Rev. Ecol. Syst.* **7**, 1–32.

Hegyi, F. (1974). A simulation model for managing jack-pine stands. *In* "Growth Models for Tree and Stand Simulation" (J. Fries, ed.), Res. Notes No. 30, pp. 74–87. Dep. For. Yield Res., R. Coll. For., Stockholm.

Heilman, P. E., and Geisel, S. P. (1963). Nitrogen requirements and the biological cycling of nitrogen in Douglas-fir in relation to the effects of nitrogen fertilization. *Plant Soil* **18**, 386–402.

Heinselman, M. L. (1973). Fire in the virgin forests of the Boundary Waters Canoe Area, Minnesota. *Quat. Res. (N.Y.)* **3**, 329–382.

Heinselman, M. L. (1981a). Fire and succession in the conifer forests of northern North America. *In* "Forest Succession: Concepts and Application" (D. C. West, H. H. Shugart, and D. B. Botkin, eds.), pp. 374–405. Springer-Verlag, Berlin and New York.

Heinselman, M. L. (1981b). Fire intensity and frequency as factors in the distribution and structure of northern ecosystems. *In* "Fire Regimes and Ecosystem Properties" (H. A. Mooney, T. M. Bonnicksen, N. L. Christensen, J. E. Lotan, and W. A. Reiners, eds.), *Gen. Tech. Rep. WO— U.S. For. Serv. [Wash. Off.]* **GTR-WO-26**, pp. 7–57.

Hellmers, H., Bonner, J. F., and Kelleher, J. M. (1955). Soil fertility: a watershed management problem in the San Gabriel Mountains of southern California. *Soil Sci.* **80**, 189–197.

Henry, J. D., and Swan, J. M. A. (1974). Reconstructing forest history from live and dead plant material—an approach to the study of forest succession in southwest New Hampshire. *Ecology* **55**, 772–783.

Hervé, A. F., Moreno, H. H., and Parada, R. M. (1974). Granitoids of the Andean range of Valdivia Province, Chile. *Pac. Geol.* **8**, 39–45.

Hesselman, H. (1917). Om våra skogsföryngringsåtgärders inverkan på salpeterbildningen i marken och dess betyldese för barrskogens föryngring. (On the effect of our regeneration measures on the formation of saltpetre in the ground and its importance in the regeneration of coniferous forests.) *Medd. Statens Skogsfoersoeksanstalt* **13/14**, 923–1076 (Engl. summ., pp. xci–cxxvi).

Heusser, C. J. (1966). Late-Pleistocene pollen diagrams from the Province of Llanquihue, southern Chile. *Proc. Am. Philos. Soc.* **110**, 269–305.

Heusser, C. J. (1974). Vegetation and climate of the southern Chilean Lake District during and since the last interglaciation. *Quat. Res. (N.Y.)* **4**, 290–315.

Hibbs, D. E. (1982). Gap dynamics in a hemlock-hardwood forest. *Can. J. For. Res.* **12**, 522–527.

Hibbs, D. E., and Fischer, B. C. (1979). Sexual and vegetative reproduction of striped maple (*Acer pensylvanicum* L.). *Bull. Torrey Bot. Club* **106**, 222–227.

Hickman, J. C. (1975). Environmental unpredictability and plastic energy allocation strategies in the annual *Polygonum cascadense*. *J. Ecol.* **63**, 689–701.

Hickman, J. C. (1979). The basic biology of plant numbers. *In* "Topics in Plant Population Biology" (O. T. Solbrig, S. Jain, G. B. Johnson, and P. H. Raven, eds.), pp. 232–263. Columbia Univ. Press, New York.

Hiebert, R. D., and Hamrick, J. L. (1983). Patterns and levels of genetic variation in Great Basin bristlecone pine, *Pinus longaeva*. *Evolution* **37**, 302–310.

Higgs A. J. (1981). Island biogeography theory and nature reserve design. *J. Biogeogr.* **8**, 117–124.

Highsmith, R. C. (1980). Passive colonization and asexual colony multiplication in the massive coral *Porites lutea* Milne Edwards and Haime. *J. Exp. Mar. Biol. Ecol.* **47**, 55–67.

Highsmith, R. C. (1982). Reproduction by fragmentation in corals. *Mar. Ecol.: Prog. Ser.* **7**, 207–226.

Highsmith, R. C., Riggs, A. C., and D'Antonio, M. (1980). Survival of hurricane-generated coral fragments and a disturbance model of reef calcification/growth rates. *Oecologia* **46**, 322–329.

Hildén, O. (1965). Habitat selection in birds. *Ann. Zool. Fenn.* **2**, 53–75.

Hill, W. G., and Robertson, A. (1966). The effect of linkage on limits to artificial selection. *Genet. Res.* **8**, 269–294.

Hirose, T. (1975). Relations between turnover rate, resource utility, and structure of some plant populations: a study in the matter budgets. *J. Fac. Sci., Univ. Tokyo, Sect. 3* **11**, Part II, 355–407.

Hixon, M. A., and Brostoff, W. N. (1983). Damselfish as keystone species in Reverse: intermediate disturbance and diversity of reef algae. *Science (Washington, D.C.)* **220**, 511–513.

Hladik, A. (1982). Dynamique d'une forêt équatoriale africaine: mesures en temps réel et comparison du potential de croissance des différentes espècies. *Acta Oecol. [Ser.]: Oecol. Gen.* **3**, 373–392.

Hodgkin, E. P. (1959). Catastrophic destruction of the littoral fauna and flora near Fremantle January 1959. *J. R. Soc. West. Aust.* **42**, 6–11.

Hodgkin, E. P. (1960). Patterns of life on rocky shores. *J. R. Soc. West. Aust.* **43**, 35–45.

Hoerner, S. F. (1965). "Fluid-dynamic Drag." (Available from S. F. Hoerner, 2 King Lane, Greenbriar, Brick Town, New Jersey 08723.)

Holdridge, L. R., Grenke, W. C., Hatheway, W. H., Liang, T., and Tosi, J. A., Jr. (1971). "Forest Environments in Tropical Life Zones." Pergamon, New York.

Holling, C. S. (1973). Resilience and stability of ecological systems. *Annu. Rev. Ecol. Syst.* **4**, 1–23.

Holmes, R. T., and Robinson, S. K. (1981). Tree species preferences of foraging insectivorous birds in a northern hardwoods forest. *Oecologia* **48**, 31–35.

Holmes, R. T., Bonney, R. E., Jr., and Pacala, S. W. (1979). Guild structure of the Hubbard Brook bird community: a multivariate approach. *Ecology* **60**, 512–520.

Holt, B. R. (1972). Effect of arrival time on recruitment, mortality and reproduction in successional plant populations. *Ecology* **53**, 668–673.

Holthuijzen, A. A. M., and Boerboom, J. H. A. (1982). The *Cecropia* seed bank in the Surinam rain forest. *Biotropica* **14**, 62–68.

Honda, H., and Fisher, J. B. (1978). Tree branch angle: maximizing effective leaf area. *Science (Washington, D.C.)* **199**, 888–889.

Honda, H., Tomlinson, P. B., and Fisher, J. B. (1981). Computer simulation of branch interaction and regulation by unequal flow rates in botanical trees. *Am. J. Bot.* **68**, 569–585.

Hool, J. N. (1966). A dynamic programming–Markov chain approach to forest production control. *For. Sci. Monogr.* No. 12, 1–26.

Hopkins, B. A. (1965). "Forest and Savanna." Heinemann, London.

Horn, H. S. (1968). The adaptive significance of colonial nesting in the Brewer's Blackbird (*Euphagus cyanocephalus*). *Ecology* **49**, 682–694.

Horn, H. S. (1971). "The Adaptive Geometry of Trees." Princeton Univ. Press, Princeton, New Jersey.

Horn, H. S. (1974). The ecology of secondary succession. *Annu. Rev. Ecol. Syst.* **5**, 25–37.

Horn, H. S. (1975). Markovian properties of forest succession. *In* "Ecology and Evolution of Communities" (M. L. Cody and J. Diamond, eds.), pp. 196–211. Belknap Press, Cambridge, Massachusetts.

Horn, H. S. (1976). Succession. *In* "Theoretical Ecology" (R. M. May, ed.), pp. 187–204. Blackwell, Oxford.

Horn, H. S. (1981a). Some causes of variety in patterns of secondary succession. *In* "Forest Succession: Concepts and Application" (D. C. West, H. H. Shugart, and D. B. Botkin, eds.), pp. 24–35. Springer-Verlag, Berlin and New York.

Horn, H. S. (1981b). Succession. *In* "Theoretical Ecology" (R. M. May, ed.), pp. 253–271. Sinauer, Sunderland, Massachusetts.

Horn, H. S., and MacArthur, R. H. (1972). Competition among fugitive species in a harlequin environment. *Ecology* **53**, 749–752.

Hornbeck, J. W., Likens, G. E., Pierce, R. S., and Bormann, F. H. (1975). Strip cutting as a means of protecting site and streamflow quality when clearcutting northern hardwoods. *In* "Forest Soils and Land Management" (B. Bernier and C. F. Winget, eds.), pp. 209–225. Presses Univ. Laval, Quebec.

Horton, J. S., and Kraebel, C. J. (1955). Development of vegetation after fire in the chamise chaparral of southern California. *Ecology* **36**, 244–262.

Horton, R. E. (1945). Erosional development of streams and their drainage basins: hydrophysical approach to quantitative morphology. *Bull. Geol. Soc. Am.* **56**, 275–370.

Hosseus, C. K. (1915). "Las Cañas de Bambu en las Cordilleras del Sud," Bol. No. 19, pp. 195–208. Minist. Agric., Buenos Aires, Argentina.

Hough, A. F. (1936). A climax forest community on East Tionesta Creek in northwestern Pennsylvania. *Ecology* **17**, 9–28.

Hough, A. F. (1949). Deer and rabbit browsing and available winter forage in Allegheny hardwood forests. *J. Wildl. Manage.* **13**, 135–141.

Hough, A. F. (1965). A twenty-year record of understory vegetation change in a virgin Pennsylvania forest. *Ecology* **46**, 370–373.

Hough, A. F., and Forbes, R. D. (1943). The ecology and silvics of forests in the high plateaus of Pennsylvania. *Ecol. Monogr.* **13**, 299–320.

Hough, F. O., and Crown, J. A. (1952). "The Campaign on New Britain," U.S. Marine Corps Monographs. U.S. Gov. Print. Off., Washington, D.C.

Howard, T. M., and Ashton, D. H. (1973). The distribution of *Nothofagus cunninghamii* rainforest. *Proc. R. Soc. Victoria* **86**, 47–75.

Howe, H. F., and Smallwood, J. (1982). Ecology of seed dispersal. *Annu. Rev. Ecol. Syst.* **13**, 201–228.

Hruby, T., and Norton, T. A. (1979). Algal colonization on rocky shores in the Firth of Clyde. *J. Ecol.* **67**, 65–77.

Hubbell, S. P. (1979). Tree dispersion, abundance and diversity in a tropical dry forest. *Science (Washington, D.C.)* **203**, 1299–1309.

Hubbell, S. P. (1980). Seed predation and the coexistence of tree species in tropical forests. *Oikos* **35**, 214–229.

Huber, O. (1978). Light compensation point of vascular plants of a tropical cloud forest and an ecological interpretation. *Photosynthetica* **12**, 382–390.

Hubert, E. E. (1918). Fungi as contributory causes of windfall in the Northwest. *J. For.* **16**, 696–714.

Huey, R. B., and Slatkin, M. (1976). Cost and benefits of lizard thermoregulation. *Q. Rev. Biol.* **51**, 363–384.

Huffaker, C. B. (1958). Experimental studies on predation:dispersion factors and predator–prey oscillations. *Hilgardia* **27**, 343–383.

Humphrey, R. R. (1962). "Range Ecology." Ronald Press, New York.

Humphrey, R. R. (1974). Fire in the deserts and desert grassland of North America. *In* "Fire and Ecosystems" (T. T. Kozlowski and C. E. Ahlegren, eds.), pp. 365–400. Academic Press, New York.

Humphreys, W. F., and Kitchener, D. J. (1982). The effect of habitat utilization on species-area curves: implications for optimal reserve area. *J. Biogeogr.* **9**, 391–396.

Hunt, R. (1978). "Plant Growth Analysis," Studies in Biology, No. 96. Arnold, London.

Huston, M. (1979). A general hypothesis of species diversity. *Am. Nat.* **113**, 81–101.

Hutchinson, G. E. (1951). Copepodology for the ornithologist. *Ecology* **32**, 571–577.

Hutchinson, G. E. (1961). The paradox of the plankton. *Am. Nat.* **95**, 137–146.

Hutnik, R. J. (1952). Reproduction on windfalls in a northern hardwood stand. *J. For.* **50**, 693–694.

Hutto, R. L. (1978). A mechanism for resource allocation among sympatric heteromyid rodent species. *Oecologia* **33**, 115–126.

Ives, R. L. (1942). The beaver-meadow complex. *J. Geomorphol.* **5**, 191–203.

Jackson, J. B. C. (1977a). Competition on marine hard substrata: the adaptive significance of solitary and colonial strategies. *Am. Nat.* **111**, 743–767.

Jackson, J. B. C. (1977b). Habitat area, colonization and development of epibenthic community structure. *In* "Biology of Benthic Organisms" (B. F. Keegan, P. O. Ceidigh, and P. J. S. Boaden, eds.), pp. 349–358. Pergamon, Oxford.

Jackson, J. B. C. (1979a). Morphological strategies of sessile animals. *In* "Biology and Systematics of Colonial Organisms" (G. P. Larwood and B. R. Rosen, eds.), pp. 499–555. Academic Press, New York.

Jackson, J. B. C. (1979b). Overgrowth competition between encrusting cheilostome ectoprocts in a Jamaican cryptic reef environment. *J. Anim. Ecol.* **48**, 805–823.

Jackson, J. B. C., and Buss, L. W. (1975). Allelopathy and spatial competition among coral reef invertebrates. *Proc. Natl. Acad. Sci. U.S.A.* **72**, 5160–5163.

Jackson, J. B. C., and Winston, J. E. (1982). Ecology of cryptic coral reef communities. I. Distribution and abundance of major groups of encrusting organisms. *J. Exp. Mar. Biol. Ecol.* **57**, 135–148.

Jackson, L. W. R. (1959). Relation of pine forest overstory opening diameter to growth of pine reproduction. *Ecology* **40**, 478–480.

Jackson, M. T. (1966). Effects of microclimate on spring flowering phenology. *Ecology* **47**, 407–415.

Jain, S. K. (1975). Patterns of survival and microevolution in plant populations. *In* "Population Genetics and Ecology" (S. Karlin and E. Nevo, eds.), pp. 49–89. Academic Press, New York.

Jain, S. K. (1976). The evolution of inbreeding in plants. *Annu. Rev. Ecol. Syst.* **7**, 469–495.

Jain, S. K. (1978). Adaptive strategies: polymorphism, plasticity and homeostasis. *In* "Topics in Plant Population Biology" (D. Solbrig, G. Johnson, S. Jain, and P. Raven, eds.), pp. 160–187. Columbia Univ. Press, New York.

Jain, S. K. (1983). Genetic characteristics of populations. *In* "Disturbance and Ecosystems" (H. A. Mooney and M. Godron, eds.), pp. 240–258. Springer-Verlag, Berlin and New York.

Jain, S. K. (1984). Some evolutionary characteristics of plant–plant interactions. *In* "Plant Population Ecology" (R. Dirzo and J. Sarukhan, eds.), pp. 128–139. Sinauer, Sunderland, Massachusetts.

Jain, S. K., and Martins, P. S. (1979). Ecological genetics of the colonizing ability of rose clover (*Trifoliums hirtum* All.). *Am. J. Bot.* **66**, 361–366.

Jain, S. K., Rai, K. N., and Singh, R. S. (1981). Population biology of *Avena*. XI. Variation in peripheral isolates of *A. barbata*. *Genetica* **56**, 213–215.

Jane, G. T., and Green, T. G. A. (1983). Episodic forest mortality in the Kaimai Ranges, North Island, New Zealand. *N.Z. J. Bot.* **21**, 21–31.

Janos, D. P. (1980). Vesicular–arbuscular mycorrhizae affect lowland tropical rain forest plant growth. *Ecology* **61**, 151–162.

Janzen, D. H. (1967). Why mountain passes are higher in the tropics. *Am. Nat.* **101**, 233–249.

Janzen, D. H. (1970). Herbivores and the number of tree species in tropical forests. *Am. Nat.* **104**, 501–528.

Janzen, D. H. (1971). Seed predation by animals. *Annu. Rev. Ecol. Syst.* **2**, 465–492.

Janzen, D. H. (1975). Intra- and interhabitat variations in *Guazuma ulmifolia* (Sterculiaceae) seed predation by *Amblycerus cistelinus* (Bruchidae) in Costa Rica. *Ecology* **56**, 1009–1013.

Jarman, P. J., and Jarman, M. V. (1979). The dynamics of ungulate social organization. *In* "Serengeti: Dynamics of an Ecosystem" (A. R. E. Sinclair and M. Norton-Griffiths, eds.), pp. 185–220. Univ. of Chicago Press, Chicago, Illinois.

Jernakoff, P. (1983). Interactions among animals and algae in an intertidal zone dominated by barnacles. Ph.D. Thesis, Univ. of Sydney.

Johansen, H. W. (1971). Effects of elevational changes on benthic algae in Prince William Sound. *In* "The Great Alaskan Earthquake of 1964: Biology," Publ. 1604, pp. 35–68. Committee on the Great Alaskan Earthquake of 1964, Natl. Res. Counc.–Natl. Acad. Sci., Washington, D.C.

Johnson, E. A. (1981). Vegetation organization and dynamics of lichen woodland communities in the Northwest Territory, Canada. *Ecology* **62**, 200–215.

Johnson, P. L., and Billings, W. D. (1962). The alpine vegetation of the Beartooth Plateau in relation to cryopedogenic processes and patterns. *Ecol. Monogr.* **32**, 102–135.

Johnson, T. K., and Jorgensen, C. D. (1981). Ability of desert rodents to find buried seeds. *J. Range Manage.* **34**, 312–314.

Johnson, W. C. (1977). A mathematical model for succession and land use for the North Carolina Piedmont. *Bull. Torrey Bot. Club* **104**, 334–346.

Johnson, W. C., and Sharpe, D. M. (1976). Forest dynamics on the northern Georgia Piedmont. *For. Sci.* **22**, 307–322.

Johnson, W. C., Burgess, R. L., and Keammerer, W. R. (1976). Forest overstory vegetation and environment on the Missouri River floodplain in North Dakota. *Ecol. Monogr.* **46**, 59–84.

Johnston, D. W., and Odum, E. P. (1956). Breeding bird populations in relation to plant succession on the Piedmont of Georgia. *Ecology* **37**, 50–62.

Jones, C. S., and Schlesinger, W. H. (1980). *Emmenanthe penduliflora:* further consideration of germination response. *Madrono* **27**, 122–125.

Jones, E. W. (1945). The structure and reproduction of the virgin forest of the north temperate zone. *New Phytol.* **44**, 130–148.

Jones, E. W. (1950). Some aspects of natural regeneration in the Benin rain forest. *Emp. For. J.* **29**, 108–124.

Jones, E. W. (1956). Ecological studies on the rain forest of Southern Nigerian. The plateau forest of the Okomu forest reserve. *J. Ecol.* **44**, 83–117.

Jones, K. B. (1981). Effects of grazing on lizard abundance and diversity in western Arizona. *Southwest. Nat.* **26**, 107–115.

Jones, R., Groves, R. H., and Specht, R. L. (1969). Growth of heath vegetation. III. Growth curves for heaths in southern Australia: a reassessment. *Aust. J. Bot.* **17**, 309–314.

Jones, W. E., and Demetropoulos, A. (1968). Exposure to wave action: measurements of an important ecological parameter on rocky shores in Anglesey. *J. Exp. Mar. Biol. Ecol.* **2**, 46–63.

Jordan, C. F. (1982). Amazon rain forests. *Am. Sci.* **70**, 394–401.

Jordan, C. F., and Farnsworth, E. G. (1980). A rain forest chronicle: perpetuation of a myth. *Biotropica* **12**, 233–234.

Jordan, J. S. (1967). Deer browsing in northern hardwoods after clear-cutting. *USDA For. Serv. Res. Pap. NE* **NE-57**.

June, S. R., and Ogden, J. (1978). Studies on the vegetation of Mount Colenso, New Zealand, 4. An assessment of the process of canopy maintenance and regeneration strategy in a Red Beech (*Nothofagus fusca*) forest. *N.Z. J. Ecol.* **1**, 7–15.

Kacser, H., and Burns, J. A. (1981). The molecular basis of dominance. *Genetics* **97**, 639–666.

Kallman, K. D., and Harrington, R. W., Jr. (1964). Evidence for the existence of homozygous clones in the self-fertilizing hermaphroditic fish, *Rivulus marmoratus* (Poey). *Biol. Bull. (Woods Hole, Mass.)* **126**, 101–114.

Kamil, A. C., and Sargent, T. D. (1981). ''Foraging Behavior. Ecological, Ethological, and Psychological Approaches.'' Garland STPM Press, New York.

Kaminsky, R. (1981). The microbial origin of the allelopathic potential of *Adenostoma fasciculatum* H. & A. *Ecol. Monogr.* **51**, 365–382.

Kareiva, P. (1982). Experimental and mathematical analyses of herbivore movement: quantifying the influence of plant spacing and quality on foraging discrimination. *Ecol. Monogr.* **52**, 261–282.

Karlson, R. H. (1978). Predation and space utilization patterns in a marine epifaunal community. *J. Exp. Mar. Biol. Ecol.* **31**, 225–239.

Karlson, R. H. (1980). Alternative competitive strategies in a periodically disturbed habitat. *Bull. Mar. Sci.* **30**, 894–900.

Karlson, R. H. (1983). Disturbance and monopolization of a spatial resource by *Zoanthus sociatus* (Coelenterata, Anthozoa). *Bull. Mar. Sci.* **33**, 118–131.

Karlson, R. H., and Jackson, J. B. C. (1981). Competitive networks and community structure: a simulation study. *Ecology* **62**, 670–678.

Karr, J. R. (1979). On the use of mist nets in the study of bird communities. *Inland Bird Banding* **51**, 1–10.

Karr, J. R. (1981a). Surveying birds with mist nets. *In* ''Estimating Numbers of Terrestrial Birds'' (C. J. Ralph and J. M. Scott, eds.), *Stud. Avian Biol.* **6**, 62–67.

Karr, J. R. (1981b). Assessment of biotic integrity using fish communities. *Fisheries* **6**, 21–27.

Karr, J. R. (1982a). Avian extinctions on Barro Colorado Island, Panama: A reassessment. *Am. Nat.* **119**, 220–239.

Karr, J. R. (1982b). Population variability and extinction in the avifauna of a tropical land bridge island. *Ecology* **63**, 1975–1978.

Karr, J. R. (1983). Commentary (Avian community ecology). *In* "Perspectives in Ornithology: Essays Presented for the Centennial of the American Ornithologists' Union" (A. H. Brush and G. A. Clark, Jr., eds.), pp. 403–410. Cambridge Univ. Press, London and New York.

Karr, J. R. (1985). Birds of Panama: Biogeography and ecological dynamics. *In* "The Botany and Natural History of Panama: La Botánica e Historia Natural de Panama" (W. G. D'Arcy and M. D. Correa, A. eds.) Missouri Bot. Gard., St. Louis.

Karr, J. R., and Freemark, K. E. (1983). Habitat selection and environmental gradients: Dynamics in the "stable" tropics. *Ecology* **64**, 1481–1494.

Karr, J. R., and Gorman, O. T. (1975). Effects of land treatment on the aquatic environment. Non-point Source Pollution Seminar—1975. *U.S. Environ. Prot. Agency, Reg. 5 [Rep.] EPA* **EPA 905/9-75-007**, pp. 120–150.

Karr, J. R., and James, F. C. (1975). Eco-morphological configurations and convergent evolution in species and communities. *In* "Ecology and Evolution of Communities" (M. L. Cody and J. M. Diamond, eds.), pp. 258–291. Harvard Univ. Press, Cambridge, Massachusetts.

Kat, P. W. (1982). The relationship between heterozygosity for enzyme loci and developmental homeostasis in peripheral populations of aquatic bivalves (Unionidae). *Am. Nat.* **119**, 824–832.

Kaufman, D. W., and Kaufman, G. A. (1982). Effect of moonlight on activity and microhabitat use by Ord's kangaroo rat (*Dipodomys ordii*). *J. Mammal.* **63**, 309–312.

Kaushik, N. K., and Hynes, H. B. N. (1971). The fate of dead leaves that fall into streams. *Arch. Hydrobiol.* **68**, 465–515.

Kawano, K., Okumura, H., and Masuda, J. (1980). The productive and reproductive biology of flowering plants. VIII. Assimilation behavior and dry matter allocation of *Hepatica nobilis* var. *japonicum* f. *variegata* (Makino) Kitam. (Ranunculaceae). *Daigaku Kyoyobu Kiyo, Shizen Kagakuhen (J. Coll. Lib. Arts, Toyama Univ., Japan)* **13**, 33–46.

Kay, A. M. (1980). The organization of sessile guilds on pier pilings. Ph.D. Thesis, Univ. of Adelaide, Adelaide, South Australia.

Kay, A. M., and Butler, A. J. (1983). "Stability" of the fouling communities on pier pilings of two piers in South Australia. *Oecologia* **56**, 70–78.

Kay, A. M., and Keough, M. J. (1981). Occupation of patches in the epifaunal communities on pier pilings and the bivalve *Pinna bicolor* at Edithburgh, South Australia. *Oecologia* **48**, 123–130.

Kayll, A. J. (1966). Some characteristics of heath fires in northeast Scotland. *J. Appl. Ecol.* **3**, 29–40.

Kazimirov, N. I., and Morozova, R. M. (1973). "Biological Cycling of Matter in Spruce Forests." Nauka, Leningrad. Cited in DeAngelis, D. L., Gardner, R. H., and Shugart, H. H. (1981). Productivity of forest ecosystems studied during the IBP: the woodlands data set. *In* "Dynamic Properties of Forest Ecosystems" (D. E. Reichle, ed.), pp. 567–672. Cambridge Univ. Press, London and New York.

Keay, R. W. (1960). Seeds in forest soils. *Niger. For. Inf. Bull.* **4**, 1–4.

Keddy, P. A. (1983). Shoreline vegetation in Axe Lake, Ontario: effects of exposure on zonation patterns. *Ecology* **64**, 331–344.

Keeley, J. E. (1979). Population differentiation along a flood frequency gradient: physiological adaptation to flooding in *Nyssa sylvatica*. *Ecol. Monogr.* **49**, 98–108.

Keeley, J. E. (1981). Reproductive cycles and fire regimes. *In* "Fire Regimes and Ecosystem Properties" (H. A. Mooney, T. M. Bonnicksen, N. L. Christensen, J. E. Lotan, and W. A. Reiners, eds.), *Gen. Tech. Rep. WO—[Wash. Off.]* **GTR-WO-26**, pp. 231–277.

Keeley, J. E., and Keeley, S. C. (1984). Postfire recovery of California coastal sage scrub. *Am. Midl. Nat.* **111**, 105–117.

Keeley, J. E., and Zedler, P. H. (1978). Reproduction of chaparral shrubs after fire: a comparison of sprouting and seeding strategies. *Am. Midl. Nat.* **99**, 142–161.

Keeley, J. E., Morton, B. A., Pedrosa, A., and Trotter, P. (1985). The role of allelopathy, heat, and charred wood in the germination of chaparral herbs and suffrutescents. *J. Ecology* (in press).

Keeley, S. C., Keeley, J. E., Hutchinson, S. M., and Johnson, A. W. (1981). Postfire succession of the herbaceous flora in southern California Chaparral. *Ecology* **62,** 1608–1621.

Keen, S. L., and Neill, W. E. (1980). Spatial relationships and some structuring processes in benthic intertidal animal communities. *J. Exp. Mar. Biol. Ecol.* **45,** 139–156.

Keeney, D. R. (1980). Prediction of soil nitrogen availability in forest ecosystems: a literature review. *For. Sci.* **26,** 159–171.

Keever, C. (1950). Causes of succession on old fields of the Piedmont, North Carolina. *Ecol. Monogr.* **20,** 229–250.

Kellman, M. C., and Adams, C. D. (1970). Milpa weeds of Cayo District, Belize (British Honduras). *Can. Geogr.* **14,** 323–343.

Kempf, J. S., and Pickett, S. T. A. (1981). The role of branch length and angle in branching pattern of forest shrubs along a successional gradient. *New Phytol.* **88,** 111–116.

Kentworthy, J. B. (1963). Temperatures in heather burning. *Nature (London)* **200,** 1226.

Keough, M. J. (1981). Community dynamics of the epifauna of the bivalve *Pinna bicolor* Gmelin. Ph.D. Thesis, Univ. of Adelaide, Adelaide, South Australia.

Keough, M. J. (1983). Patterns of recruitment of sessile invertebrates in two subtidal habitats. *J. Exp. Mar. Biol. Ecol.* **66,** 213–245.

Keough, M. J. (1984a). Dynamics of the epifauna of the bivalve *Pinna bicolor:* interactions among recruitment, predation, and competition. *Ecology* **65,** 677–688.

Keough, M. J. (1984b). Effects of patch size on the abundance of sessile marine invertebrates. *Ecology* **65,** 423–437.

Keough, M. J., and Butler, A. J. (1983). Temporal changes in speices number in an assemblage of sessile marine invertebrates. *J. Biogeogr.* **10,** 317–330.

Keough, M. J., and Downes, B. J. (1982). Recruitment of marine invertebrates: the role of active larval choices and early mortality. *Oecologia* **54,** 348–352.

Kerster, H. W. (1968). Population age structure in the prairie forb, *Liatris aspera. BioScience* **18,** 430–432.

Kessell, S. R. (1976). Gradient modeling: A new approach to fire modeling and wilderness resource management. *Environ. Manage.* **1,** 39–48.

Kessell, S. R. (1979). "Gradient Modeling: Resource and Fire Management." Springer-Verlag, Berlin and New York.

Kimura, M., and Ohta, T. (1971). "Theoretical Aspects of Population Genetics." Princeton Univ. Press, Princeton, New Jersey.

Kinerson, R. S. (1975). Relationships between plant surface area and respiration in loblolly pine. *J. Appl. Ecol.* **12,** 965–971.

Kinerson, R. S., Ralston, C. W., and Wells, G. G. (1977). Carbon cycling in a loblolly pine plantation. *Oecologia* **29,** 1–10.

King, D. A. (1979). The comparative geometry of successional vs. climax trees. Ph.D. Thesis, Univ. of Wisconsin, Madison.

King, D. A. (1981). Tree dimensions: maximizing the rate of height growth in dense stands. *Oecologia* **51,** 351–356.

King, T. J. (1977). The plant ecology of ant-hills in calcareous grasslands. I. Patterns of species in relation to ant-hills in southern England. *J. Ecol.* **65,** 235–256.

Kira, T. (1975). Primary production of forests. *In* "Photosynthesis and Productivity in Different Environments" (J. P. Cooper, ed.), pp. 5–40. Cambridge Univ. Press, London and New York.

Kira, T. (1978). Community architecture and organic matter dynamics in tropical lowland rain forests of Southern Asia, with special reference to Pasoh Forest, West Malaysia. *In* "Tropical Trees as Living Systems" (P. B. Tomlinson and M. H. Zimmerman, eds.), pp. 561–590. Cambridge Univ. Press, London and New York.

Kira, T., and Shidei, T. (1967). Primary production and turnover of organic matter in different forest ecosystems of the western Pacific. *Nippon Seitai Gakkaishi (Jpn. J. Ecol.)* **17,** 70–87.

Kirkland, G. L. (1977). Responses of small mammals to the clearcutting of northern Appalachian forests. *J. Mammal.* **58,** 600–609.

Kitchener, D. J., Dell, J., Muir, B. G., and Palmer, M. (1982). Birds in western Australian wheatbelt reserves—implications for conservation. *Biol. Conserv.* **22,** 127–163.

Kivilaan, A., and Bandurski, R. S. (1973). The ninety-year period for Dr. Beal's seed viability experiment. *Am. J. Bot.* **60,** 140–145.

Knapp, R., ed. (1974). "Vegetation Dynamics." Junk, The Hague.

Knight, D. H. (1975). A phytosociological analysis of species rich tropical forest on Barro Colorado Island, Panama. *Ecol. Monogr.* **45,** 259–284.

Knight, F. B., and Heikkenen, H. J. (1980). "Principles of Forest Entomology," 5th Ed. McGraw-Hill, New York.

Knowlton, N., Lang, J. C., Rooney, M. C., and Clifford, P. (1981). Evidence for delayed mortality in hurricane-damaged Jamaican staghorn corals. *Nature (London)* **294,** 251–252.

Koehl, M. A. R. (1977). Effects of sea anemones on the flow forces they encounter. *J. Exp. Biol.* **69,** 87–105.

Koehl, M. A. R. (1979). Stiffness or extensibility of intertidal algae: a comparative study of modes of withstanding wave action. *J. Biomech.* **12,** 634.

Koehl, M. A. R. (1982). The interaction of moving water and sessile organisms. *Sci. Am.* **247,** 124–134.

Koehl, M. A. R., and Wainright, S. A. (1977). Mechanical adaptation of a giant kelp. *Limnol. Oceanogr.* **22,** 1067–1071.

Koehn, R. K., and Shumway, S. E. (1982). A genetic/physiological explanation for differential growth rate among indivdiuals of the American oyster, *Crassostrea virginica* (Gmelin). *Mar. Biol. Lett.* **3,** 35–42.

Koehn, R. K., Turano, F. J., and Mitton, J. B. (1973). Population genetics of marine pelecypods. II. Genetic differences in microhabitats of *Modiolus demissus. Evolution* **27,** 100–105.

Koehn, R. K., Milkman, R., and Mitton, J. B. (1976). Population genetics of marine pelecypods. IV. Selection, migration, genetic differentiation in the blue mussel, *Mytilus edulis. Evolution* **30,** 2–32.

Kojis, B. L., and Quinn, N. J. (1981). Aspects of sexual reproduction and larval development in the shallow water hermatypic coral, *Goniastrea australensis* (Edwards and Haime, 1857). *Bull. Mar. Sci.* **31,** 558–573.

Kojis, B. L., and Quinn, N. J. (1983). Reproductive strategies in four species of *Porites* (Scleractinia). *Proc. Int. Coral Reef Symp., 4th, Manila, 1981* **2,** 145–151.

Kologiski, R. L. (1977). The phytosociology of the Green Swamp, North Carolina. *N.C. Agric. Exp. Stn., Tech. Bull.* No. 250.

Komarek, E. V. (1968). Lightning and lightning fires as ecological forces. *Proc. Annu. Tall Timbers Fire Ecol. Conf.* **8,** 161–207.

Komarek, E. V. (1974). Effects of fire on temperate forests and related ecosystems: Southeastern United States. *In* "Fire and Ecosystems" (T. T. Kozlowski and C. E. Ahlgren, eds.), pp. 251–277. Academic Press, New York.

Kozlowski, T. T., and Ahlgren, C. E., eds. (1974). "Fire and Ecosystems." Academic Press, New York.

Kozlowski, T. T., and Ward, R. C. (1957). Seasonal height growth of deciduous trees. *For. Sci.* **3,** 168–174.

Kramer, F. (1933). De natuurlijke verjonging in het Goenoeng-Gedehcomplex. *Tectona* **26,** 156–185.

Kramer, P. J. (1943). Amount and duration of growth of various species of tree seedlings. *Plant Physiol.* **18,** 239–251.

Krantz, G. W., and Mellott, J. L. (1972). Studies on phoretic specificity in *Macrocheles mycotrupetes* and *M. peltotrupetes* Krantz and Mellott (Acari: Macrochelidae), associates of geotrupine Scarabaeidae. *Acarologia* **14**, 317–344.

Krebs, J. E. (1975). A comparison of soils under agriculture and forests in San Carlos, Costa Rica. *In* "Tropical Ecological Systems" (F. B. Golley and E. Medina, eds.), pp. 381–390. Springer-Verlag, Berlin and New York.

Krebs, J. R., Ryan, J. C., and Charnov, E. L. (1974). Hunting by expectation or optimal foraging? A study of patch use by chickadees. *Anim. Behav.* **22**, 953–964.

Krebs, J. R., Stephens, D. W., and Sutherland, W. J. (1983). Perspectives in optimal foraging. *In* "Perspectives in Ornithology" (A. H. Brush and G. A. Clark, Jr., eds.), pp. 165–216. Cambridge Univ. Press, London and New York.

Krefting, L. W., and Roe, E. I. (1949). The role of some birds and mammals in seed germination. *Ecol. Monogr.* **19**, 269–286.

Kruger, F. J. (1977). Ecology of Cape fynbos in relation to fire. *In* "Environmental Consequences of Fire and Fuel Management in Mediterranean Ecosystems" (H. A. Mooney and C. E. Conrad, eds.), *Gen. Tech. Rep. WO—U.S. For. Ser.* [*Wash. Off.*] **GTR-WO-3**, pp. 230–244.

Kruger, F. J. (1982). Prescribing fire frequencies in Cape fynbos in relation to plant demography. *In* "Dynamics and Management of Mediterranean-Type Ecosystems" (C. E. Conrad and W. C. Oechel, eds.), *USDA For. Ser. Gen. Tech. Rep. PSW* **PSW-58**, pp. 483–489.

Kumerow, J. (1981). Structure of roots and root systems. *In* "Ecosystems of the World. Vol. 11: Mediterranean-type Shrublands" (F. di Castri, D. W. Goodall, and R. L. Specht, eds.), pp. 269–288. Elsevier, Amsterdam.

Kushlan, J. A. (1974). Effects of a natural fish kill on the water quality, plankton, and fish population of a pond in the Big Cypress Swamp, Florida. *Trans. Am. Fish. Soc.* **103**, 235–243.

Kushlan, J. A. (1976). Wading bird predation in a seasonally fluctuating pond. *Auk* **93**, 464–476.

Kushlan, J. A. (1979). Design and management of continental wildlife reserves: Lessons from the Everglades. *Biol. Conserv.* **15**, 281–290.

LaBue, J., and Darnell, R. M. (1959). Effect of habitat disturbance on a small mammal population. *J. Mammal.* **40**, 425–437.

Laessle, A. M. (1965). Spacing and competition in natural stands of sand pine. *Ecology* **46**, 65–72.

Lambert, R. L., Lang, G. E., and Reiners, W. A. (1980). Loss of mass and chemical change in decaying boles of a subalpine balsam fir forest. *Ecology* **61**, 1460–1473.

Lang, G. (1970). Die Vegetation der Brindabella Range bei Canberra. *Math.-Naturwiss. Kl., Akad. Wiss. Lit., Mainz* No. 1.

Lang, G. E., and Knight, D. E. (1983). Tree growth, mortality, recruitment, and canopy gap formation during a 10-year period in a tropical moist forest. *Ecology* **64**, 1075–1080.

Lang, J. C. (1971). Interspecific aggression by scleractinian corals. 1. The rediscovery of *Scoiymia cubensis* (Milne-Edwards and Haime). *Bull. Mar. Sci.* **21**, 952–959.

Lang, J. C. (1973). Interspecific aggression by scleractinian corals. 2. Why the race is not only to the swift. *Bull. Mar. Sci.* **23**, 260–279.

Lang, J. C. (1974). Biological zonation at the base of a reef. *Am. Sci.* **62**, 272–281.

Langford, A. N., and Buell, M. F. (1969). Integration, identity, and stability in the plant association. *Adv. Ecol. Res.* **6**, 83–135.

Latter, B. D. H., and Robertson, A. (1962). The effect of inbreeding and artificial selection on reproductive fitness. *Genet. Res.* **3**, 110–138.

Lauer, W. (1961). Wandlungen im Landschaftsbild des südchilenischen Seengebietes seit Ende der spanischen Kolonialzeit. *Schr. Geogr. Inst. Univ. Kiel* **20**, 227–276.

Laufersweiler, J. D. (1955). Changes with age in the proportion of the dominants in a beech–maple forest in central Ohio. *Ohio J. Sci.* **55**, 73–80.

Law, R. (1981). The dynamics of a colonizing species of *Poa annua*. *Ecology* **62**, 1267–1277.

Lawhon, D. K., and Hafner, M. S. (1981). Tactile discriminatory ability and foraging strategies in kangaroo rats and pocket mice (Rodentia: Heteromyidae). *Oecologia* **50**, 303–309.

Lawrence, G. E. (1966). Ecology of vertebrate animals in relation to chaparral fire in the Sierra Nevada foothills. *Ecology* **47,** 278–291.

Laws, R. M., Parker, I. S. C., and Johnstone, R. C. B. (1975). "Elephants and their Habitats: The Ecology of Elephants in North Bunyoro, Uganda." Oxford Univ. Press, London and New York.

Lawton, J. H. (1983). Plant architecture and the diversity of phytophagous insects. *Annu. Rev. Entomol.* **28,** 23–39.

Lawton, J. H., and Hassell, M. P. (1981). Asymmetrical competition in insects. *Nature (London)* **289,** 793–795.

Lawton, J. H., and McNeill, S. (1979). Between the devil and the deep blue sea: on the problem of being a herbivore. *In* "Population Dynamics" (R. M. Anderson, B. D. Turner, and L. R. Taylor, eds.), *Symp. Br. Ecol. Soc.* **20,** 223–244.

Lawton, R. O. (1982). Wind stress and elfin stature in a montane rain forest tree: an adaptive explanation. *Am. J. Bot.* **69,** 1224–1230.

Lawton, R. O., and Dryer, V. (1980). The vegetation of the Monteverde cloud forest reserve. *Brenesia* **18,** 101–116.

Leak, W. B. (1963). Delayed germination of white ash seeds under forest conditions. *J. For.* **61,** 768–772.

Leak, W. B., and Filip, S. M. (1977). Thirty-eight years of group selection in New England northern hardwoods. *J. For.* **75,** 641–643.

Leak, W. B., and Martin, C. W. (1975). Relationship of stand age to streamwater nitrate in New Hampshire. *USDA For. Serv. Northeast. For. Exp. Stn. Res. Note NE* **NE-211.**

Leary, R. F., Allendorf, F. W., and Knudsen, K. L. (1983). Developmental stability and enzyme heterozygosity in rainbow trout. *Nature (London)* **301,** 71–72.

Lebednik, P. A. (1973). Ecological effects of intertidal uplifting from nuclear testing. *Mar. Biol. (Berlin)* **20,** 197–207.

Lebron, M. L. (1979). An autecological study of *Palicourea riparia* in Puerto Rico. *Oecologia* **42,** 31–46.

LeCren, E. D., Kipling, C., and McCormack, J. C. (1972). Windermere: effects of exploitation and eutrophication on the salmonid community. *J. Fish. Res. Board Can.* **29,** 819–832.

Ledig, T. F., and Conkle, M. T. (1983). Genetic diversity and genetic structure in a narrow endemic, Torrey pine (*Pinus torregyana* Parry ex Carn.). *Evolution* **37,** 79–85.

Ledig, T. F., Guries, R. P., and Bonefield, B. A. (1983). The relation of growth to heterozygosity in pitch pine. *Evolution* **37,** 1227–1238.

Lee, R. (1978). "Forest Microclimatology." Columbia Univ. Press, New York.

Leigh, E. G., Jr. (1975). Structure and climate in tropical rain forest. *Annu. Rev. Ecol. Syst.* **6,** 67–86.

Leigh, E. G., Jr., Rand, A. S., and Windsor, D. M., eds. (1982). "The Ecology of a Tropical Forest: Seasonal Rhythms and Long-term Changes." Smithsonian Inst. Press, Washington, D.C.

Lemen, C. A. (1978). Seed size selection in heteromyids: a second look. *Oecologia* **35,** 13–19.

Lemon, P. C. (1961). Forest ecology of ice storms. *Bull. Torrey Bot. Club* **88,** 21–29.

Leopold, L. B. (1971). Trees and streams: the efficiency of branching patterns. *J. Theor. Biol.* **31,** 339–354.

Lerner, I. M. (1954). "Genetic Homeostasis." Oliver & Boyd, Edinburgh.

Leslie, J. F., and Vrijenhoek, R. C. (1977). Genetic analysis of natural populations of *Poeciliopsis monacha:* allozyme inheritance and pattern of mating. *J. Hered.* **68,** 301–306.

Leslie, J. F., and Vrijenhoek, R. C. (1980). Consideration of Muller's ratchet mechanism through studies of genetic linkage and genomic compatibilities in clonally reproducing *Poecilsiopsis. Evolution* **34,** 1105–1115.

Levin, D. A. (1981). Dispersal versus gene flow in plants. *Ann. Mo. Bot. Gard.* **68,** 233–253.

Levin, D. A., and Kerster, H. W. (1974). Gene flow in seed plants. *Evol. Biol.* **7,** 139–220.

Levin, D. A., and Wilson, J. B. (1978). The genetic implications of ecological adaptations in plants. *In* "Structure and Functioning of Plant Populations" (A. H. J. Freysen and J. W. Woldendorp, eds.), pp. 75–98. North-Holland Publ., Amsterdam.

Levin, S. A. (1974). Dispersion and population interactions. *Am. Nat.* **108**, 207–228.

Levin, S. A. (1976). Population dynamics in heterogeneous environments. *Annu. Rev. Ecol. Syst.* **7**, 287–310.

Levin, S. A. (1978). Pattern formation in ecological communities. *In* "Spatial Pattern in Phytoplankton Communities" (J. H. Steele, ed.), pp. 433–465. Plenum, New York.

Levin, S. A., and Paine, R. T. (1974). Disturbance, patch formation, and community structure. *Proc. Natl. Acad. Sci. U.S.A.* **71**, 2744–2747.

Levings, S. C., and Garrity, S. D. (1983). Diel and tidal movement of two co-occuring neritid snails; differences in grazing patterns on a tropical rocky shore. *J. Exp. Mar. Biol. Ecol.* **67**, 261–278.

Levins, R. (1968). "Evolution in Changing Environments." Princeton Univ. Press, Princeton, New Jersey.

Levins, R. (1973). Extinction. *In* "Some Mathematical Questions in Biology: Lectures on Mathematics in the Life Sciences," (R. Gerstenhaber, ed.). *Am. Math. Soc.* **2**, 75–108.

Levins, R., and Culver, D. (1975). Regional coexistence of species and competition between rare species. *Proc. Natl. Acad. Sci. U.S.A.* **68**, 1246–1248.

Leviten, P. J., and Kohn, A. J. (1980). Microhabitat resource use, activity pattern, and episodic catastrophe: *Conus* on tropical intertidal reef rock benches. *Ecol. Monogr.* **50**, 55–75.

Levyns, M. R. (1966). *Haemanthus canaliculatus,* a new fire-lily from the Western Province. *J. S. Afr. Bot.* **32**, 73–75.

Lewis, A. C. (1979). Feeding preference for diseased and wilted sunflower in the grasshopper, *Melanoplus differentialis. Entomol. Exp. Appl.* **26**, 202–207.

Lewis, J. R. (1954). Observations on a high-level population of limpets. *J. Anim. Ecol.* **23**, 85–100.

Lewis, J. R. (1964). "The Ecology of Rocky Shores." English Univ. Press, London.

Lewis, J. R., and Bowman, R. S. (1975). Local habitat-induced variations in the population dynamics of *Patella vulgata* L. *J. Exp. Mar. Biol. Ecol.* **17**, 165–203.

Lewontin, R. C. (1974). "The Genetic Basis of Evolutionary Change." Columbia Univ. Press, New York.

Lewontin, R. C. (1979). Fitness, survival, and optimality. *In* "Analysis of Ecological Systems" (D. H. Horn, R. Mitchell, and G. R. Stairs, eds.), pp. 3–21. Ohio State Univ. Press, Columbus.

Lewontin, R. C., and Krakauer, J. (1973). Distribution of gene-frequency as a test of the theory of the selective neutralism of polymorphisms. *Genetics* **74**, 175–195.

Lidicker, W. Z., Jr. (1975). The role of dispersal in the demography of small mammals. *In* "Small Mammals: Their Productivity and Population Dynamics" (F. B. Golley, K. Petrusewicz, and L. Ryszkowski, eds.), pp. 103–128. Cambridge Univ. Press, London and New York.

Liew, T. C. (1973). Occurence of seeds in virgin forest top soil with particular reference to secondary species in Sabah. *Malays. For.* **36**, 185–193.

Liew, T. C., and Wong, F. O. (1973). Density, recruitment, mortality and growth of dipterocarp seedlings in virgin and logged-over forest in Sabah. *Malays. For.* **36**, 3–15.

Lindenmuth, A. W., and Davis, J. R. (1973). Predicting fire spread in Arizona's oak chaparral. *USDA For. Ser., Rocky Mountain For. Range Exp. Stn. Res. Pap.* **RM-101.**

Linhart, Y. B. (1974). Intra-population differentiation in annual plants. I. *Veronica peregrina* L. raised under non-competitive conditions. *Evolution* **28**, 232–243.

Linhart, Y. B. (1976). Evolutionary studies of plant populations in Vernal Pools. *In* "Vernal Pools—Their Ecology and Conservation" (S. Jain, ed.), Publ. No. 9, pp. 40–46. Inst. Ecol., Univ. of California, Davis.

Linhart, Y. B., Mitton, J. B., Sturgeon, K. B., and Davis, M. L. (1981). Genetic variation in space and time in a population of ponderosa pine. *Heredity* **46**, 407–426.

Little, S., and Somes, H. A. (1965). Atlantic white cedar being eliminated by excessive animal damage in South Jersey. *USDA For. Ser. Res. Note NE* **NE-33.**

Littler, M. M., and Littler, D. S. (1980). The evolution of thallus form and survival strategies in benthic marine macroalgae: field and laboratory tests of a functional form model. *Am. Nat.* **116**, 25–44.

Littler, M. M., Martz, D. R., and Littler, D. S. (1983). Effects of recurrent sand deposition on rocky intertidal organisms: importance of substrate heterogeneity in a fluctuating environment. *Mar. Ecol.: Prog. Ser.* **11,** 129–139.

Logan, K. T. (1965). Growth of tree seedlings as affected by light intensity. I. White birch, yellow birch, sugar maple and silver maple. *Can., For. Branch, Dep. Publ.* No. 1121.

Logan, K. T. (1970). Adaptations of the photosynthetic apparatus of sun- and shade-grown yellow birch (*Betula alleghaniensis* Britt.). *Can. J. Bot.* **48,** 1681–1688.

Logan, K. T., and Krotkov, G. (1968). Adaptations of the photosynthetic mechanism of sugar maple (*Acer saccharum*) seedlings grown in various light intensities. *Physiol. Plant.* **22,** 104–116.

Loiselle, B. A., and Hoppes, W. G. (1983). Nest predation in insular and mainland lowland rainforest in Panama. *Condor* **85,** 93–95.

Loope, L. L., and Gruell, G. E. (1973). The ecological role of fire in the Jackson Hole area, northwestern Wyoming. *Quat. Res.* (*N.Y.*) **3,** 425–443.

López Q., M. M., and Vázquez-Yanes, C. (1976). Estudios sobre germinación de semillas en condiciones naturales controladas. *In* "Regeneración des Selvas" (A. Gómez-Pompa, C. Vázquez-Yanes, S. del Amo, R. and A. Butanda C., eds.), pp. 250–262. Compañia Editorial Continental, Mexico City, Mexico.

Lorimer, C. G. (1980). Age structure and disturbance history of a southern Appalachian virgin forest. *Ecology* **61,** 1169–1184.

Lotan, J. E. (1975). The role of cone serotiny in lodgepole pine forests. *In* "Management of Lodgepole Pine Ecosystems Symposium Proceedings" (D. M. Baumgartner, ed.), pp. 471–495. Washington State Univ. Coop. Ext. Serv., Pullman.

Loucks, O. L. (1970). Evolution of diversity, efficiency, and community stability. *Am. Zool.* **10,** 17–25.

Loucks, O. L. (1975). Analysis of perturbations in ecosystems. *In* "The Study of Species Transients, their Characteristics and Significance of Natural Resource Systems" (O. L. Loucks, ed.), pp. 4–7. Inst. Ecol., Indianapolis, Indiana.

Loucks, O. L., Ek, A. R., Johnson, W. C., and Monserud, R. A. (1980). Growth, aging and succession. *In* "Dynamic Properties of Forest Ecosystems" (D. E. Reichle, ed.), pp. 37–85. Cambridge Univ. Press, London and New York.

Loveless, A. R. (1961). Nutritional interpretation of sclerophylly based on differences in chemical composition of sclerophyllous and mesophytic leaves. *Ann. Bot.* (*London*) **25,** 168–184.

Loveless, A. R. (1962). Further evidence to support a nutritional interpretation of sclerophylly. *Ann. Bot.* (*London*) **26,** 551–561.

Lovett, G. M., Reiners, W. A., and Olson, R. K. (1982). Cloud droplet deposition in subalpine fir forests. *Science* **218,** (*Washington, D.C.*) 1303–1304.

Lowe-McConnell, R. H. (1975). "Fish Communities in Tropical Freshwaters." Longmans, Green, New York.

Loya, Y. (1976). Recolonization of red sea corals affected by natural catastrophes and man-made perturbations. *Ecology* **57,** 278–289.

Lubchenco, J. (1978). Plant species diversity in a marine intertidal community: importance of herbivore food preference and algal competitive abilities. *Am. Nat.* **112,** 23–39.

Lubchenco, J. (1980). Algal zonation in the New England rocky intertidal community: an experimental analysis. *Ecology* **61,** 333–344.

Lubchenco, J. (1983). *Littorina* and *Fucus:* effects of herbivores, substratum heterogeneity, and plant escapes during succession. *Ecology* **64,** 1116–1123.

Lubchenco, J., and Gaines, S. C. (1981). A unified approach to marine plant–herbivore interactions. I. Populations and communities. *Annu. Rev. Ecol. Syst.* **12,** 405–437.

Lubchenco, J., and Menge, B. A. (1978). Community development and persistence in a low rocky intertidal zone. *Ecol. Monogr.* **48,** 67–94.

Ludlow, W. B., and Vázquez-Yanes, C. (1976). Germinación de semillas de *Piper hispidum* S.W. bajo differentes condiciones de iluminación. *In* "Regeneración des Selvas" (A. Gómez-Pompa, C. Vázquez-Yanes, S. Del Amo R. and A. Butanda C., eds.), pp. 263–278. Compañia Editorial Continental, Mexico City, Mexico.

Ludwig, J. A., and Whitford, W. G. (1981). Short-term water and energy flow in arid ecosystems. *In* "Arid-land Ecosystems: Structure, Functioning and Management" (D. W. Goodall, R. A. Perry, and K. M. W. Howes, eds.), Vol. 2, pp. 271–299. Cambridge Univ. Press, London and New York.

Lugo, A. (1970). Photosynthetic studies on four species of rain forest seedlings. *In* "A Tropical Rainforest" (H. T. Odum and R. F. Pigeon, eds.), pp. 181–102. USAEC, Oak Ridge, Tennessee.

Lundberg, A., Alatalo, R. V., Carlson, A., and Ulfstrand, S. (1981). Biometry, habitat distribution and breeding success in the Pied Flycatcher *Ficedula hypoleuca*. *Ornis Scand.* **12,** 68–79.

Lustick, S. I. (1983). Cost–benefit of thermoregulation in birds: influences of posture, microhabitat selection, and color. *In* "Behavioral Energetics: The Cost of Survival in Vertebrates" (W. P. Aspey and S. I. Lustick, eds.), pp. 265–294. Ohio State Univ. Press, Columbus.

Lutz, H. J. (1930a). Original forest composition in northwestern Pennsylvania as indicated by early land survey notes. *J. For.* **28,** 1098–1103.

Lutz, H. J. (1930b). The vegetation of Heart's Content, a virgin forest in northwestern Pennsylvania. *Ecology* **11,** 1–29.

Lutz, H. J., and Chandler, R. F. (1946). "Forest Soils." Wiley, New York.

Lutz, H. J., and McComb, A. L. (1935). Origin of white pine in virgin forest stands of northwestern Pennsylvania as indicated by stem and basal branch features. *Ecology* **16,** 252–256.

Lyford, W. H., and MacLean, D. W. (1966). Mound and pit microrelief in relation to soil disturbance and tree distribution in New Brunswick, Canada. *Harv. For. Pap.* No. 15.

Mabry, T. J., and Difeo, D. R., Jr. (1973). The role of secondary plant chemistry in the evolution of mediterranean scrub vegetation. *In* "Mediterranean Type Ecosystems" (F. di Castri and H. A. Mooney, eds.), pp. 121–155. Springer-Verlag, Berlin and New York.

MacArthur, R. H., and MacArthur, J. W. (1961). On bird species diversity. *Ecology* **42,** 594–598.

MacArthur, R. H., and Pianka, E. R. (1966). On optimal use of a patchy environment. *Am. Nat.* **100,** 603–609.

MacArthur, R. H., and Wilson, E. O. (1967). "The Theory of Island Biogeography." Princeton Univ. Press, Princeton, New Jersey.

McCauley, O. D., and Trimble, G. R. (1975). Site quality in Appalachian hardwoods: the biological and economic response under selection silviculture. *USDA For. Serv. Res. Pap. NE (U.S.)* **NE-312.**

M'Closkey, R. T. (1980). Spatial patterns in sizes of seeds collected by four species of heteromyid rodents. *Ecology* **61,** 486–489.

M'Closkey, R. T. (1981). Microhabitat use in coexisting desert rodents—the role of population density. *Oecologia* **50,** 310–315.

McColl, J. G., and Grigal, D. F. (1977). Nutrient changes following a forest wildfire in Minnesota: effects in watersheds with differing soils. *Oikos* **28,** 105–112.

McCormick, J. F., and Platt, R. B. (1980). Recovery of an Appalachian forest following the chestnut blight or Catherine Keever—you were right! *Am. Midl. Nat.* **104,** 264–273.

McCoy, E. D. (1983). The application of island-biogeographic theory to patches of habitat: How much land is enough? *Biol. Conserv.* **25,** 53–61.

McCune, B. (1982). Site, history and forest dynamics in the Bitterroot Canyons, Montana. Ph.D. Thesis, Univ. of Wisconsin, Madison.

McDonnell, M. J., and Stiles, E. W. (1983). The structural complexity of old field vegetation and the recruitment of bird-dispersed plant species. *Oecologia* **56,** 109–116.

McDowell, W. H., and Fisher, S. G. (1976). Autumnal processing of dissolved organic matter in a small woodland stream ecosystem. *Ecology* **57,** 561–569.

McGee, C. E. (1976). Maximum soil temperatures on clearcut forest land in western North Carolina. *USDA For. Ser., Southeast. For. Exp. Stn., Res. Note SE* **SE-237.**

McGill, W. B., and Cole, C. V. (1981). Comparative aspects of organic C, N, and P cycling through soil organic matter during pedogensis. *Geoderma* **26,** 267–289.

McIntosh, R. P. (1980). The relationship between succession and the recovery process in ecosystems. *In*

"The Recovery Process in Damaged Ecosystems" (J. Cairns, ed.), pp. 11–62. Ann Arbor Sci. Publ., Ann Arbor, Michigan.

McIntosh, R. P. (1981). Succession and ecological theory. In "Forest Succession: Concepts and Application" (D. C. West, H. H. Shugart, and D. B. Botkin, eds.), pp. 10–23. Springer-Verlag, Berlin and New York.

McKay, F. E. (1971). Behavioral aspects of population dynamics in unisexual–bisexual *Poeciliopsis* (Pisces: Poeciliidae). *Ecology* **52**, 778–790.

McLean, J. H. (1962). Sublittoral ecology of kelp beds of the open coast area near Carmel, California. *Biol. Bull. (Woods Hole, Mass.)* **122**, 95–114.

McLintock, T F. (1959). Soil moisture pattern in northern coniferous forest. *USDA For. Serv., Northeast. For. Exp. Stn., Stn. Pap.* No. 128.

MacMahon, J. A. (1980). Ecosystems over time: Succession and other types of change. In "Forests: Fresh Perspectives from Ecosystem Analysis" (R. H. Waring, ed.), *Proc. Annu. Biol. Colloq. [Ore. State Univ.]* **40**, 27–58.

MacMahon, J. A. (1981). Successional processes: Comparisons among biomes with special reference to probably roles of influences on animals. In "Forest Succession: Concepts and Application" (D. C. West, H. H. Shugart, and D. B. Botkin, eds.), pp. 277–321. Springer-Verlag, Berlin and New York.

MacMahon, J. A., Phillips, D. L., Robinson, J. V., and Schimpf, D. J. (1978). Levels of organization: An organism-centered approach. *BioScience* **28**, 700–704.

McMillen, G. G., and McClendon, J. H. (1979). Leaf angle: an adaptive feature of sun and shade leaves. *Ann. Bot. (London)* **48**, 437–442.

McMurtie, R. (1978). Persistence and stability of single-species and prey–predator systems in spatially heterogeneous environments. *Math. Biosci.* **39**, 11–51.

McNair, J. N. (1982). Optimal giving-up and the marginal value theorem. *Am. Nat.* **119**, 511–529.

McNaughton, S. J. (1979). Grassland–herbivore dynamics. In "Serengeti: Dynamics of an Ecosystem" (A. R. E. Sinclair and M. Norton-Griffiths, eds.), pp. 46–81. Univ. of Chicago Press, Chicago, Illinois.

McNaughton, S. J. (1983). Serengeti grassland ecology: The role of composite environmental factors and contingency in community organization. *Ecol. Monogr.* **53**, 291–320.

McPherson, J. K., and Muller, C. H. (1969). Allelopathic effect of *Adenastoma* fasciculatum, "chamise," in the California chaparral. *Ecol. Monogr.* **39**, 177–198.

McPherson, J. K., Chou, C. H., and Muller, C. H. (1971). Allelopathic constituents of the chaparral shrub *Adenostoma fasciculatum. Phytochemistry* **10**, 2925–2933.

McWilliam, J. R., Schroeder, H. E., Marshall, D. R., and Oram, R. N. (1971). Genetic sterility of Australian phalaris under domestication. *Aust. J. Agric. Res.* **22**, 895–908.

Maguire, D. A., and Forman, R. T. R. (1983). Herb cover effects on tree seedling patterns in a mature hemlock-hardwood forest. *Ecology* **64**, 1367–1380.

Mahoney, S. A. (1976). Thermal and ecological energetics of the White-crowned Sparrow (*Zonotrichia leucophrys*) using the equivalent black-body temperature. Ph.D. Thesis, Washington State Univ., Pullman.

Mahoney, S. A., and King, J. R. (1977). The use of the equivalent black-body temperature in the thermal energetics of small birds. *J. Therm. Biol.* **2**, 115–120.

Main, A. R. (1981). Fire tolerance of heathland animals. In "Ecosystems of the World. Vol. 9b: Heathlands and Related Shrublands" (R. L. Specht, ed.), pp. 85–90. Elsevier, Amsterdam.

Malecot, G. (1955). Remarks on decrease of relationship with distance, following paper by M. Kimura. *Cold Spring Harbor Symp. Quant. Biol.* **20**, 52–53.

March, W. J., and Skeen, J. N. (1976). Global radiation beneath the canopy and in a clearing of a suburban hardwood forest. *Agric. Meteorol.* **16**, 321–327.

Mares, M. A., and Williams, D. F. (1977). Experimental support for food particle size resource allocation in heteromyid rodents. *Ecology* **58**, 1186–1190.

Margaris, N. S. (1976). Structure and dynamics in a phryganic (East Mediterranean) ecosystem. *J. Biogeogr.* **3**, 249–259.

Margaris, N. S. (1981). Adaptive strategies in plants dominating Mediterranean-type ecosystems. *In* "Mediterranean Type Shrublands" (F. di Castri, D. W. Goodall, and R. L. Specht, eds.), pp. 257–267. Elsevier, Amsterdam.

Margules, C., Higgs, A. J., and Rafe, R. W. (1982). Modern biogeographic theory: Are there any lessons for nature reserve design? *Biol. Conserv.* **24**, 115–128.

Marion, G. M. (1979). Biomass and nutrient removal in long-rotation stands. *In* "Impact of Intensive Harvesting on Forest Nutrient Cycling" (A. L. Leaf, ed.), pp. 98–110. USDA For. Serv., Northeast. For. Exp. Stn., Broomall, Pennsylvania.

Markham, J. W. (1973). Observations on the ecology of *Laminaria sinclairii* on three northern Oregon beaches. *J. Phycol.* **9**, 336–341.

Marks, P. L. (1974). The role of pin cherry (*Prunus pensylvanica* L.) in the maintenance of stability in northern hardwood ecosystems. *Ecol. Mongr.* **44**, 73–88.

Marks, P. L. (1975). On the relation between extension growth and successional status of deciduous trees of the northeastern United States. *Bull. Torrey Bot. Club.* **102**, 172–177.

Marks, P. L., and Bormann, F. H. (1972). Revegetation following forest cutting: Mechanisms for return to steady-state nutrient cycling. *Science (Washington, D.C.)* **176**, 914–915.

Marquis, D. A. (1965). Scarify soil during logging to increase birch reproduction. *North. Logger* **14**, 24, 42.

Marquis, D. A. (1973). The effect of environmental factors on advance regeneration of Allegheny hardwoods. Ph.D. Thesis, Yale Univ., New Haven, Connecticut.

Marquis, D. A. (1974). The impact of deer browsing on Allegheny hardwood regeneration. *USDA For. Ser. Res. Pap. NE* **NE-308.**

Marquis, D. A. (1975a). Seed storage and germination under northern hardwood forests. *Can. J. For. Res.* **5**, 478–484.

Marquis, D. A. (1975b). The Allegheny hardwood forests of Pennsylvania. *USDA For. Serv. Gen. Tech. Rep. NE* **NE-15.**

Marquis, D. A.(1981). Effect of deer browsing on timber production in Allegheny hardwood forests of Northwestern Pennsylvania. *For. Serv. Res. Pap. NE (U.S.)* **NE-475.**

Marquis, D. A., and Brenneman, R. (1981). The impact of deer on forest vegetation in Pennsylvania. *USDA For. Serv. Gen. Tech. Rep. NE* **NE-65.**

Marquis, D. A., Bjorkbom, J. C., and Yelenosky, G. (1964). Effect of seedbed condition and light exposure on paper birch regeneration. *J. For.* **62**, 876–881.

Marshall, L. G. (1981). The great American interchange—An invasion induced crisis for South American mammals. *In* "Biotic Crises in Ecological and Evolutionary Time" (M. H. Nitecki, ed.), pp. 133–229. Academic Press, New York.

Marshall, S. M., and Stephenson, T. A.(1933). The breeding of reef animals. Part 1: The corals. *Sci. Rep. Great Barrier Reef Exped., 1928–1929* **3**, 219–245.

Martin, C. (1923). "Landeskunde von Chile." Friederichsen, Hamburg.

Martin, R. E., Cushwa, C. T., and Miller, R. L. (1969). Fire as a physical factor in wildland management. *Proc. Annu. Tall Timbers Fire Ecol. Conf.* **9**, 271–288.

Martin, W. E. (1959). The vegetation of Island Beach State Park, New Jersey. *Ecol. Monogr.* **29**, 1–46.

Martindale, S. (1983). Foraging patterns of nesting Gila Woodpeckers. *Ecology* **64**, 888–898.

Maruyama, T., and Kimura, M. (1980). Genetic variability and effective population size when local extinction and recolonization of subpopulations are frequent. *Proc. Natl. Acad. Sci. U.S.A.* **77**, 6710–6714.

Matson, P. A., and Boone, R. (1984). Nitrogen mineralization and natural disturbance: wave-form dieback of mountain hemlock in the Oregon Cascades. *Ecology* (in press).

Matson, P. A., and Vitousek, P. M. (1981). Nitrification potentials following clearcutting in the Hoosier National Forest, Indiana. *For. Sci.* **27**, 781–791.

Matson, P. A., and Waring, R. H. (1984). Effects of nutrient and light limitation on mountain hemlock: susceptibility to laminated root rot. *Ecology* (in press).

Matthews, R. W., and Matthews, J. R. (1978). "Insect Behavior." Wiley, New York.

Mattson, W. J. (1980). Herbivory in relation to plant nitrogen content. *Annu. Rev. Ecol. Syst.* **11**, 119–161.

Mattson, W. J., and Addy, N. D. (1975). Phytophagous insects as regulators of forest primary production. *Science (Washington, D.C.)* **190**, 515–522.

May, R. M. (1973). "Stability and Complexity in Model Ecosystems." Princeton Univ. Press, Princeton, New Jersey.

May, R. M. (1976). Patterns in multi-species communities. *In* "Theoretical Ecology" (R. H. May, ed.), pp. 142–162. Saunders, Philadelphia, Pennsylvania.

May, R. M. (1981). Modeling recolonization by neotropical migrants in habitats with changing patch structure, with notes on the age structure of populations. *In* "Forest Island Dynamics in Man-dominated Landscapes" (R. L. Burgess and D. M. Sharpe, eds.), pp. 207–213. Springer-Verlag, Berlin and New York.

Maynard Smith, J. (1978). "The Evolution of Sex." Cambridge Univ. Press, London and New York.

Maynard Smith, J., ed. (1983). "Evolution Now." Freeman, San Francisco, California.

Melillo, J. M., Aber, J. D., and Muratore, J. M. (1982). Nitrogen and lignin control of hardwood leaf litter decomposition dynamics. *Ecology* **63**, 621–626.

Menge, B. A. (1976). Organization of the New England rocky intertidal community: role of predation, competition, and environmental heterogeneity. *Ecol. Monogr.* **46**, 355–393.

Menge, B. A. (1978a). Predation intensity in a rocky intertidal community: relation between predator foraging activity and environmental harshness. *Oecologia* **34**, 1–16.

Menge, B. A. (1978b). Predation intensity in a rocky intertidal community: effect of an algal canopy, wave action and dessication on predator feeding rates. *Oecologia* **34**, 17–35.

Menge, B. A. (1979). Coexistence between the seastars *Asterias vulgaris* and *A. forbesi* in a heterogeneous environment: a non-equilibrium explanation. *Oecologia* **41**, 245–272.

Menge, J. L. (1975). Effect of herbivores on community structure of the New England rocky intertidal region: distribution, abundance and diversity of algae. Ph.D. Thesis, Harvard Univ., Cambridge, Massachusetts.

Menges, E. S., and Locks, O. L. (1980). Modeling a disease-caused patch disturbance: oak wilt in the midwestern United States. *Ecology* **65**, 487–498.

Merz, R. W., and Boyce, S. G. (1958). Reproduction of upland hardwoods in southeastern Ohio. *U.S. For. Serv., Cent. States For. Exp. Stn., Tech. Pap.* No. 155.

Messina, F. J. (1981). Plant protection as a consequence of an antmembracid mutualism: interactions on goldenrod (*Solidago* sp.). *Ecology* **62**, 1433–1440.

Metzger, F., and Schultz, J. (1981). Spring ground layer vegetation 50 years after harvesting in northern hardwoods forests. *Am. Midl. Nat.* **105**, 44–50.

Mielke, D. L., Shugart, H. H., and West, D. C. (1977). User's manual for FORAR, a stand model for upland forests of southern Arkansas. *Oak Ridge Natl. Lab.* [*Rep.*] *ORNL-TM (U.S.)* **ORNL-TM-5767.**

Miles, J. (1979). "Vegetation Dynamics." Chapman & Hall, London.

Miller, H. G. (1969). Nitrogen nutrition on the sands of Culbin Forest, Morayshire. *J. Sci. Food Agric.* **20**, 417–419.

Miller, H. G. (1981). Forest fertilization: some guiding concepts. *Forestry* **54**, 157–167.

Miller, H. G., Cooper, J. M., and Miller, J. D. (1976). Effect of nitrogen supply on nutrients in litterfall and crown leading in a stand of Corsican pine. *J. Appl. Ecol.* **13**, 233–248.

Miller, J. C. (1980). Niche relationships among parasitic insects occurring in a temporary habitat. *Ecology* **61**, 270–275.

Miller, P. C. (1969a). Tests of solar radiation models in three forest canopies. *Ecology* **50**, 878–885.

Miller, P. C. (1969b). Solar radiation profiles in openings in canopies of aspen and oak. *Science (Washington, D.C.)* **164**, 308–309.

Miller, P. C. (1981). Similarities and limitations of resource utilization in Mediterranean type eco-systems. *In* ''Resource Use by Chaparral and Matorral'' (P. C. Miller, ed.), p. 369–407. Spring-er-Verlag, Berlin and New York.

Miller, T. E. (1982). Community diversity and interactions between the size and frequency of distur-bance. *Am. Nat.* **120**, 533–536.

Milton, K. (1982). Dietary quality and population regulation in a howler monkey population. *In* ''The Ecology of a Tropical Forest: Seasonal Rhythms and Long-term Changes'' (E. G. Leigh, Jr., A. S. Rand, and D. M. Windsor, eds.), pp. 273–289. Smithsonian Inst. Press, Washington, D.C.

Minckler, L. S. (1961). Measuring light in uneven-aged hardwood stands. *U.S. For. Serv. Cent. States For. Exp. Stn., Tech. Pap.* No. 184.

Minckler, L. S., and Woerheide, J. D. (1965). Reproduction of hardwoods 10 years after cutting as affected by site and opening size. *J. For.* **63**, 103–107.

Minckler, L. S., Woerheide, J. D., and Schlesinger, R. C. (1973). Light, soil moisture, and tree reproduction in hardwood forest openings. *USDA For. Serv., N. C. For. Exp. Stn. Res. Pap. NC* **NC-89.**

Minnich, R. A. (1983). Fire mosaics in southern California and northern Baja California. *Science (Washington, D.C.)* **219**, 1287–1294.

Mitchell, K. J. (1975). Dynamics and simulated yield of Douglas-fir. *For. Sci. Monogr.* No. 17.

Mitchell, R. G., and Martin, R. E. (1980). Fire and insects in pine culture of the Pacific Northwest. *Proc. Conf. Fire For. Meteorol.* **6**, 182–190.

Mitter, C., Futuyma, D. J., Schneider, J. C., and Hare, J. D. (1979). Genetic variation and host–plant relations in a parthenogenetic moth. *Evolution* **33**, 777–790.

Mitton, J. B. (1978). Relationship between heterozygosity for enzyme loci and variation of mor-phological characters in natural populations. *Nature (London)* **273**, 661–662.

Mitton, J. B., and Pierce, B. A. (1980). The distribution of individual heterozygosity in natural popula-tions. *Genetics* **95**, 1043–1054.

Mitton, J. B., Knowles, P., Sturgeon, K. B., Linhart, Y. B., and Davis, M. (1981). Associations between heterozygosity and growth rate variables in three western forest trees. *In* ''Isozymes of North American Trees and Forest Insects'' (M. T. Conkle, ed.), pp. 27–34. Pac. Southwest For. Range Exp. Stn., Berkeley, California.

Mohler, C. L., Marks, P. L., and Sprugel, D. G. (1978). Stand structure and allometry of trees during self-thinning of pure stands. *J. Ecol.* **66**, 599–614.

Moller, C. M., Muller, D., and Nielsen, J. (1954). Graphic presentation of dry matter production in European beech. *Forstl. Forsoegsvaes. Dan.* **21**, 327–335

Molloy, B. P. J. (1969). Evidence for post-glacial climatic changes in New Zealand. *J. Hydrol. (N.Z.)* **8**, 56–67.

Mooney, H. A. (1972). The carbon balance of plants. *Annu. Rev. Ecol. Syst.* **3**, 315–346.

Mooney, H. A., ed. (1977). ''Convergent Evolution in Chile and California: Mediterranean Climate Ecosystems,'' US/IBP Synthesis Series, No. 5. Dowden, Hutchinson & Ross, Stroudsburg, Pennsylvania.

Mooney, H. A., and Dunn, E. L. (1970). Convergent evolution of mediterranean climate evergreen sclerophyll shrubs. *Evolution* **24**, 292–303.

Mooney, H. A., and Godron, M., eds. (1983) ''Disturbance and Ecosystems.'' Springer-Verlag, Berlin and New York.

Mooney, H. A., and Rundel, P. W. (1979). Nutrient relations of the evergreen shrub, *Adenostoma fasciculatum,* in the California chaparral. *Bot. Gaz.* **140**, 109–113.

Mooney, H. A., Bonnicksen, T. M., Christensen, N. L., Lotan, J. E., and Reiners, W. A., eds. (1981). Fire regimes and ecosystem properties. *Gen. Tech. Rep. WO—U.S. For. Serv. [Wash. Off.]* **GTR-WO-26.**

Moran, G. F., and Marshall, D. R. (1978). Allozyme uniformity within and variation between races of the colonizing species *Xanthium strumarium* L. (Noogora burr). *Aust. J. Biol. Sci.* **31**, 283–291.

Moreno C., P. (1976). Latencia y viabilidad de semillas de vegetación primaria. *In* "Regeneración des Selvas" (A. Gómez-Pompa, C. Vázquez-Yanes, S. del Amo R., and A. Butanda C., eds.), pp. 527–548. Compañia Editorial Continental, Mexico City, Mexico.

Morey, H. F. (1936). Age–size relationships of Heart's Content, a virgin forest in northwestern Pennsylvania. *Ecology* **17**, 251–257.

Morisita, M. (1959). Measuring of the dispersion of individuals and analysis of the distributional patterns. *Mem. Fac. Sci., Kyushu Univ., Ser. E* **2**, 215–235.

Morison, S. E. (1950). "Breaking the Bismarcks Barrier. Vol. VI: History of United States Naval Operations in World War II." Little, Brown, Boston, Massachusetts.

Morris, R. F. (1963). The dynamics of epidemic spruce budworm populations. *Mem. Entomol. Soc. Can.* **31**.

Morton, E. S. (1980). Adaptations to seasonal changes by migrant land birds in the Panama Canal Zone. *In* "Migrant Birds in the Neotropics: Ecology, Behavior, Distribution, and Abundance" (A. Keast and E. S. Morton, eds.), Smithsonian Inst. Press, Washington, D.C.

Motro, V., and Thomson, G. (1982). On heterozygosity and the effective size of populations subject to size changes. *Evolution (Lawrence Kans.)* **36**, 1059–1066.

Mount, A. B. (1979). Natural regeneration processes in Tasmanian forests. *Search* **10**, 180–186.

Moyer, J. T., Emerson, W. K., and Ross, M. (1982). Massive destruction of scleractinian corals by the muricid gastropod, *Drupella,* in Japan and the Philippines. *Nautilus* **96**, 69–82.

Mugaas, J. N., and King, J. R. (1981). Annual variation of daily energy expenditure by the Black-billed Magpie. *Stud. Avian Biol.* **5**, 1–78.

Mukai, T., Watanabe, T. K., and Yamaguchi, O. (1974). The genetic structure of natural populations of *Drosophila melanogaster*. XII. Linkage disequilibrium in a large local population. *Genetics* **77**, 771–793.

Muller, C. H. (1940). Plant succession in the *Larrea–Flourensia* climax. *Ecology* **21**, 206–212.

Muller, C. H. (1952). Plant succession in arctic heath and tundra in northern Scandinavia. *Bull. Torrey Bot. Club* **79**, 296–309.

Muller, C. H. (1966). The role of chemical inhibition (allelopathy) in vegetational composition. *Bull. Torrey Bot. Club* **93**, 332–351.

Muller, C. H., Hanawalt, R. B., and McPherson, J. K. (1968). Allelophathic control of herb growth in the fire cycle of California chaparral. *Bull. Torrey Bot. Club* **95**, 225–231.

Muller, H. J. (1964). The relation of mutation to mutational advance. *Mutat. Res.* **1**, 2–9.

Muller, R. N. (1978). The phenology, growth, and ecosystem dynamics of *Erythronium americanum* in the northern hardwood forest. *Ecol. Monogr.* **48**, 1–20.

Müller-Using, B. (1973). Untersuchungen Über Die Verjungung Von *Nothofagus alpina* (Poepp. et Endl.) Oerst Und Ihrer Wichtigsten Begleitbaumarten In Der Chilenischen Anden-Und Kusten-Kordillere. Ph. D. Thesis, Univ. of Munich.

Mullette, K. J., and Bamber, R. K. (1978). Studies of the lignotubers of *Eucalyptus gummifera* (Gaertn. & Hochr.). III. Inheritance and chemical composition. *Aust. J. Bot.* **26**, 23–28.

Mulvey, M., and Vrijenhoek, R. C. (1981). Multiple paternity in the hermaphroditic snail, *Biomphalaria obstructa*. *J. Hered.* **72**, 308–312.

Muñoz, P., C. (1966). "Sinopsis de la Flora Chilena," 2nd Ed. Univ. de Chile, Santiago.

Munro, D. D.(1974). Forest growth models: a prognosis. *In* "Growth Models for Tree and Stand Simulation" (J. Fries, ed.), Research Notes, No. 30, pp. 7–21. Dep. For. Yield Res., R. Coll. For., Stockholm.

Munteanu, N., and Maly, E. J. (1981). The effect of current on the distribution of diatoms settling on submerged glass slides. *Hydrobiologia* **78**, 273–282.

Murdoch, W. W. (1977). Stabilizing effects of spatial heterogeneity in predator–prey systems. *Theor. Popul. Biol.* **11**, 252–273.

Murray, J. (1964). Multiple mating and effective population size in *Cepaea nemoralis*. *Evolution* **18**, 283–291.

Mutch, R. W. (1970). Wildland fires and ecosystems—a hypothesis. *Ecology* **51**, 1046–1051.

Myrold, D. D., Gosz, J. R., and Tiedje, J. M. (1982). Moisture gradient effects on denitrification rates within a non-aggrading aspen site. *Bull. Ecol. Soc. Am.* **63,** 95.

Naka, K. (1982). Community dynamics of evergreen broadleaf forests in southwestern Japan. I. Wind damaged trees and canopy gaps in an evergreen oak forest. *Bot. Mag.* **95,** 385–399.

Nakashizuka, T., and Numata, M. (1982a). Regeneration process of climax beech forests. I. Structure of a beech forest with the undergrowth of Sasa. *Nippon Seitai Gakkaishi (Jpn. J. Ecol.)* **32,** 57–67.

Nakashizuka, T., and Numata, M. (1982b). Regeneration process of climax beech forests. II. Structure of a forest under the influences of grazing. *Nippon Seitai Gakkaishi, (Jpn. J. Ecol.)* **32,** 473–482.

Nanson, G. C., and Beach, H. F. (1977). Forest succession and sedimentation on a meandering river flood plain, northeast British Columbia, Canada. *J. Biogeogr.* **4,** 229–252.

Naveh, Z. (1975). The evolutionary significance of fire in the Mediterranean region. *Vegetatio* **29,** 199–208.

Nei, M., Maruyama, T., and Chakraborty, R. (1975). The bottleneck effect and genetic variability in populations. *Evolution* **29,** 1–10.

Neilson, R. P., and Wullstein, L. H. (1983). Biogeography of two southwest American oaks in relation to atmospheric dynamics. *J. Biogeogr.* **10,** 275–297.

Nelson, T. C., and Zillgitt, W. M. (1969). "A Forest Atlas of the South." USDA For. Serv., South. For. Exp. Stn., New Orleans, Louisiana, and Southeast. For. Exp. Stn., Ashville, North Carolina.

Newton, I. (1972). "Finches." Collins, London.

Ng, F. S. P. (1978). Strategies of establishment in Malayan forest trees. *In* "Tropical Trees as Living Systems" (P. B. Tomlinson and M. N. Zimmerman, eds.), pp. 129–162. Cambridge Univ. Press, London and New York.

Nicholson, D. I. (1965). A review of natural regeneration in the dipterocarp forest of Sabah. *Malay. For.* **28,** 4–24.

Nicotri, M. E. (1977). Grazing effects of four marine intertidal herbivores on the microflora. *Ecology* **58,** 1020–1032.

Nienhuis, P. H. (1969). The significance of the substratum for intertidal algal growth on the artificial rocky shore of the Netherlands. *Int. Rev. Gesamten Hydrobiol.* **54,** 207–215.

Nisbet, R. M., and Gurney, W. S. C. (1982). "Modelling Fluctuating Populations." Wiley, New York.

Nitzberg, M. E., and Keeley, J. E. (1985). Investigations into the role of charred wood in the germination of two chaparral herbs. Manuscript.

Noble, I. R. (1981). Predicting successional change. *In* "Fire Regimes and Ecosystem Properties" (H. A. Mooney, T. M. Bonnicksen, N. L. Christensen, J. E. Lothan, and W. A. Reiners, eds.), *U.S.D.A. For. Serv. Gen. Tech. Rep. WO—U.S. For. Serv. [Wash. Off.]* **GTR-WO-26,** pp. 278–300.

Noble, I. R., and Slatyer, R. O. (1980). The use of vital attributes to predict successional changes in plant communities subject to recurrent disturbances. *Vegetatio* **43,** 5–21.

Norton, T. A., and Fetter, R. (1981). The settlement of *Sargassum muticum* propagules in stationary and flowing water. *J. Mar. Biol. Assoc. U.K.* **61,** 929–940.

Norton-Griffiths, M. (1979). The influence of grazing, browsing, and fire on the vegetation dynamics of the Serengeti. *In* "Serengeti: Dynamics of an Ecosystem" (A. R. E. Sinclair and M. Norton-Griffiths, eds.), pp. 310–352. Univ. of Chicago Press, Chicago, Illinois.

Norton-Griffiths, M., Herlocker, D., and Pennycuick, L. (1975). The patterns of rainfall in the Serengeti ecosystem, Tanzania. *East Afr. Wildl. J.* **13,** 347–374.

Noyes, R. F., Barrett, G. W., and Taylor, D. H. (1982). Social structure of feral house mouse (*Mus musculus* L.) populations: effects of resource partitioning. *Behav. Ecol. Sociobiol.* **10,** 157–163.

Noy-Meir, I. (1973). Desert ecosystems: environment and producers. *Annu. Rev. Ecol. Syst.* **4,** 25–51.

Noy-Meir, I. (1974). Desert ecosystems: higher trophic levels. *Annu. Rev. Ecol. Syst.* **5,** 195–213.

Noy-Meir, I. (1981). Spatial effects in modelling of arid ecosystems. *In* "Arid-land Ecosystems: Structure, Functioning and Management" (D. W. Goodall, R. A. Perry, and K. M. W. Howes, eds.), pp. 411–432. Cambridge Univ. Press, London and New York.

Nye, P. H., and Greenland, D. J. (1960). "The Soil under Shifting Cultivation," Tech. Rep. No. 51. Commonw. Bur. Soils, Farnham Royal, England.

Nye, P. H., and Greenland, D. J. (1964). Changes in the soil after clearing tropical forest. *Plant Soil* **21**, 101–112.

Nye, P. H., and Tinker, P. B. (1977). "Solute Movement in the Soil–Root System." Univ. of California Press, Berkeley.

Oberdorfer, E. (1960). "Pflanzensoziologische Studien in Chile—Ein Vergleich mit Europa." Cramer, Weinheim.

Odum, E. P. (1969). The strategy of ecosystem development. *Science (Washington, D.C.)* **164**, 262–270.

Odum, E. P. (1971). "Fundamentals of Ecology." Saunders, Philadelphia, Pennsylvania.

Odum, E. P. (1975). "Ecology," 2nd Ed. Holt, Rinehart, New York.

O'Farrell, M. J. (1980). Spatial relationships of rodents in a sagebrush community. *J. Mammal.* **61**, 589–605.

Ohta, T. (1982). Linkage disequilibrium due to random genetic drift in finite subdivided populations. *Proc. Natl. Acad. Sci. U.S.A.* **70**, 1940–1944.

Okia, N. O. (1976). Birds of the understory of lake–shore forests on the Entebbe Peninsula, Uganda. *Ibis* **118**, 1–13.

Oldeman, R. A. A. (1972). L'architecture de la végétation ripicole forestière des fleuves et criques guyanais. *Adansonia* **12**, 253–265.

Oldeman, R. A. A. (1975). "Bioarquitectura de las Vegetationes y Método Práctico para su Observación." Nota Técnica No. E.1. Min. Agr. y Ganado, Depto. de Regionalización, Quito, Ecuador.

Oldeman, R. A. A. (1978). Architecture and energy exchange of dicotyledonous trees in the forest. *In* "Tropical Trees as Living Systems" (P. B. Tomlinson and M. H. Zimmerman, eds.), pp. 535–560. Cambridge Univ. Press, London and New York.

O'Leary, J. F., and Minnich, R. A. (1981). Postfire recovery of creosote bush scrub vegetation in the western Colorado desert. *Madrono* **28**, 61–66.

Oliver, C. D. (1981). Forest development in North America following major disturbances. *For. Ecol. Manage.* **3**, 153–168.

Oliver, C. D. (1982). Stand development—its uses and methods of study. *In* "Forest Succession and Stand Development Research in the Northwest" (J. E. Means, ed.), pp. 100–111. Oregon State Univ. Press, Corvallis.

Oliver, C. D., and Stephens, E. P. (1977). Reconstruction of a mixed-species forest in central New England. *Ecology* **58**, 562–572.

Olson, J. S. (1958). Rates of succession and soil changes on southern Lake Michigan sand dunes. *Bot. Gaz.* **119**, 125–170.

Olson, J. S. (1963). Analog computer models for the movement of isotopes through ecosystems. *Radioecol., Proc. Natl. Symp., 1st, Fort Collins, Colo., 1961.*

Oohata, S., and Shidei, T. (1971). Studies on the branching structure of trees. I. Bifurcation ratio of trees in Horton's law. *Nippon Seitai Gakkaishi (Jpn. J. Ecol.)* **21**, 7–14.

Oosting, H. J., and Bourdeau, P. F. (1955). Virgin hemlock segregates in the Joyce Kilmer Memorial Forest of western North Carolina. *Bot. Gaz.* **116**, 340–359.

Orians, G. H. (1982). The influence of tree falls in tropical forests on tree species richness. *Trop. Ecol.* **23**, 255–279.

Orians, G. H., and Pearson, N. E. (1979). On the theory of central place foraging. *In* "Analysis of Ecological Systems" (D. J. Horn, G. R. Stairs, and R. D. Mitchell, eds.), pp. 155–177. Ohio State Univ. Press, Columbus.

Orians, G. H., and Solbrig, O. T. (1977). A cost–income model of leaves and roots with special reference to arid and semi-arid areas. *Am. Nat.* **111**, 677–690.

Osman, R. W. (1977). The establishment and development of a marine epifaunal community. *Ecol. Monogr.* **47**, 37–63.

Ostarello, G. L. (1976). Larval dispersal in the subtidal hydrocoral *Allopora californica* Verrill (1866). *In* "Coelenterate Ecology and Behavior" (G. O. Mackie, ed.), pp. 331–337. Plenum, New York.

Pacala, S. W., and Roughgarden, J. (1982). Spatial heterogeneity and interspecific competition. *Theor. Popul. Biol.* **21**, 92–113.

Paijmans, K. (1970). An analysis of four tropical rainforest sites in New Guinea. *J. Ecol.* **58**, 77–101.

Paine, R. T. (1966). Food web complexity and species diversity. *Am. Nat.* **100**, 65–75.

Paine, R. T. (1974). Intertidal community structure: experimental studies on the relationship between a dominant competitor and its principal predator. *Oecologia* **15**, 93–120.

Paine, R. T. (1976). Biological observations on a subtidal *Mytilus californianus* bed. *Veliger* **19**, 125–130.

Paine, R. T. (1977). Controlled manipulations in the marine intertidal zone and their contributions to ecological theory. *Spec. Publ., Acad. Nat. Sci. Philadelphia* No. 12, 245–270.

Paine, R. T. (1979). Disaster, catastrophe and local persistence of the sea palm *Postelsia palmaeformis*. *Science (Washington, D.C.)* **205**, 685–687.

Paine, R. T., and Levin, S. A. (1981). Intertidal landscapes: disturbance and the dynamics of pattern. *Ecol. Monogr.* **51**, 145–178.

Paine, R. T., and Suchanek, T. H. (1983). Convergence of ecological processes between independently evolved competitive dominants: a tunicate–mussel comparison. *Evolution* **37**, 821–831.

Paine, R. T., and Vadas, R. L. (1969). The effects of grazing by sea urchins, *Strongylocentrotus* spp., on benthic algal populations. *Limnol. Oceanogr.* **14**, 710–719.

Palumbi, S. R., and Jackson, J. B. C. (1982). Ecology of cryptic coral reef communities. II. Recovery from small disturbances by encrusting bryozoa: the influence of "host" species and lesion size. *J. Exp. Mar. Biol. Ecol.* **64**, 103–115.

Parkhurst, D. F., and Loucks, O. L. (1972). Optimal leaf size in relation to environment. *J. Ecol.* **60**, 505–537.

Parrish, J. A. D., and Bazzaz, F. A. (1982). Responses of plants from three successional communities to a nutrient gradient. *J. Ecol.* **70**, 233–248.

Parsons, D. J. (1976). The role of fire in natural communities: An example from the Southern Sierra Nevada, California. *Environ. Conserv.* **3**, 91–99.

Patric, J. H., and Hanes, T. L. (1964). Chaparral succession in a San Gabriel mountain area of southern California. *Ecology* **45**, 353–360.

Peace, W. J. H., and Grubb, P. J. (1982). Interaction of light and mineral nutrient supply in the growth of *Impatiens parviflora*. *New Phytol.* **90**, 127–150.

Pearson, J. A. (1982). Biomass distribution and ecosystem development in lodgepole pine forests of the Medicine Bow Mountains, Wyoming. Ph.D. Thesis, Univ. of Wyoming, Laramie.

Pearson, R. G. (1981). Recovery and recolonization of coral reefs. *Mar. Ecol.: Prog. Ser.* **4**, 105–122.

Peet, R. K. (1981). Changes in biomass and production during secondary forest succession. *In* "Forest Succession: Concepts and Application" (D. C. West, H. H. Shugart, and B. D. Botkin, eds.), pp. 324–338. Springer-Verlag, Berlin and New York.

Peet, R. K., and Christensen, N. L. (1980). Succession: a population process. *Vegetatio* **43**, 131–140.

Peet, R. K., and Loucks, O. L. (1977). A gradient analysis of southern Wisconsin forests. *Ecology* **58**, 485–499.

Peet, R. K., Glenn-Lewin, D. C., and Wolf, J. W. (1983). Predictions of man's impact on vegetation. *In* "Man's Impact on Vegetation" (W. Holzner, M. J. A. Werger, and I. Ikusiam, eds.), pp. 41–53. Junk, The Hague.

Penning de Vries, F. W. T. (1972). Respiration and growth. *In* "Crop Processes in Controlled Environments" (A. R. Rees, K. E. Cockshull, D. W. Hard, and R. G. Hurd, eds.), pp. 327–346. Academic Press, New York.

Penning de Vries, F. W. T. (1975). Use of assimilates in higher plants. *In* "Photosynthesis and Production in Different Environments" (J. P. Cooper, ed.), pp. 459–480. Cambridge Univ. Press, London and New York.

Pennycuick, C. J. (1979). Energy costs of locomotion and the concept of "foraging radius". *In*

"Serengeti: Dynamics of an Ecosystem" (A. R. E. Sinclair and M. Norton-Griffiths, eds.), pp. 164–184. Univ. of Chicago Press, Chicago, Illinois.

Peterson, D. L., and Bazzaz, F. A. (1978). Life cycle characteristics of *Aster pilosus* in early successional habitats. *Ecology* **59**, 1005–1013.

Peterson, D. L., and Rolfe, G. L. (1982). Nutrient dynamics of herbaceous vegetation in upland and floodplain forest communities. *Am. Midl. Nat.* **107**, 325–339.

Phares, R. E., and Williams, R. D. (1971). Crown release promotes faster diameter growth of pole-size black walnut. *Res. Note NC (U.S. For. Serv.)* **NC-124.**

Philpot, C. W. (1969). Seasonal changes in heat control and ether extractive content of chamise. *USDA For. Serv. Intermountain For. Range Exp. Stn. Res. Pap INT* **INT-61.**

Philpot, C. W. (1970). Influence of mineral content on pyrolysis of plant material. *For. Sci.* **16**, 416–417.

Phipps, R. L. (1979). Simulation of wetlands forest vegetation dynamics. *Ecol. Model.* **7**, 257–288.

Pickett, S. T. A. (1976). Succession: an evolutionary interpretation. *Am. Nat.* **110**, 107–119.

Pickett, S. T. A. (1980). Non-equilibrium coexistence of plants. *Bull. Torrey Bot. Club* **107**, 238–248.

Pickett, S. T. A. (1983). Differential adaptation of tropical species to canopy gaps and its role in community dynamics. *Trop. Ecol.* **24**, 68–84.

Pickett, S. T. A., and Kempf, J. S. (1980). Branching patterns in forest shrubs and understory trees in relation to habitat. *New Phytol.* **86**, 219–228.

Pickett, S. T. A., and Thompson, J. N. (1978). Patch dynamics and the design of nature reserves. *Biol. Conserv.* **13**, 27–37.

Pierce, B. A., and Mitton, J. B. (1982). Allozyme heterozygosity and growth in the tiger salamander, *Ambystoma tigrinum*. *J. Hered.* **73**, 250–253.

Pinero, D., and Sarukhán, J. (1982). Reproductive behavior and its individual variability in a tropical palm, *Astrocaryum mexicanum*. *J. Ecol.* **70**, 461–472.

Pinto, A. E. (1970). Phenological studies of trees at El Verde. *In* "A Tropical Rainforest" (H. Odum and R. F. Pigeon, eds.), pp. D237–D239. USAEC, Washington, D.C.

Pitcher, C. R. (1981). Some mutualistic aspects of the ecology of the scallop *Chlamys asperrimus* (Lamarck) and its epizooic sponges. B.S. Thesis, Univ. of Adelaide, Adelaide, South Australia.

Pitelka, L. F., Stanton, D. S., and Peckenham, M. O. (1980). Effects of light and density on resource allocation in a forest herb, *Aster acuminatus* (Compositae). *Am. J. Bot.* **67**, 942–948.

Pitman, G. B., Larsson, S., and Tenow, O. (1982). Stem growth efficiency: an index of susceptibility to bark beetle and sawfly attack. *In* "Carbon Uptake and Allocation in Subalpine Ecosystems as a Key to Management" (R. H. Waring, ed.), Proc. Int. Union For. Res. Organ. Workshop, pp. 52–56. Oregon State Univ. For. Res. Lab., Corvallis.

Platt, W. J. (1975). The colonization and formation of equilibrium plant species associations on badger disturbances in a tall-grass prairie. *Ecol. Monogr.* **45**, 285–305.

Platt, W. J. (1976). The natural history of a fugitive prairie plant (*Mirabilis hirsuta* (Pursh.) MacM.). *Oecologia* **22**, 389–410.

Platt, W. J., and Weis, I. M. (1977). Resource partitioning and competition within a guild of fugitive prairie plants. *Am. Nat.* **111**, 479–513.

Pleasants, J. M., and Zimmerman, M. (1979). Patchiness in the dispersion of nectar resources: evidence for hot and cold spots. *Oecologia* **41**, 283–288.

Plumb, M. L. (1979). The dynamics of short-lived species in a sand prairie. Ph.D. Thesis, Univ. of Wisconsin, Madison.

Pond, F. W., and Cable, D. R. (1962). Recovery of vegetation following wildfire on a chaparral area in Arizona. *U.S. For. Ser., Rocky Mt. For. Range Exp. Stn., Res. Notes* No. 72.

Pons, T. L. (1977). An ecophysiological study in the field layer of ash coppice. II. Experiments with *Geum urbanum* and *Cirsium palustre* in different light intensities. *Acta Bot. Neerl.* **26**, 29–42.

Pontailler, J. Y. (1979). La régénération du hêtre en forêt de fontainbleau, ses relations avec les conditions hydriques stationnelles. Ph.D. Thesis, Univ. Paris, Sud, Cent. D'Orsay.

Poore, M. E. D. (1964). Integration of the plant community. *J. Anim. Ecol.* **33**, Suppl., 213–226.

Poore, M. E. D. (1968). Studies in Malaysian rainforest. *J. Ecol.* **56**, 143–189.

Porter, J. W. (1974). Community structure of coral reefs on opposite sides of the Isthmus of Panama. *Science (Washington, D.C.)* **186**, 543–545.

Porter, J. W., Battey, J. F., and Smith, G. J. (1982). Perturbation and change in coral communities. *Proc. Natl. Acad. Sci. U.S.A.* **79**, 1678–1681.

Porter, W. P., and James, F. C. (1979). Behavioral implications of mechanistic ecology. II: the African rainbow lizard, *Agama agama. Copeia, 1979* pp. 594–619.

Porter, W. P., Mitchell, J. W., Bockman, W. A., and DeWitt, C. B. (1973). Behavioral implications of mechanistic ecology. *Oecologia* **13**, 1–54.

Potts, D. C. (1981). Crown-of-thorns starfish—man-induced pest or natural phenomenon? *In* "The Ecology of Pests" (R. L. Kitching and R. E. Jones, eds.), pp. 55–86.

Powell, N. A. (1968). Studies on bryozoa of the Bay of Fundy region. Part II. Bryozoa from fifty fathoms, Bay of Fundy. *Cah. Biol. Mar.* **9**, 247–259.

Power, M. E. (1984). Depth distribution of armored catfishes: Predator-induced resource avoidance. *Ecology* **65**, 523–528.

Powers, R. F. (1980). Mineralizable soil nitrogen as an index of nitrogen availability to forest trees. *Soil Sci. Soc. Am. J.* **44**, 1314–1320.

Preston, F. W. (1962). The canonical distribution of commonness and rarity: Part I. *Ecology* **43**, 185–215.

Price, M. V. (1978). Seed dispersion preferences of coexisting desert rodent species. *J. Mammal.* **59**, 624–626.

Price, P. W. (1975). "Insect Ecology." Wiley, New York.

Price, P. W. (1980). "Evolutionary Biology of Parasites." Princeton Univ. Press, Princeton, New Jersey.

Priestley, C. H. B. (1959). Heat conduction and temperature profiles in air and soil. *J. Aust. Inst. Agric. Sci.* **25**, 94–107.

Pritchett, W. L. (1979). "Properties and Management of Forest Soils." Wiley, New York.

Pritchett, W. L., and Wells, C. G. (1978). Harvesting and site preparation increase nutrient mobilization. *In* "Proceedings: A Symposium on Principles of Maintaining Productivity on Prepared Sites" (T. Tippin, ed.), pp. 98–110. Atlanta, Georgia.

Pryor, L. D., and Moore, R. M. (1954). Plant communities. *In* "Canberra, A Nation's Capital" (H. L. White, ed.), pp. 162–177. Angus & Robertson, Sydney.

Pulliam, H. R., and Brand, M. R. (1975). The production and utilization of seeds in plains grassland of southeastern Arizona. *Ecology* **56**, 1158–1166.

Pulliam, H. R., and Mills, G. S. (1977). The use of space by wintering sparrows. *Ecology* **58**, 1393–1399.

Putz, F. E. (1980). Lianas vs. trees. *Biotropica* **12**, 224–225.

Putz, F. E. (1983). Treefall pits and mounds, buried seeds, and the importance of soil disturbance to pioneer trees on Barro Colorado Island, Panama. *Ecology* **64**, 1069–1074.

Putz, F. E. (1984). The natural history of lianas on Barro Colorado Island, Panama. *Ecology,* **65**, 1713–1724.

Putz, F. E., and Milton, K. (1982). Tree mortality rates on Barro Colorado Island. *In* "The Ecology of a Tropical Forest" (E. G. Leigh, A. S. Rand, and D. M. Windsor, eds.), pp. 95–100. Smithsonian Inst. Press, Washington, D.C.

Putz, F. E., Coley, P. D., Lu, K., Montalvo, A., and Aiello, A. (1983). Uprooting and snapping of trees, structural causes and ecological consequences. *Can. J. For. Res.* **13**, 1011–1020.

Pyke, G. H., Pulliam, H. R., and Charnov, E. L. (1977). Optimal foraging: a selective review of theory and tests. *Q. Rev. Biol.* **52**, 137–154.

Pyne, S. J. (1982). "Fire in America." Princeton Univ. Press, Princeton, N.J.

Quarterman, E. (1970). Germination of seeds of certain tropical species. *In* "A Tropical Rainforest" (H. T. Odum and R. F. Pigeon, eds.), pp. D173–D175. USAEC, Washington, D.C.

Quick, C. R. (1935). Notes on the germination of *Ceanothus* seeds. *Madrono* **3**, 135–140.

Quinn, J. F. (1979). Disturbance, predation and diversity in the rocky intertidal zone. Ph.D. Thesis, Univ. of Washington, Seattle.

Quinn, J. F. (1982). Competitive hierarchies in marine benthic communities. *Oecologia* **54**, 129–135.

Quinn, R. D. (1979). Effects of fire on small mammals in the chaparral. *In.* "Cal-Neva Wildlife Transactions," pp. 125–133. Cited in Wirtz (1982).

Rabinowitz, D. (1981). Buried viable seeds in a North American tall-grass prairie: the resemblance of their abundance and composition to dispersing seeds. *Oikos* **36**, 191–195.

Radtke, K. W.-H., Arndt, A. M., and Wakimoto, R. H. (1982). Fire history at the Santa Monica Mountains. *In* "Dynamics and Management of Mediterranean-Type Ecosystems" (C. E. Conrad and W. C. Oechel, eds.), *USDA For. Ser. Pac. Southwest For. Range Exp. Stn. Gen. Tech. Rep. PSW* **PSW-58**, pp. 438–443.

Raffa, K. F., and Berryman, A. A. (1980). Flight responses and host selection by bark beetles. *In* "Dispersal of Forest Insects: Evaluation, Theory and Management Implications" (A. A. Berryman and L. Safranyik, eds.), Proc. Int. Union For. Res. Organ. Conf., pp. 213–233. Washington State Univ. Coop. Ext. Serv., Pullman.

Rai, K. N., and Jain, S. K. (1982). Population biology of *Avena*. IX. Gene flow and neighborhood size in relation to microgeographic variation in *A. barbata*. *Oecologia* **53**, 399–405.

Raison, R. J. (1979). Modification of the soil environment by vegetation fires, with particular reference to nitrogen transformations: a review. *Plant Soil* **51**, 73–108.

Rand, A. S. (1976). Report. *In* "Environmental Sciences Program Data: 1975." Smithsonian Trop. Res. Inst., Panama.

Rand, A. S., and Rand, W. M. (1982). Variation in rainfall on Barro Colorado Island. *In* "The Ecology of a Tropical Forest: Seasonal Rhythms and Long-term Changes" (E. G. Leigh, Jr., A. S. Rand, and D. M. Windsor, eds.), pp. 47–59. Smithsonian Inst. Press, Washington, D.C.

Rand, A. S., Guerrero, S., and Andrews, R. M. (1983). The ecological effects of malaria on populations of the lizard *Anolis limifrons* on Barro Colorado Island, Panama. *In* "Advances in Herpetology and Evolutionary Biology: Essays in Honor of Ernest E. Williams" (A. G. J. Rhodin and K. Miyata, eds.), pp. 455–471. Mus. Comp. Zool., Cambridge, Massachusetts.

Randall, J. E. (1967). Food habits of reef fishes of the West Indies. *Stud. Trop. Oceanogr.* **5**, 665–847.

Randall, J. E. (1974). The effect of fishes on coral fishes. *Proc. Int. Symp. Coral Reefs, 2nd, Great Barrier Reef, 1973* **1**, 159–166.

Rapp, M., and Lossaint, P. (1981). Some aspects of mineral cycling in the Garrique of southern France. *In* "Ecosystems of the World. Vol. 11: Mediterranean-Type Shrublands" (F. di Castri, D. W. Goodall, and R. L. Specht, eds.), pp. 289–307. Elsevier, Amsterdam.

Raup, H. M. (1957). Vegetational adjustment to the instability of the site. *Proc. Pap. Tech. Meet., Int. Union Conserv. Nat. Nat. Resour., 6th, Edinburgh* pp. 36–48.

Raup, H. M. (1975). Species versatility in shore habitats. *J. Arnold Arbor., Harv. Univ.* **55**, 126–165.

Raven, P. H. (1973). The evolution of mediterranean floras. *In* "Mediterranean Type Ecosystems: Origin and Structure" (F. di Castri and H. A. Mooney, eds.), pp. 213–224. Springer-Verlag, Berlin and New York.

Reddingius, J., and Den Boer, P. J. (1970). Simulation experiments illustrating stabilization of animal number by spreading of risk. *Oecologia* **5**, 240–248.

Reiche, K. F. (1934). "Geografía Botánica de Chile," Vol. 1. Imprenta Univ., Santiago, Chile.

Reichle, D. E., ed. (1981). "Dynamic Properties of Forest Ecosystems." Cambridge Univ. Press, London and New York.

Reichman, O. J. (1975). Relation of desert rodent diets to available resources. *J. Mammal.* **56**, 731–751.

Reichman, O. J. (1979). Desert granivore foraging and its impact on seed densities and distributions. *Ecology* **60**, 1085–1092.

Reichman, O. J., and Oberstein, D. (1977). Selection of seed distribution types by *Dipodomys merriami* and *Perognathus amplus*. *Ecology* **58**, 636–643.

Reiners, N. M., and Reiners, W. A. (1965). Natural harvesting of trees. *William L. Hutcheson Mem. For. Bull.* **2**, 9–17.

Reiners, W. A. (1983). Disturbance and basic properties of ecosystem energetics. *In* "Disturbance and Ecosystems: Components of Response" (H. A. Money and M. Godron, eds.), pp. 83–98. Springer-Verlag, Berlin and New York.

Reiners, W. A., and Lang, G. E. (1979). Vegetational patterns and processes in the balsam fir zone, White Mountains, New Hampshire. *Ecology* **60**, 403–417.

Rex, M. A. (1981). Community structure in the deep sea benthos. *Annu. Rev. Ecol. Syst.* **12**, 331–353.

Rice, J., Ohmart, R. D., and Anderson, B. W. (1983). Habitat selection attributes of an avian community: a discriminant analysis investigation. *Ecol. Monogr.* **53**, 263–290.

Rice, K. J. (1984). Plant life history variation and colonization dynamics in a variable environment: the population biology of two sympatric winter annuals, *Erodium botrys* and *E. brachycarpum*. Ph.D. Thesis, Univ. California, Davis.

Richards, P., and Williamson, G. B. (1975). Treefalls and patterns of understory species in a wet lowland tropical forest. *Ecology* **56**, 1226–1229.

Richards, P. W. (1952). "The Tropical Rain Forest." Cambridge Univ. Press, London and New York.

Richardson, C. A., Dustan, P., and Lang, J. C. (1979). Maintenance of living space by sweeper tentacles of *Montastrea cavernosa*, a Caribbean reef coral. *Mar. Biol. (Berlin)* **55**, 181–186.

Richerson, P., Armstrong, R., and Goldman, C. (1970). Contemporaneous disequilibrium: a new hypothesis to explain the "paradox of the plankton." *Proc. Natl. Acad. Sci. U.S.A.* **67**, 1710–1714.

Richter, D. D., Ralston, C. W., and Harms, W. R. (1982). Prescribed fire: effects on water quality and forest nutrient cycling. *Science (Washington, D.C.)* **215**, 661–663.

Ricker, W. E. (1963). Big effects from small causes: two examples from fish population dynamics. *J. Fish. R. Board Can.* **20**, 257–284.

Ricklefs, R. E. (1977). Environmental heterogeneity and plant species diversity: a hypothesis. *Am. Nat.* **111**, 376–381.

Riera, B. (1984). Observations sur les chablis, piste de St. Elie en Guyane. Manuscript.

Ries, R. E., and Fisser, H. G. (1979). Influence of environmental factors upon sagebrush and grass production in Wyoming. *Agro-Ecosystems* **5**, 41–55.

Risser, P. G., Birney, E. C., Blocker, H. D., May, S. W. D., Parton, W. J., and Wiens, J. A. (1981). "The True Prairie Ecosystem." Dowden, Hutchinson & Ross, Stroudsburg, Pennsylvania.

Robbins, R. G. (1958). Direct effect of the 1855 earthquake on the vegetation of the Orongorongo Valley, Wellington. *Trans. R. Soc. N.Z.* **85**, 205–212.

Robinson, S. K., and Holmes, R. T. (1982). Foraging behavior of forest birds: the relationships among search tactics, diet, and habitat structure. *Ecology* **63**, 1918–1931.

Robles, C. D. (1982). Disturbance and predation in an assemblage of herbivorous diptera and algae on rocky shores. *Oecologia* **54**, 23–31.

Robles, C. D., and Cubit, J. (1981). Influence of biotic factors in an upper intertidal community: dipteran larvae grazing on algae. *Ecology* **62**, 1536–1547.

Rodin, L. E., and Bazilevich, N. I. (1967). "Production and Mineral Cycling in Terresrial Vegetation." Oliver & Boyd, Edinburgh.

Roff, D. A. (1978). Size and survival in a stochastic environment. *Oecologia* **36**, 163–172.

Rogers, D. S. (1979). Patterns of colonization following gopher disturbance and cultivation in a sand prairie region. M.S. Thesis. Univ. of Wisconsin, Madison.

Romell, L. G. (1935). Ecological problems of the humus layer in the forest. *Mem.—N.Y. Agric. Exp. Stn. (Ithaca)* No. 170.

Romme, W. H. (1982). Fire and landscape diversity in subalpine forests of Yellowstone National Park. *Ecol. Monogr.* **52**, 199–221.

Romme, W. H., and Knight, D. H. (1981). Fire frequency and subalpine forest succession along a topographic gradient in Wyoming. *Ecology* **62**, 319–326.

Romme, W. H., and Knight, D. H. (1982). Landscape diversity: the concept applied to Yellowstone Park, *BioScience* **32**, 664–670.

Romme, W. H., and Martin, W. H. (1982). Natural disturbance by tree-falls in old-growth mixed mesophytic forest: Lilley Cornett Woods, Kentucky. *In* "Central Hardwood Forest Conference IV Proceedings" (R. N. Muller, ed.), pp. 367–383. Univ. of Kentucky, Lexington.

Root, R. B. (1973). Organization of a plant–arthropod association in simple and diverse habitats: the fauna of collards (*Brassica oleracea*). *Ecol. Monogr.* **43**, 95–124.

Rosenzweig, M. L., and Abramsky, Z. (1980). Microtine cycles: the role of habitat heterogeneity. *Oikos* **34**, 141–146.

Rosenzweig, M. L., Smigel, B., and Kraft, A. (1975). Patterns of food, space and diversity. *In* "Rodents in Desert Environments" (I. Prakash and P. K. Ghosh, eds.), pp. 241–268. Junk, The Hague.

Ross, B. A., Bray, J. R., and Marshall, W. H. (1970). Effects of long-term exclusion on a *Pinus resinosa* forest in north-central Minnesota. *Ecology* **51**, 1088–1093.

Rosswall, T. (1976). The internal cycle between vegetation, microorganisms, and soils. *In* "Nitrogen, Phosphorus, and Sulfur–Global Cycles" (B. H. Svennson and R. Soderlund, eds.), *Ecol. Bull.* No. 22, 157–167.

Rotenberry, J. T., and Wiens, J. A. (1978). Nongame bird communities in northwestern rangelands. *In* "Proceedings of the Workshop on Nongame Bird Habitat Management in the Coniferous Forests of the Western United States" (R. M. DeGraaf, ed.), *USDA For. Serv. Pac. Northwest For. Range Exp. Stn. Gen. Tech. Rep. PNW* **PNW-64**, pp. 32–46.

Rotenberry, J. T., and Wiens, J. A. (1980). Habitat structure, patchiness, and avian communities in North American steppe vegetation: a multivariate analysis. *Ecology* **61**, 1228–1250.

Rotenberry, J. T., Hinds, W. T., and Thorp, J. M. (1976). Microclimatic patterns on the Arid Lands Ecology Reserve. *Northwest Sci.* **50**, 122–130.

Roth, R. R. (1976). Spatial heterogeneity and bird species diversity. *Ecology* **57**, 773–782.

Rothermal, R. C. (1972). A mathematical model for predicting fire spread in wildland fuels. *USDA For. Serv. Intermountain For. Range Exp. Stn. Res. Pap. INT* **INT-115.**

Rothermal, R. C., and Philpot, C. W. (1973). Predicting changes in chaparral flammability. *J. For.* **71**, 640–643.

Roughgarden, J. (1972). Evolution of niche width. *Am. Nat.* **106**, 683–718.

Roughgarden, J. (1978). Influence of competition on patchiness in a random environment. *Theor. Popul. Biol.* **14**, 185–203.

Roughgarden, J. (1979). "Theory of Population Genetics and Evolutionary Ecology: An Introduction." Macmillan, New York.

Rowe, J. S. (1961). Critique of some vegetational concepts as applied to forests of northwestern Alberta. *Can. J. Bot.* **39**, 1007–1017.

Rowe, J. S., Bersteinsson, J. L., Padbury, G. A., and Hermesh, G. A.(1974). Fire studies in the Mackenzie Valley. *Can., Dep. Indian North. Aff., Publ.* no. QS-1567-000-EE-A1.

Rowley, I. (1974). "Bird Life." Collins, London.

Royama, T. (1970). Factors governing the hunting behavior and selection of food by the Great Tit (*Parus major* L.). *J. Anim. Ecol.* **39**, 619–668.

Rubin, J. A. (1982). The degree of intransitivity and its measurement in an assemblage of encrusting cheilostome bryozoans. *J. Exp. Mar. Biol. Ecol.* **60**, 119–128.

Rudstam, L. (1983). Cisco in Wisconsin Lakes: A long term comparison of their population structure and an analysis of their vertical distribution. M.S. Thesis, Univ. of Wisconsin, Madison.

Rundel, P. W. (1973). The relationship between basal fire scars and crown damage in giant sequoia. *Ecology* **54**, 210–213.

Rundel, P. W. (1981a). Fire as an ecological factor. *Encycl. Plant Physiol., New Ser.* **12A**, 501–538.

Rundel, P. W. (1981b). Structural and chemical components of flammability. *In* "Fire Regimes and Ecosystem Properties" (H. A. Mooney, T. M. Bonnicksen, N. L. Christensen, J. E. Lotan, and W. A. Reiners, eds.), *Gen. Tech. Rep. WO—U.S. For. Serv.* [*Wash. Off.*] **GTR-WO-26,** pp. 183–207.

Rundel, P. W., and Parsons, D. J. (1979). Structural changes in chamise (*Adenostoma fasciculatum*) along a fire-induced age gradient. *J. Range Manage.* **32**, 462–466.

Rundel, P. W., and Parsons, D. J. (1980). Nutrient changes in two chaparral shrubs along a fire-induced age gradient. *Am. J. Bot.* **67,** 51–58.

Runkle, J. R. (1981). Gap regeneration in some old-growth forest of the eastern United States. *Ecology* **62,** 1041–1051.

Runkle, J. R. (1982). Patterns of disturbance in some old-growth mesic forests of eastern North America. *Ecology* **63,** 1533–1546.

Runkle, J. R. (1984). Development of woody vegetation in treefall gaps in a beech-sugar maple forest. *Holarctic Ecol.* **7,** 157–164.

Russ, G. R. (1980). Effects of predation by fishes, competition, and structural complexity of the substratum on the establishment of a marine epifaunal community. *J. Exp. Mar. Biol. Ecol.* **42,** 55–70.

Russ, G. R. (1982). Overgrowth in marine epifaunal community: competitive hierarchies and competitive networks. *Oecologia* **53,** 12–19.

Russell, J. K. (1982). Timing of reproduction in coatis (*Nasua narica*) in relation to fluctuations in food resources. *In* "The Ecology of a Tropical Forest: Seasonal Rhythms and Long-term Changes" (E. G. Leigh, Jr., A. S. Rand, and D. M. Windsor, eds.), pp. 413–431. Smithsonian Inst. Press, Washington, D.C.

Russell, R. P., and Parsons, R. F. (1978). Effects of time since fire on heath floristics at Wilson's Promontory, southern Australia. *Aust. J. Bot.* **26,** 53–61.

Rutzler, K. (1970). Spatial competition among Porifera: solution by epizooism. *Oecologia* **5,** 85–95.

Ryland, J. S. (1974). Behavior, settlement, and metamorphosis of bryozoan larvae: a review. *Thalassia Jugosl.* **10,** 239–262.

Ryland, J. S. (1976). Physiology and ecology of marine bryozoans. *Adv. Mar. Biol.* **14,** 286–443.

Saint-Amand, P. (1962). The great earthquakes of May 1960 in Chile. *Smithson. Inst., Annu. Rep., 1962* pp. 337–363.

Sale, P. F. (1977). Maintenance of high diversity in coral reef communities. *Amer. Nat.* **111,** 337–359.

Sale, P. F. (1979). Recruitment, loss, and coexistence in a guild of territorial coral reef fishes. *Oecologia* **42,** 159–177.

Salisbury, E. J. (1942). "The Reproductive Capacity of Plants." Bell, London.

Salman, S. D. (1982). Seasonal and short-term variations in abundance of barnacle larval near southwest of the Isle of Man. *Estaurine Coastal Shelf Sci.* **15,** 241–253.

Sammarco, P. W., and Carleton, J. H. (1983). Damselfish territoriality and coral community structure: reduced grazing, coral recruitment, and effects on coral spat. *Proc. Int. Coral Reef Symp., 4th, Manila, 1981* **2,** 525–535.

Samollow, P. B., and Soulé, M. E. (1983). A case of stress related heterozygote superiority in nature. *Evolution* **37,** 646–648.

Sampson, A. W. (1944). Plant succession on burned chaparral lands in northern California. *Bull.— Calif. Agric. Exp. Stn.* No. 685.

Sanchez, P. A. (1976). "Properties and Management of Soils in the Tropics." Wiley, New York.

Sanders, H. L. (1969). Benthic marine diversity and the stability–time hypothesis. *Brookhaven Symp. Biol.* No. 22, 71–81.

Sanderson, G. C., and Bellrose, F. C. (1969). Wildlife habitat management and wetlands. *An. Acad. Bras. Cienc.* **41,** Suppl., 153–204.

Santelices, B., Castilla, J. C., Cancino, J., and Schmiede, P. (1980). Comparative ecology of *Lessonia nigrescens* and *Durvillaea antarctica* (Phaeophyta) in central Chile. *Mar. Biol. (Berlin)* **59,** 119–132.

Santos, P. F., and Whitford, W. G. (1981). The effects of microarthropods on litter decomposition in a Chihuahuan desert ecosystem. *Ecology* **62,** 654–663.

Sara, M. (1970). Competition and cooperation in sponge populations. *Symp. Zool. Soc. London* No. 25, 273–284.

Sarukhán, J. (1964). Estudio sucesional de una area talada en Tuxtapec, Oax. Inst. Nac. Invest. For. Mex., Publ. Espec. **3,** 107–172.

Sarukhán, J. (1978). Studies on the demography of tropical trees. *In* "Tropical Trees as Living Systems" (P. B. Tomlinson and M. Zimmerman, eds.), pp. 163–184. Cambridge Univ. Press, London and New York.

Schaal, B. A. (1975). Population structure and local differentiation in *Liatris cylindracea*. *Am. Nat.* **109,** 511–528.

Schaal, B. A., and Levin, D. A. (1976). The demographic genetics of *Liatris cylindracea* Michx. (Compositae). *Am. Nat.* **110,** 191–206.

Schaffer, W. M., and Gadgil, M. D. (1975). Selection for optimal life histories in plants. *In* "The Ecology and Evolution of Communities" (M. Cody and J. Diamond, eds.), pp. 142–157. Harvard Univ. Press, Cambridge, Massachusetts.

Scheer, G. (1959). Die Formenvielfalt der Riffkorallen. *Zool. Abt. Ber., 1958/1959* pp. 50–67.

Scheltema, R. S. (1971). The dispersal of the larvae of shoalwater benthic invertebrate species over long distances by ocean currents. *Eur. Mar. Biol. Symp. [Proc.], 4th, Bangor, Wales, 1969* pp. 7–28.

Schemske, D. W., and Brokaw, N. (1981). Treefalls and the distribution of understory birds in a tropical forest. *Ecology* **62,** 938–945.

Schindler, D. W. Kling, H., Schmidt, R. V., Prokopowich, J., Frost, V. E., Reid, R. A., and Capel, M. (1973). Enrichment of Lake 227 by addition of phosphate and nitrate: The second, third, and fourth years of enrichment 1970, 1971, and 1972. *J. Fish. Res. Board Can.* **30,** 1415–1440.

Schindler, D. W., Newbury, R. W., Beaty, K. G., Prokopowich, J., Ruszczynski, T., and Dalton, A. (1980). Effects of a windstorm and forest fire on chemical losses from forested watersheds and on the quality of receiving streams. *Can. J. Fish. Aquat. Sci.* **37,** 328–334.

Schlesinger, R. C. (1976). Sixteen years of selection silviculture in upland hardwood stands. *U.S. For. Serv., Res. Pap. NC* **NC-125.**

Schlesinger, W. H., Gray, J. T., Gill, D. S., and Mahall, B. E. (1982). *Ceanothus megacarpus* chaparral: a synthesis of ecosystem processes during development and annual growth. *Bot. Rev.* **48,** 78–117.

Schlessman, M. A. (1980). Systematics of the tuberous species of *Lomatium* (Umbelliferae). Ph.D. Thesis, Univ. of Washington, Seattle.

Schlosser, I. J. (1982). Fish community structure and function along two habitat gradients in a headwater stream. *Ecol. Monogr.* **52,** 395–414.

Schlosser, I. J. (1985). Patterns of flow regime, juvenile abundance and the assemblage structure of stream fishes. *Ecology* (in press).

Schluter, D. (1981). Does the theory of optimal diets apply in complex environments? *Am. Nat.* **118,** 139–147.

Schluter, D. (1982). Seed and patch selection by Galapagos ground finches: relation to foraging efficiency and food supply. *Ecology* **63,** 1106–1120.

Schmithüsen, J. (1956). Die räumliche Ordung der chilenischen Vegetation. *Bonner Geogr. Abh.* **17,** 1–89.

Schmithüsen, J. (1960). Die Nadelhölzer in den Waldgesellshaften der südlichen Anden. *Vegetatio* **10,** 313–327.

Schoener, A., and Schoener, T. W. (1981). The dynamics of the species–area relationship in marine fouling systems: 1. Biological correlates of changes in the species–area slope. *Am. Nat.* **118,** 339–360.

Schoener, T. W., and Schoener, A. (1983). The time to extinction of a colonizing propagule of lizards increases with island area. *Nature (London)* **302,** 332–334.

Schonbeck, M., and Norton, T. A. (1978). Factors controlling the upper limits of fucoid algae on the shore. *J. Exp. Mar. Biol. Ecol.* **31,** 303–313.

Schonbeck, M. W., and Norton, T. A. (1980). Factors controlling the lower limits of fucoid algae on the shore. *J. Exp. Mar. Biol. Ecol.* **43,** 131–150.

Schowalter, T. D. (1981). Insect herbivore relationship to the state of the host plant: biotic regulation of ecosystem nutrient cycling through ecological succession. *Oikos* **37,** 126–130.

Schowalter, T. D., and Whitford, W. G. (1979). Territorial behavior of *Bootettix argentatus* Bruner (Orthoptera: Acrididae). *Am. Midl. Nat.* **102**, 182–184.

Schowalter, T. D., Coulson, R. N., and Crossley, D. A., Jr. (1981a). Role of southern pine beetle and fire in maintenance of structure and function of the southeastern coniferous forest. *Environ. Entomol.* **10**, 821–825.

Schowalter, T. D., Pope, D. N., Coulson, R. N., and Fargo, W. S. (1981b). Patterns of southern pine beetle (*Dendroctonus frontalis* Zimm.) infestation enlargement. *For. Sci.* **27**, 837–849.

Schowalter, T. D., Webb, J. W., and Crossley, D. A., Jr. (1981c). Community structure and nutrient content of canopy arthropods in clearcut and uncut forest ecosystems. *Ecology* **62**, 1010–1019.

Schroder, G. D., and Rosenzweig, M. L. (1975). Perturbation analysis of competition and overlap in habitat utilization between *Dipodomys ordii* and *Dipodomys merriami*. *Oecologia* **19**, 9–28.

Schroeder, P. M., Dular, R., and Hayden, B. P. (1976). Vegetation changes associated with barrier dune construction on the Outer Banks of North Carolina. *Environ. Manage.* **1**, 105–114.

Schulz, J. P. (1960). Ecological studies on rain forest in northern Suriname. *Verh. K. Ned. Akad. Wet., Afd. Natuurkd., Reeks 2* **53**, No. 1.

Schultz, R. J. (1969). Hybridization, unisexuality, and polyploidy in the teleost Poeciliopsis (Poeciliidae) and other vertebrates. *Am. Nat.* **103**, 605–619.

Schultz, R. J. (1977). Evolution and ecology of unisexual fishes. *In* "Evolutionary Biology" (M. K. Hecht, W. C. Steere, and B. Wallace, eds.), Vol. 10, pp. 277–331. Plenum, New York.

Schwaegerle, K. E., and Schaal, B. A. (1979). Genetic variability and founder effect in the pitcher plant, *Sarracemia purpurea* L. *Evolution* **33**, 1210–1218.

Schwenke, H. (1971). Water movement: plants. *In* "Marine Ecology. Vol. 1: Environmental Factors," Part 2 (O. Kinne, ed.), pp. 1091–1121. Wiley (Interscience), New York.

Seapy, R. R., and Littler, M. M. (1982). Population and species diversity fluctuations in a rocky intertidal community relative to severe aerial exposure and sediment burial. *Mar. Biol. (Berlin)* **71**, 87–96.

Seastedt, T. R., and Crossley, D. A., Jr. (1981). Microarthropod response following cable logging and clear-cutting in the southern Appalachians. *Ecology* **62**, 126–135.

Seed, R. (1969). The ecology of *Mytilus edulis* L. (Lamellibranchiata) on exposed rocky shores. 2. Growth and mortality. *Oecologia* **3**, 317–350.

Selander, R. K., and Kaufman, D. (1973). Genic variability and strategies of adaptation in animals. *Proc. Natl. Acad. Sci. U.S.A.* **70**, 1875–1877.

Selander, R. K., and Whittam, T. S. (1983). Protein polymorphism and the genetic structure of populations. *In* "Evolution of Genes and Proteins" (M. Nei and R. K. Koehn, eds.), pp. 89–114. Sinauer, Sunderland, Massachusetts.

Sexton, O. J., Ortleb, E. P., Hathaway, L. M., Ballinger, R. E., and Licht, P. (1971). Reproductive cycles of three species of anoline lizards from the Isthmus of Panama. *Ecology* **52**, 201–215.

Sexton, O. J., Bauman, J., and Ortleb, E. (1972). Seasonal food habits of *Anolis limifrons*. *Ecology* **53**, 182–186.

Shafi, M. I., and Yarranton, G. A. (1973). Diversity, floristic richness, and species evenness during secondary (post-fire) succession. *Ecology* **54**, 897–902.

Shafizadeh, F. (1968). Pyrolysis and combustion of cellulosic fuels. *Adv. Carbohydr. Chem.* **23**, 1–54.

Shafizadeh, F., Chin, P. P. S., and DeGroot, W. F. (1977). Effective heat content of forest fuels. *For. Sci.* **23**, 81–89.

Shaver, G. (1981). Mineral nutrition and leaf longevity in an evergreen shrub, *Ledum palustre* ssp. *decumbens*. *Oecologia* **49**, 362–365.

Shepherd, S. A., and Womersley, H. B. S.(1976). The subtidal algae and seagrass ecology of St. Francis Island, South Australia. *Trans. R. Soc. South Aust.* **100**, 177–191.

Sheppard, C. R. C. (1979). Interspecific aggression between reef corals with reference to their distribution. *Mar. Ecol.: Prog. Ser.* **1**, 237–247.

Sheppard, C. R. C. (1981). Illumination and the coral community beneath tabular *Acropora* species. *Mar. Biol. (Berlin)* **64**, 53–58.

Shields, W. (1982). "Philopatry, Inbreeding, and the Evolution of Sex." State Univ. of New York Press, Albany.

Shmida, A., and Whittaker, R. H. (1981). Pattern and biological microsite effects in two shrub communities, southern California. *Ecology* **62**, 234–251.

Shugart, H. H., and Noble, J. R. (1981). A computer model of succession and fire response of the high-altitude *Eucalyptus* forest of the Brindabella Range, Australian Capital Territory. *Aust. J. Ecol.* **6**, 149–164.

Shugart, H. H., and West, D. C. (1977). Development of an Appalachian deciduous forest succession model and its application to assessment of the impact of the chestnut blight. *J. Environ. Manage.* **5**, 161–179.

Shugart, H. H., and West, D. C. (1980). Forest succession models. *BioScience* **30**, 308–313.

Shugart, H. H., and West, D. C. (1981). Long-term dynamics of forest ecosystems. *Am. Sci.* **69**, 647–652.

Shugart, H. H., Crow, T. R., and Hett, J. M. (1973). Forest succession models: A rationale and methodology for modeling forest succession over large regions. *For. Sci.* **19**, 203–212.

Shugart, H. H., Hopkins, M. S., Burgess, I. P., and Mostlock, A. T. (1981a). The development of a succession model for subtropical rainforest and its application to assess the effects of timber harvest at Wiangaree State Forest, New South Wales. *J. Environ. Manage.* **11**, 243–265.

Shugart, H. H., West, D. C., and Emanuel, W. R. (1981b). Patterns and dynamics of forests: an application of simulation models. *In* "Forest Succession: Concepts and Applications" (D. C. West, H. H. Shugart, and D. B. Botkin, eds.), pp. 74–94. Springer-Verlag, Berlin and New York.

Shumway, J., and Atkinson, W. A. (1978). Predicting nitrogen fertilizer response in unthinned stands of Douglas-fir. *Commun. Soil Sci. Plant Anal.* **9**, 529–539.

Siccama, T. G., Weir, G., and Wallace, K. (1976). Ice damage in a mixed nardwood forest in Connecticut in relation to *Vitis* infestation. *Bull. Torrey Bot. Club* **103**, 180–183.

Siegel, S. (1956). "Nonparametric Statistics for the Behavioral Sciences." McGraw-Hill, New York.

Sigafoos, R. S. (1952). Frost action as a primary physical factor in tundra plant communities. *Ecology* **33**, 480–487.

Silberbauer-Gottsberger, I., Morawetz, W., and Gottsberger, G. (1977). Frost damage of Cerrado plants in Botucatu, Brazil, as related to the geographical distribution of species. *Biotropica* **9**, 253–261.

Simberloff, D., and Abele, L. G. (1982). Refuge design and island biogeographic theory: effects of fragmentation. *Am. Nat.* **120**, 41–50.

Sinclair, A. R. E., and Norton-Griffiths, M., eds. (1979). "Serengeti: Dynamics of an Ecosystem." Univ. of Chicago Press, Chicago, Illinois.

Singh, S. M., and Zouros, E. (1978). Genetic variation associated with growth rate in the American oyster (*Crassostrea virginica*). *Evolution* **32**, 342–353.

Sirèn, G. (1955). The development of spruce forest on raw humus sites in northern Finland and its ecology. *Acta For. Fenn.* **62**.

Sivaramakrishnan, V. R. (1951). Studies on early development and regeneration in some Indian sponges. *Proc.—Indian Acad. Sci., Sect. B* **34B**, 273–310.

Skeen, J. N. (1976). Regeneration and survival of woody species in a naturally-created forest opening. *Bull. Torrey Bot. Club* **103**, 259–265.

Skeen, J. N., and March, W. T. (1977). The effect of summer thunderstorms on the near-ground temperature regimen within a suburban forest. *J. Tenn. Acad. Sci.* **52**, 95–99.

Skeen, J. N., Carter, M. E. B., and Ragsdale, H. L. (1980). Yellow-poplar: the Piedmont case. *Bull. Torrey Bot. Club* **107**, 1–6.

Slatkin, M. (1974). Competition and regional coexistence. *Ecology* **55**, 128–134.

Slatkin, M. (1981a). Fixation probabilities and fixation times in a subdivided population. *Evolution* **35**, 477–488.

Slatkin, M. (1981b). Populational heritability. *Evolution* **35**, 859–871.

Slobodkin, L. B., and Sanders, H. L. (1969). On the contribution of environmental predictability to species diversity. *Brookhaven Symp. Biol.* No. 22, 82–95.

Smith, A. T., and Vrieze, J. M. (1979). Population structure of Everglades rodents: responses to a patchy environment. *J. Mammal.* **60**, 778–794.

Smith, C. C. (1970). Coevolution of pine squirrel (*Tamiasciurus*) and conifers. *Ecol. Monogr.* **40**, 349–371.

Smith, D. M. (1962). "The Practice of Silviculture," 7th Ed. Wiley, New York.

Smith, H. (1972). Light quality and germination: ecological implications. *In* "Seed Ecology" (W. Heydecker, ed.), pp. 219–231. Pennsylvania State Univ. Press, University Park.

Smith, H. C. (1983). Growth of Appalachian hardwoods kept free to grow from 2 to 12 years after clearcutting. *For. Serv. Res. Pap. NE (U.S.)* **NE-528.**

Smith, O. L. (1980). The influence of environmental gradients on ecosystem stability. *Am. Nat.* **116**, 1–24.

Smith, S. H. (1968). Species succession and fishery exploitation in the Great Lakes. *J. Fish. Res. Board Can.* **25**, 667–693.

Smith, T. M., Shugart, H. H., and West, D. C. (1981). The use of forest simulation models to integrate timber harvest and nongame bird habitat management. pp. 501–510. 46th North American Wildlife and Natural Resources Conf. 1981. Wildlife Management Institute, Washington, D.C.

Smith, W. H. (1981). "Air Pollution and Forests: Interactions Between Air Contaminants and Forest Ecosystems." Springer-Verlag, Berlin and New York.

Smythe, N. (1970). Relationships between fruiting seasons and dispersal methods in a neotropical forest. *Am. Nat.* **104**, 25–35.

Snaydon, R. W. (1980). Plant demography in agricultural systems. *In* "Demography and Evolution in Plant Populations" (O. T. Solbrig, ed.), pp. 131–160. California Press, Berkeley.

Solbrig, O. T., ed. (1980). "Demography and Evolution in Plant Populations." Univ. of California Press, Berkeley.

Solbrig, O. T., and Simpson, B. B. (1974). Components of regulation of a population of dandelions in Michigan. *J. Ecol.* **62**, 473–486.

Solbrig, O. T., Newell, S. J., and Kincaid, D. T. (1980). The population biology of the genus *Viola.* I. The demography of *Viola sororia. J. Ecol.* **68**, 521–546.

Solomon, A. M., and Shugart, H. H. (1984). Integrating forest stand simulations with paleoecological records to examine long-term forest dynamics. *Proc. Europ. Sci. Found. Workshop For. Dynam., Uppsala, Sweden 1983.* In press. Elsevier, North Holland.

Soulé, M. E. (1976). Allozyme variation: its determination in space and time. *In* "Molecular Evolution" (F. J. Ayala, ed.), pp. 60–77. Sinauer, Sunderland, Massachusetts.

Soulé, M. E. (1979). Heterozygosity and developmental stability: another look. *Evolution* **33**, 396–401.

Soulé, M. E., and Wilcox, B. A., eds. (1980). "Conservation Biology: An Evolutionary–Ecological Perspective." Sinauer, Sunderland, Massachusetts.

Sousa, W. P. (1977). Disturbance and ecological succession in marine intertidal boulder fields. Ph.D. Thesis, Univ. of California, Santa Barbara.

Sousa, W. P. (1979a). Experimental investigations of disturbance and ecological succession in a rocky intertidal algal community. *Ecol. Monogr.* **49**, 227–254.

Sousa, W. P. (1979b). Disturbance in marine intertidal boulder fields: the nonequilibrium maintenance of species diversity. *Ecology* **60**, 1225–1239.

Sousa, W. P. (1980). The responses of a community to disturbance: the importance of successional age and species' life histories. *Oecologia* **45**, 72–81.

Sousa, W. P. (1984). Intertidal mosaics: propagule availability, and spatially variable patterns of succession. *Ecology* **65**, 1918–1935.

Sousa, W. P., Schroeter, S. C., and Gaines, S. D. (1981). Latitudinal variation in intertidal algal community structure: the influence of grazing and vegetative propagation. *Oecologia* **48**, 297–307.

Southwood, T. R. E., and Comins, H. N. (1976). A synoptic population model. *J. Anim. Ecol.* **45**, 949–965.

Sower, L. L. (1980). Control of moth reproduction by disruption of the pheromone communication: problems and promise. *In* "Advances in Invertebrate Reproduction" (W. H. Clark, Jr. and T. S. Adams, eds.), pp. 197–212. Elsevier/North-Holland, New York.

Sparling, J. H. (1967). Assimilation rates of some woodland herbs in Ontario. *Bot. Gaz.* **128**, 160–168.

Specht, R. L. (1969). A comparison of schlerophyllous vegetation characteristic of mediterranean type climates in France, California, and southern Australia. II. Dry matter, energy, and nutrient accumulation. *Aust. J. Bot.* **17**, 293–308.

Specht, R. L. (1979). Heathlands and related shrublands of the world. *In* "Ecosystems of the World. Vol. 9a: Heathlands and Related Shrublands" (R. L. Specht, ed.), pp. 1–18. Elsevier, Amsterdam.

Specht, R. L. (1981). Primary production Mediterranean-climate ecosystems regenerating after fire. *In* "Mediterranean-Type Shrublands" (F. di Castri, D. W. Goodall, and R. L. Specht, eds.), pp. 257–267. Elsevier, Amsterdam.

Specht, R. L., and Rayson, P. (1957). Dark Island heath (Ninety-Mile Plain, South Australia). I. Definition of the ecosystem. *Aust. J. Bot.* **5**, 52–85.

Specht, R. L., Rayson, P., and Jackman, M. E. (1958). Dark Island Heath (Ninety-Mile Plain, South Australia). VI. Pyric succession: changes in composition, coverage, dry weight, and mineral nutrient status. *Aust. J. Bot.* **6**, 59–88.

Springett, B. P. (1978). On the ecological role of insects in Australian eucalypt forests. *Aust. J. Ecol.* **3**, 129–139.

Sprugel, D. G. (1976). Dynamic structure of wave-generated *Abies balsamea* forests in the northeastern United States. *J. Ecol.* **64**, 889–911.

Sprugel, D. G. (1984). Density, biomass, productivity, and nutrient cycling changes during stand development in wave-regenerated balsam fir forests. *Ecol. Monogr.* **54**, 165–186.

Sprugel, D. G., and Bormann, F. H. (1981). Natural disturbance and the steady state in high-altitude balsam fir forests. *Science (Washington, D.C.)* **211**, 390–393.

Spurr, S. H. (1956). Natural restocking of forests following the 1938 hurricane in central New England. *Ecology* **30**, 350–358.

Spurr, S. H., and Barnes, B. V. (1973). "Forest Ecology," 2nd Ed. Ronald Press, New York.

Spurr, S. H., and Barnes, B. V. (1980). "Forest Ecology," 3rd Ed. Ronald Press, New York.

Squires, V. R. (1975). Environmental heterogeneity as a factor in group size determination among grazing sheep. *Proc. N.Z. Soc. Anim. Prod.* **35**, 184–190.

Stark, N. (1973). Nutrient cycling in a desert ecosystem. *Bull. Ecol. Soc. Am.* **54**, 21.

Stark, N. (1977). Fire and nutrient cycling in a Douglas-fir/larch forest. *Ecology* **58**, 16–30.

Stark, N. (1978). Man, tropical forests, and the biological life of a soil. *Biotropica* **10**, 1–10.

States, J. B. (1976). Local adaptations in chipmunk (*Eutamias amoenus*) populations and evolutionary potential at species' borders. *Ecol. Monogr.* **46**, 221–256.

Stearns, S. C. (1976). Life-history tactics: a review of ideas. *Q. Rev. Biol.* **51**, 3–47.

Stearns, S. C. (1977). The evolution of life history traits: A critique of the theory and a review of the data. *Annu. Rev. Ecol. Syst.* **8**, 145–171.

Stearns, S. C., and Crandall, R. E. (1981). Bet-hedging and persistence as adaptations of colonizers. *Evol. Today, Proc. Int. Congr. Syst. Evol., 2nd, Vancouver, B.C., 1980* pp. 371–384.

Stebbing, A. R. D.(1973). Competition for space between epiphytes of *Fucus serratus* L. *J. Mar. Biol. Assoc. U.K.* **53**, 247–261.

Stebbins, G. L. (1957). Self-fertilization and population variability in higher plants. *Am. Nat.* **41**, 337–354.

Stebbins, G. L. (1971). "Chromosomal Evolution in Higher Plants." Arnold, London.

Stebbins, G. L. (1972). Evolution and diversity of arid-land shrubs. *In* "Wildland Shrubs—Their Biology and Utilization" (C. M. McKell, J. P. Blasidell, and J. R. Goodin, eds.), *USDA For. Ser. Intermountain For. Range Exp. Stn. Gen. Tech. Rep. INT* **INT-1**, pp. 111–120.

Steingraeber, D. A. (1980). The analyses of branching patterns in trees. Ph.D. Thesis, Univ. of Wisconsin, Madison.

Steingraeber, D. A., Kascht, L. J., and Franck, D. H. (1979). Variation of shoot morphology and bifurcation ratio in sugar maple (*Acer sacchrum*) saplings. *Am. J. Bot.* **66**, 441–445.

Stenseth, N. C., and Hansson, L. (1981). The importance of population dynamics in heterogeneous landscapes: management of vertebrate pests and some other animals. *Agro-Ecosystems* **7**, 187–211.

Stephenson, T. A., and Stephenson, A. (1954). Life between tide-marks in North America. IIIB. Nova Scotia and Prince Edward Island: the geographical features of the region. *J. Ecol.* **12**, 46–70.

Stephenson, T. A., and Stephenson, A. (1972). "Life Between Tidemarks on Rocky Shores." Freeman, San Francisco, California.

Stergios, B. G. (1976). Achene production, dispersal, seed germination, and seedling establishment of *Hieracium aurantiacum* in an abandoned field community. *Can. J. Bot.* **54**, 1189–1197.

Stewart, J. W. B., Cole, C. V., and Maynard, D. G. (1983). Interactions of the biogeochemical cycles in grassland ecosystems. *In* "The Biogeochemical Cycles and Their Interactions" (B. Bolin and R. B. Cook, eds.). Wiley, New York.

Stiles, F. G. (1975). Ecology, flowering phenology, and hummingbird pollination of some Costa Rican *Heliconia* species. *Ecology* **56**, 285–301.

Stimson, J. S. (1973). The role of territory in the ecology of the intertidal limpet *Lottia gigantea* (Gray). *Ecology* **54**, 1020–1030.

Stimson, J. S. (1978). Mode and timing of reproduction in some common hermatypic corals of Hawaii and Enewetak. *Mar. Biol. (Berlin)* **48**, 173–184.

Stoeckler, J. H., Strothman, R. O., and Krefting, L. W. (1957). Effect of deer browsing on hardwood regeneration in the northern hardwood-hemlock type in northeastern Wisconsin. *J. Wildl. Manage.* **21**, 75–80.

Stone, E. C. (1951). The stimulative effect of fire on the flowering of the golden brodiaea (*Brodiaea ixioides* Wats. var. *lugens*. Jepson). *Ecology* **32**, 534–537.

Stone, E. C., and Juhren, G. (1951). The effect of fire on the germination of the seed of *Rhus ovata* Wats. *Am. J. Bot.* **38**, 368–372.

Stone, E. C., and Juhren, G. (1953). Fire stimulated germination. *Calif. Agric.* **7**, 13–14.

Stone, E. L. (1975). Windthrow influences on spatial heterogeneity in a forest soil. *Mitt. Schweiz. Anst. Forstl. Versuchswes.* **51**, 77–87.

Storr, J. F. (1976). Ecological factors controlling sponge distribution in the Gulf of Mexico and the resulting zonation. *In* "Aspects of Sponge Biology" (F. W. Harrison and R. R. Cowden, eds.), pp. 261–276. Academic Press, New York.

Stoszek, K. J., Mika, P. G., Moore, J. A., and Osborne, H. L. (1981). Relationships of Douglas-fir tussock moth defoliation to site and stand characteristics in northern Idaho. *For. Sci.* **27**, 431–442.

Strahler, A. M. (1957). Quantitative analysis of watershed geomorphology. *Trans. Am. Geophys. Union* **38**, 913–920.

Strong, D. R. (1977). Epiphyte loads, treefalls, and perennial forest disruption: a mechanism for maintaining higher tree species richness in the tropics without animals. *J. Biogeogr.* **4**, 215–218.

Strothman, R. O. (1967). The influence of light and moisture on the growth of red pine seedlings in Minnesota. *For. Sci.* **13**, 182–191.

Sturgeon, K. B. (1979). Monoterpene variation in ponderosa pine xylem resin related to western pine beetle predation. *Evolution* **33**, 803–814.

Suchanek, T. H. (1978). The ecology of *Mytilus edulis* L. in exposed rocky intertidal communities. *J. Exp. Mar. Biol. Ecol.* **31**, 105–120.

Suchanek, T. H. (1979). The *Mytilus californianus* community: studies on the composition, structure, organization, and dynamics of a mussel bed. Ph.D. Dissertation. Univ. of Washington, Seattle.

Suchanek, T. H. (1981). The role of disturbance in the evolution of life history strategies in the intertidal mussels *Mytilus edulis* and *Mytilus californianus*. *Oecologia* **50**, 143–152.

Sugihara, G. (1981). $S = CA^2$, $z = \frac{1}{4}$: A reply to Connor and McCoy. *Am. Nat.* **117**, 790–793.

Sutherland, J. P. (1970). Dynamics of high and low populations of the limpet, *Acmaea scabra* (Gould). *Ecol. Monogr.* **40**, 169–188.

Sutherland, J. P. (1974). Multiple stable points in natural communities. *Am. Nat.* **108**, 859–872.

Sutherland, J. P. (1976). Life histories and the dynamics of fouling communities. *In* "Ecology of Fouling Communities" (J. D. Costlow, ed.), pp. 137–153. U.S. Nav. Inst. Press, Annapolis, Maryland.

Sutherland, J. P. (1980). Dynamics of the epibenthic community on roots of the mangrove *Rhizophora mangle*, at Bahia de Buche, Venezuela. *Mar. Biol. (Berlin)* **58**, 75–84.

Sutherland, J. P. (1981). The fouling community at Beaufort, North Carolina: a study in stability. *Am. Nat.* **118**, 499–519.

Sutherland, J. P., and Karlson, R. H. (1977). Development and stability of the fouling community at Beaufort, North Carolina. *Ecol. Monogr.* **47**, 425–446.

Swain, A. M. (1973). A history of fire and vegetation in northeastern Minnesota as recorded in lake sediments. *Quat. Res. (N.Y.)* **3**, 383–396.

Swain, A. M. (1978). Environmental changes during the past 2000 years in North Central Wisconsin: Analysis of pollen, charcoal, and seeds from varved lake sediments. *Quat. Res. (N.Y.)* **10**, 55–68.

Swank, W. T., Waide, J. B., Crossley, D. A., Jr., and Todd, R. L. (1981). Insect defoliation enhances nitrate export from forest ecosystems. *Oecologia* **51**, 297–299.

Sweeney, J. R. (1956). Responses of vegetation to fire. *Univ. Calif. Publ. Bot.* **28**, 141–216.

Swift, L. W., Swank, W. T., Mankin, J. B., Luxmoore, R. J., and Goldstein, R. A. (1975). Simulation of evapotranspiration and drainage from mature and clearcut deciduous forests and young pine plantations. *Water Resour. Res.* **11**, 667–673.

Switzer, G. L., and Nelson, L. E. (1972). Nutrient accumulation and cycling in loblolly pine (*Pinus taeda* L.) plantation ecosystems: the first twenty years. *Soil Sci. Soc. Am. Proc.* **36**, 143–147.

Tadaki, Y., Sato, A., Sakurai, S., Takeuchi, I., and Kawahara, T. (1977). Studies on the production structure of forests. XVIII. Structure and production of subalpine "dead trees strips" *Abies* forests near Mt. Asahi. *Nippon Seitai Gakkaishi (Jpn. J. Ecol.)* **27**, 83–90.

Talbot, G. H., Russell, B. C., and Anderson, G. R. V. (1978). Coral reef fish communities: unstable, high-diversity systems? *Ecol. Monogr.* **48**, 425–440.

Tansley, A. G. (1935). The use and abuse of vegetational concepts and terms. *Ecology* **16**, 284–307.

Taylor, D. L. (1973). Some ecological implications of forest fire control in Yellowstone National Park, Wyoming. *Ecology* **54**, 1394–1396.

Taylor, L. R., and Taylor, R. A. J. (1977). Aggregation, migration and population mechanics. *Nature (London)* **265**, 415–421.

Taylor, L. R., and Woiwod, I. P. (1980). Temporal stability as a density-dependent species characteristic. *J. Anim. Ecol.* **49**, 209–224.

Taylor, L. R., and Woiwod, I. P. (1982). Comparative synoptic dynamics. I. Relationships between inter- and intra-specific spatial and temporal variance/mean population parameters. *J. Anim. Ecol.* **51**, 879–906.

Taylor, L. R., Woiwod, I. P., and Perry, J. N. (1978). The density-dependence of spatial behaviour and the rarity of randomness. *J. Anim. Ecol.* **47**, 383–406.

Taylor, L. R., Woiwod, I. P., and Perry, J. N. (1980). Variance and the large scale spatial stability of aphids, moths and birds. *J. Anim. Ecol.* **49**, 831–854.

Taylor, P. R., and Littler, M. M. (1982). The roles of compensatory mortality, physical disturbance, and substrate retention in the development and organization of a sand-influenced, rocky-intertidal community. *Ecology* **63**, 135–146.

Taylor, R. A. J., and Taylor, L. R. (1979). A behavioral model for the evolution of spatial dynamics. *In* "Population Dynamics" (R. M. Anderson, B. D. Turner, and L. R. Taylor, eds.), *Symp. Br. Ecol. Soc.* **20**, 1–27.

Taylor, R. J., and Pearcy, R. W. (1976). Seasonal patterns of the CO_2 exchange characteristics of the understory plants from a deciduous forest. *Can. J. Bot.* **54**, 1094–1103.

Taylor, S. E. (1975). Optimal leaf form. *In* "Perspectives of Biophysical Ecology" (D. M. Gates and R. B. Schmerl, eds.), pp. 73–86. Springer-Verlag, Berlin and New York.

Terres, J. K. (1980). "Encyclopedia of North American Birds." Knopf, New York.

Tharp, M. L. (1978). Modeling major perturbations on a forest ecosystem. M.S. Thesis, Univ. of Tennessee, Knoxville.

Thibault, R. E. (1974). The ecology of unisexual and bisexual fishes of the genus *Poeciliopsis:* A study of niche relationships. Ph.D. Thesis, Univ. of Connecticut, Storrs.

Thiery, R. G. (1982). Environmental instability and community diversity. *Biol. Rev. Cambridge Philos. Soc.* **57**, 691–710.

Thom, B. G. (1967). Mangrove ecology and geomorphology: Tabasco, Mexico. *J. Ecol.* **55**, 301–343.

Thomas, A. G., and Dale, H. M. (1974). Zonation and regulation of old pasture populations of *Hieracium floribundum. Can J. Bot.* **52**, 1451–1458.

Thomasson, K. (1963). Araucanian Lakes: Plankton studies in North Patagonia with notes on terrestrial vegetation. *Acta Phytogeogr. Suec.* **47**, 1–139.

Thompson, J. N. (1978). Within-patch structure and dynamics in *Pastinaca sativa* and resource availability to a specialized herbivore. *Ecology* **59**, 443–448.

Thompson, J. N. (1980). Treefalls and colonization patterns of temperate forest herbs. *Am. Midl. Nat.* **104**, 176–184.

Thompson, J. N. (1982). "Interaction and Coevolution." Wiley (Interscience), New York.

Thompson, J. N. (1983a). Partitioning of variance in demography: within-patch differences in herbivory, survival, and flowering of *Lomatium farinosum* (Umbelliferae). *Oikos* **40**, 315–317.

Thompson, J. N. (1983b). Selection pressures on phytophagous insects on small hostplants. *Oikos* **40**, 438–444.

Thompson, J. N. (1983c). The use of ephemeral plant parts on small hostplants: how *Depressaria leptotaeniae* (Lep., Oecophoridae) feeds on *Lomatium dissectum* (Umbelliferae). *J. Anim. Ecol.* **52**, 281–291.

Thompson, J. N. (1983d). Selection of plant parts by *Depressaria multifidae* (Lep., Oecophoridae) on its seasonally-restricted hostplant, *Lomatium grayi* (Umbelliferae). *Ecol. Entomol.* **8**, 203–211.

Thompson, J. N., and Moody, M. E. (1985). Assessing probability of interaction in size-structured populations: *Depressaria multifidae* (Lep., Oecophoridae) attack on *Lomatium grayi* (Umbelliferae). *Ecology* (in press).

Thompson, J. N., and P. W. Price. (1977). Plant plasticity, phenology and herbivore dispersion: wild parsnip and the parsnip webworm. *Ecology* **58**, 1112–1119.

Thompson, J. N., and Willson, M. F. (1978). Disturbance and dispersal of fleshy fruits. *Science (Washington, D.C.)* **200**, 1161–1163.

Thompson, J. N., and Willson, M. F. (1979). Evolution of temperate fruit/bird interactions: phenological strategies. *Evolution* **33**, 973–982.

Thompson, S. D. (1982a). Microhabitat utilization and foraging behavior of bipedal and quadrupedal heteromyid rodents. *Ecology* **63**, 1303–1312.

Thompson, S. D. (1982b). Structure and species composition of desert heteromyid rodent species assemblages: effects of a simple habitat manipulation. *Ecology* **63**, 1313–1321.

Thorson, G. (1950). Reproductive and larval ecology of marine bottom invertebrates. *Biol. Rev. Cambridge Philos. Soc.* **25**, 1–45.

Tierson, W. C., Patric, E. F., and Behrend, D. F. (1966). Influence of white-tailed deer on the logged northern hardwood forest. *J. For.* **64**, 801–805.

Tilman, D. (1978). Cherries, ants and tent caterpillars: timing of nectar production in relation to susceptibility of caterpillars to ant predation. *Ecology* **59**, 686–692.

Tilman, D. (1982). "Resource Competition and Community Structure." Princeton Univ. Press, Princeton, New Jersey.

Tinkle, D. W., and Selander, R. K. (1973). Age-dependent allozymic variation in a natural population of lizards. *Biochem. Genet.* **8**, 231–237.

Titmus, G. (1983). Are animal populations really aggregated? *Oikos* **40**, 64–68.

Toft, C. A. (1980). Seasonal variation in populations of Panamanian litter frogs and their prey: a comparison of wetter and drier sites. *Oecologia* **47**, 34–38.

Toft, C. A., Rand, A. S., and Clark, M. (1982). Population dynamics and seasonal recruitment in *Bufo typhonius* and *Colostethus nubicola* (Anura). *In* "The Ecology of a Tropical Forest: Seasonal

Rhythms and Long-Term Changes'' (E. G. Leigh, Jr., A. S. Rand, and D. M. Windsor, eds.), pp. 397–403. Smithsonian Inst. Press, Washington, D.C.

Tomanek, J. (1960). Mikroklimatische Verhaltnisse in Lochhiebe. *Verh. Ganzstaatl. Bioklimatol. Konf., 2nd, Tschechoslowak. Acad. d. Wiss., Prague 1958* pp. 297–313.

Tomlinson, J. (1966). The advantage of hermaphroditism and parthenogenesis. *J. Theor. Biol.* **11,** 54– 58.

Tomoff, C. S. (1974). Avian species diversity in desert scrub. *Ecology* **55,** 396–403.

Tonn, W. M., and Magnuson, J. J. (1982). Patterns in the species composition and richness of fish assemblages in northern Wisconsin lakes. *Ecology* **63,** 1149–1166.

Torquebiau, E. (1981). Analyse architecturale de la forêt de Los Tuxtlas (Veracruz), Mexique. Ph.D. Thesis, Univ. Sci. Tech. Languedoc, Acad. de Montpellier, Montpellier, France.

Toth, L. A., Dudley, D. R., Karr, J. R., and Gorman, O. T. (1982). Natural and man-induced variability in a silverjaw minnow (*Ericymba buccata*) population. *Am. Midl. Nat.* **107,** 284–293.

Townsley, S. J., Trott, L., and Trott, E. (1962). A preliminary report on the rehabilitation of the littoral marine community on a new lava flow at Kapoho, Hawaii. *Ecology* **43,** 728–730.

Tracey, M. L., Bellet, N. F., and Graven, C. B. (1975). Excess of allozyme homozygosity and breeding population structure in the mussel *Mytilus californianus*. *Mar. Biol. (Berlin)* **32,** 303–311.

Traubaud, L. (1974). Experimental study on the effects of prescribed burning on a *Quercus coccifera* L. garrique: early results. *Proc. Annu. Tall Timbers Fire Ecol. Conf.* **13,** 97–129.

Traubaud, L. (1975). Les températures de feux de végétation. *Dewxince Symp. Eur. Combust., Orleans, Fr.* **1,** 210–214.

Traubaud, L. (1976). In flammabilite et combustibilite des principales espèces des garrigues de la région méditerranéenne. *Acta Oecologica* [Ser.]*Oecol. Plant.* **11,** 117–136.

Traubaud, L. (1977). Comparison between the effect of prescribed fires and wild fires on the global quantitative evolution of the kermes scrub oak (*Quercus coccifera* L.) garriques. *In* "Environmental Consequences of Fire and Fuel Management in Mediterranean Ecosystems'' (H. A. Mooney and C. E. Conrad, eds.), *Gen. Tech. Rep. WO—U.S. For. Serv. [Wash. Off.]* **GTR-WO-3,** pp. 271–282.

Traubaud, L. (1981). Man and fire: impacts on Mediterranean vegetation. *In* "Mediterranean-Type Shrublands'' (F. di Castri, D. W. Goodall, and R. L. Specht, eds.), pp. 523–537. Elsevier, Amsterdam.

Trejo P., L. (1976). Diseminación de semillas por aves en "Los Tuxtlas'', Ver. *In* "Regeneración des Selvas'' (A. Gómez-Pompa, C. Vázquez-Yanes, S. del Amo, R. and A. Butanda C., eds.), pp. 447–470. Compañia Editorial Continental, Mexico City, Mexico.

Trimble, G. R. (1965). Species composition changes under individual tree selection cutting in cove hardwoods. *USDA For. Serv. Northeast. For. Exp. Stn. Res. Note NE* **NE-30.**

Trimble, G. R. (1970). 20 years of intensive uneven-aged management: effect on growth, yield, and species composition in two hardwood stands in West Virginia. *USDA For. Serv. Res. Northeast. For. Exp. Stn. Pap. NE* **NE-154.**

Trimble, G. R., and Tryon, E. H. (1966). Crown encroachment into openings cut in Appalachian hardwood stands. *J. For.* **62,** 104–108.

Trombulak, S. C., and Kenagy, G. J. (1980). Effects of seed distribution and competitors on seed harvesting efficiency in heteromyid rodents. *Oecologia* **44,** 342–346.

Tryon, E. H., and Trimble, G. R. (1969). Effect of distance from stand border on height of hardwood reproduction in openings. *Proc. W. Va. Acad. Sci.* **41,** 125–133.

Tsuchiya, M. (1983). Mass mortality in a population of the mussel *Mytilus edulis* L. caused by high temperature on rocky shores. *J. Exp. Mar. Biol. Ecol.* **66,** 101–111.

Tubbs, C. H. (1969). Natural regeneration of yellow birch in the Lake States. *In* "Birch Symposium Proceedings'' (E. H. Lason, ed.), pp. 74–78. Northeast. For. Exp. Stn., Upper Darby, Pennsylvania.

Tubbs, C. H. (1977). Natural regeneration of northern hardwoods in the northern Great Lakes Region. *USDA For. Serv., North Central For. Exp. Stn. Res. Pap. NC* **NC-50.**

Tunnicliffe, V. (1981). Breakage and propagation of the stony coral *Acropora cervicornis. Proc. Natl. Acad. Sci. U.S.A.* **78**, 2427–2431.

Turberville, H. W., and Hough, A. F. (1939). Errors in age counts of suppressed trees. *J. For.* **37**, 417–418.

Turkington, R., and Harper, J. L. (1979). The growth distribution and neighbour relationships of *Trifolium repens* in a permanent pasture. IV. Fine-scale biotic differentiation. *J. Ecol.* **67**, 245–254.

Turner, J. (1977). Effect of nitrogen availability on nitrogen cycling in a Douglas fir stand. *For. Sci.* **23**, 307–316.

Turner, J., and Long, J. N. (1975). Accumulation of organic matter in a series of Douglas-fir stands. *Can. J. For. Res.* **5**, 681–690.

Turner, M. E., Stephens, J. C., and Anderson, W. W. (1982). Homozygosity and patch structure in plant populations as a result of nearest-neighbor pollination. *Proc. Natl. Acad. Sci. U.S.A.* **79**, 203–207.

Tyson, P. D. (1980). Temporal and spatial variation of rainfall anomalies in Africa south of latitude 22° during the period of meteorological record. *Clim. Change* **2**, 363–371.

Uhl, C. (1982a). Recovery following disturbances of different intensities in the Amazon rain forest of Venezuela. *Interciencia (Caracas)*

Uhl, C. (1982b). Tree dynamics in a species rich tierra firme forest in Amazonia, Venezuela. *Acta Cient. Venez.* **33**, 72–77.

Uhl, C., and Murphy, P. G. (1982). Composition, structure, and regeneration of a tierra firme forest in the Amazonian Basin of Venezuela. *Trop. Ecol.* **22**, 219–237.

Uhl, C., Clark, K., Clark, H., and Murphy, P. (1981). Early plant succession after cutting and burning in the upper Rio Negro region of the Amazon Basin. *J. Ecol.* **69**, 631–649.

Uhl, C., Jordan, C., Clark, K., Clark, H., and Herrera, R. (1982). Ecosystem recovery in Amazon caatinga forest after cutting, cutting and burning, and bulldozer clearing treatments. *Oikos* **38**, 313–320.

Underwood, A. J. (1980). The effects of grazing by gastropods and physical factors on the upper limits of distribution of intertidal macroalgae. *Oecologia* **46**, 201–213.

Underwood, A. J. (1981). Structure of a rocky intertidal community in New South Wales: patterns of vertical distribution and seasonal changes. *J. Exp. Mar. Biol. Ecol.* **51**, 57–85.

Underwood, A. J., and Denley, E. J. (1984). Paradigms, explanations, and generalizations in models for the structure of intertidal communities on rocky shores. *In* ''Ecological Communities: Conceptual Issues and the Evidence'' (D. R. Strong, Jr., D. Simberloff, L. G. Abele, and A. B. Thistle, eds.), pp. 151–180. Princeton Univ. Press, Princeton, New Jersey.

Underwood, A. J., and Jernakoff, P. (1981). Effects of interactions between algae and grazing gastropods on the structure of a low-shore intertidal algal community. *Oecologia* **48**, 221–233.

Underwood, A. J., Denley, E. J., and Moran, M. J. (1983). Experimental analyses of the structure and dynamics of mid-shore rocky intertidal communities in New South Wales. *Oecologia* **56**, 202–219.

U.S. Forest Service (1973). Silvicultural systems for the major forest types of the United States. *U.S. Dep. Agric., Agric. Handb.* No. 445.

U.S. Forest Service (1978). Uneven-aged silviculture and management in the United States. *Gen. Tech. Rep. WO—U.S. For. Serv.* [*Wash. Off.*] **GTR-WO-24.**

Usher, M. B. (1979). Markovian approaches to ecological succession. *J. Anim. Ecol.* **48**, 413–426.

van Andel, J., and Jager, J. C. (1981). Analysis of growth and nutrition of six plant species of woodland clearings. *J. Ecol.* **69**, 871–882.

van Blaricom, G. R. (1982). Experimental analysis of structural regulation in marine sand communities exposed to oceanic swell. *Ecol. Monogr.* **52**, 283–305.

Vance, R. R. (1973). A mutualistic interaction between a sessile marine clam and its epibionts. *Ecology* **59**, 679–685.

Vance, R. R. (1979). Effects of grazing by the sea urchin, *Centrostephanus coronatus*, on prey community composition. *Ecology* **60**, 537–546.

Van Cleve, K., and Viereck, L. A. (1981). Forest succession in relation to nutrient cycling in the boreal forest of Alaska. *In* "Forest Succession: Concepts and Application" (D. C. West, H. H. Shugart, and D. B. Botkin, eds.), pp. 185–211. Springer-Verlag, Berlin and New York.

Van den Driessche, R. (1974). Prediction of mineral nutrient status of trees by foliar analysis. *Bot. Rev.* **40**, 347–394.

Vandermeer, J. H. (1970). The community matrix and the number of species in a community. *Am. Nat.* **104**, 73–83.

van der Meijden, E. (1979). Herbivore exploitation of a fugitive plant species: local survival and extinction of the cinnabar moth and ragwort in a heterogeneous environment. *Oecologia* **42**, 307–323.

Van Hook, R. I., Jr., Nielsen, M. G., and Shugart, H. H. (1980). Energy and nitrogen relations for a *Macrosiphum liriodendri* (Homoptera: Aphididae) population in an east Tennessee *Liriodendron tulipifera* stand. *Ecology* **61**, 960–975.

Van Horne, B. (1983). Density as a misleading indicator of habitat quality. *J. Wildl. Manage.* **47**, 893–901.

van Steenis, C. G. G. J. (1958). Rejuvenation as a factor for judging the status of vegetation types: The biological nomad theory. *In* "Proceedings of the Kandy Symposium," pp. 212–215. UNESCO, Paris.

Van Valen, L. (1962). A study of fluctuating asymmetry. *Evolution* **16**, 125–142.

Van Valen, L. (1973). A new evolutionary law. *Evol. Theory* **1**, 1–30.

van Wilgen, B. W. (1981). Some effects of fire frequency on fynbos plant community composition and structure at Jonkershoek, Stellenbosch. *S. Afr. For. J.* **118**, 42–55.

van Wilgen, B. W. (1982). Some effects of post-fire age on the above-ground plant biomass of fynbos (macchia) vegetation in South Africa. *J. Ecol.* **70**, 217–225.

van Wilgen, B. W., and le Maitre, D. C. (1981). Preliminary estimates of nutrient levels in fynbos vegetation and the role of fire in nutrient cycling. *S. Afr. For. J.* **119**, 24–28.

Vázquez-Yanes, C. (1976a). Estudios sobre la ecofisiología de la germinación en una zona cálido-húmeda de México. *In* "Regeneración des Selvas" (A. Gómez-Pompa, C. Vázquez-Yanes, S. Del Amo, R. and A. Butanda, C. eds.), pp. 279–387. Compañia Editorial Continental, Mexico City, Mexico.

Vázquez-Yanes, C. (1976b). Seed dormancy and germination in secondary vegetation tropical plants: the role of light. *Comp. Physiol. Ecol.* **57**, 30–34.

Vázquez-Yanes, C. (1977). Germination of a pioneer tree (*Trema guineensis*, Ficahlo) from equatorial Africa. *Turrialba* **27**, 301–302.

Vázquez-Yanes, C. (1980a). Light quality and seed germination in *Cecropia obtusifolia* and *Piper auritum* from a tropical rain forest in Mexico. *Phyton (Buenos Aires)* **38**, 33–35.

Vázquez-Yanes, C. (1980b). Notas sobre la autoecología de los árboles pioneros de rapido crecimiento de la selva tropical lluviosa. *Trop. Ecol.* **21**, 103–112.

Vázquez-Yanes, C., and Orozco-Segovia, A. (1982a). Germination of a tropical rain forest shrub, *Piper hispidum* Sw. (Piperaceae) under different light qualities. *Phyton (Buenos Aires)* **42**, 143–149.

Vázquez-Yanes, C., and Orozco-Segovia, A. (1982b). Seed germination of a tropical rain forest pioneer tree *Heliocarpus donnell-smithii* in response to diurnal fluctuation of temperature. *Physiol. Plant.* **56**, 295–298.

Vázques-Yanes, C., and Smith, H. (1982). Phytochrome control of seed germination in the tropical rain forest pioneer trees *Cecropia obtusifolia* and *Piper auritum* and its ecological significance. *New Phytol.* **92**, 477–485.

Vázquez-Yanes, C., Orozco-Segovia, A., François, G., and Trejo P., L. (1975). Observations on seed dispersal by bats in a tropical humid region in Veracruz, Mexico. *Biotropica* **7**, 73–76.

Veblen, T. T. (1979). Structure and dynamics of *Nothofagus* forests near timberline in south-central Chile. *Ecology* **60**, 937–945.

Veblen, T. T. (1982). Growth patterns of *Chusquea* bamboos in the understory of Chilean *Nothofagus* forests and their influences in forest dynamics. *Bull. Torrey Bot. Club* **109**, 474–487.

Veblen, T. T., and Ashton, D. H. (1978). Catastrophic influences on the vegetation of the Valdivian Andes, Chile. *Vegetatio* **36,** 149–167.

Veblen, T. T., and Stewart, G. H. (1982). On the conifer regeneration gap in New Zealand: the dynamics of *Libocedrus bidwillii* stands on South Island. *J. Ecol.* **70,** 413–436.

Veblen, T. T., Ashton, D. H., and Schlegel, F. M. (1979). Tree regeneration strategies in a lowland *Nothofagus*-dominated forest in south-central Chile. *J. Biogeogr.* **6,** 329–340.

Veblen, T. T., Schlegel, F. M., and Escobar R., B. (1980). Structure and dynamics of old-growth *Nothofagus* forests in the Valdivian Andes, Chile. *J. Ecol.* **68,** 1–31.

Veblen, T. T., Donoso Z., C., Schlegel, F. M., and Escobar R., B. (1981). Forest dynamics in south-central Chile. *J. Biogeogr.* **8,** 211–247.

Veblen, T. T., Schlegel, F. M., and Oltremari, J. V. (1983). Temperate broad-leaved evergreen forests of South America. *In* "Temperate Broad-Leaved Evergreen Forests" (J. D. Ovington, ed.), pp. 5–31. Elsevier, Amsterdam.

Veres, J. S., and Pickett, S. T. A. (1982). Branching patterns of *Lindera benzoin* beneath gaps and closed canopies. *New Phytol.* **91,** 767–772.

Vezina, P. E., and Grandter, M. M. (1965). Phenological observations of spring geophytes in Quebec. *Ecology* **46,** 869–872.

Viereck, L. A., and Schandelmeier, L. A. (1980). Effects of fire in Alaska and adjacent Canada—a literature review. *BLM—Alaska Tech. Rep.* No. 6.

Visher, S. S. (1949). American dry seasons: their intensity and frequency. *Ecology* **30,** 365–370.

Vitousek, P. M. (1977). The regulation of element concentrations in mountain streams in the northeastern United States. *Ecol. Monogr.* **47,** 65–87.

Vitousek, P. M. (1982). Nutrient cycling and nutrient use efficiency. *Am. Nat.* **119,** 553–572.

Vitousek, P. M. (1983). Nitrogen turnover in a ragweed-dominated first-year old field in southern Indiana. *Am. Midl. Nat.* **110,** 46–53.

Vitousek, P. M. (1984). Litterfall, nutrient cycling and nutrient limitation in tropical forests. *Ecology* **65,** 285–298.

Vitousek, P. M., and Matson, P. A. (1984). Mechanisms of nitrogen retention in forest ecosystems: A field experiment. *Science* **255,** 51–52.

Vitousek, P. M., and Melillo, J. M. (1979). Nitrate losses from disturbed forests: Patterns and mechanisms. *For. Sci.* **25,** 605–619.

Vitousek, P. M., and Reiners, W. A. (1975). Ecosystem succession and nutrient retention: A hypothesis. *BioScience* **25,** 376–381.

Vitousek, P. M., Gosz, J. R., Grier, C. C., Melillo, J. M., and Reiners, W. A. (1982). A comparative analysis of potential nitrification and nitrate mobility in forest ecosystems. *Ecol. Monogr.* **52,** 155–177.

Vogl, R. J. (1973). Ecology of knobcone pine in the Santa Ana Mountains, California. *Ecol. Monogr.* **43,** 125–143.

Vogl, R. J. (1974). Effects of fire on grasslands. *In* "Fire and Ecosystems" (T. T. Kozlowski and C. E. Ahlgren, eds.), pp. 139–194. Academic Press, New York.

Vogl, R. J. (1980). The ecological factors that produce perturbation-dependent ecosystems. *In* "The Recovery Process in Damaged Ecosystems" (J. Cairns, Jr., ed.), pp. 63–94. Ann Arbor Sci. Publ., Ann Arbor, Michigan.

Vogt, K. A., Grier, C. C., Meier, C. C., and Keyes, M. R. (1983). Organic matter and nutrient dynamics in forest floors of young and mature *Abies amabilis* stands in western Washington as affected by fine root input. *Ecol. Monogr.* **53,** 139–157.

Voroney, R. P., Van Veen, J. A., and Paul, E. A. (1981). Organic C dynamics in grassland soils. 2. Model validation and simulation of the long-term effects of cultivation and rainfall erosion. *Can. J. Soil. Sci.* **61,** 211–224.

Vrijenhoek, R. C. (1972). Genetic relationships of unisexual hybrid fishes to their progenitors using lactate dehydrogenase isozymes as gene markers (Poeciliopsis, Poeciliidae). *Am. Nat.* **106,** 754–766.

Vrijenhoek, R. C. (1978). Coexistence of clones in a heterogeneous environment. *Science (Washington, D.C.)* **199,** 549–552.

Vrijenhoek, R. C. (1979). Genetics of a sexually reproducing fish in a highly fluctuating environment. *Am. Nat.* **113,** 17–29.

Vrijenhoek, R. C.(1984a). The evolution of clonal diversity in *Poeciliopsis*. *In* ''Evolutionary Genetics of Fishes'' (B. J. Turner, ed.), pp. 218–231. Plenum, New York.

Vrijenhoek, R. C. (1984b). Ecological differentiation among clones: the frozen niche variation model. *In* ''Population Biology and Evolution'' (K. Wöhrmann and V. Loeschcke, eds.), pp. 217–231. Springer-Verlag, Berlin and New York.

Vrijenhoek, R. C. (1985). Homozygosity and interstrain variation in the self-fertilizing hermaphroditic fish *Rivulus marmoratus* Poey. *J. Hered.* (in press).

Vrijenhoek, R. C., and Lerman, S. (1982). Heterozygosity and developmental stability under sexual and asexual breeding systems. *Evolution* **36,** 768–776.

Wade, M. J., and McCauley, D. E. (1980). Group selection: the phenotypic and genotypic differentiation of small populations. *Evolution* **34,** 799–812.

Waggoner, P. E., and Stephens, G. R. (1970). Transition probabilities for a forest. *Nature (London)* **225,** 1160–1161.

Wagner, T. L., Feldman, R. M., Gagne, J. A., Cover, J. D., Coulson, R. N., and Schoolfield, R. M. (1981). Factors affecting gallery construction, oviposition, and reemergence of *Dendroctonus frontalis* in the laboratory. *Ann. Entomol. Soc. Am.* **74,** 255–273.

Waldschmidt, S., and Tracy, C. R. (1983). Interactions between a lizard and its thermal environment: implications for sprint performance and space utilization in the lizard *Uta stansburiana. Ecology* **64,** 476–484.

Walker, B. H. (1981). Is succession a viable concept in African savanna ecosystems? *In* ''Forest Succession: Concepts and Applications'' (D. C. West, H. H. Shugart, and D. B. Botkin, eds.), pp. 431–447. Springer-Verlag, Berlin and New York.

Walker, J., Thompson, C. H., Fergus, I. F., and Tunstall, B. R. (1981). Plant succession and soil development in coastal sand dunes of subtropical eastern Australia. *In* ''Forest Succession: Concepts and Application'' (D. C. West, H. H. Shugart, and D. B. Botkin, eds.), pp. 107–131. Springer-Verlag, Berlin and New York.

Walker, T. W., and Syers, J. K. (1976). The fate of phosphorous during pedogenesis. *Geoderma* **15,** 1–19.

Wallace, B. (1981). ''Basic Population Genetics.'' Columbia Univ. Press, New York.

Wallace, B., and Srb, A. M. (1964). ''Adaptation,'' 2nd Ed. Prentice-Hall, Englewood Cliffs, New Jersey.

Wallace, B., and Vetukhiv, M. (1955). Adaptative organization of gene pools of *Drosophila* populations. *Cold Spring Harbor Symp. Quant. Biol.* **20,** 303–309.

Wallace, C. C., and Bull, G. D. (1983). Patterns of juvenile coral recruitment on a reef front during a spring–summer spawning period. *Proc. Int. Coral Reef Symp., 4th, Manila, 1981* **2,** 345–350.

Wardle, J. A. (1970). The ecology of *Nothofagus solandri*. I. The distribution and relationship with other major forest and scrub species. *N.Z. J. Bot.* **8,** 494–531.

Wardle, J. A., and Guest, R. (1977). Forests of the Waitaki and Lake Hawea Catchments. *N.Z. J. For. Sci.* **7,** 44–67.

Waring, R. H. (1982). Coupling stress physiology with ecosystem analysis. *In* ''Carbon Uptake and Allocation in Subalpine Ecosystems as a Key to Management'' (R. H. Waring, ed.), Proc. Int. Union For. Res. Organ. Workshop, pp. 5–8. Oregon State Univ. For. Res. Lab., Corvallis.

Waring, R. H., Thies, W. G., and Muscato, D. (1980). Stem growth per unit of leaf area: a measure of tree vigor. *For. Sci.* **26,** 112–117.

Warren Wilson, J. (1981). Analyses of growth, photosynthesis and light interception for single plants and stands. *Ann. Bot. (London)* **48,** 507–512.

Waser, P. M. (1981). Sociality or territorial defense? The influence of resource renewal. *Behav. Evol. Sociobiol.* **8,** 231–237.

Waser, P. M., and Wiley, R. H. (1980). Mechanisms and evolution of spacing in animals. *In* "Handbook of Behavioral Neurobiology" (P. Marler and J. G. Vandenberg, eds.), Vol. 3, pp. 159–223. Plenum, New York.

Watt, A. S. (1925). On the ecology of the British beechwoods with special reference to their regeneration. Part II, Sections II and III. The development and structure of beech communities on the Sussex Downs. *J. Ecol.* **13,** 27–73.

Watt, A. S. (1947). Pattern and process in the plant community. *J. Ecol.* **35,** 1–22.

Watt, W. B. (1977). Adaptation at specific loci. I. Natural selection on phosphoglucose isomerase of *Colias* butterflies: biochemical and population aspects. *Genetics* **87,** 177–194.

Weaver, J. E. (1968). "Prairie Plants and their Environment: A 50 Year Study in the Midwest." Univ. of Nebraska Press, Lincoln.

Weaver, J. E., and Albertson, F. W. (1943). Resurvey of grasses, forbs, and underground plant parts at the end of the great drought. *Ecol. Monogr.* **13,** 63–117.

Webb, L. J. (1958). Cyclones as an ecological factor in tropical lowland rain forest, North Queensland. *Aust. J. Bot.* **6,** 220–228.

Webb, L. J., Tracey, J. G., and Williams, W. T. (1972). Regeneration and pattern in the subtropical rain forest. *J. Ecol.* **60,** 675–696.

Webb, W. L., King, R. T., and Patric, E. F. (1956). Effect of white-tailed deer on a mature northern forest. *J. For.* **54,** 391–398.

Webster, J. R., Waide, J. B., and Patten, B. C. (1974). Nutrient recycling and the stability of ecosystems. *In* "Mineral Cycling in Southeastern Ecosystems" (F. G. Howell, J. B. Gentry, and M. H. Smith, eds.), *AEC Symp. Ser.* (*U.S. At. Energy Comm.* **CONF-740513**), pp. 1–27.

Weetman, G. F. (1962a). "Nitrogen Relations in a Black Spruce (*Picea mariana*) Stand Subject to Various Fertilizer and Soil Treatments," Woodlands Res. Index No. 129. Pulp Pap. Res. Inst. Can., Montreal.

Weetman, G. F. (1962b). "Establishment Report on a Humus Decomposition Experiment," Woodlands Res. Index No. 134. Pulp Pap. Inst. Can., Montreal.

Wein, R. W.(1974). Vegetation recovery in arctic tundra and forest-tundra after fire. *Can., Dep. Indian North. Aff., Publ.* No. QS. 8037-000-EE-A1.

Wein, R. W., and Bliss, L. C. (1973). Changes in arctic *Eriophorum* tussock communities following fire. *Ecology* **54,** 845–852.

Wein, R. W., and Bliss, L. C. (1974). Primary production in arctic cottongrass tussock tundra communities. *Arct. Alp. Res.* **6,** 261–274.

Weiner, J., and Conte, P. T. (1981). Dispersal and neighborhood effects in an annual plant competition model. *Ecol. Modell.* **13,** 131–147.

Weinstein, D. A., and Shugart, H. H. (1983). Ecological modeling of landscape dynamics. *In* "Disturbance and Ecosystems" (H. Mooney and M. Godron, eds.), pp. 29–45. Springer-Verlag, Berlin and New York.

Weinstein, D. A., Shugart, H. H., and West, D. C. (1982). The long-term nutrient retention properties of forest ecosystems: A simulation investigation. *Oak Ridge Natl. Lab.* [*Rep.*] *ORNL-TM (U.S.)* **ORNL-TM-8472.**

Weischet, W. (1960). "Contribuciones al Estudio de las Transformaciones Geográficas en la Parte Septentrional del Sur de Chile por Efecto del Sismo, del 22 de Mayo de 1960." Inst. Geol., Univ. de Chile, Santiago.

Weller, M. W. (1981). "Freshwater Marshes: Ecology and Wildlife Management." Univ. of Minnesota Press, St. Paul.

Weller, M. W., and Spatcher, C. E. (1965). Role of habitat in the distribution and abundance of marsh birds. *Iowa State Univ., Agric. Home Econ. Exp. Stn., Spec. Rep.* No. 43.

Wellington, G. M. (1980). Reversal of digestive interactions between Pacific reef corals: mediation by sweeper tentacles. *Oecologia* **47,** 340–343.

Wellington, W. G. (1980). Dispersal and population change. *In* "Dispersal of Forest Insects: Evaluation, Theory and Management Implications'' (A. A. Berryman and L. Safranyik, eds.), Proc. Int. Union For. Res. Organ. Conf., pp. 11–24. Washington State Univ. Coop. Ext. Serv., Pullman.

Wells, C. B., Campbell, R. E., DeBano, L. F., Fredriksen, E. L., Froelich, R. C., and Dunn, P. H. (1979). Effects of fire on soil. *Gen. Tech. Rep. WO—U.S. For. Ser. [Wash. Off.]* **GTR-WO-7.**

Wells, H. G., Wells, N. J., and Gray, I. E. (1964). The calico scallop community in North Carolina. *Bull. Mar. Sci.* **14,** 561–593.

Wells, P. V. (1962). Vegetation in relation to geological substratum and fire in the San Luis Obispo Quadrangle, California. *Ecol. Monogr.* **32,** 79–103.

Wells, P. V. (1969). The relation between mode of reproduction and extent of speciation in woody genera of the California chaparral. *Evolution* **23,** 264–267.

Wendel, G. W. (1972). Longevity of black cherry seed in the forest floor. *USDA For. Serv. Northeast. For. Exp. Stn. Res. Note NE* **NE-149.**

Wendel, G. W. (1977). Longevity of black cherry, wild grape, and sassafras seed in the forest floor. USDA *For. Serv. Northeast. For. Exp. Stn. Res. Pap. NE (U.S.)* **NE-375.**

Wendel, G. W., Storey, T. G., and Dyram, G. M. (1962). Forest fuels on organic and associated soils in the Coastal Plain of North Carolina. *U.S.D.A. For. Ser., Southeast. For. Exp. Stn., Stn. Pap.* No. 144.

Went, F. W., Juhren, G., and Juhren, M. C. (1952). Fire and biotic factors affecting germination. *Ecology* **33,** 351–364.

Werner, P. A. (1979). Competition and coexistence of similar species. *In* ''Topics in Plant Population Biology'' (O. T. Solbrig, S. Jain, G. B. Johnson, and P. H. Raven, eds.), pp. 287–312. Columbia Univ. Press, New York.

Werner, P. A., and Caswell, H. (1977). Population growth rates and age versus stage-distribution models for teasel (*Dipsacus sylvestris* (Huds.). *Ecology* **58,** 1103–1111.

West, N. E. (1983). ''Temperate Deserts and Semi-Deserts. Vol. 5. Ecosystems of the World'' (D. Goodall, ed.). Elsevier, New York.

West, D. C., Shugart, H. H., and D. B. Botkin, eds. (1981). ''Forest Succession: Concepts and Application.'' Springer-Verlag, Berlin and New York.

Western, D., and Ssemakula, J. (1981). The future of the savannah ecosystems: ecological islands or faunal enclaves? *Afr. J. Ecol.* **19,** 7–19.

Westman, W. E. (1979). A potential role of coastal sage scrub understories in the recovery of chaparral after fire. *Madrono* **26,** 64–68.

Westman, W. E. (1981a). Factors influencing the distribution of species of Californian coastal sage scrub. *Ecology* **62,** 429–455.

Westman, W. E. (1981b). Diversity relations and succession in Californian coastal sage scrub. *Ecology* **62,** 170–184.

Westman, W. E., O'Leary, J. F., and Malanson, G. P. (1981). The effects of fire intensity, aspect, and substrate on post-fire growth of Californian coastal sage scrub. *In* ''Components of Productivity of Mediterranean-Climate Regions—Basic and Applied'' (N. S. Margaris and H. A. Mooney, eds.), pp. 151–179. Junk, The Hague.

Wethey, D. S. (1979). Demographic variation in intertidal barnacles. Ph.D. Thesis, Univ. of Michigan, Ann Arbor.

Whigham, D. (1974). An ecological life history study of *Uvularia perfoliata* L. *Am. Midl. Nat.* **91,** 343–359.

Whitcomb, R. F., Lynch, J. F., Opler, P. A., and Robbins, C. S. (1976). Island biogeography and conservation: strategy and limitations. *Science (Washington, D.C.)* **193,** 1030–1032.

Whitcomb, R. F. Lynch, J. F., Klimkiewicz, M. K., Robbins, C. S., Whitcomb, B. L., and Bystrak, D. (1981). Effects of forest fragmentation on avifauna of the eastern deciduous forest. *In* ''Forest Island Dynamics in Man-dominated Landscapes'' (R. L. Burgess and D. M. Sharpe, eds.), pp. 125–205. Springer-Verlag, Berlin and New York.

White, P. S. (1979). Pattern, process, and natural disturbance in vegetation. *Bot. Rev.* **45,** 229–299.

White, P. S., and Bratton, S. P. (1980). After preservation: philosophical and practical problems of change. *Biol. Conserv.* **18**, 241–255.

Whitham, T. G. (1978). Habitat selection by *Pemphigus* aphids in response to resource limitation and competition. *Ecology* **59**, 1164–1176.

Whitham, T. G. (1981). Individual trees as heterogeneous environments: adaptation to herbivory or epigenetic noise? *In* "Insect Life History Patterns: Habitat and Geographic Variation" (R. F. Denno and H. Dingle, eds.), pp. 9–27. Springer-Verlag, Berlin and New York.

Whitmore, T. C. (1966). A study of light conditions in forests in Ecuador with some suggestions for further studies in tropical forests. *In* "Light as an Ecological Factor" (R. Bainbridge, C. C. Evans, and O. Rackham, eds.), pp. 235–247. Blackwell, Oxford.

Whitmore, T. C. (1974). "Change with Time and the Role of Cyclones in Tropical Rain Forest on Kolombangara, Solomon Islands," Inst. Pap. No. 46. Commonw. For. Inst., Oxford.

Whitmore, T. C. (1975). "Tropical Rainforests of the Far East." Oxford Univ. Press (Clarendon), London and New York.

Whitmore, T. C. (1978). Gaps in the forest canopy. In "Tropical Trees as Living Systems" (P. B. Tomlinson and M. H. Zimmermann, eds.), pp. 639–655. Cambridge Univ. Press, London and New York.

Whitmore, T. C. (1982). On pattern and process in forests. In "The Plant Community as a Working Mechanism" (E. J. Newman, ed.), pp. 45–57. Blackwell, Oxford.

Whitney, G. G. (1976). The bifurcation ratio as an indicator of adaptive strategy on woody plant species. *Bull. Torrey Bot. Club* **103**, 67–72.

Whittaker, R. H. (1953). A consideration of climax theory: the climax as a population and pattern. *Ecol. Monogr.* **23**, 41–78.

Whittaker, R. H. (1956). Vegetation of the Great Smoky Mountains. *Ecol. Monogr.* **26**, 1–80.

Whittaker, R. H. (1975). "Communities and Ecosystems," 2nd Ed. Macmillan, New York.

Whittaker, R. H., and Feeny, P. P. (1971). Allelochemics; chemical interaction between species. *Science (Washington, D.C.)***171**, 757–770.

Whittaker, R. H., and Levin, S. A. (1977). The role of mosaic phenomena in natural communities. *Theor. Popul. Biol.* **12**, 117–139.

Whittaker, R. H., and Woodwell, G. M. (1967). Surface area relations of woody plants and forest communities. *Am. J. Bot.* **54**, 931–939.

Whittaker, R. H., and Woodwell, G. M. (1968). Dimension and production relations of trees and shrubs in the Brookhaven forest, New York. *J. Ecol.* **56**, 1–25.

Whittaker, R. H., Bormann, F. H., Likens, G. E., and Siccama, T. G.(1974). The Hubbard Brook Ecosystem Study: forest biomass production. *Ecol. Monogr.* **44**, 233–252.

Wicklow, D. T. (1977). Germination response in *Emmenanthe penduliflora* (Hyrophyllaceae). *Ecology* **58**, 201–205.

Wiens, J. A. (1973). Interterritorial habitat variation in Grasshopper and Savannah sparrows. *Ecology* **54**, 877–884.

Wiens, J. A. (1974a). Climatic instability and the "ecological saturation" of bird communities in North American grasslands. *Condor* **76**, 385–400.

Wiens, J. A. (1974b). Habitat heterogeneity and avian community structure in North American grasslands. *Am. Midl. Nat.* **91**, 195–213.

Wiens, J. A. (1976). Population responses to patchy environments. *Annu. Rev. Ecol. Syst.* **7**, 81–120.

Wiens, J. A. (1977). On competition and variable environments. *Am. Sci.* **65**, 590–597.

Wiens, J. A. (1981). Scale problems in avian censusing. *Stud. Avian Biol.* **6**, 513–521.

Wiens, J. A. (1983). Avian community ecology: an iconoclastic view. *In* "Perspectives in Ornithology" (A. H. Brush and G. A. Clark, Jr., eds.), pp. 355–403. Cambridge Univ. Press, London and New York.

Wiens, J. A. (1984). Resource systems, populations, and communities. *In* "A New Ecology: Novel Approaches to Interactive Systems" (P. W. Price, C. N. Slobodchikoff, and W. S. Gaud, eds.), pp. 397–436. Wiley, New York.

Wiens, J. A., and Johnston, R. F. (1977). Adaptive correlates of granivory in birds. *In* "Granivorous Birds in Ecosystems" (J. Pinowski and S. C. Kendeigh, eds.), pp. 301–340. Cambridge Univ. Press, London and New York.

Wiens, J. A., and Rotenberry, J. T. (1981a). Censusing and the evaluation of avian habitat occupancy. *Stud. Avian Biol.* **6**, 522–532.

Wiens, J. A., and Rotenberry, J. T. (1981b). Habitat associations and community structure of birds in shrubsteppe environments. *Ecol. Monogr.* **51**, 21–41.

Wiklander, G. (1981). Rapporteur's comment on clearcutting. *In* "Nitrogen Cycling in Terrestrial Ecosystems: Processes, Ecosystem Strategies, and Management Implications" (F. E. Clark and T. H. Rosswall, eds.), *Ecol. Bull.* No. 33, 642–647.

Wilbur, R. B., and Christensen, N. L. (1983). Effects of fire on nutrient availability in a North Carolina coastal plain pocosin. *Am. Midl. Nat.* **110**, 54–63.

Wilkins, C. W. (1977). A stochastic analysis of the effect of fire on remote vegetation. Ph.D. Thesis, Univ. of Adelaide, Adelaide, South Australia.

Williamson, G. B. (1975). Pattern and seral composition in an old-growth beech–maple forest. *Ecology* **56**, 727–731.

Willson, M. F., and Melampy, M. N. (1983). The effect of bicolored fruit displays on fruit removal by avian frugivores. *Oikos* **41**, 27–31.

Wilson, E. O., and Willis, E. O. (1975). Applied biogeography. *In* "Ecology and Evolution of Communities" (M. L. Cody and J. M. Diamond, eds.), pp. 522–534. Belknap Press, Cambridge, Massachusetts.

Winn, A. A., and Pitelka, L. F. (1981). Some effects of density on the reproductive patterns and patch dynamics of *Aster acuminatus*. *Bull. Torrey Bot. Club* **108**, 438–445.

Wirtz, W. O., II (1977). Vertebrate post-fire succession. *In* "Environmental Consequences of Fire and Fuel Management in Mediterranean Ecosystems" (H. A. Mooney and C. E. Conrad, eds.), *Gen. Tech. Rep. WO—U.S. For. Serv. [Wash. Off.]* **GTR-WO-3**, pp. 46–57.

Wirtz, W. O., II (1982). Postfire community structure of birds and rodents in southern California chaparral. *In* "Dynamics and Management of Mediterranean-Type Ecosystems" (C. E. Conrad and W. C. Oechel, eds.), *USDA For. Ser. Pac. Southwest For. Range Exp. Stn. Gen. Tech. Rep. PSW* **PSW-58**, pp. 241–246.

Witcosky, J. J. (1981). Insects associated with black stain root disease of Douglas-fir in western Oregon. M.S. Thesis, Oregon State Univ., Corvallis.

Wolcott, T. G. (1973). Physiological ecology and intertidal zonation in limpets (*Acmaea*): a critical look at "limiting factors". *Biol. Bull. (Woods Hole, Mass.)* **145**, 389–422.

Wolff, J. O. (1980). The role of habitat patchiness in the population dynamics of snowshoe hares. *Ecol. Monogr.* **50**, 111–130.

Wood, V., and Seed, R. (1980). The effects of shore level on the epifaunal communities associated with *Fucus serratus* (L.) in the Menai Strait, North Wales. *Cah. Biol. Mar.* **21**, 135–154.

Woodley, J. D., Chonesky, E. A., Clifford, P. A., Jackson, J. B. C., Kaufman, L. S., Knowlton, N., Lang, J. C., Pearson, M. P., Porter, J. W., Rooney, M. C., Rylaarsdam, K. W., Tunnicliffe, V. J., Dallmeyer, M. D., Wahle, C. M., Wulff, J. L., Curtis, A. S. G., Jupp, B. P., Koehl, M. A. R., Neigel, J., and Sides, E. M. (1981). Hurricane Allen's impact on Jamaican coral reefs. *Science (Washington, D.C.)* **214**, 749–755.

Woodmansee, R. G. (1978). Additions and losses of nitrogen in grassland ecosystems. *BioScience* **28**, 488–453.

Woods, F. W., and Shanks, R. E.(1959). Natural replacement of chestnut by other species in the Great Smoky Mountains National Park. *Ecology* **40**, 349–361.

Woods, K. D., and Whittaker, R. H. (1981). Canopy-understory interaction and the internal dynamics of mature hardwood and hemlock-hardwood forests. *In* "Forest Succession: Concepts and Application" (D. C. West, H. H. Shugart and D. B. Botkin, eds.), pp. 305–323. Springer-Verlag, New York.

Woodwell, G. M. (1958). Factors controlling growth of pond pine seedlings in organic soils of the Carolinas. *Ecol. Monogr.* **28,** 219–236.

Woodwell, G. M. (1974). Success, succession, and Adam Smith. *BioScience* **24,** 81–87.

Woodwell, G. M., Whittaker, R. H., Reiners, W. A., Likens, G. E., Delwiche, C. C., and Botkin, D. B. (1978). The biota and the world carbon budget. *Science (Washington, D.C.)* **199,** 141–146.

Worthington, A. (1982). Population sizes and breeding rhythms of two species of manakins in relation to food supply. *In* "The Ecology of a Tropical Forest: Seasonal Rhythms and Long-term Changes" (E. G. Leigh, Jr., A. S. Rand, and D. M. Windsor, eds.), pp. 213–225. Smithsonian Inst. Press, Washington, D.C.

Wright, C., and Mella, A. (1963). Modifications to the soil pattern of south-central Chile resulting from seismic and associated phenomena during the period May to August 1960. *Bull. Seismol. Soc. Am.* **53,** 1367–1402.

Wright, H. A. (1972). Shrub response of fire. *In* "Wildland Shrubs—Their Biology and Utilization" (C. M. McKell, J. P. Blaisdell, and J. R. Goodwin, eds.), *USDA For. Serv. Intermountain For. Range Exp. Stn. Gen. Tech. Rep. INT* **INT-1,** pp. 204–217.

Wright, H. E., Jr. (1974). Landscape development, forest fires, and wilderness management. *Science (Washington, D.C.)* **186,** 487–495.

Wright, S. (1929). Evolution in a Mendelian population. *Anat. Rec.* **44,** 287.

Wright, S. (1931). Evolution in Mendelian populations. *Genetics* **10,** 97–159.

Wright, S. (1932). The roles of mutation, inbreeding, crossbreeding and selection in evolution. *Proc. Int. Congr. Genet., 6th.* **1,** 356–366.

Wright, S. (1940). Breeding structure of populations in relation to speciation. *Am. Nat.* **74,** 232–248.

Wright, S. (1969). "Evolution and the Genetics of Populations. Vol. II: The Theory of Gene Frequencies." Univ. of Chicago Press, Chicago, Illinois.

Wright, S. (1970). Random drift and the shifting balance theory of evolution. *In* "Mathematical Topics in Population Genetics" (K. Kojima, ed.), pp. 1–30. Springer-Verlag, Berlin and New York.

Wright, S. (1977). "Evolution and the Genetics of Populations. Vol. III: Experimental Results and Evolutionary Deductions." Univ. of Chicago Press, Chicago, Illinois.

Wright, S. (1978). "Evolution and the Genetics of Populations. Vol. IV: Variability within and Among Natural Populations." Univ. of Chicago Press, Chicago, Illinois.

Wright, S. (1982). The shifting balance theory and macroevolution. *Annu. Rev. Genet.* **16,** 1–19.

Wright, S. J. (1979). Competition between insectivorous lizards and birds in Central Panama. *Am. Zool.* **19,** 1145–1156.

Wu, K. K., and Jain, S. K. (1979). Population regulation in *Bromus rubens* and *B. mollis:* Life cycle components and competition. *Oecologia* **39,** 337–357.

Wyatt-Smith, J. (1954). Storm forest in Kelantan. *Malay. For.* **17,** 5–111.

Yamaguchi, M. (1975). Sea level fluctuations and mass mortalities of reef animals in Guam, Mariana Islands. *Micronesica* **11,** 227–243.

Yeaton, R. I. (1978). A cyclical relationship between *Larrea tridentata* and *Opuntia leptocaulis* in the northern Chihuahuan desert. *J. Ecol.* **66,** 651–656.

Yoda, K. (1983). Community respiration in a lowland rain forest in Pasoh, peninsular Malaysia. *Nippon Seitai Gakkaishi (Jpn. J. Ecol.)* **33,** 183–197.

Yoda, K., Shinozaki, K., Ogawa, H., Hozumi, K., and Kira, T. (1965). Estimation of the total amount of respiration in woody organs of trees and forest communities. *J. Biol., Osaka City Univ.* **16,** 15–26.

Young, C. M. (1982). Larval behavior, predation and early post-settling mortality as determinates of spatial distribution in subtidal solitary ascidians of the San Juan Islands, Washington. Ph.D. Thesis, Univ. of Alberta, Edmonton.

Young, C. M., and Chia, F.-S. (1984). Abundance and distribution of pelagic larvae, as influenced by predatory, behavioral and hydrographic factors. *In* "Reproduction of Marine Invertebrates. Vol. 9: General Concepts" (A. C. Giese and J. S. Pearse, eds.) Academic Press, New York. In press.

Young, E. G., Jr., and Murray, J. C. (1966). Heterosis and inbreeding depression in diploid and tetraploid cottons. *Crop Sci.* **6,** 436–438.

Zackrisson, O. (1977). Influence of forest fires on the North Swedish boreal forest. *Oikos* **29,** 22–32.

Zangerl, A. R., and Bazzaz, F. A. (1983). Responses of an early and a late successional species of *Polygonum* to variations in resource availability. *Oecologia* **56,** 397–404.

Zangerl, A. R., Pickett, S. T. A., and Bazzaz, F. A. (1977). Some hypotheses on variation in plant population and an experimental approach. *Biologist* **59,** 113–122.

Zaret, T. M. (1979). Predation in freshwater fish communities. *In* "Predator–prey Systems in Fisheries Management" (R. H. Stroud and H. Clepper, eds.), pp. 135–144. Sport Fish. Inst., Washington, D.C.

Zedler, J., and Loucks, O. L. (1969). Differential burning response of *Poa pratensis* fields and *Andropogon scoparius* prairies in central Wisconsin. *Am. Midl. Nat.* **81,** 341–352.

Zedler, P. H. (1981). Vegetation change in chaparral and desert communities in San Diego County, California. *In* "Forest Succession: Concepts and Application" (D. C. West, H. H. Shugart, and D. B. Botkin, eds.), pp. 406–430. Springer-Verlag, Berlin and New York.

Zedler, P. H., and Goff, F. G. (1973). Size-association analysis of forest succesional trends in Wisconsin. *Ecol. Monogr.* **43,** 79–94.

Zedler, P. H., Gautier, C. R., and McMaster, G. S. (1983). Vegetation change in response to extreme events: the effects of a short interval between fires in California chapparal and coastal shrub. *Ecology* **64,** 809–818.

Zlotin, R. I., and Khodashova, K. S. (1980). "The Role of Animals in Biological Cycling of Forest–Steppe Ecosystems" (N. R. French, ed.), Engl. Ed. Dowden, Hutchinson & Ross, Stroudsburg, Pennsylvania.

Zouros, E., Singh, S. M., and Miles, H. E. (1980). Growth rate in oysters: an overdominant phenotype and its possible explanations. *Evolution* **34,** 856–867.

Zucker, W. V. (1982). How aphids choose leaves: the roles of phenolics in host selection by a galling aphid. *Ecology* **63,** 972–981.

Index